Digital Control
Engineering

Digital Control
Engineering
Analysis and Design

M. Sami Fadali
Antonio Visioli

AMSTERDAM • BOSTON • HEIDELBERG • LONDON
NEW YORK • OXFORD • PARIS • SAN DIEGO
SAN FRANCISCO • SINGAPORE • SYDNEY • TOKYO

Academic Press is an imprint of Elsevier

Academic Press is an imprint of Elsevier
30 Corporate Drive, Suite 400
Burlington, MA 01803

This book is printed on acid-free paper. ∞

Library of Congress Cataloging-in-Publication Data
Application submitted

ISBN 13: 978-0-12-374498-2

For information on all Academic Press publications,
visit our Website at *www.books.elsevier.com*

Printed in the United States
09 10 11 12 13   10 9 8 7 6 5 4 3 2 1

Contents

Preface

Approach

Control systems are an integral part of everyday life in today's society. They control our appliances, our entertainment centers, our cars, and our office environments; they control our industrial processes and our transportation systems; they control our exploration of land, sea, air, and space. Almost all of these application use digital controllers implemented with computers, microprocessors, or digital electronics. Every electrical, chemical, or mechanical engineering senior or graduate student should therefore be familiar with the basic theory of digital controllers.

This text is designed for a senior or combined senior/graduate-level course in digital controls in departments of mechanical, electrical, or chemical engineering. Although other texts are available on digital controls, most do not provide a satisfactory format for a senior/graduate-level class. Some texts have very few examples to support the theory, and some were written before the wide availability of computer-aided-design (CAD) packages. Others make some use of CAD packages but do not fully exploit their capabilities. Most available texts are based on the assumption that students must complete several courses in systems and control theory before they can be exposed to digital control. We disagree with this assumption, and we firmly believe that students can learn digital control after a one-semester course covering the basics of analog control. As with other topics that started at the graduate level—linear algebra and Fourier analysis to name a few—the time has come for digital control to become an integral part of the undergraduate curriculum.

Features

To meet the needs of the typical senior/graduate-level course, this text includes the following features:

Numerous examples. The book includes a large number of examples. Typically, only one or two examples can be covered in the classroom because of time limitations. The student can use the remaining examples for self-study. The experience of the authors is that students need more examples to experiment with so as to gain a better understanding of the theory. The examples are varied to bring out subtleties of the theory that students may overlook.

Extensive use of CAD packages. The book makes extensive use of CAD packages. It goes beyond the occasional reference to specific commands to the integration of these commands into the modeling, design, and analysis of digital control systems. For example, root locus design procedures given in most digital control texts are not CAD procedures and instead emphasize paper-and-pencil design. The use of CAD packages, such as MATLAB®, frees students from the drudgery of mundane calculations and allows them to ponder more subtle aspects of control system analysis and design. The availability of a simulation tool like Simulink® allows the student to simulate closed-loop control systems including aspects neglected in design such as nonlinearities and disturbances.

Coverage of background material. The book itself contains review material from linear systems and classical control. Some background material is included in appendices that could either be reviewed in class or consulted by the student as necessary. The review material, which is often neglected in digital control texts, is essential for the understanding of digital control system analysis and design. For example, the behavior of discrete-time systems in the time domain and in the frequency domain is a standard topic in linear systems texts but often receives brief coverage. Root locus design is almost identical for analog systems in the s-domain and digital systems in the z-domain. The topic is covered much more extensively in classical control texts and inadequately in digital control texts. The digital control student is expected to recall this material or rely on other sources. Often, instructors are obliged to compile their own review material, and the continuity of the course is adversely affected.

Inclusion of advanced topics. In addition to the basic topics required for a one-semester senior/graduate class, the text includes some advanced material to make it suitable for an introductory graduate-level class or for two quarters at the senior/graduate level. We would also hope that the students in a single-semester course would acquire enough background and interest to read the additional chapters on their own. Examples of optional topics are state-space methods, which may receive brief coverage in a one-semester course, and nonlinear discrete-time systems, which may not be covered.

Standard mathematics prerequisites. The mathematics background required for understanding most of the book does not exceed what can be reasonably expected from the average electrical, chemical, or mechanical engineering senior. This background includes three semesters of calculus, differential equations, and basic linear algebra. Some texts on digital control require more mathematical maturity and are therefore beyond the reach of the typical senior.

On the other hand, the text does include optional topics for the more advanced student. The rest of the text does not require knowledge of this optional material so that it can be easily skipped if necessary.

Senior system theory prerequisites. The control and system theory background required for understanding the book does not exceed material typically covered in one semester of linear systems and one semester of control systems. Thus, the students should be familiar with Laplace transforms, the frequency domain, and the root locus. They need not be familiar with the behavior of discrete-time systems in the frequency and time domain or have extensive experience with compensator design in the *s*-domain. For an audience with an extensive background in these topics, some topics can be skipped and the material can be covered at a faster rate.

Coverage of theory and applications. The book has two authors: the first is primarily interested in control theory and the second is primarily interested in practical applications and hardware implementation. Even though some control theorists have sufficient familiarity with practical issues such as hardware implementation and industrial applications to touch on the subject in their texts, the material included is often deficient because of the rapid advances in the area and the limited knowledge that theorists have of the subject.

It became clear to the first author that to have a suitable text for his course and similar courses, he needed to find a partner to satisfactorily complete the text. He gradually collected material for the text and started looking for a qualified and interested partner. Finally, he found a co-author who shared his interest in digital control and the belief that it can be presented at a level amenable to the average undergraduate engineering student.

For about 10 years, Dr. Antonio Visioli has been teaching an introductory and a laboratory course on automatic control, as well as a course on control systems technology. Further, his research interests are in the fields of industrial regulators and robotics. Although he contributed to the material presented throughout the text, his major contribution was adding material related to the practical design and implementation of digital control systems. This material is rarely covered in control systems texts but is an essential prerequisite for applying digital control theory in practice.

The text is written to be as self-contained as possible. However, the reader is expected to have completed a semester of linear systems and classical control. Throughout the text, extensive use is made of the numerical computation and computer-aided-design package MATLAB. As with all computational tools, the enormous capabilities of MATLAB are no substitute for a sound understanding of the theory presented in the text. As an example of the inappropriate use of supporting technology, we recall the story of the driver who followed the instructions of his GPS system and drove into the path of an oncoming train![1] The reader must use MATLAB as a tool to support the theory without blindly accepting its computational results.

Organization of Text

The text begins with an introduction to digital control and the reasons for its popularity. It also provides a few examples of applications of digital control from the engineering literature.

Chapter 2 considers discrete-time models and their analysis using the **z-transform**. We review the z-transform, its properties, and its use to solve difference equations. The chapter also reviews the properties of the **frequency response** of discrete-time systems. After a brief discussion of the **sampling theorem**, we are able to provide rules of thumb for **selecting the sampling rate** for a given signal or for given system dynamics. This material is often covered in linear systems courses, and much of it can be skipped or covered quickly in a digital control course. However, the material is included because it serves as a foundation for much of the material in the text.

Chapter 3 derives simple mathematical models for linear discrete-time systems. We derive models for the analog-to-digital converter (ADC), the digital-to-analog converter (DAC), and for an analog system with a DAC and an ADC. We include systems with time delays that are not an integer multiple of the sampling period. These transfer functions are particularly important because many applications include an analog plant with DAC and ADC. Nevertheless, there are situations where different configurations are used. We therefore include an analysis of a variety of configurations with samplers. We also characterize the **steady-state tracking error** of discrete-time systems and define error constants for the unity feedback case. These error constants play an analogous role to the error constants for analog systems. Using our analysis of more complex configurations, we are able to obtain the **error due to a disturbance input**.

In Chapter 4, we present stability tests for input-output systems. We examine the definitions of **input-output stability** and **internal stability** and derive conditions for each. By transforming the characteristic polynomial of a discrete-time system, we are able to test it using the standard **Routh-Hurwitz criterion** for analog systems. We use the **Jury criterion**, which allows us to directly test the stability of a discrete-time system. Finally, we present the **Nyquist criterion** for the z-domain and use it to determine closed-loop stability of discrete-time systems.

Chapter 5 introduces analog s-domain design of proportional (P), proportional-plus-integral (PI), proportional-plus-derivative (PD), and proportional-plus-integral-

---

[1] The story was reported in the *Chicago Sun-Times*, on January 4, 2008. The driver, a computer consultant, escaped just in time before the train slammed into his car at 60 mph in Bedford Hills, New York.

plus-derivative (PID) control using MATLAB. We use MATLAB as an integral part of the design process, although many steps of the design can be competed using a scientific calculator. It would seem that a chapter on analog design does not belong to a text on digital control. This is false. Analog control can be used as a first step toward obtaining a digital control. In addition, direct digital control design in the z-domain is similar in many ways to s-domain design.

Digital controller design is topic of Chapter 6. It begins with proportional control design then examines digital controllers based on analog design. The direct design of digital controllers is considered next. We consider root locus design in the z-plane for PI and PID controllers. We also consider a synthesis approach due to Ragazzini that allows us to specify the desired closed-loop transfer function. As a special case, we consider the design of deadbeat controllers that allow us to exactly track an input at the sampling points after a few sampling points. For completeness, we also examine frequency response design in the w-plane. This approach requires more experience because values of the stability margins must be significantly larger than in the more familiar analog design. As with analog design, MATLAB is an integral part of the design process for all digital control approaches.

Chapter 7 covers state-space models and state-space realizations. First, we discuss analog state-space equations and their solutions. We include nonlinear analog equations and their linearization to obtain linear state-space equations. We then show that the solution of the analog state equations over a sampling period yields a discrete-time state-space model. Properties of the solution of the analog state equation can thus be used to analyze the discrete-time state equation. The discrete-time state equation is a recursion for which we obtain a solution by induction. In Chapter 8, we consider important properties of state–space models: stability, **controllability**, and **observability**. As in Chapter 4, we consider internal stability and input-output stability, but the treatment is based on the properties of the state-space model rather than those of the transfer function. Controllability is a property that characterizes our ability to drive the system from an arbitrary initial state to an arbitrary final state in finite time. Observability characterizes our ability to calculate the initial state of the system using its input and output measurements. Both are structural properties of the system that are independent of its stability. Next, we consider **realizations** of discrete-time systems. These are ways of implementing discrete-time systems through their state-space equations using summers and delays.

Chapter 9 covers the design of controllers for state-space models. We show that the system dynamics can be arbitrarily chosen using state feedback if the system is controllable. If the state is not available for feedback, we can design a state estimator or **observer** to estimate it from the output measurements. These are dynamic systems that mimic the system but include corrective feedback to account for errors that are inevitable in any implementation. We give two types of observers. The first is a simpler but more computationally costly full-order observer that estimates the entire state vector. The second is a reduced-order observer with the order reduced by virtue of the fact that the measurements are available and need not be estimated. Either observer can be used to provide an estimate of the state for feedback control, or for other purposes. Control schemes based on state estimates are said to use **observer state feedback**.

Chapter 10 deals with the optimal control of digital control systems. We consider the problem of unconstrained optimization, followed by constrained optimization, then generalize to dynamic optimization as constrained by the system dynamics. We are particularly interested in the linear quadratic regulator where optimization results are easy to interpret and the prerequisite mathematics background is minimal. We consider both the finite time and steady-state regulator and discuss conditions for the existence of the steady-state solution. The first 10 chapters are mostly restricted to linear discrete-time systems. Chapter 11 examines the far more complex behavior of nonlinear discrete-time systems. It begins with equilibrium points and their stability. It shows how equivalent discrete-time models can be easily obtained for some forms of nonlinear analog systems using **global** or **extended linearization**. It provides stability theorems and instability theorems using Lyapunov stability theory. The theory gives sufficient conditions for nonlinear systems, and failure of either the stability or instability tests is inconclusive. For linear systems, Lyapunov stability yields necessary and sufficient conditions. Lyapunov stability theory also allows us to design controllers by selecting a control that yields a closed-loop system that meets the Lyapunov stability conditions. For the classes of nonlinear systems for which extended linearization is straightforward, linear design methodologies can yield nonlinear controllers.

Chapter 12 deals with practical issues that must be addressed for the successful implementation of digital controllers. In particular, the hardware and software requirements for the correct implementation of a digital control system are analyzed. We discuss the choice of the sampling frequency in the presence of anti-aliasing filters and the effects of quantization, rounding, and truncation errors. We also discuss **bumpless switching** from automatic to manual control, avoiding discontinuities in the control input. Our discussion naturally leads to approaches for the effective implementation of a PID controller. Finally, we consider nonuniform sampling, where the sampling frequency is changed during control operation, and multirate sampling, where samples of the process outputs are available at a slower rate than the controller sampling rate.

Supporting Material

The following resources are available to instructors adopting this text for use in their courses. Please visit *www.elsevierdirect9780123744982.com* to register for access to these materials:

Instructor solutions manual. Fully typeset solutions to the end-of-chapter problems in the text.

PowerPoint images. Electronic images of the figures and tables from the book, useful for creating lectures.

ACKNOWLEDGMENTS

We would like to thank the anonymous reviewers who provided excellent suggestions for improving the text. We would also like to thank Dr. Qing-Chang Zhong of the University of Liverpool who suggested the cooperation between the two authors that led to the completion of this text. We would also like to thank Joseph P. Hayton, Maria Alonso, Mia Kheyfetz, Marilyn Rash, and the Elsevier staff for their help in producing the text. Finally, we would like to thank our wives Betsy Fadali and Silvia Visioli for their support and love throughout the months of writing this book.

Introduction to Digital Control

In most modern engineering systems, there is a need to control the evolution with time of one or more of the system variables. Controllers are required to ensure satisfactory transient and steady-state behavior for these engineering systems. To guarantee satisfactory performance in the presence of disturbances and model uncertainty, most controllers in use today employ some form of negative feedback. A sensor is needed to measure the controlled variable and compare its behavior to a reference signal. Control action is based on an error signal defined as the difference between the reference and the actual values.

The controller that manipulates the error signal to determine the desired control action has classically been an analog system, which includes electrical, fluid, pneumatic, or mechanical components. These systems all have *analog* inputs and outputs (i.e., their input and output signals are defined over a continuous time interval and have values that are defined over a continuous range of amplitudes). In the past few decades, analog controllers have often been replaced by *digital* controllers whose inputs and outputs are defined at discrete time instances. The digital controllers are in the form of digital circuits, digital computers, or microprocessors.

Intuitively, one would think that controllers that continuously monitor the output of a system would be superior to those that base their control on sampled values of the output. It would seem that control variables (controller outputs) that change continuously would achieve better control than those that change periodically. This is in fact true! Had all other factors been identical for digital and analog control, analog control would be superior to digital control. What then is the reason behind the change from analog to digital that has occurred over the past few decades?

Objectives

After completing this chapter, the reader will be able to do the following:

1. Explain the reasons for the popularity of digital control systems.
2. Draw a block diagram for digital control of a given analog control system.
3. Explain the structure and components of a typical digital control system.

1.1 WHY DIGITAL CONTROL?

Digital control offers distinct advantages over analog control that explain its popularity. Here are some of its many advantages:

Accuracy. Digital signals are represented in terms of zeros and ones with typically 12 bits or more to represent a single number. This involves a very small error as compared to analog signals where noise and power supply drift are always present.

Implementation errors. Digital processing of control signals involves addition and multiplication by stored numerical values. The errors that result from digital representation and arithmetic are negligible. By contrast, the processing of analog signals is performed using components such as resistors and capacitors with actual values that vary significantly from the nominal design values.

Flexibility. An analog controller is difficult to modify or redesign once implemented in hardware. A digital controller is implemented in firmware or software, and its modification is possible without a complete replacement of the original controller. Furthermore, the structure of the digital controller need not follow one of the simple forms that are typically used in analog control. More complex controller structures involve a few extra arithmetic operations and are easily realizable.

Speed. The speed of computer hardware has increased exponentially since the 1980s. This increase in processing speed has made it possible to sample and process control signals at very high speeds. Because the interval between samples, the sampling period, can be made very small, digital controllers achieve performance that is essentially the same as that based on continuous monitoring of the controlled variable.

Cost. Although the prices of most goods and services have steadily increased, the cost of digital circuitry continues to decrease. Advances in very large scale integration (VLSI) technology have made it possible to manufacture better, faster, and more reliable integrated circuits and to offer them to the consumer at a lower price. This has made the use of digital controllers more economical even for small, low-cost applications.

1.2 THE STRUCTURE OF A DIGITAL CONTROL SYSTEM

To control a physical system or process using a digital controller, the controller must receive measurements from the system, process them, and then send control signals to the actuator that effects the control action. In almost all applications, both the plant and the actuator are analog systems. This is a situation

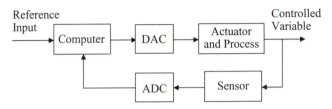

FIGURE 1.1

Configuration of a digital control system.

where the controller and the controlled do not "speak the same language" and some form of translation is required. The translation from controller language (digital) to physical process language (analog) is performed by a digital-to-analog converter, or DAC. The translation from process language to digital controller language is performed by an analog-to-digital converter, or ADC. A sensor is needed to monitor the controlled variable for feedback control. The combination of the elements discussed here in a control loop is shown in Figure 1.1. Variations on this control configuration are possible. For example, the system could have several reference inputs and controlled variables, each with a loop similar to that of Figure 1.1. The system could also include an inner loop with digital or analog control.

1.3 EXAMPLES OF DIGITAL CONTROL SYSTEMS

In this section, we briefly discuss examples of control systems where digital implementation is now the norm. There are many other examples of industrial processes that are digitally controlled, and the reader is encouraged to seek other examples from the literature.

1.3.1 Closed-Loop Drug Delivery System

Several chronic diseases require the regulation of the patient's blood levels of a specific drug or hormone. For example, some diseases involve the failure of the body's natural closed-loop control of blood levels of nutrients. Most prominent among these is the disease diabetes, where the production of the hormone insulin that controls blood glucose levels is impaired.

To design a closed-loop drug delivery system, a sensor is utilized to measure the levels of the regulated drug or nutrient in the blood. This measurement is converted to digital form and fed to the control computer, which drives a pump that injects the drug into the patient's blood. A block diagram of the drug delivery system is shown in Figure 1.2. Refer to Carson and Deutsch (1992) for a more detailed example of a drug delivery system.

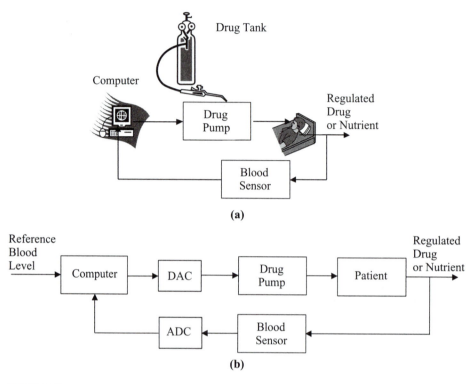

FIGURE 1.2

Drug delivery digital control system. (a) Schematic of a drug delivery system. (b) Block diagram of a drug delivery system.

1.3.2 Computer Control of an Aircraft Turbojet Engine

To achieve the high performance required for today's aircraft, turbojet engines employ sophisticated computer control strategies. A simplified block diagram for turbojet computer control is shown in Figure 1.3. The control requires feedback of the engine state (speed, temperature, and pressure), measurements of the aircraft state (speed and direction), and pilot command.

1.3.3 Control of a Robotic Manipulator

Robotic manipulators are capable of performing repetitive tasks at speeds and accuracies that far exceed those of human operators. They are now widely used in manufacturing processes such as spot welding and painting. To perform their tasks accurately and reliably, manipulator hand (or end-effector) positions and velocities are controlled digitally. Each motion or degree of freedom (D.O.F.) of the manipulator is positioned using a separate position control system. All the

(a)

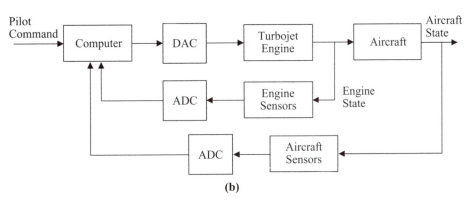

(b)

FIGURE 1.3

Turbojet engine control system. (a) F-22 military fighter aircraft. (b) Block diagram of an engine control system.

motions are coordinated by a supervisory computer to achieve the desired speed and positioning of the end-effector. The computer also provides an interface between the robot and the operator that allows programming the lower-level controllers and directing their actions. The control algorithms are downloaded from the supervisory computer to the control computers, which are typically specialized microprocessors known as digital signal processing (DSP) chips. The DSP chips execute the control algorithms and provide closed-loop control for the manipulator. A simple robotic manipulator is shown in Figure 1.4a, and a block diagram of its digital control system is shown in Figure 1.4b. For simplicity, only one motion control loop is shown in Figure 1.4, but there are actually n loops for an n-D.O.F. manipulator.

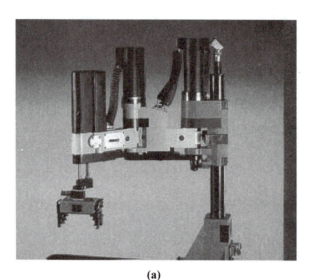

(a)

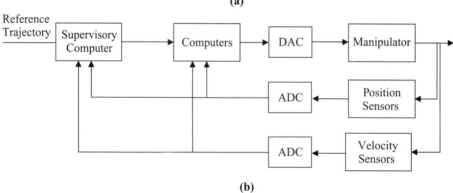

(b)

FIGURE 1.4

Robotic manipulator control system. (a) 3-D.O.F. robotic manipulator. (b) Block diagram of a manipulator control system.

RESOURCES

Carson, E. R., and T. Deutsch, A spectrum of approaches for controlling diabetes, *Control Syst. Mag.,* 12(6):25-31, 1992.

Chen, C. T., *Analog and Digital Control System Design,* Saunders–HBJ, 1993.

Koivo, A. J., *Fundamentals for Control of Robotic Manipulators,* Wiley, 1989.

Shaffer, P. L., A multiprocessor implementation of a real-time control of turbojet engine, *Control Syst. Mag.,* 10(4):38-42, 1990.

PROBLEMS

1.1 A fluid level control system includes a tank, a level sensor, a fluid source, and an actuator to control fluid inflow. Consult any classical control text[1] to obtain a block diagram of an analog fluid control system. Modify the block diagram to show how the fluid level could be digitally controlled.

1.2 If the temperature of the fluid in Problem 1.1 is to be regulated together with its level, modify the analog control system to achieve the additional control. (*Hint:* An additional actuator and sensor are needed.) Obtain a block diagram for the two-input-two-output control system with digital control.

1.3 Position control servos are discussed extensively in classical control texts. Draw a block diagram for a direct current motor position control system after consulting your classical control text. Modify the block diagram to obtain a digital position control servo.

1.4 Repeat Problem 1.3 for a velocity control servo.

1.5 A ballistic missile is required to follow a predetermined flight path by adjusting its angle of attack α (the angle between its axis and its velocity vector v). The angle of attack is controlled by adjusting the thrust angle δ (angle between the thrust direction and the axis of the missile). Draw a block diagram for a digital control system for the angle of attack including a gyroscope to measure the angle α and a motor to adjust the thrust angle δ.

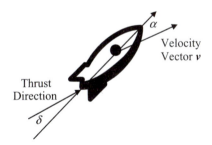

FIGURE P1.5

Missile angle-of-attack control.

1.6 A system is proposed to remotely control a missile from an earth station. Because of cost and technical constraints, the missile coordinates would be measured every 20 seconds for a missile speed of up to 500 m/s. Is such a control scheme feasible? What would the designers need to do to eliminate potential problems?

---

[1]See, for example, J. Van deVegte, *Feedback Control Systems*, Prentice Hall, 1994.

1.7 The control of the recording head of a dual actuator hard disk drive (HDD) requires two types of actuators to achieve the required a high real density. The first is a coarse voice coil motor (VCM) with a large stroke but slow dynamics, and the second is a fine piezoelectric transducer (PZT) with a small stroke and fast dynamics. A sensor measures the head position and the position error is fed to a separate controller for each actuator. Draw a block diagram for a dual actuator digital control system for the HDD.[2]

---

[2]J. Ding, F. Marcassa, S.-C. Wu, and M. Tomizuka, Multirate control for computational saving, *IEEE Trans. Control Systems Tech.*, *14*(1):165-169, 2006.

Discrete-Time Systems

Digital control involves systems whose control is updated at discrete time instants. Discrete-time models provide mathematical relations between the system variables at these time instants. In this chapter, we develop the mathematical properties of discrete-time models that are used throughout the remainder of the text. For most readers, this material provides a concise review of material covered in basic courses on control and system theory. However, the material is self-contained, and familiarity with discrete-time systems is not required. We begin with an example that illustrates how discrete-time models arise from analog systems under digital control.

Objectives

After completing this chapter, the reader will be able to do the following:

1. Explain why difference equations result from digital control of analog systems.
2. Obtain the z-transform of a given time sequence and the time sequence corresponding to a function of z.
3. Solve linear time-invariant (LTI) difference equations using the z-transform.
4. Obtain the z-transfer function of an LTI system.
5. Obtain the time response of an LTI system using its transfer function or impulse response sequence.
6. Obtain the modified z-transform for a sampled time function.
7. Select a suitable sampling period for a given LTI system based on its dynamics.

2.1 ANALOG SYSTEMS WITH PIECEWISE CONSTANT INPUTS

In most engineering applications, it is necessary to control a physical system or **plant** so that it behaves according to given design specifications. Typically, the plant is analog, the control is piecewise constant, and the control action is updated periodically. This arrangement results in an overall system that is conveniently described by a discrete-time model. We demonstrate this concept using a simple example.

EXAMPLE 2.1

Consider the tank control system of Figure 2.1. In the figure, lowercase letters denote perturbations from fixed steady-state values. The variables are defined as

- H = steady-state fluid height in the tank
- h = height perturbation from the nominal value
- Q = steady-state flow through the tank
- q_i = inflow perturbation from the nominal value
- q_o = outflow perturbation from the nominal value

It is necessary to maintain a constant fluid level by adjusting the fluid flow rate into the tank. Obtain an analog mathematical model of the tank, and use it to obtain a discrete-time model for the system with piecewise constant inflow q_i and output h.

Solution

Although the fluid system is nonlinear, a linear model can satisfactorily describe the system under the assumption that fluid level is regulated around a constant value. The linearized model for the outflow valve is analogous to an electrical resistor and is given by

$$h = R q_o$$

where h is the perturbation in tank level from nominal, q_o is the perturbation in the outflow from the tank from a nominal level Q, and R is the fluid resistance of the valve.

Assuming an incompressible fluid, the principle of conservation of mass reduces to the volumetric balance: rate of fluid volume increase = rate of volume fluid in—rate of volume fluid out:

$$\frac{dC(h+H)}{dt} = (q_i + Q) - (q_o + Q)$$

where C is the area of the tank or its fluid capacitance. The term H is a constant and its derivative is zero, and the term Q cancels so that the remaining terms only involve perturba-

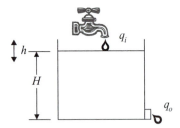

FIGURE 2.1

Fluid level control system.

tions. Substituting for the outflow q_0 from the linearized valve equation into the volumetric fluid balance gives the analog mathematical model

$$\frac{dh}{dt} + \frac{h}{\tau} = \frac{q_i}{C}$$

where $\tau = RC$ is the fluid time constant for the tank. The solution of this differential equation is

$$h(t) = e^{-(t-t_0)/\tau} h(t_0) + \frac{1}{C} \int_{t_0}^{t} e^{-(t-\lambda)/\tau} q_i(\lambda) d\lambda$$

Let q_i be constant over each sampling period T, that is, $q_i(t) = q_i(k)$ = constant for t in the interval $[k\ T, (k + 1)\ T]$. Then we can solve the analog equation over any sampling period to obtain

$$h(k+1) = e^{-T/\tau} h(k) + R\left[1 - e^{-T/\tau}\right] q_i(k)$$

where the variables at time kT are denoted by the argument k. This is the desired discrete-time model describing the system with piecewise constant control. Details of the solution are left as an exercise (Problem 2.1).

The discrete-time model obtained in Example 2.1 is known as a difference equation. Because the model involves a linear time-invariant analog plant, the equation is linear time invariant. Next, we briefly discuss difference equations; then we introduce a transform used to solve them.

2.2 DIFFERENCE EQUATIONS

Difference equations arise in problems where the independent variable, usually time, is assumed to have a discrete set of possible values. The nonlinear difference equation

$$y(k+n) = f[y(k+n-1), y(k+n-2), \ldots, y(k+1), y(k), u(k+n), \atop u(k+n-1), \ldots, u(k+1), u(k)] \tag{2.1}$$

with forcing function $u(k)$ is said to be of order n because the difference between the highest and lowest time arguments of $y(.)$ and $u(.)$ is n. The equations we deal with in this text are almost exclusively linear and are of the form

$$y(k+n) + a_{n-1} y(k+n-1) + \ldots + a_1 y(k+1) + a_0 y(k) \atop = b_n u(k+n) + b_{n-1} u(k+n-1) + \ldots + b_1 u(k+1) + b_0 u(k) \tag{2.2}$$

We further assume that the coefficients a_i, b_i, $i = 0, 1, 2, \ldots$, are constant. The difference equation is then referred to as linear time invariant, or LTI. If the forcing function $u(k)$ is equal to zero, the equation is said to be *homogeneous*.

EXAMPLE 2.2

For each of the following difference equations, determine the order of the equation. Is the equation (a) linear, (b) time invariant, or (c) homogeneous?

1. $y(k + 2) + 0.8y(k + 1) + 0.07y(k) = u(k)$
2. $y(k + 4) + \sin(0.4k)y(k + 1) + 0.3y(k) = 0$
3. $y(k + 1) = -0.1y^2(k)$

Solution

1. The equation is second order. All terms enter the equation linearly and have constant coefficients. The equation is therefore LTI. A forcing function appears in the equation, so it is nonhomogeneous.
2. The equation is fourth order. The second coefficient is time dependent but all the terms are linear and there is no forcing function. The equation is therefore linear time varying and homogeneous.
3. The equation is first order. The right-hand side (RHS) is a nonlinear function of $y(k)$ but does not include a forcing function or terms that depend on time explicitly. The equation is therefore nonlinear, time invariant, and homogeneous.

Difference equations can be solved using classical methods analogous to those available for differential equations. Alternatively, z-transforms provide a convenient approach for solving LTI equations, as discussed in the next section.

2.3 THE z-TRANSFORM

The z-transform is an important tool in the analysis and design of discrete-time systems. It simplifies the solution of discrete-time problems by converting LTI difference equations to algebraic equations and convolution to multiplication. Thus, it plays a role similar to that served by Laplace transforms in continuous-time problems. Because we are primarily interested in application to digital control systems, this brief introduction to the z-transform is restricted to **causal signals** (i.e., signals with zero values for negative time) and the one-sided z-transform.

The following are two alternative definitions of the z-transform.

Definition 2.1: Given the causal sequence $\{u_0, u_1, u_2, \ldots, u_k, \ldots\}$, its z-transform is defined as

$$U(z) = u_0 + u_1 z^{-1} + u_2 z^{-2} + \ldots + u_k z^{-k}$$

$$= \sum_{k=0}^{\infty} u_k z^{-k} \tag{2.3}$$

∎

The variable z^{-1} in the preceding equation can be regarded as a time delay operator. The z-transform of a given sequence can be easily obtained as in the following example.

Definition 2.2: Given the impulse train representation of a discrete-time signal,

$$u^*(t) = u_0\delta(t) + u_1\delta(t-T) + u_2\delta(t-2T) + \ldots + u_k\delta(t-kT) + \ldots$$

$$= \sum_{k=0}^{\infty} u_k\delta(t-kT) \tag{2.4}$$

the Laplace transform of (2.4) is

$$U^*(s) = u_0 + u_1 e^{-sT} + u_2 e^{-2sT} + \ldots + u_k e^{-ksT} + \ldots$$

$$= \sum_{k=0}^{\infty} u_k \left(e^{-sT}\right)^k \tag{2.5}$$

Let z be defined by

$$z = e^{sT} \tag{2.6}$$

Then substituting from (2.6) in (2.5) yields the z-transform expression (2.3).   ∎

EXAMPLE 2.3

Obtain the z-transform of the sequence $\{u_k\}_{k=0}^{\infty} = \{1, 1, 3, 2, 0, 4, 0, 0, 0, \ldots\}$.

Solution

Applying Definition 2.1 gives $U(z) = 1 + 3z^{-1} + 2z^{-2} + 4z^{-4}$.

Although the preceding two definitions yield the same transform, each has its advantages and disadvantages. The first definition allows us to avoid the use of impulses and the Laplace transform. The second allows us to treat z as a complex variable and to use some of the familiar properties of the Laplace transform (such as linearity).

Clearly, it is possible to use Laplace transformation to study discrete time, continuous time, and mixed systems. However, the z-transform offers significant simplification in notation for discrete-time systems and greatly simplifies their analysis and design.

2.3.1 *z*-Transforms of Standard Discrete-Time Signals

Having defined the z-transform, we now obtain the z-transforms of commonly used discrete-time signals such as the sampled step, exponential, and the discrete-time impulse. The following identities are used repeatedly to derive several important results:

$$\sum_{k=0}^{n} a^k = \frac{1-a^{n+1}}{1-a}, \quad a \neq 1$$

$$\sum_{k=0}^{\infty} a^k = \frac{1}{1-a}, \quad |a| < 1 \tag{2.7}$$

EXAMPLE 2.4: UNIT IMPULSE

Consider the discrete-time impulse (Figure 2.2)

$$u(k) = \delta(k) = \begin{cases} 1, & k = 0 \\ 0, & k \neq 0 \end{cases}$$

Applying Definition 2.1 gives the z-transform

$$U(z) = 1$$

Alternatively, one may consider the impulse-sampled version of the delta function $u^*(t)$ = $\delta(t)$. This has the Laplace transform

$$U^*(s) = 1$$

Substitution from (2.6) has no effect. Thus, the z-transform obtained using Definition 2.2 is identical to that obtained using Definition 2.1.

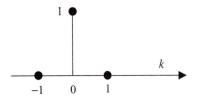

FIGURE 2.2

Discrete-time impulse.

EXAMPLE 2.5: SAMPLED STEP

Consider the sequence $\{u_k\}_{k=0}^{\infty} = \{1, 1, 1, 1, 1, 1, \ldots\}$. Definition 2.1 gives the z-transform

$$U(z) = 1 + z^{-1} + z^{-2} + z^{-3} + \ldots + z^{-k} + \ldots$$

$$= \sum_{k=0}^{\infty} z^{-k}$$

Using the identity (2.7) gives the following closed-form expression for the z-transform:

$$U(z) = \frac{1}{1 - z^{-1}}$$

$$= \frac{z}{z - 1}$$

Note that (2.7) is only valid for $|z| < 1$. This implies that the z-transform expression we obtain has a region of convergence outside which it is not valid. The region of convergence must be clearly given when using the more general two-sided transform with functions that are nonzero for negative time. However, for the one-sided z-transform and time functions that are zero for negative time, we can essentially extend regions of convergence and use the z-transform in the entire z-plane.[1] (See Figure 2.3.)

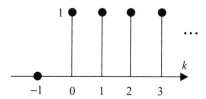

FIGURE 2.3

Sampled unit step.

EXAMPLE 2.6: EXPONENTIAL

Let

$$u(k) = \begin{cases} a^k, & k \geq 0 \\ 0, & k < 0 \end{cases}$$

Then

$$U(z) = 1 + az^{-1} + a^2 z^{-2} + \ldots + a^k z^{-k} + \ldots$$

Using (2.7), we obtain

$$U(z) = \frac{1}{1-(a/z)}$$

$$= \frac{z}{z-a}$$

As in Example 2.5, we can use the transform in the entire z-plane in spite of the validity condition for (2.7) because our time function is zero for negative time. (See Figure 2.4.)

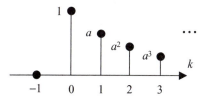

FIGURE 2.4

Sampled exponential.

[1]The idea of extending the definition of a complex function to the entire complex plane is known as **analytic continuation**. For a discussion of this topic, consult any text on complex analysis.

2.3.2 Properties of the z-Transform

The z-transform can be derived from the Laplace transform as shown in Definition 2.2. Hence, it shares several useful properties with the Laplace transform, which can be stated without proof. These properties can also be easily proved directly and the proofs are left as an exercise for the reader. Proofs are provided for properties that do not obviously follow from the Laplace transform.

Linearity
This equation follows directly from the linearity of the Laplace transform.

$$\mathscr{Z}\{\alpha f_1(k) + \beta f_2(k)\} = \alpha F_1(z) + \beta F_2(z) \tag{2.8}$$

EXAMPLE 2.7

Find the z-transform of the causal sequence

$$f(k) = 2 \times 1(k) + 4\delta(k), \quad k = 0, 1, 2, \ldots$$

Solution
Using linearity, the transform of the sequence is

$$F(z) = \mathscr{Z}\{2 \times 1(k) + 4\delta(k)\} = 2\,\mathscr{Z}\{1(k)\} + 4\mathscr{Z}\{\delta(k)\} = \frac{2z}{z-1} + 4 = \frac{6z-4}{z-1}$$

Time Delay
This equation follows from the time delay property of the Laplace transform and equation (2.6).

$$\mathscr{Z}\{f(k-n)\} = z^{-n}F(z) \tag{2.9}$$

EXAMPLE 2.8

Find the z-transform of the causal sequence

$$f(k) = \begin{cases} 4, & k = 2, 3, \ldots \\ 0, & \text{otherwise} \end{cases}$$

Solution
The given sequence is a sampled step starting at $k = 2$ rather than $k = 0$ (i.e., it is delayed by two sampling periods). Using the delay property, we have

$$F(z) = \mathscr{Z}\{4 \times 1(k-2)\} = 4\,z^{-2}\mathscr{Z}\{1(k)\} = z^{-2}\frac{4z}{z-1} = \frac{4}{z(z-1)}$$

Time Advance

$$\mathscr{Z}\{f(k+1)\} = zF(z) - zf(0)$$
$$\mathscr{Z}\{f(k+n)\} = z^n F(z) - z^n f(0) - z^{n-1} f(1) - \ldots - z f(n-1) \qquad \text{(2.10)}$$

PROOF. Only the first part of the theorem is proved here. The second part can be easily proved by induction. We begin by applying the z-transform Definition 2.1 to a discrete-time function advanced by one sampling interval. This gives

$$\mathscr{Z}\{f(k+1)\} = \sum_{k=0}^{\infty} f(k+1) z^{-k}$$

$$= z \sum_{k=0}^{\infty} f(k+1) z^{-(k+1)}$$

Now add and subtract the initial condition f(0) to obtain

$$\mathscr{Z}\{f(k+1)\} = z \left\{ \left[f(0) + \sum_{k=0}^{\infty} f(k+1) z^{-(k+1)} \right] - f(0) \right\}$$

Next, change the index of summation to $m = k + 1$ and rewrite the z-transform as

$$\mathscr{Z}\{f(k+1)\} = z \left\{ \left[\sum_{m=0}^{\infty} f(m) z^{-m} \right] - f(0) \right\}$$

$$= zF(z) - zf(0) \qquad \blacksquare$$

EXAMPLE 2.9

Using the time advance property, find the z-transform of the causal sequence

$$\{f(k)\} = \{4, 8, 16, \ldots\}$$

Solution
The sequence can be written as

$$f(k) = 2^{k+2} = g(k+2), \quad k = 0, 1, 2, \ldots$$

where g(k) is the exponential time function

$$g(k) = 2^k, \quad k = 0, 1, 2, \ldots$$

Using the time advance property, we write the transform

$$F(z) = z^2 G(z) - z^2 g(0) - z g(1) = z^2 \frac{z}{z-2} - z^2 - 2z = \frac{4z}{z-2}$$

Clearly, the solution can be obtained directly by rewriting the sequence as

$$\{f(k)\} = 4\{1, 2, 4, \ldots\}$$

and using the linearity of the z-transform.

Multiplication by Exponential

$$\mathscr{Z}\left\{a^{-k}f(k)\right\} = F(az) \tag{2.11}$$

PROOF

$$LHS = \sum_{k=0}^{\infty} a^{-k}f(k)z^{-k} = \sum_{k=0}^{\infty} f(k)(az)^{-k} = F(az) \qquad \blacksquare$$

EXAMPLE 2.10

Find the z-transform of the exponential sequence

$$f(k) = e^{-\alpha kT}, \quad k = 0, 1, 2, \ldots$$

Solution

Recall that the z-transform of a sampled step is

$$F(z) = \left(1 - z^{-1}\right)^{-1}$$

and observe that $f(k)$ can be rewritten as

$$f(k) = \left(e^{\alpha T}\right)^{-k} \times 1, \quad k = 0, 1, 2, \ldots$$

Then apply the multiplication by exponential property to obtain

$$\mathscr{Z}\left\{\left(e^{\alpha T}\right)^{-k}f(k)\right\} = \left[1 - \left(e^{\alpha T}z\right)^{-1}\right]^{-1} = \frac{z}{z - e^{-\alpha T}}$$

This is the same as the answer obtained in Example 2.6.

Complex Differentiation

$$\mathscr{Z}\left\{k^{m}f(k)\right\} = \left(-z\frac{d}{dz}\right)^{m}F(z) \tag{2.12}$$

PROOF. To prove the property by induction, we first establish its validity for $m = 1$. Then we assume its validity for any m and prove it for $m + 1$. This establishes its validity for $1 + 1 = 2$, then $2 + 1 = 3$, and so on.

For $m = 1$, we have

$$\mathscr{Z}\left\{kf(k)\right\} = \sum_{k=0}^{\infty} kf(k)z^{-k} = \sum_{k=0}^{\infty} f(k)\left(-z\frac{d}{dz}\right)z^{-k}$$

$$= \left(-z\frac{d}{dz}\right)\sum_{k=0}^{\infty} f(k)z^{-k} = \left(-z\frac{d}{dz}\right)F(z)$$

Next, let the statement be true for any m and define the sequence

$$f_m(k) = k^m f(k), \quad k = 0, 1, 2, \ldots$$

and obtain the transform

$$
\begin{aligned}
\mathscr{Z}\{k\, f_m(k)\} &= \sum_{k=0}^{\infty} k\, f_m(k)\, z^{-k} \\
&= \sum_{k=0}^{\infty} f_m(k)\left(-z\frac{d}{dz}\right) z^{-k} \\
&= \left(-z\frac{d}{dz}\right) \sum_{k=0}^{\infty} f_m(k)\, z^{-k} = \left(-z\frac{d}{dz}\right) F_m(z)
\end{aligned}
$$

Substituting for $F_m(z)$, we obtain the result

$$\mathscr{Z}\{k\, f_m(k)\} = \left(-z\frac{d}{dz}\right)^{m+1} F(z) \qquad \blacksquare$$

EXAMPLE 2.11

Find the z-transform of the sampled ramp sequence

$$f(k) = k, \quad k = 0, 1, 2, \ldots$$

Solution

Recall that the z-transform of a sampled step is

$$F(z) = \frac{z}{z-1}$$

and observe that $f(k)$ can be rewritten as

$$f(k) = k \times 1, \quad k = 0, 1, 2, \ldots$$

Then apply the complex differentiation property to obtain

$$\mathscr{Z}\{k \times 1\} = \left(-z\frac{d}{dz}\right)\left(\frac{z}{z-1}\right) = (-z)\frac{(z-1)-z}{(z-1)^2} = \frac{z}{(z-1)^2}$$

2.3.3 Inversion of the z-Transform

Because the purpose of z-transformation is often to simplify the solution of time domain problems, it is essential to inverse-transform z-domain functions. As in the case of Laplace transforms, a complex integral can be used for inverse transformation. This integral is difficult to use and is rarely needed in engineering applications. Two simpler approaches for inverse z-transformation are discussed in this section.

Long Division

This approach is based on Definition 2.1, which relates a time sequence to its z-transform directly. We first use long division to obtain as many terms as desired of the z-transform expansion; then we use the coefficients of the expansion to write the time sequence. The following two steps give the inverse z-transform of a function $F(z)$:

1. Using long division, expand $F(z)$ as a series to obtain

$$F_i(z) = f_0 + f_1 z^{-1} + \ldots + f_i z^{-i} = \sum_{k=0}^{i} f_k z^{-k}$$

2. Write the inverse transform as the sequence

$$\{f_0, f_1, \ldots, f_i, \ldots\}$$

The number of terms obtained by long division i is selected to yield a sufficient number of points in the time sequence.

EXAMPLE 2.12

Obtain the inverse z-transform of the function $F(z) = \dfrac{z+1}{z^2 + 0.2z + 0.1}$

Solution

1. Long Division

$$
\begin{array}{r}
z^{-1} + 0.8z^{-2} - 0.26z^{-3} + \ldots \ldots \\
z^2 + 0.2z + 0.1 \overline{\smash{)}\, z + 1} \\
z + 0.2 + 0.1z^{-1} \\
\overline{0.8 - 0.10z^{-1}} \\
0.8 + 0.16z^{-1} + 0.08z^{-2} \\
\overline{-0.26z^{-1} - \ldots}
\end{array}
$$

Thus, $F_t(z) = 0 + z^{-1} + 0.8z^{-2} - 0.26z^{-3}$

2. Inverse Transformation

$$\{f_k\} = \{0, 1, 0.8, -0.26, \ldots\}$$

Partial Fraction Expansion

This method is almost identical to that used in inverting Laplace transforms. However, because most z-functions have the term z in their numerator, it is often convenient to expand $F(z)/z$ rather than $F(z)$. As with Laplace transforms, partial fraction expansion allows us to write the function as the sum of simpler functions that are the *z*-transforms of known discrete-time functions. The time functions are available in *z*-transform tables such as the table provided in Appendix I.

The procedure for inverse *z*-transformation is

1. Find the partial fraction expansion of $F(z)/z$ or $F(z)$.
2. Obtain the inverse transform $f(k)$ using the *z*-transform tables.

We consider three types of *z*-domain functions $F(z)$: functions with simple (nonrepeated) real poles, functions with complex conjugate and real poles, and functions with repeated poles. We discuss examples that demonstrate partial fraction expansion and inverse *z*-transformation in each case.

Case 1

SIMPLE REAL ROOTS

The most convenient method to obtain the partial fraction expansion of a function with simple real roots is the method of residues. The residue of a complex function $F(z)$ at a simple pole z_i is given by

$$A_i = (z - z_i) F(z)]_{z \to z_i} \qquad (2.13)$$

This is the partial fraction coefficient of the ith term of the expansion

$$F(z) = \sum_{i=1}^{n} \frac{A_i}{z - z_i} \qquad (2.14)$$

Because most terms in the *z*-transform tables include a z in the numerator (see Appendix I), it is often convenient to expand $F(z)/z$ and then to multiply both sides by z to obtain an expansion whose terms have a z in the numerator. Except for functions that already have a z in the numerator, this approach is slightly longer but has the advantage of simplifying inverse transformation. Both methods are examined through the following example.

EXAMPLE 2.13

Obtain the inverse *z*-transform of the function $F(z) = \dfrac{z+1}{z^2 + 0.3z + 0.02}$.

Solution
It is instructive to solve this problem using two different methods. First we divide by z; then we obtain the partial fraction expansion.

1. Partial Fraction Expansion

Dividing the function by z, we expand as

$$\frac{F(z)}{z} = \frac{z+1}{z(z^2+0.3z+0.02)}$$

$$= \frac{A}{z} + \frac{B}{z+0.1} + \frac{C}{z+0.2}$$

where the partial fraction coefficients are given by

$$A = z\,\frac{F(z)}{z}\bigg]_{z=0} = F(0) = \frac{1}{0.02} = 50$$

$$B = (z+0.1)\frac{F(z)}{z}\bigg]_{z=-0.1} = \frac{1-0.1}{(-0.1)(0.1)} = -90$$

$$C = (z+0.2)\frac{F(z)}{z}\bigg]_{z=-0.2} = \frac{1-0.2}{(-0.2)(-0.1)} = 40$$

Thus, the partial fraction expansion is

$$F(z) = \frac{50z}{z} - \frac{90z}{z+0.1} + \frac{40z}{z+0.2}$$

2. Table Lookup

$$f(k) = \begin{cases} 50\delta(k) - 90(-0.1)^k + 40(-0.2)^k, & k \geq 0 \\ 0, & k < 0 \end{cases}$$

Note that $f(0) = 0$ so the time sequence can be rewritten as

$$f(k) = \begin{cases} -90(-0.1)^k + 40(-0.2)^k, & k \geq 1 \\ 0, & k < 1 \end{cases}$$

Now, we solve the same problem without dividing by z.

1. Partial Fraction Expansion

We obtain the partial fraction expansion directly

$$F(z) = \frac{z+1}{z^2+0.3z+0.02}$$

$$= \frac{A}{z+0.1} + \frac{B}{z+0.2}$$

where the partial fraction coefficients are given by

$$A = (z+0.1)F(z)]_{z=-0.1} = \frac{1-0.1}{0.1} = 9$$

$$B = (z+0.2)F(z)]_{z=-0.2} = \frac{1-0.2}{-0.1} = -8$$

Thus, the partial fraction expansion is

$$F(z) = \frac{9}{z+0.1} - \frac{8}{z+0.2}$$

2. Table Lookup
Standard z-transform tables do not include the terms in the expansion of $F(z)$. However, $F(z)$ can be written as

$$F(z) = \frac{9z}{z+0.1} z^{-1} - \frac{8z}{z+0.2} z^{-1}$$

Then we use the delay theorem to obtain the inverse transform

$$f(k) = \begin{cases} 9(-0.1)^{k-1} - 8(-0.2)^{k-1}, & k \geq 1 \\ 0, & k < 1 \end{cases}$$

Verify that this is the answer obtained earlier when dividing by z written in a different form (observe the exponent in the preceding expression).

Although it is clearly easier to obtain the partial fraction expansion without dividing by z, inverse transforming requires some experience. There are situations where division by z may actually simplify the calculations as seen in the following example.

EXAMPLE 2.14

Find the inverse z-transform of the function

$$F(z) = \frac{z}{(z+0.1)(z+0.2)(z+0.3)}$$

Solution
1. Partial Fraction Expansion
Dividing by z simplifies the numerator and gives the expansion

$$\frac{F(z)}{z} = \frac{1}{(z+0.1)(z+0.2)(z+0.3)}$$

$$= \frac{A}{z+0.1} + \frac{B}{z+0.2} + \frac{C}{z+0.3}$$

where the partial fraction coefficients are

$$A = (z + 0.1) \frac{F(z)}{z} \bigg]_{z=-0.1} = \frac{1}{(0.1)(0.2)} = 50$$

$$B = (z + 0.2) \frac{F(z)}{z} \bigg]_{z=-0.2} = \frac{1}{(-0.1)(0.1)} = -100$$

$$C = (z + 0.3) \frac{F(z)}{z} \bigg]_{z=-0.3} = \frac{1}{(-0.2)(-0.1)} = 50$$

Thus, the partial fraction expansion is

$$F(z) = \frac{50z}{z + 0.1} - \frac{100z}{z + 0.2} + \frac{50z}{z + 0.3}$$

2. Table Lookup

$$f(k) = \begin{cases} 50(-0.1)^k - 100(-0.2)^k + 50(-0.3)^k, & k \geq 0 \\ 0, & k < 0 \end{cases}$$

Case 2

COMPLEX CONJUGATE AND SIMPLE REAL ROOTS

For a function $F(z)$ with real and complex poles, the partial fraction expansion includes terms with real roots and others with complex roots. Assuming that $F(z)$ has real coefficients, then its complex roots occur in complex conjugate pairs and can be combined to yield a function with real coefficients and a quadratic denominator. To inverse-transform such a function, use the following z-transforms:

$$\mathcal{Z}\{e^{-\alpha k} \sin(k\omega_d)\} = \frac{e^{-\alpha} \sin(\omega_d) z}{z^2 - 2e^{-\alpha} \cos(\omega_d) z + e^{-2\alpha}} \tag{2.15}$$

$$\mathcal{Z}\{e^{-\alpha k} \cos(k\omega_d)\} = \frac{z[z - e^{-\alpha} \cos(\omega_d)]}{z^2 - 2e^{-\alpha} \cos(\omega_d) z + e^{-2\alpha}} \tag{2.16}$$

The denominators of the two transforms are identical and have complex conjugate roots. The numerators can be scaled and combined to give the desired inverse transform.

To obtain the partial fraction expansion, we use the residues method shown in Case 1. With complex conjugate poles, we obtain the partial fraction expansion

$$F(z) = \frac{Az}{z - p} + \frac{A^*z}{z - p^*} \tag{2.17}$$

We then inverse z-transform to obtain

$$f(k) = Ap^k + A^*p^{*k}$$

$$= |A||p|^k \left[e^{j(\theta_p k + \theta_A)} + e^{-j(\theta_p k + \theta_A)} \right]$$

where θ_p and θ_A are the angle of the pole p and the angle of the partial fraction coefficient A, respectively. We use the exponential expression for the cosine function to obtain

$$f(k) = 2|A||p|^k \cos(\theta_p k + \theta_A) \qquad (2.18)$$

Most modern calculators can perform complex arithmetic, and the residues method is preferable in most cases. Alternatively, by equating coefficients, we can avoid the use of complex arithmetic entirely but the calculations can be quite tedious. The following example demonstrates the two methods.

EXAMPLE 2.15

Find the inverse z-transform of the function

$$F(z) = \frac{z^3 + 2z + 1}{(z - 0.1)(z^2 + z + 0.5)}$$

Solution: Equating Coefficients

1. Partial Fraction Expansion
Dividing the function by z gives

$$\frac{F(z)}{z} = \frac{z^3 + 2z + 1}{z(z - 0.1)(z^2 + z + 0.5)}$$

$$= \frac{A_1}{z} + \frac{A_2}{z - 0.1} + \frac{Az + B}{z^2 + z + 0.5}$$

The first two coefficients can be easily evaluated as before. Thus,

$$A_1 = F(0) = -20$$

$$A_2 = (z - 0.1)\frac{F(z)}{z} \cong 19.689$$

To evaluate the remaining coefficients, we multiply the equation by the denominator and equate coefficients to obtain

$$z^3: \quad A_1 + A_2 + A = 1$$

$$z^1: \quad 0.4\,A_1 + 0.5A_2 - 0.1B = 2$$

where the coefficients of the third- and first-order terms yield separate equations in A and B. Because A_1 and A_2 have already been evaluated, we can solve each of the two equations for one of the remaining unknowns to obtain

$$A \cong 1.311 \quad B \cong -1.557$$

Had we chosen to equate coefficients without first evaluating A_1 and A_2, we would have faced that considerably harder task of solving four equations in four unknowns. The remaining coefficients can be used to check our calculations

$$z^0: \quad -0.05 A_1 = 0.05(20) = 1$$

$$z^2: \quad 0.9 A_1 + A_2 - 0.1A + B = 0.9(-20) + 19.689 - 0.1(1.311) - 1.557 \cong 0$$

The results of these checks are approximate, because approximations were made in the calculations of the coefficients. The partial fraction expansion is

$$F(z) = -20 + \frac{19.689\, z}{z - 0.1} + \frac{1.311 z^2 - 1.557 z}{z^2 + z + 0.5}$$

2. Table Lookup

The first two terms of the partial fraction expansion can be easily found in the z-transform tables. The third term resembles the transforms of a sinusoid multiplied by an exponential if rewritten as

$$\frac{1.311 z^2 - 1.557 z}{z^2 - 2(-0.5) z + 0.5} = \frac{1.311 z \left[z - e^{-\alpha} \cos(\omega_d) \right] - C z e^{-\alpha} \sin(\omega_d)}{z^2 - 2 e^{-\alpha} \cos(\omega_d) z + e^{-2\alpha}}$$

Starting with the constant term in the denominator, we equate coefficients to obtain

$$e^{-\alpha} = \sqrt{0.5} = 0.707$$

Next, the denominator z^1 term gives

$$\cos(\omega_d) = -0.5/e^{-\alpha} = -\sqrt{0.5} = -0.707$$

Thus, $\omega_d = 3\pi/4$, an angle in the second quadrant, with $\sin(\omega_d) = 0.707$.
    Finally, we equate the coefficients of z^1 in the numerator to obtain

$$-1.311 e^{-\alpha} \cos(\omega_d) - C e^{-\alpha} \sin(\omega_d) = -0.5(C - 1.311) = -1.557$$

and solve for $C = 4.426$. Referring to the z-transform tables, we obtain the inverse transform

$$f(k) = -20\delta(k) + 19.689(0.1)^k + (0.707)^k [1.311 \cos(3\pi k/4) - 4.426 \sin(3\pi k/4)]$$

for positive time k. The sinusoidal terms can be combined using the trigonometric identities

$$\sin(A - B) = \sin(A)\cos(B) - \sin(B)\cos(A)$$

$$\sin^{-1}(1.311/4.616) = 0.288$$

and the constant $4.616 = \sqrt{(1.311)^2 + (4.426)^2}$. This gives

$$f(k) = -20\delta(k) + 19.689(0.1)^k - 4.616(0.707)^k \sin(3\pi k/4 - 0.288) \big]$$

Residues

1. Partial Fraction Expansion
Dividing by z gives

$$\frac{F(z)}{z} = \frac{z^3 + 2z + 1}{z(z-0.1)\left[(z+0.5)^2 + 0.5^2\right]}$$

$$= \frac{A_1}{z} + \frac{A_2}{z-0.1} + \frac{A_3}{z+0.5-j0.5} + \frac{A_3^*}{z+0.5+j0.5}$$

The partial fraction expansion can be obtained as in the first approach

$$\left.\frac{z+1}{\cdots\,0.5+j0.5)}\right|_{z=-0.5+j0.5} \cong 0.656 + j2.213$$

$$\cdots + \frac{(0.656+j2.213)z}{z+0.5-j0.5} + \frac{(0.656-j2.213)z}{z+0.5+j0.5}$$

Cartesian to polar form:

$$.656 + j2.213 = 2.308e^{j1.283}$$

$$\cdots 9(0.1)^k + 4.616(0.707)^k \cos(3\pi k/4 + 1.283)\Big]$$

ned earlier because $1.283 - \pi/2 = -0.288$.

Case 3

REPEATED ROOTS

repeated root of multiplicity r, r partial fraction with the repeated root. The partial fraction expan-

$$\frac{\cdots z)}{\prod z - z_j} = \sum_{i=1}^{r} \frac{A_{1i}}{(z-z_1)^{r+1-i}} + \sum_{j=r+1}^{n} \frac{A_j}{z-z_j} \quad (2.19)$$

eated roots are governed by

$$\left.\frac{\cdots}{\cdots}(z-z_1)^r F(z)\right]_{z\to z_1}, \quad i=1,2,\ldots,r \quad (2.20)$$

ple or complex conjugate roots can be obtained

EXAMPLE 2.16

Obtain the inverse z-transform of the function

$$F(z) = \frac{1}{z^2(z - 0.5)}$$

Solution

1. Partial Fraction Expansion
Dividing by z gives

$$\frac{F(z)}{z} = \frac{1}{z^3(z - 0.5)} = \frac{A_{11}}{z^3} + \frac{A_{12}}{z^2} + \frac{A_{13}}{z} + \frac{A_4}{z - 0.5}$$

where

$$A_{11} = z^3 \frac{F(z)}{z}\Big|_{z=0} = \frac{1}{z - 0.5}\Big|_{z=0} = -2$$

$$A_{12} = \frac{1}{1!}\frac{d}{dz} z^3 \frac{F(z)}{z}\Big|_{z=0} = \frac{d}{dz}\frac{1}{z - 0.5}\Big|_{z=0} = \frac{-1}{(z - 0.5)^2}\Big|_{z=0} = -4$$

$$A_{13} = \frac{1}{2!}\frac{d^2}{dz^2} z^3 \frac{F(z)}{z}\Big|_{z=0}$$

$$= \left(\frac{1}{2}\right)\frac{d}{dz}\frac{-1}{(z - 0.5)^2}\Big|_{z=0} = \left(\frac{1}{2}\right)\frac{(-1)(-2)}{(z - 0.5)^3}\Big|_{z=0} = -8$$

$$A_4 = (z - 0.5)\frac{F(z)}{z}\Big|_{z=0.5} = \frac{1}{z^3}\Big|_{z=0.5} = 8$$

Thus, we have the partial fraction expansion

$$F(z) = \frac{1}{z^2(z - 0.5)} = \frac{8z}{z - 0.5} - 2z^{-2} - 4z^{-1} - 8$$

2. Table Lookup
The z-transform tables and Definition 2.1 yield

$$f(k) = \begin{cases} 8(0.5)^k - 2\delta(k - 2) - 4\delta(k - 1) - 8\delta(k), & k \geq 0 \\ 0, & k < 0 \end{cases}$$

Evaluating $f(k)$ at $k = 0, 1, 2$ yields

$$f(0) = 8 - 8 = 0$$

$$f(1) = 8(0.5) - 4 = 0$$

$$f(2) = 8(0.5)^2 - 2 = 0$$

We can therefore rewrite the inverse transform as

$$f(k) = \begin{cases} (0.5)^{k-3}, & k \geq 3 \\ 0, & k < 3 \end{cases}$$

Note that the solution can be obtained directly using the delay theorem without the need for partial fraction expansion because $F(z)$ can be written as

$$F(z) = \frac{z}{z - 0.5} z^{-3}$$

The delay theorem and the inverse transform of an exponential yield the solution obtained earlier.

2.3.4 The Final Value Theorem

The final value theorem allows us to calculate the limit of a sequence as k tends to infinity, if one exists, from the z-transform of the sequence. If one is only interested in the final value of the sequence, this constitutes a significant short cut. The main pitfall of the theorem is that there are important cases where the limit does not exist. The two main case are

1. An unbounded sequence
2. An oscillatory sequence

The reader is cautioned against blindly using the final value theorem, because this can yield misleading results.

Theorem 2.1: The Final Value Theorem. If a sequence approaches a constant limit as k tends to infinity, then the limit is given by

$$f(\infty) = \lim_{k \to \infty} f(k) = \lim_{z \to 1} \left(\frac{z-1}{z} \right) F(z) = \lim_{z \to 1} (z-1) F(z) \qquad \text{(2.21)}$$

PROOF. Let $f(k)$ have a constant limit as k tends to infinity; then the sequence can be expressed as the sum

$$f(k) = f(\infty) + g(k), \quad k = 0, 1, 2, \ldots$$

with $g(k)$ a sequence that decays to zero as k tends to infinity; that is,

$$\lim_{k \to \infty} f(k) = f(\infty)$$

The z-transform of the preceding expression is

$$F(z) = \frac{f(\infty) z}{z - 1} + G(z)$$

The final value $f(\infty)$ is the partial fraction coefficient obtained by expanding $F(z)/z$ as follows:

$$f(\infty) = \lim_{z \to 1}(z - 1)\frac{F(z)}{z} = \lim_{z \to 1}(z - 1)F(z)$$ ∎

EXAMPLE 2.17

Verify the final value theorem using the z-transform of a decaying exponential sequence and its limit as k tends to infinity.

Solution
The z-transform pair of an exponential sequence is

$$\{e^{-akT}\} \xleftarrow{\ \mathcal{Z}\ } \frac{z}{z - e^{-aT}}$$

with $a > 0$. The limit as k tends to infinity in the time domain is

$$f(\infty) = \lim_{k \to \infty} e^{-akT} = 0$$

The final value theorem gives

$$f(\infty) = \lim_{z \to 1}\left(\frac{z - 1}{z}\right)\left(\frac{z}{z - e^{-aT}}\right) = 0$$

EXAMPLE 2.18

Obtain the final value for the sequence whose z-transform is

$$F(z) = \frac{z^2(z - a)}{(z - 1)(z - b)(z - c)}$$

What can you conclude concerning the constants b and c if it is known that the limit exists?

Solution
Applying the final value theorem, we have

$$f(\infty) = \lim_{z \to 1}\frac{z(z - a)}{(z - b)(z - c)} = \frac{1 - a}{(1 - b)(1 - c)}$$

To inverse z-transform the given function, one would have to obtain its partial fraction expansion, which would include three terms: the transform of the sampled step, the transform of the exponential $(b)^k$, and the transform of the exponential $(c)^k$. Therefore, the conditions for the sequence to converge to a constant limit and for the validity of the final value theorem are $|b| < 1$ and $|c| < 1$.

2.4 COMPUTER-AIDED DESIGN

In this text, we make extensive use of computer-aided design (CAD) and analysis of control systems. We use MATLAB,[2] a powerful package with numerous useful commands. For the reader's convenience, we list some MATLAB commands after covering the relevant theory. The reader is assumed to be familiar with the CAD package but not with the digital system commands. We adopt the notation of bolding all user commands throughout the text. Readers using other CAD packages will find similar commands for digital control system analysis and design.

MATLAB typically handles coefficients as vectors with the coefficients listed in descending order. The function $G(z)$ with numerator $5(z + 3)$ and denominator $z^3 + 0.1z^2 + 0.4z$ is represented as the numerator polynomial

>> **num = 5*[1, 3]**

and the denominator polynomial

>> **den = [1, 0.1, 0.4, 0]**

Multiplication of polynomials is equivalent to the convolution of their vectors of coefficients and is performed using the command

>> **denp = conv(den1, den2)**

where **denp** is the product of **den1** and **den2**.

The partial fraction coefficients are obtained using the command

>> **[r, p, k] = residue(num, den)**

where **p** represents the poles, **r** their residues, and **k** the coefficients of the polynomial resulting from dividing the numerator by the denominator. If the highest power in the numerator is smaller than the highest power in the denominator, **k** is zero. This is the usual case encountered in digital control problems.

MATLAB allows the user to sample a function and z-transform it with the commands

>> **g = tf(num, den)**

>> **gd = c2d(g, 0.1, 'imp')**

Other useful MATLAB commands are available with the symbolic manipulation toolbox.

ztrans   z-transform

iztrans   inverse z-transform

---

[2]MATLAB® is a copyright of MathWorks Inc., of Natick, Massachusetts.

To use these commands, we must first define symbolic variables such as **z**, **g**, and **k** with the command

>> **syms z g k**

Powerful commands for symbolic manipulations are also available through packages such as MAPLE, MATHEMATICA, and MACSYMA.

2.5 z-TRANSFORM SOLUTION OF DIFFERENCE EQUATIONS

By a process analogous to Laplace transform solution of differential equations, one can easily solve linear difference equations. The equations are first transformed to the z-domain (i.e., both the right- and left-hand side of the equation are z-transformed). Then the variable of interest is solved for and z-transformed. To transform the difference equation, we typically use the time delay or the time advance property. Inverse z-transformation is performed using the methods of Section 2.3.

EXAMPLE 2.19

Solve the linear difference equation

$$x(k+2) - (3/2)x(k+1) + (1/2)x(k) = 1(k)$$

with the initial conditions $x(0) = 1$, $x(1) = 5/2$.

Solution

1. z-Transform
We begin by z-transforming the difference equation using (2.10) to obtain

$$[z^2 X(z) - z^2 x(0) - zx(1)] - (3/2)[zX(z) - zx(0)] + (1/2)X(z) = z/(z-1)$$

2. Solve for X(z)
Then we substitute the initial conditions and rearrange terms to obtain

$$[z^2 - (3/2)z + (1/2)]X(z) = z/(z-1) + z^2 + (5/2 - 3/2)z$$

Then

$$X(z) = \frac{z[1 + (z+1)(z-1)]}{(z-1)(z-1)(z-0.5)} = \frac{z^3}{(z-1)^2 (z-0.5)}$$

3. Partial Fraction Expansion
The partial fraction of $X(z)/z$ is

$$\frac{X(z)}{z} = \frac{z^2}{(z-1)^2 (z-0.5)} = \frac{A_{11}}{(z-1)^2} + \frac{A_{12}}{z-1} + \frac{A_3}{z-0.5}$$

where

$$A_{11} = (z-1)^2 \frac{X(z)}{z}\bigg|_{z=1} = \frac{z^2}{z-0.5}\bigg|_{z=1} = \frac{1}{1-0.5} = 2$$

$$A_3 = (z-0.5)\frac{X(z)}{z}\bigg|_{z=0.5} = \frac{z^2}{(z-1)^2}\bigg|_{z=0.5} = \frac{(0.5)^2}{(0.5-1)^2} = 1$$

To obtain the remaining coefficient, we multiply by the denominator and get the equation

$$z^2 = A_{11}(z-0.5) + A_{12}(z-0.5)(z-1) + A_3(z-1)^2$$

Equating the coefficient of z^2 gives

$$z^2 : 1 = A_{12} + A_3 = A_{12} + 1 \quad \text{i.e., } A_{12} = 0$$

Thus, the partial fraction expansion in this special case includes two terms only. We now have

$$X(z) = \frac{2z}{(z-1)^2} + \frac{z}{z-0.5}$$

4. Inverse z-Transformation
From the z-transform tables, the inverse z-transform of $X(z)$ is

$$x(k) = 2k + (0.5)^k$$

2.6 THE TIME RESPONSE OF A DISCRETE-TIME SYSTEM

The time response of a discrete-time linear system is the solution of the difference equation governing the system. For the *linear time-invariant* (LTI) case, the response due to the initial conditions and the response due to the input can be obtained separately and then added to obtain the overall response of the system. The response due to the input, or the forced response, is the convolution summation of its input and its response to a unit impulse. In this section, we derive this result and examine its implications.

2.6.1 Convolution Summation

The response of a discrete-time system to a unit impulse is known as the *impulse response sequence*. The impulse response sequence can be used to represent the response of a linear discrete-time system to an arbitrary input sequence

$$\{u(k)\} = \{u(0), u(1), \ldots, u(i), \ldots\} \qquad \textbf{(2.22)}$$

To derive this relationship, we first represent the input sequence in terms of discrete impulses as follows

$$u(k) = u(0)\delta(k) + u(1)\delta(k-1) + u(2)\delta(k-2) + \ldots + u(i)\delta(k-i) + \ldots$$

$$= \sum_{i=0}^{\infty} u(i)\delta(k-i) \tag{2.23}$$

For a linear system, the principle of superposition applies and the system output due to the input is the following sum of impulse response sequences:

$$\{y(l)\} = \{h(l)\}u(0) + \{h(l-1)\}u(1) + \{h(l-2)\}u(2) + \ldots + \{h(l-i)\}u(i) + \ldots$$

Hence, the output at time k is given by

$$y(k) = h(k) * u(k) = \sum_{i=0}^{\infty} h(k-i)u(i) \tag{2.24}$$

where (*) denotes the convolution operation.

For a causal system, the response due to an impulse at time i is an impulse response starting at time i and the delayed response $h(k-i)$ satisfies (Figure 2.5)

$$h(k-i) = 0, \quad i > k \tag{2.25}$$

In other words, a causal system is one whose impulse response is a causal time sequence. Thus, (2.24) reduces to

$$y(k) = u(0)h(k) + u(1)h(k-1) + u(2)h(k-2) + \ldots + u(k)h(0)$$

$$= \sum_{i=0}^{k} u(i)h(k-i) \tag{2.26}$$

A simple change of summation variable ($j = k - i$) transforms (2.26) to

$$y(k) = u(k)h(0) + u(k-1)h(1) + u(k-2)h(2) + \ldots + u(0)h(k)$$

$$= \sum_{j=0}^{k} u(k-j)h(j) \tag{2.27}$$

Equation (2.24) is the convolution summation for a noncausal system, whose impulse response is nonzero for negative time, and it reduces to (2.26) for a causal

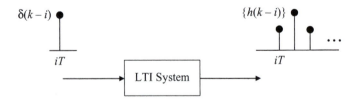

FIGURE 2.5

Response of a causal LTI discrete-time system to an impulse at iT.

system. The summations for time-varying systems are similar, but the impulse response at time i is $h(k, i)$. Here, we restrict our analysis to LTI systems. We can now summarize the result obtained in the following theorem.

Theorem 2.2: Response of an LTI System. The response of an LTI discrete-time system to an arbitrary input sequence is given by the convolution summation of the input sequence and the impulse response sequence of the system.   ■

To better understand the operations involved in convolution summation, we evaluate one point in the output sequence using (2.24). For example,

$$y(2) = \sum_{i=0}^{2} u(i) h(2-i)$$
$$= u(0) h(2) + u(1) h(1) + u(2) h(0)$$

From Table 2.1 and Figure 2.6, one can see the output corresponding to various components of the input of (2.23) and how they contribute to $y(2)$. Note that future input values do not contribute because the system is causal.

Table 2.1 Input Components and Corresponding Output Components

Input	Response	Figure 2.6 Color
$u(0)\ \delta(k)$	$u(0)\ \{h(k)\}$	White
$u(1)\ \delta(k-1)$	$u(1)\ \{h(k-1)\}$	Gray
$u(2)\ \delta(k-2)$	$u(2)\ \{h(k-2)\}$	Black

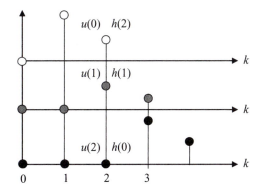

FIGURE 2.6

Output at $k = 2$.

2.6.2 The Convolution Theorem

The convolution summation is considerably simpler than the convolution integral that characterizes the response of linear continuous-time systems. Nevertheless, it is a fairly complex operation, especially if the output sequence is required over a long time period. The following theorem shows how the convolution summation can be avoided by z-transformation.

Theorem 2.3: The Convolution Theorem. The z-transform of the convolution of two time sequences is equal to the product of their z-transforms.

PROOF. z-transforming (2.24) gives

$$
\begin{aligned}
Y(z) &= \sum_{k=0}^{\infty} y(k) z^{-k} \\
&= \sum_{k=0}^{\infty} \left[\sum_{i=0}^{\infty} u(i) h(k-i) \right] z^{-k}
\end{aligned}
\tag{2.28}
$$

Interchange the order of summation and substitute $j = k - i$ to obtain

$$
Y(z) = \sum_{i=0}^{\infty} \sum_{j=-i}^{\infty} u(i) h(j) z^{-(i+j)}
\tag{2.29}
$$

Using the causality property, (2.24) reduces (2.29) to

$$
Y(z) = \left[\sum_{i=0}^{\infty} u(i) z^{-i} \right] \left[\sum_{j=0}^{\infty} h(j) z^{-j} \right]
\tag{2.30}
$$

Therefore,

$$
Y(z) = H(z) U(z)
\tag{2.31}
$$

∎

The function $H(z)$ of (2.31) is known as the **z-transfer function** or simply the **transfer function**. It plays an important role in obtaining the response of an LTI system to any input, as explained later. Note that the transfer function and impulse response sequence are z-transform pairs.

Applying the convolution theorem to the response of an LTI system allows us to use the z-transform to find the output of a system without convolution by doing the following:

1. z-transforming the input
2. Multiplying the z-transform of the input and the z-transfer function
3. Inverse z-transforming to obtain the output temporal sequence

An added advantage of this approach is that the output can often be obtained in closed form. The preceding procedure is demonstrated in the example that follows.

EXAMPLE 2.20

Given the discrete-time system

$$y(k+1) - 0.5y(k) = u(k), \; y(0) = 0$$

find the impulse response of the system $h(k)$:

1. From the difference equation
2. Using z-transformation

Solution

Let $u(k) = \delta(k)$. Then

$$y(1)$$

$$y(2) = 0.5y(1) = 0.5$$

$$y(3) = 0.5y(2) = (0.5)^2$$

$$\text{i.e., } h(i) = \begin{cases} (0.5)^{i-1}, & i = 1, 2, 3, \ldots \\ 0, & i < 1 \end{cases}$$

Alternatively, z-transforming the difference equation yields the transfer function

$$H(z) = \frac{Y(z)}{U(z)} = \frac{1}{z - 0.5}$$

$$= \frac{Y(z)}{U(z)} = \frac{1}{z - 0.5}$$

Inverse-transforming with the delay theorem gives the impulse response

$$h(i) = \begin{cases} (0.5)^{i-1}, & i = 1, 2, 3, \ldots \\ 0, & i < 1 \end{cases}$$

Observe that the response decays exponentially because the pole has magnitude less than unity. In Chapter 4, we discuss this property and relate to the stability of discrete-time systems.

EXAMPLE 2.21

Given the discrete-time system

$$y(k+1) - y(k) = u(k+1)$$

find the system transfer function and its response to a sampled unit step.

Solution

The transfer function corresponding to the difference equation is

$$H(z) = \frac{z}{z - 1}$$

We multiply the transfer function by the sampled unit step's z-transform to obtain

$$Y(z) = \left(\frac{z}{z-1}\right) \times \left(\frac{z}{z-1}\right) = \left(\frac{z}{z-1}\right)^2 = z\frac{z}{(z-1)^2}$$

The z-transform of a unit ramp is

$$F(z) = \frac{z}{(z-1)^2}$$

Then, using the time advance property of the z-transform, we have the inverse transform

$$y(i) = \begin{cases} i+1, & i = 0, 1, 2, 3, \ldots \\ 0, & i < 0 \end{cases}$$

It is obvious from this simple example that z-transforming yields the response of a system in closed form more easily than direct evaluation. For higher-order difference equations, obtaining the response in closed form directly may be impossible, whereas z-transforming to obtain the response remains a relatively simple task.

2.7 THE MODIFIED z-TRANSFORM

Sampling and z-transformation capture the values of a continuous-time function at the sampling points only. To evaluate the time function between sampling points, we need to delay the sampled waveform by a fraction of a sampling interval before sampling. We can then vary the sampling points by changing the delay period. The z-transform associated with the delayed waveform is known as the modified z-transform.

We consider a causal continuous-time function $y(t)$ sampled every T seconds. Next, we insert a delay $T_d < T$ before the sampler as shown in Figure 2.7. The output of the delay element is the waveform

$$y_d(t) = \begin{cases} y(t - T_d), & t \geq 0 \\ 0, & t < 0 \end{cases} \tag{2.32}$$

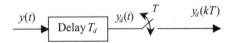

FIGURE 2.7

Sampling of a delayed signal.

Note that delaying a causal sequence always results in an initial zero value. To avoid inappropriate initial values, we rewrite the delay as

$$T_d = T - mT, \quad 0 \le m < 1$$
$$m = 1 - T_d/T \tag{2.33}$$

For example, a time delay of 0.2 s with a sampling period T of 1 s corresponds to $m = 0.8$—that is, a time advance of 0.8 of a sampling period and a time delay of one sampling period. If $y_{-1}(t + mT)$ is defined as $y(t + mT)$ delayed by one complete sampling period, then, based on (2.32), $y_d(t)$ is given by

$$y_d(t) = y(t - T + mT) = y_{-1}(t + mT) \tag{2.34}$$

We now sample the delayed waveform with sampling period T to obtain

$$y_d(kT) = y_{-1}(kT + mT), \quad k = 0, 1, 2, \ldots \tag{2.35}$$

From the delay theorem, we know the z-transform of $y_{-1}(t)$

$$Y_{-1}(z) = z^{-1}Y(z) \tag{2.36}$$

We need to determine the effect of the time advance by mT to obtain the z-transform of $y_d(t)$. We determine this effect by considering specific examples. At this point, it suffices to write

$$Y(z, m) = \mathcal{Z}_m\{y(kT)\} = z^{-1}\mathcal{Z}\{y(kT + mT)\} \tag{2.37}$$

where $\mathcal{Z}_m\{\bullet\}$ denotes the modified z-transform.

EXAMPLE 2.22: STEP

The step function has fixed amplitude for all time arguments. Thus, shifting it or delaying it does not change the sampled values. We conclude that the modified z-transform of a sampled step is the same as its z-transform, times z^{-1} for all values of the time advance mT—that is, $1/(1-z^{-1})$.

EXAMPLE 2.23: EXPONENTIAL

We consider the exponential waveform

$$y(t) = e^{-pt} \tag{2.38}$$

The effect of a time advance mT on the sampled values for an exponential decay is shown in Figure 2.8. The sampled values are given by

$$y(kT + mT) = e^{-p(k+m)T} = e^{-pmT}e^{-pkT}, \quad k = 0, 1, 2, \ldots \tag{2.39}$$

We observe that the time advance results in a scaling of the waveform by the factor e^{-pmT}. By the linearity of the z-transform, we have the following:

$$\mathscr{Z}\{y(kT+mT)\} = e^{-pmT}\,\frac{z}{z-e^{-pT}} \tag{2.40}$$

Using (2.37), we have the modified z-transform

$$Y(z,m) = \frac{e^{-pmT}}{z-e^{-pT}} \tag{2.41}$$

For example, for $p = 4$ and $T = 0.2$ s, to delay by $0.7\ T$, we let $m = 0.3$ and calculate e^{-pmT} $= e^{-0.24} = 0.787$ and $e^{-pT} = e^{-0.8} = 0.449$. We have the modified z-transform

$$Y(z,m) = \frac{0.787}{z-0.449}$$

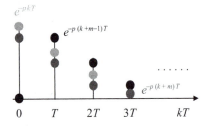

FIGURE 2.8

Effect of time advance on sampling an exponential decay.

The modified z-transforms of other important functions, such as the ramp and the sinusoid, can be obtained following the procedure presented earlier. The derivations of these modified z-transforms are left as exercises.

2.8 FREQUENCY RESPONSE OF DISCRETE-TIME SYSTEMS

In this section, we discuss the steady-state response of a discrete-time system to a sampled sinusoidal input.[3] It is shown that, as in the continuous-time case, the response is a sinusoid of the same frequency as the input with frequency-dependent phase shift and magnitude scaling. The scale factor and

---

[3]Unlike continuous sinusoids, sampled sinusoids are only periodic if the ratio of the period of the waveform and the sampling period is a rational number (equal to a ratio of integers). However, the continuous envelope of the sampled form is clearly always periodic. See the text by Oppenheim et al., 1997, p. 26) for more details.

phase shift define a complex function of frequency known as the frequency response.

We first obtain the frequency response using impulse sampling and the Laplace transform to exploit the well known relationship between the transfer function $H_a(s)$ and the frequency response $H_a(j\omega)$

$$H_a(j\omega) = H_a(s)|_{s=j\omega} \tag{2.42}$$

The impulse-sampled representation of a discrete-time waveform is

$$u^*(t) = \sum_{k=0}^{\infty} u(kT)\delta(t - kT) \tag{2.43}$$

where $u(kT)$ is the value at time kT, and $\delta(t - kT)$ denotes a Dirac delta at time kT. The representation (2.43) allows Laplace transformation to obtain

$$U^*(s) = \sum_{k=0}^{\infty} u(kT)e^{-kTs} \tag{2.44}$$

It is now possible to define a transfer function for sampled inputs as

$$H^*(s) = \frac{Y^*(s)}{U^*(s)}\bigg|_{\text{zero initial conditions}} \tag{2.45}$$

Then, using (2.42), we obtain

$$H^*(j\omega) = H^*(s)|_{s=j\omega} \tag{2.46}$$

To rewrite (2.46) in terms of the complex variable $z = e^{sT}$, we use the equation

$$H(z) = H^*(s)\bigg|_{s=\frac{1}{T}\ln(z)} \tag{2.47}$$

Thus, the frequency response is given by

$$\begin{aligned} H^*(j\omega) &= H(z)|_{z=e^{j\omega T}} \\ &= H(e^{j\omega T}) \end{aligned} \tag{2.48}$$

Equation (2.48) can also be verified without the use of impulse sampling by considering the sampled complex exponential

$$\begin{aligned} u(kT) &= u_0 e^{jk\omega_0 T} \\ &= u_0[\cos(k\omega_0 T) + j\sin(k\omega_0 T)], \quad k = 0, 1, 2, \dots \end{aligned} \tag{2.49}$$

This eventually yields the sinusoidal response while avoiding its second-order z-transform. The z-transform of the chosen input sequence is the first-order function

$$U(z) = u_0 \frac{z}{z - e^{j\omega_0 T}} \tag{2.50}$$

Assume the system z-transfer function to be

$$H(z) = \frac{N(z)}{\prod_{i=1}^{n}(z - p_i)} \tag{2.51}$$

where $N(z)$ is a numerator polynomial of order n or less, and the system poles p_i are assumed to lie inside the unit circle.

The system output due to the input of (2.49) has the z-transform

$$Y(z) = \left[\frac{N(z)}{\prod_{i=1}^{n}(z - p_i)} \right] u_0 \frac{z}{z - e^{j\omega_0 T}} \tag{2.52}$$

This can be expanded into the partial fractions

$$Y(z) = \frac{Az}{z - e^{j\omega_0 T}} + \sum_{i=1}^{n} \frac{B_i z}{z - p_i} \tag{2.53}$$

Then inverse z-transforming gives the output

$$y(kT) = Ae^{jk\omega_0 T} + \sum_{i=1}^{n} B_i p_i^k, \quad k = 0, 1, 2, \ldots \tag{2.54}$$

The assumption of poles inside the unit circle implies that, for sufficiently large k, the output reduces to

$$y_{ss}(kT) = Ae^{jk\omega_0 T}, \quad k \text{ large} \tag{2.55}$$

where $y_{ss}(kT)$ denotes the steady-state output.

The term A is the partial fraction coefficient

$$A = \frac{Y(z)}{z}(z - e^{j\omega_0 T}) \bigg|_{z = e^{j\omega_0 T}} \tag{2.56}$$

$$= H(e^{j\omega_0 T}) u_0$$

Thus, we write the steady-state output in the form

$$y_{ss}(kT) = |H(e^{j\omega_0 T})| u_0 e^{j[k\omega_0 T + \angle H(e^{j\omega_0 T})]}, \quad k \text{ large} \tag{2.57}$$

The real part of this response is the response due to a sampled cosine input, and the imaginary part is the response of a sampled sine. The sampled cosine response is

$$y_{ss}(kT) = |H(e^{j\omega_0 T})| u_0 \cos[k\omega_0 T + \angle H(e^{j\omega_0 T})], \quad k \text{ large} \tag{2.58}$$

The sampled sine response is similar to (2.58) with the cosine replaced by sine.

Equations (2.57) and (2.58) show that the response to a sampled sinusoid is a sinusoid of the same frequency scaled and phase-shifted by the magnitude and angle

$$|H(e^{j\omega_0 T})| \angle H(e^{j\omega_0 T}) \tag{2.59}$$

respectively. This is the frequency response function obtained earlier using impulse sampling. Thus, one can use complex arithmetic to determine the steady-state response due to a sampled sinusoid without the need for z-transformation.

EXAMPLE 2.24

Find the steady-state response of the system

$$H(z) = \frac{1}{(z - 0.1)(z - .5)}$$

due to the sampled sinusoid $u(kT) = 3 \cos(0.2\,k)$.

Solution
Using (2.58) gives the response

$$y_{ss}(kT) = \left| H\left(e^{j0.2}\right) \right| 3 \cos\left(0.2k + \angle H\left(e^{j0.2}\right)\right), k \text{ large}$$

$$= \left| \frac{1}{\left(e^{j0.2} - 0.1\right)\left(e^{j0.2} - 0.5\right)} \right| 3 \cos\left(0.2k + \angle \frac{1}{\left(e^{j0.2} - 0.1\right)\left(e^{j0.2} - 0.5\right)}\right)$$

$$= 6.4 \cos(0.2k - 0.614)$$

2.8.1 Properties of the Frequency Response of Discrete-Time Systems

Using (2.48), the following frequency response properties can be derived:

1. *DC gain:* The DC gain is equal to $H(1)$.

PROOF. From (2.48),

$$H\left(e^{j\omega T}\right)\Big|_{\omega \to 0} = H(z)\Big|_{z \to 1}$$
$$= H(1) \qquad\blacksquare$$

2. *Periodic nature:* The frequency response is a periodic function of frequency with period $\omega_s = 2\pi/T$ rad/s.

PROOF. The complex exponential

$$e^{j\omega T} = \cos(\omega T) + j\sin(\omega T)$$

is periodic with period $\omega_s = 2\pi/T$ rad/s. Because $H(e^{j\omega T})$ is a single-valued function of its argument, it follows that it also is periodic and that it has the same repetition frequency. $\blacksquare$

3. *Symmetry:* For transfer functions with real coefficients, the magnitude of the transfer function is an even function of frequency and its phase is an odd function of frequency.

PROOF. For negative frequencies, the transfer function is

$$H\left(e^{-j\omega T}\right) = H\left(\overline{e^{j\omega T}}\right)$$

For real coefficients, we have

$$\overline{H\left(e^{j\omega T}\right)} = H\left(\overline{e^{j\omega T}}\right)$$

Combining the last two equations gives

$$H\left(e^{-j\omega T}\right) = \overline{H\left(e^{j\omega T}\right)}$$

Equivalently, we have

$$\left|H\left(e^{-j\omega T}\right)\right| = \left|H\left(e^{j\omega T}\right)\right|$$
$$\angle H\left(e^{-j\omega T}\right) = -\angle H\left(e^{j\omega T}\right) \qquad \blacksquare$$

Hence, it is only necessary to obtain $H(e^{j\omega T})$ for frequencies ω in the range from DC to $\omega_s/2$. The frequency response for negative frequencies can be obtained by symmetry, and for frequencies above $\omega_s/2$ the frequency response is periodically repeated. If the frequency response has negligible amplitudes at frequencies above $\omega_s/2$, the repeated frequency response cycles do not overlap. The overall effect of sampling for such systems is to produce a periodic repetition of the frequency response of a continuous-time system.

Because the frequency response functions of physical systems are not band-limited, overlapping of the repeated frequency response cycles, known as *folding*, occurs. The frequency $\omega_s/2$ is known as the *folding frequency*. Folding results in distortion of the frequency response and should be minimized. This can be accomplished by proper choice of the sampling frequency $\omega_s/2$ or filtering. Figure 2.9 shows the frequency response of a second-order underdamped digital system.

2.8.2 MATLAB Commands for the Discrete-Time Frequency Response

The MATLAB commands **bode**, **nyquist**, and **nichols** calculate and plot the frequency response of a discrete-time system. For a sampling period of 0.2 s and a transfer function with numerator **num** and denominator **den**, the three commands have the form

```
>> g = tf(num, den, 0.2)

>> bode(g)

>> nyquist(g)

>> nichols(g)
```

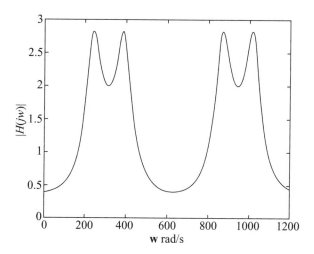

FIGURE 2.9

Frequency response of a digital system.

MATLAB does not allow the user to select the grid for automatically generated plots. However, all the commands have alternative forms that allow the user to obtain the frequency response data for later plotting.

The commands **bode** and **nichols** have the alternative form

>> [M, P, w] = bode(g, w)

>> [M, P, w] = nichols(g, w)

where **w** is predefined frequency grid, **M** is the magnitude, and **P** is the phase of the frequency response. MATLAB selects the frequency grid if none is given and returns the same list of outputs. The frequency vector can also be eliminated from the output or replaced by a scalar for single-frequency computations. The command **nyquist** can take similar forms to those just described but yields the real and imaginary parts of the frequency response as follows

>> [Real, Imag, w] = nyquist(g, w)

As with all MATLAB commands, printing the output is suppressed if any of the frequency response commands is followed by a semicolon. The output can then be used with the command **plot** to obtain user selected plot specifications. For example, a plot of the actual frequency response points without connections is obtained with the command

>> plot(Real(:), Imag(:), '*')

where the locations of the data points are indicated with the "*". The command

>> subplot(2, 3, 4)

creates a 2-row, 3-column grid and draw axes at the first position of the second row (the first three plots are in the first row and 4 is the plot number). The next plot command superimposes the plot on these axes. For other plots, the subplot and plot commands are repeated with the appropriate arguments. For example, a plot in the first row and second column of the grid is obtained with the command

$$>> \text{subplot}(2, 3, 2)$$

2.9 THE SAMPLING THEOREM

Sampling is necessary for the processing of analog data using digital elements. Successful digital data processing requires that the samples reflect the nature of the analog signal and that analog signals be recoverable, at least in theory, from a sequence of samples. Figure 2.10 shows two distinct waveforms with identical samples. Obviously, faster sampling of the two waveforms would produce distinguishable sequences. Thus, it is obvious that sufficiently fast sampling is a prerequisite for successful digital data processing. The sampling theorem gives a lower bound on the sampling rate necessary for a given **band-limited** signal (i.e., a signal with a known finite bandwidth).

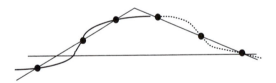

FIGURE 2.10

Two different waveforms with identical samples.

Theorem 2.4: The Sampling Theorem. The band-limited signal with

$$f(t) \xleftrightarrow{\mathscr{F}} F(j\omega), \ F(j\omega) \neq 0, \ -\omega_m \leq \omega \leq \omega_m \tag{2.60}$$
$$F(j\omega) = 0, \ \text{elsewhere}$$

with F denoting the Fourier transform, can be reconstructed from the discrete-time waveform

$$f^*(t) = \sum_{k=-\infty}^{\infty} f(t)\delta(t - kT) \tag{2.61}$$

if and only if the sampling angular frequency $\omega_s = 2\pi/T$ satisfies the condition

$$\omega_s > 2\omega_m \qquad\qquad (2.62)$$

The spectrum of the continuous-time waveform can be recovered using an ideal low-pass filter of bandwidth ω_b in the range

$$\omega_m < \omega_b < \omega_s/2 \qquad\qquad (2.63)$$

PROOF. Consider the unit impulse train

$$\delta_T(t) = \sum_{k=-\infty}^{\infty} \delta(t - kT) \qquad\qquad (2.64)$$

and its Fourier transform

$$\delta_T(\omega) = \frac{2\pi}{T} \sum_{n=-\infty}^{\infty} \delta(\omega - n\omega_s) \qquad\qquad (2.65)$$

Impulse sampling is achieved by multiplying the waveforms $f(t)$ and $\delta_T(t)$. By the frequency convolution theorem, the spectrum of the product of the two waveforms is given by the convolution of their two spectra; that is,

$$\Im\{\delta_T(t) \times f(t)\} = \frac{1}{2\pi}\delta_T(j\omega) * F(j\omega)$$

$$= \left[\frac{1}{T}\sum_{n=-\infty}^{\infty} \delta(\omega - n\omega_s)\right] * F(j\omega)$$

$$= \frac{1}{T}\sum_{n=-\infty}^{\infty} F(\omega - n\omega_s)$$

where ω_m is the bandwidth of the signal. Therefore, the spectrum of the sampled waveform is a periodic function of frequency ω_s. Assuming that $f(t)$ is a real valued function, then it is well known that the magnitude $|F(j\omega)|$ is an even function of frequency, whereas the phase $F(j\omega)$ is an odd function. For a band-limited function, the amplitude and phase in the frequency range 0 to $\omega_s/2$ can be recovered by an ideal low-pass filter as shown in Figure 2.11. ■

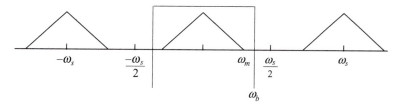

FIGURE 2.11

Sampling theorem.

2.9.1 Selection of the Sampling Frequency

In practice, finite bandwidth is an idealization associated with infinite-duration signals, whereas finite duration implies infinite bandwidth. To show this, assume that a given signal is to be band limited. Band limiting is equivalent to multiplication by a pulse in the frequency domain. By the convolution theorem, multiplication in the frequency domain is equivalent to convolution of the inverse Fourier transforms. Hence, the inverse transform of the band-limited function is the convolution of the original time function with the sinc function, a function of infinite duration. We conclude that a band-limited function is of infinite duration.

A time-limited function is the product of a function of infinite duration and a pulse. The frequency convolution theorem states that multiplication in the time domain is equivalent to convolution of the Fourier transforms in the frequency domain. Thus, the spectrum of a time-limited function is the convolution of the spectrum of the function of infinite duration with a sinc function, a function of infinite bandwidth. Hence, the Fourier transform of a time-limited function has infinite bandwidth. Because all measurements are made over a finite time period, infinite bandwidths are unavoidable. Nevertheless, a given signal often has a finite "effective bandwidth" beyond which its spectral components are negligible. This allows us to treat physical signals as band limited and choose a suitable sampling rate for them based on the sampling theorem.

In practice, the sampling rate chosen is often larger than the lower bound specified in the sampling theorem. A rule of thumb is to choose ω_s as

$$\omega_s = k\omega_m, \quad 5 \le k \le 10 \tag{2.66}$$

The choice of the constant k depends on the application. In many applications, the upper bound on the sampling frequency is well below the capabilities of state-of-the-art hardware. A closed-loop control system cannot have a sampling period below the minimum time required for the output measurement; that is, the sampling frequency is upper-bounded by the **sensor delay**.[4] For example, oxygen sensors used in automotive air/fuel ratio control have a sensor delay of about 20 ms, which corresponds to a sampling frequency upper bound of 50 Hz. Another limitation is the computational time needed to update the control. This is becoming less restrictive with the availability of faster microprocessors but must be considered in sampling rate selection.

In digital control, the sampling frequency must be chosen so that samples provide a good representation of the analog physical variables. A more detailed discussion of the practical issues that must be considered when choosing the sampling frequency is given in Chapter 12. Here, we only discuss choosing the sampling period based on the sampling theorem.

---

[4]It is possible to have the sensor delay as an integer multiple of the sampling period if a state estimator is used, as discussed in Franklin et al. (1998).

For a linear system, the output of the system has a spectrum given by the product of the frequency response and input spectrum. Because the input is not known a priori, we must base our choice of sampling frequency on the frequency response.

The frequency response of a first-order system is

$$H(j\omega) = \frac{K}{j\omega/\omega_b + 1} \tag{2.67}$$

where K is the DC gain and ω_b is the system bandwidth. The frequency response amplitude drops below the DC level by a factor of about 10 at the frequency $7\omega_b$. If we consider $\omega_m = 7\omega_b$, the sampling frequency is chosen as

$$\omega_s = k\omega_b, \quad 35 \le k \le 70 \tag{2.68}$$

For a second-order system with frequency response

$$H(j\omega) = \frac{K}{j2\zeta\omega/\omega_n + 1 - (\omega/\omega_n)^2} \tag{2.69}$$

and the bandwidth of the system is approximated by the damped natural frequency

$$\omega_d = \omega_n \sqrt{1 - \zeta^2} \tag{2.70}$$

Using a frequency of $7\omega_d$ as the maximum significant frequency, we choose the sampling frequency as

$$\omega_s = k\omega_d, \quad 35 \le k \le 70 \tag{2.71}$$

In addition, the impulse response of a second-order system is of the form

$$y(t) = Ae^{-\zeta\omega_n t} \sin(\omega_d t + \phi) \tag{2.72}$$

where A is a constant amplitude, and ϕ is a phase angle. Thus, the choice of sampling frequency of (2.71) is sufficiently fast for oscillations of frequency ω_d and time to first peak π/ω_d.

EXAMPLE 2.25

Given a first-order system of bandwidth 10 rad/s, select a suitable sampling frequency and find the corresponding sampling period.

Solution

A suitable choice of sampling frequency is $\omega_s = 60$, $\omega_b = 600$ rad/s. The corresponding sampling period is approximately $T = 2\pi/\omega_s \cong 0.01$ s.

EXAMPLE 2.26

A closed-loop control system must be designed for a steady-state error not to exceed 5 percent, a damping ratio of about 0.7, and an undamped natural frequency of 10 rad/s. Select a suitable sampling period for the system if the system has a sensor delay of

    1. 0.02 s
    2. 0.03 s

Solution

Let the sampling frequency be

$$\omega_s \geq 35\omega_d$$
$$= 35\omega_n \sqrt{1 - \zeta^2}$$
$$= 350\sqrt{1 - 0.49}$$
$$= 249.95 \text{ rad/s}$$

The corresponding sampling period is $T = 2\pi/\omega_s \leq 0.025$ s.

    1. A suitable choice is $T = 20$ ms because this is equal to the sensor delay.
    2. We are forced to choose $T = 30$ ms, which is equal to the sensor delay.

RESOURCES

Chen, C.-T., *System and Signal Analysis,* Saunders, 1989.

Feuer, A., and G. C. Goodwin, *Sampling in Digital Signal Processing and Control,* Birkhauser, 1996.

Franklin, G. F., J. D. Powell, and M. L. Workman, *Digital Control of Dynamic Systems,* Addison-Wesley, 1998.

Goldberg, S., *Introduction to Difference Equations,* Dover, 1986.

Jacquot, R. G., *Modern Digital Control Systems,* Marcel Dekker, 1981.

Kuo, B. C., *Digital Control Systems,* Saunders, 1992.

Mickens, R. E., *Difference Equations,* Van Nostrand Reinhold, 1987.

Oppenheim, A. V., A. S. Willsky, and S. H. Nawab, *Signals and Systems,* Prentice Hall, 1997.

PROBLEMS

2.1 Derive the discrete-time model of Example 2.1 from the solution of the system differential equation with initial time kT and final time $(k + 1)T$.

2.2 For each of the following equations, determine the order of the equation and then test it for (i) linearity, (ii) time invariance, (iii) homogeneity.
(a) $y(k + 2) = y(k + 1) y(k) + u(k)$
(b) $y(k + 3) + 2 y(k) = 0$
(c) $y(k + 4) + y(k - 1) = u(k)$
(d) $y(k + 5) = y(k + 4) + u(k + 1) - u(k)$
(e) $y(k + 2) = y(k) u(k)$

2.3 Find the transforms of the following sequences using Definition 2.1.
(a) $\{0, 1, 2, 4, 0, 0, \ldots\}$
(b) $\{0, 0, 0, 1, 1, 1, 0, 0, 0, \ldots\}$
(c) $\{0, 2^{-0.5}, 1, 2^{-0.5}, 0, 0, 0, \ldots\}$

2.4 Obtain closed forms of the transforms of Problem 2.3 using the table of z-transforms and the time delay property.

2.5 Prove the linearity and time delay properties of the z-transform from basic principles.

2.6 Use the linearity of the z-transform and the transform of the exponential function to obtain the transforms of the discrete-time functions.
(a) $\sin(k\omega T)$
(b) $\cos(k\omega T)$

2.7 Use the multiplication by exponential property to obtain the transforms of the discrete-time functions.
(a) $e^{-\alpha kT}\sin(k\omega T)$
(b) $e^{-\alpha kT}\cos(k\omega T)$

2.8 Find the inverse transforms of the following functions using Definition 2.1 and, if necessary, long division.
(a) $F(z) = 1 + 3z^{-1} + 4z^{-2}$
(b) $F(z) = 5z^{-1} + 4z^{-5}$
(c) $F(z) = \dfrac{z}{z^2 + 0.3z + 0.02}$
(d) $F(z) = \dfrac{z - 0.1}{z^2 + 0.04z + 0.25}$

2.9 For Problems 2.8(c) and (d), find the inverse transforms of the functions using partial fraction expansion and table lookup.

2.10 Solve the following difference equations.
(a) $y(k + 1) - 0.8 y(k) = 0$, $y(0) = 1$
(b) $y(k + 1) - 0.8 y(k) = 1(k)$, $y(0) = 0$

(c) $y(k + 1) - 0.8\ y(k) = 1(k),\ y(0) = 1$

(d) $y(k + 2) + 0.7\ y(k + 1) + 0.06\ y(k) = \delta(k),\ y(0) = 0,\ y(1) = 2$

2.11 Find the transfer functions corresponding to the difference equations of Problem 2.2 with input $u(k)$ and output $y(k)$. If no transfer function is defined, explain why.

2.12 Test the linearity with respect to the input of the systems for which you found transfer functions in Problem 2.11.

2.13 If the rational functions of Problems 2.8.(c) and (d) are transfer functions of LTI systems, find the difference equation governing each system.

2.14 We can use z-transforms to find the sum of integers raised to various powers. This is accomplished by first recognizing that the sum is the solution of the difference equation

$$f(k) = f(k-1) + a(k)$$

where $a(k)$ is the kth term in the summation. Evaluate the following summations using z-transforms.

(a) $\displaystyle\sum_{k=1}^{n} k$

(b) $\displaystyle\sum_{k=1}^{n} k^2$

2.15 Find the impulse response functions for the systems governed by the following difference equations.

(a) $y(k + 1) - 0.5\ y(k) = u(k)$

(b) $y(k + 2) - 0.1\ y(k + 1) + 0.8\ y(k) = u(k)$

2.16 Find the final value for the functions if it exists.

(a) $F(z) = \dfrac{z}{z^2 - 1.2z + 0.2}$

(b) $F(z) = \dfrac{z}{z^2 + 0.3z + 2}$

2.17 Find the steady-state response of the systems resulting from the sinusoidal input $u(k) = 0.5\ \sin(0.4\ k)$.

(a) $H(z) = \dfrac{z}{z - 0.4}$

(b) $H(z) = \dfrac{z}{z^2 + 0.4z + 0.03}$

2.18 Find the frequency response of a noncausal system whose impulse response sequence is given by

$$\{u(k), u(k) = u(k + K),\ k = -\infty, \ldots, \infty\}$$

Hint: The impulse response sequence is periodic with period K and can be expressed as

$$u^*(t) = \sum_{l=0}^{K-1} \sum_{m=-\infty}^{\infty} u(l + mK)\delta(t - l - mK)$$

2.19 The well-known Shannon reconstruction theorem states that any band-limited signal $u(t)$ with bandwidth $\omega_s/2$ can be exactly reconstructed from its samples at a rate $\omega_s = 2\pi/T$. The reconstruction is given by

$$u(t) = \sum_{k=-\infty}^{\infty} u(k) \frac{\sin\left[\dfrac{\omega_s}{2}(t - kT)\right]}{\dfrac{\omega_s}{2}(t - kT)}$$

Use the convolution theorem to justify the preceding expression.

2.20 Obtain the convolution of the two sequences $\{1, 1, 1\}$ and $\{1, 2, 3\}$.
(a) Directly
(b) Using z-transformation

2.21 Obtain the modified z-transforms for the functions of Problems (2.6) and (2.7).

2.22 Using the modified z-transform, examine the intersample behavior of the functions $h(k)$ of Problem 2.15. Use delays of (1) $0.3T$, (2) $0.5T$, and (3) $0.8T$. Attempt to obtain the modified z-transform for Problem 2.16 and explain why it is not defined.

2.23 The following open-loop systems are to be digitally feedback-controlled. Select a suitable sampling period for each if the closed-loop system is to be designed for the given specifications.

(a) $G_{ol}(s) = \dfrac{1}{s+3}$ Time constant = 0.1 s

(b) $G_{ol}(s) = \dfrac{1}{s^2 + 4s + 3}$ Undamped natural frequency = 5 rad/s, damping

ratio = 0.7

2.24 Repeat problem 2.23 if the systems have the following sensor delays.
(a) 0.025 s
(b) 0.03 s

COMPUTER EXERCISES

2.25 Consider the closed-loop system of Problem 2.23(a).
(a) Find the impulse response of the **closed-loop** transfer function, and obtain the impulse response sequence for a sampled system output.

(b) Obtain the z-transfer function by z-transforming the impulse response sequence.

(c) Using MATLAB, obtain the frequency response plots for sampling frequencies $\omega_s = k\omega_b$, $k = 5, 35, 70$.

(d) Comment on the choices of sampling periods of part (b).

2.26 Repeat Problem 2.25 for the second-order closed-loop system of Problem 2.23(b) with plots for sampling frequencies $\omega_s = k\omega_d$, $k = 5, 35, 70$.

2.27 Use MATLAB with a sampling period of 1 s and a delay of 0.5 s to verify the results of Problem 2.17 for $\omega = 5$ rad/s and $\alpha = 2$ s^{-1}.

2.28 The following difference equation describes the evolution of the expected price of a commodity[5]

$$p_e(k+1) = (1-\gamma)\, p_e(k) + \gamma p(k)$$

where $p_e(k)$ is the expected price after k quarters, $p(k)$ is the actual price after k quarters, and γ is a constant.

(a) Simulate the system with $\gamma = 0.5$ and a fixed actual price of one unit, and plot the actual and expected prices. Discuss the accuracy of the model prediction.

(b) Repeat part (a) for an exponentially decaying price $p(k) = (0.4)^k$.

(c) Discuss the predictions of the model referring to your simulation results.

---

[5]D. N. Gujarate, *Basic Econometrics*. McGraw-Hill, p. 547, 1988.

Modeling of Digital Control Systems

3

As in the case of analog control, mathematical models are needed for the analysis and design of digital control systems. A common configuration for digital control systems is shown in Figure 3.1. The configuration includes a digital-to-analog converter (DAC), an analog subsystem, and an analog-to-digital converter (ADC). The DAC converts numbers calculated by a microprocessor or computer into analog electrical signals that can be amplified and used to control an analog plant. The analog subsystem includes the plant as well as the amplifiers and actuators necessary to drive it. The output of the plant is periodically measured and converted to a number that can be fed back to the computer using an ADC. In this chapter, we develop models for the various components of this digital control configuration. Many other configurations that include the same components can be similarly analyzed. We begin by developing models for the ADC and DAC, then for the combination of DAC, analog subsystem, and ADC.

Objectives

After completing this chapter, the reader will be able to do the following:

1. Obtain the transfer function of an analog system with analog-to-digital and digital-to-analog converters including systems with a time delay.
2. Find the closed-loop transfer function for a digital control system.
3. Find the steady-state tracking error for a closed-loop control system.
4. Find the steady-state error caused by a disturbance input for a closed-loop control system.

3.1 ADC MODEL

Assume that

- ADC outputs are exactly equal in magnitude to their inputs (i.e., quantization errors are negligible).

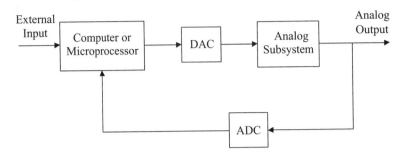

FIGURE 3.1

Common digital control system configuration.

- The ADC yields a digital output instantaneously.
- Sampling is perfectly uniform (i.e., occurs at a fixed rate).

Then the ADC can be modeled as an ideal sampler with sampling period T as shown in Figure 3.2.

Clearly, the preceding assumptions are idealizations that can only be approximately true in practice. Quantization errors are typically small but nonzero; variations in sampling rate occur but are negligible, and physical ADCs have a finite conversion time. Nevertheless, the ideal sampler model is acceptable for most engineering applications.

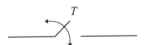

FIGURE 3.2

Ideal sampler model of an ADC.

3.2 DAC MODEL

Assume that

- DAC outputs are exactly equal in magnitude to their inputs.
- The DAC yields an analog output instantaneously.
- DAC outputs are constant over each sampling period.

Then the input-output relationship of the DAC is given by

$$\{u(k)\} \xrightarrow{\;ZOH\;} u(t) = u(k), \; kT \leq t < (k+1)T, \; k = 0, 1, 2, \ldots \qquad \textbf{(3.1)}$$

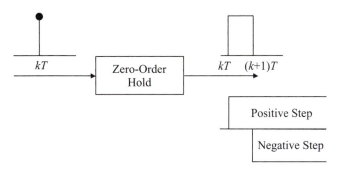

FIGURE 3.3

Model of a DAC as a zero-order hold.

where $\{u(k)\}$ is the input sequence. This equation describes a **zero-order hold** (ZOH), shown in Figure 3.3. Other functions may also be used to construct an analog signal from a sequence of numbers. For example, a **first-order hold** constructs analog signals in terms of straight lines, whereas a **second-order hold** constructs them in terms of parabolas.

In practice, the DAC requires a short but nonzero interval to yield an output; its output is not exactly equal in magnitude to its input and may vary slightly over a sampling period. But the model of (3.1) is sufficiently accurate for most engineering applications. The zero-order hold is the most commonly used DAC model and is adopted in most digital control texts. Analyses involving other hold circuits are similar, as seen from Problem 3.1.

3.3 THE TRANSFER FUNCTION OF THE ZOH

To obtain the transfer function of the ZOH, we replace the number or discrete impulse shown in Figure 3.3 by an impulse $\delta(t)$. The transfer function can then be obtained by Laplace transformation of the impulse response. As shown in the figure, the impulse response is a unit pulse of width T. A pulse can be represented as a positive step at time zero followed by a negative step at time T. Using the Laplace transform of a unit step and the time delay theorem for Laplace transforms,

$$\mathscr{L}\{\mathbf{1}(t)\} = \frac{1}{s}$$
$$\mathscr{L}\{\mathbf{1}(t-T)\} = \frac{e^{-sT}}{s}$$

(3.2)

where $\mathbf{1}(t)$ denotes a unit step.

Thus, the transfer function of the ZOH is

$$G_{ZOH}(s) = \frac{1 - e^{-sT}}{s} \tag{3.3}$$

Next, we consider the frequency response of the ZOH:

$$G_{ZOH}(j\omega) = \frac{1 - e^{-j\omega T}}{j\omega} \tag{3.4}$$

We rewrite the frequency response in the form

$$G_{ZOH}(j\omega) = \frac{e^{-j\omega\frac{T}{2}}}{\omega}\left(\frac{e^{j\omega\frac{T}{2}} - e^{-j\omega\frac{T}{2}}}{j}\right)$$

$$= \frac{e^{-j\omega\frac{T}{2}}}{\omega}\left(2\sin\left(\omega\frac{T}{2}\right)\right) = Te^{-j\omega\frac{T}{2}}\frac{\sin\left(\omega\frac{T}{2}\right)}{\omega\frac{T}{2}}$$

We now have

$$|G_{ZOH}(j\omega)|\angle G_{ZOH}(j\omega) = T\left|\text{sinc}\left(\frac{\omega T}{2}\right)\right|\angle - \omega\frac{T}{2} \tag{3.5}$$

The angle of frequency response of the ZOH hold is seen to decrease linearly with frequency, whereas the magnitude is proportional to the sinc function. As shown in Figure 3.4, the magnitude is oscillatory with its peak magnitude equal to the sampling period and occurring at the zero frequency.

3.4 EFFECT OF THE SAMPLER ON THE TRANSFER FUNCTION OF A CASCADE

In a discrete-time system including several analog subsystems in cascade and several samplers, the location of the sampler plays an important role in determining the overall transfer function. Assuming that interconnection does not change the mathematical models of the subsystems, the Laplace transform of the output of the system of Figure 3.5 is given by

$$Y(s) = H_2(s)X(s)$$
$$= H_2(s)H_1(s)U(s) \tag{3.6}$$

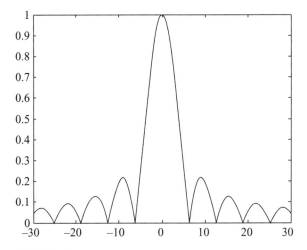

FIGURE 3.4

Magnitude of the frequency response of the zero-order hold with $T = 1$ s.

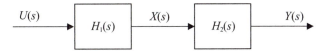

FIGURE 3.5

Cascade of two analog systems.

Inverse Laplace transforming gives the time response

$$y(t) = \int_0^t h_2(t - \tau) x(\tau) d\tau$$
$$= \int_0^t h_2(t - \tau) \left[\int_0^\tau h_1(\tau - \lambda) u(\lambda) d\lambda \right] d\tau$$

(3.7)

Changing the order and variables of integration, we obtain

$$y(t) = \int_0^t u(t - \tau) \left[\int_0^\tau h_1(\tau - \lambda) h_2(\lambda) d\lambda \right] d\tau$$
$$= \int_0^t u(t - \tau) h_{eq}(\tau) d\tau$$

(3.8)

where $h_{eq}(t) = \int_0^t h_1(t - \tau) h_2(\lambda) d\tau$.

Thus, the equivalent impulse response for the cascade is given by the convolution of the cascaded impulse responses. The same conclusion can be reached by

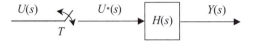

FIGURE 3.6

Analog system with sampled input.

inverse-transforming the product of the s-domain transfer functions. The time domain expression shows more clearly that cascading results in a new form for the impulse response. So if the output of the system is sampled to obtain

$$y(iT) = \int_o^{iT} u(iT - \tau) h_{eq}(\tau) d\tau, i = 1, 2, \ldots \tag{3.9}$$

it is not possible to separate the three time functions that are convolved to produce it.

By contrast, convolving an impulse-sampled function $u^*(t)$ with a continuous-time signal as shown in Figure 3.6 results in repetitions of the continuous-time function each of which is displaced to the location of an impulse in the train. Unlike the earlier situation, the resultant time function is not entirely new and there is hope of separating the functions that produced it. For a linear time-invariant (LTI) system with impulse-sampled input, the output is given by

$$y(t) = \int_0^t h(t - \tau) u(\tau) d\tau$$
$$= \int_0^t h(t - \tau) \left[\sum_{k=0}^{\infty} u(kT) \delta(\tau - kT) \right] d\tau \tag{3.10}$$

Changing the order of summation and integration gives

$$y(t) = \sum_{k=0}^{\infty} u(kT) \int_0^t h(t - \tau) \delta(\tau - kT) d\tau$$
$$= \sum_{k=0}^{\infty} u(kT) h(t - kT) \tag{3.11}$$

Sampling the output yields the convolution summation

$$y(iT) = \sum_{k=0}^{\infty} u(kT) h(iT - kT), i = 0, 1, 2, 3, \ldots \tag{3.12}$$

As discussed earlier, the convolution summation has the z-transform

$$Y(z) = H(z) U(z) \tag{3.13}$$

or in s-domain notation

$$Y^*(s) = H^*(s) U^*(s) \tag{3.14}$$

If a single block is an equivalent transfer function for the cascade of Figure 3.5, then its components cannot be separated after sampling. However, if the cascade is separated by samplers, then each block has a sampled output and input as well as a z-domain transfer function. For n blocks not separated by samplers, we use the notation

$$Y(z) = H(z)U(z)$$
$$= (H_1 H_2 \dots H_n)(z)U(z)$$

(3.15)

as opposed to n blocks separated by samplers where

$$Y(z) = H(z)U(z)$$
$$= H_1(z)H_2(z)\dots H_n(z)U(z)$$

(3.16)

EXAMPLE 3.1

Find the equivalent sampled impulse response sequence and the equivalent z-transfer function for the cascade of the two analog systems with sampled input

$$H_1(s) = \frac{1}{s+2} \quad H_2(s) = \frac{2}{s+4}$$

1. If the systems are directly connected.
2. If the systems are separated by a sampler.

Solution

1. In the absence of samplers between the systems, the overall transfer function is

$$H(s) = \frac{2}{(s+2)(s+4)}$$
$$= \frac{1}{s+2} - \frac{1}{s+4}$$

The impulse response of the cascade is

$$h(t) = e^{-2t} - e^{-4t}$$

and the sampled impulse response is

$$h(kT) = e^{-2kT} - e^{-4kT}, \ k = 0, 1, 2, \dots$$

Thus, the z-domain transfer function is

$$H(z) = \frac{z}{z - e^{-2T}} - \frac{z}{z - e^{-4T}} = \frac{(e^{-2T} - e^{-4T})z}{(z - e^{-2T})(z - e^{-4T})}$$

2. If the analog systems are separated by a sampler, then each has a z-domain transfer function and the transfer functions are given by

$$H_1(z) = \frac{z}{z - e^{-2T}} \quad H_2(z) = \frac{2z}{z - e^{-4T}}$$

The overall transfer function for the cascade is

$$H(z) = \frac{2z^2}{(z - e^{-2T})(z - e^{-4T})}$$

The partial fraction expansion of the transfer function is

$$H(z) = \frac{2}{e^{-2T} - e^{-4T}} \left[\frac{e^{-2T} z}{z - e^{-2T}} - \frac{e^{-4T} z}{z - e^{-4T}} \right]$$

Inverse z-transforming gives the impulse response sequence

$$h(kT) = \frac{2}{e^{-2T} - e^{-4T}} \left[e^{-2T} e^{-2kT} - e^{-4T} e^{-4kT} \right]$$

$$= \frac{2}{e^{-2T} - e^{-4T}} \left[e^{-2(k+1)T} - e^{-4(k+1)T} \right], \, k = 0, 1, 2, \ldots$$

Example 3.1 clearly shows the effect of placing a sampler between analog blocks on the impulse responses and the corresponding z-domain transfer function.

3.5 DAC, ANALOG SUBSYSTEM, AND ADC COMBINATION TRANSFER FUNCTION

The cascade of a DAC, analog subsystem, and ADC, shown in Figure 3.7, appears frequently in digital control systems (see Figure 3.1, for example). Because both the input and the output of the cascade are sampled, it is possible to obtain its z-domain transfer function in terms of the transfer functions of the individual subsystems. The transfer function is derived using the discussion of cascades given in Section 3.4.

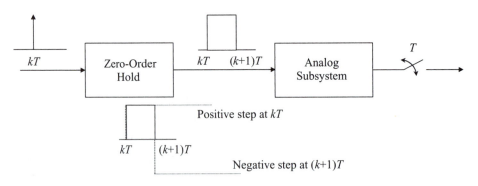

FIGURE 3.7

Cascade of a DAC, analog subsystem, and ADC.

Using the DAC model of Section 3.3, and assuming that the transfer function of the analog subsystem is $G(s)$, the transfer function of the DAC and analog subsystem cascade is

$$G_{ZA}(s) = G(s)G_{ZOH}(s)$$
$$= (1 - e^{-sT})\frac{G(s)}{s} \tag{3.17}$$

The corresponding impulse response is

$$g_{ZA}(t) = g(t)*g_{ZOH}(t)$$
$$= g_s(t) - g_s(t-T) \tag{3.18}$$
$$g_s(t) = \mathscr{L}^{-1}\left\{\frac{G(s)}{s}\right\}$$

The impulse response of (3.18) is the analog system step response minus a second step response delayed by one sampling period. This response is shown in Figure 3.8 for a second-order underdamped analog subsystem. The analog response of (3.18) is sampled to give the sampled impulse response

$$g_{ZA}(kT) = g_s(kT) - g_s(kT-T) \tag{3.19}$$

By z-transforming, we obtain the z-transfer function of the DAC (zero-order hold), analog subsystem, and ADC (ideal sampler) cascade

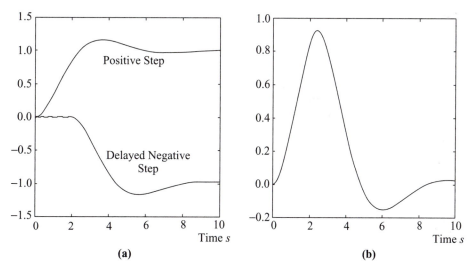

(a)   **(b)**

FIGURE 3.8

Impulse response of a DAC and analog subsystem. (a) Response of an analog system to step inputs. (b) Response of an analog system to a unit pulse input.

$$G_{ZAS}(z) = (1 - z^{-1}) \mathcal{Z}\{g_s^*(t)\}$$

$$= (1 - z^{-1}) \mathcal{Z}\left\{\mathcal{L}^{-1}\left[\frac{G(s)}{s}\right]_*\right\}$$ **(3.20)**

The cumbersome notation in (3.20) is to emphasize that sampling of a time function is necessary before z-transformation. Having made this point, the equation can be rewritten more concisely as

$$G_{ZAS}(z) = (1 - z^{-1}) \mathcal{Z}\left\{\frac{G(s)}{s}\right\}$$ **(3.21)**

EXAMPLE 3.2

Find $G_{ZAS}(z)$ for the cruise control system for the vehicle shown in Figure 3.9, where u is the input force, v is the velocity of the car, and b is the viscous friction coefficient.

Solution

We first draw a schematic to represent the cruise control system as shown in Figure 3.10. Using Newton's law, we obtain the following model:

$$M\dot{v}(t) + bv(t) = u(t)$$

which corresponds to the following transfer function:

$$G(s) = \frac{V(s)}{U(s)} = \frac{1}{Ms + b}$$

FIGURE 3.9

Automotive vehicle.

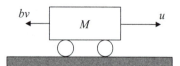

FIGURE 3.10

Schematic representation of a cruise control system for an automotive vehicle.

We rewrite the transfer function in the form

$$G(s) = \frac{K}{\tau s + 1} = \frac{K/\tau}{s + 1/\tau}$$

where $K = 1/b$ and $\tau = M/b$. The corresponding partial fraction expansion is

$$\frac{G(s)}{s} = \left(\frac{K}{\tau}\right)\left(\frac{A_1}{s} + \frac{A_2}{s + 1/\tau}\right)$$

where

$$A_1 = \left.\frac{1}{s + 1/\tau}\right|_{s=0} = \tau \quad A_2 = \left.\frac{1}{s}\right|_{s=-1/\tau} = -\tau$$

Using (3.21) and the z-transform table (see Appendix I), the desired z-domain transfer function is

$$G_{ZAS}(z) = (1 - z^{-1})\mathcal{Z}\left\{\left(\frac{K}{\tau}\right)\left[\frac{\tau}{s} - \frac{\tau}{s + 1/\tau}\right]\right\}$$

$$= K\left[1 + \frac{z-1}{z - e^{-T/\tau}}\right]$$

We simplify to obtain the transfer function

$$G_{ZAS}(z) = K\left[\frac{2z - (1 + e^{-T/\tau})}{z - e^{-T/\tau}}\right]$$

EXAMPLE 3.3

Find $G_{ZAS}(z)$ for the vehicle position control system shown in Figure 3.10, where u is the input force, y is the position of the car, and b is the viscous friction coefficient.

Solution

As with the previous example, we obtain the following equation of motion:

$$M\ddot{y}(t) + b\dot{y}(t) = u(t)$$

and the corresponding transfer function

$$G(s) = \frac{Y(s)}{U(s)} = \frac{1}{s(Ms + b)}$$

We rewrite the transfer function in terms of the system time constant

$$G(s) = \frac{K}{s(\tau s + 1)} = \frac{K/\tau}{s(s + 1/\tau)}$$

where $K = 1/b$ and $\tau = M/b$. The corresponding partial fraction expansion is

$$\frac{G(s)}{s} = \left(\frac{K}{\tau}\right)\left(\frac{A_{11}}{s^2} + \frac{A_{12}}{s} + \frac{A_2}{s+1/\tau}\right)$$

where

$$A_{11} = \frac{1}{s+1/\tau}\bigg|_{s=0} = \tau \quad A_{12} = \frac{d}{ds}\left[\frac{1}{s+1/\tau}\right]\bigg|_{s=0} = -\tau^2$$

$$A_2 = \frac{1}{s^2}\bigg|_{s=-1/\tau} = \tau^2$$

Using (3.21), the desired z-domain transfer function is

$$G_{ZAS}(z) = (1-z^{-1})\mathscr{Z}\left\{K\left[\frac{1}{s^2} - \frac{\tau}{s} + \frac{\tau}{s+1/\tau}\right]\right\}$$

$$= K\left[\frac{1}{z-1} - \tau + \frac{\tau(z-1)}{z-e^{-T/\tau}}\right]$$

which can be simplified to

$$G_{ZAS}(z) = K\left[\frac{(1-\tau+\tau e^{-T/\tau})z + [\tau - e^{-T/\tau}(\tau+1)]}{(z-1)(z-e^{-T/\tau})}\right]$$

EXAMPLE 3.4

Find $G_{ZAS}(z)$ for the series **R-L** circuit shown in Figure 3.11 with the inductor voltage as output.

Solution
Using the voltage divider rule gives

$$\frac{V_o}{V_{in}} = \frac{Ls}{R+Ls} = \frac{(L/R)s}{1+(L/R)s} = \frac{\tau s}{1+\tau s} \quad \tau = \frac{L}{R}$$

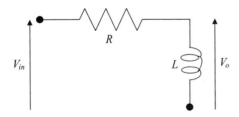

FIGURE 3.11

Series R-L circuit.

Hence, using (3.21), we obtain

$$G_{ZAS}(z) = (1 - z^{-1})\mathcal{Z}\left\{\frac{1}{s + 1/\tau}\right\}$$

$$= \frac{z-1}{z} \times \frac{z}{z - e^{-T/\tau}} = \frac{z-1}{z - e^{-T/\tau}}$$

EXAMPLE 3.5

Find the z-domain transfer function of the furnace sketched in Figure 3.12, where the inside temperature T_i is the controlled variable, T_w is the wall temperature, and T_o is the outside temperature. Assume perfect insulation so that there is no heat transfer between the wall and the environment. Assume also that heating is provided by a resistor and that the control variable u has the dimension of temperature with the inclusion of an amplifier with gain K.

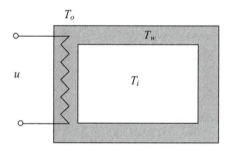

FIGURE 3.12

Schematic of a furnace.

Solution

The system can be modeled by means of the following differential equations:

$$\dot{T}_w(t) = g_{rw}(Ku(t) - T_w(t)) + g_{iw}(T_i(t) - T_w(t))$$

$$\dot{T}_i(t) = g_{iw}(T_w(t) - T_i(t))$$

where g_{rw} and g_{iw} are the heat transfer coefficients. By applying the Laplace transform and after some trivial calculations, we obtain the following transfer function

$$G(s) = \frac{T_i(s)}{U(s)} = \frac{Kg_{rw}g_{iw}}{s^2 + (2g_{iw} + g_{rw})s + g_{rw}g_{iw}}$$

Note that the two poles are real and therefore the transfer function can be rewritten as

$$G(s) = \frac{Y(s)}{U(s)} = \frac{K}{(s + p_1)(s + p_2)} \tag{3.22}$$

with appropriate values of p_1 and p_2. The corresponding partial fraction expansion is

$$\frac{G(s)}{s} = K\left(\frac{A_1}{s} + \frac{A_2}{s + p_1} + \frac{A_3}{s + p_2} \right)$$

where

$$A_1 = \frac{1}{(s + p_1)(s + p_2)}\bigg|_{s=0} = \frac{1}{p_1 p_2}$$

$$A_2 = \frac{1}{s(s + p_2)}\bigg|_{s=-p_1} = -\frac{1}{p_1(p_2 - p_1)}$$

$$A_3 = \frac{1}{s(s + p_1)}\bigg|_{s=-p_2} = \frac{1}{p_2(p_2 - p_1)}$$

Using (3.21), the desired z-domain transfer function is

$$G_{ZAS}(z) = (1 - z^{-1})\mathscr{Z}\left\{ K\left[\frac{1}{p_1 p_2 s^2} - \frac{p_1 + p_2}{p_1^2 p_2^2 s} + \frac{1}{p_1^2(p_2 - p_1)(s + p_1)} - \frac{1}{p_2^2(p_2 - p_1)}\frac{}{(s + p_2)} \right] \right\}$$

$$= K\left[\frac{1}{p_1 p_2(z - 1)} - \frac{p_1 + p_2}{p_1^2 p_2^2} + \frac{1}{p_1^2(p_2 - p_1)}\frac{}{(z - 1)(z - e^{-p_1 T})} - \frac{1}{p_2^2(p_2 - p_1)}\frac{}{(z - 1)(z - e^{-p_2 T})} \right]$$

which can be rewritten as

$$G_{ZAS}(z) = K\left[\frac{(p_1 e^{-p_2 T} - p_2 e^{-p_1 T} + p_2 - p_1)z + \left(\begin{array}{c} p_1 e^{-p_1 T} - p_2 e^{-p_2 T} + p_2 e^{-(p_1 + p_2)T} \\ - p_1 e^{-(p_1 + p_2)T} \end{array} \right)}{p_1 p_2(p_2 - p_1)(z - e^{-p_1 T})(z - e^{-p_2 T})} \right]$$

EXAMPLE 3.6

Find the z-domain transfer function of an armature-controlled DC motor.

Solution

The system can be modeled by means of the following differential equations:

$$J\ddot{\theta}(t) + b\dot{\theta}(t) = K_t i(t)$$

$$L\frac{di(t)}{dt} + Ri(t) = u(t) - K_e \dot{\theta}(t)$$

$$y(t) = \theta(t)$$

where θ is the position of the shaft (i.e., the output y of the system), i is the armature current, u is the source voltage (i.e., the input of the system), J is the moment of inertia of the motor, b is the viscous friction coefficient, K_t is the armature constant, K_e torque is the motor constant, R torque is the electric resistance, and L is the electric inductance. By applying the Laplace transform and after some trivial calculations, we obtain the following transfer function:

$$G(s) = \frac{Y(s)}{U(s)} = \frac{K_t}{s[(Js+b)(Ls+R)+K_t K_e]}$$

which can be rewritten as

$$G(s) = \frac{Y(s)}{U(s)} = \frac{K}{s(s+p_1)(s+p_2)}$$

with appropriate values of K, p_1, and p_2. The corresponding partial fraction expansion is

$$\frac{G(s)}{s} = K\left(\frac{A_{11}}{s^2} + \frac{A_{12}}{s} + \frac{A_2}{s+p_1} + \frac{A_3}{s+p_2}\right)$$

where

$$A_{11} = \frac{1}{(s+p_1)(s+p_2)}\bigg|_{s=0} = \frac{1}{p_1 p_2}$$

$$A_{12} = \frac{d}{ds}\left[\frac{1}{(s+p_1)(s+p_2)}\right]\bigg|_{s=0} = -\frac{p_1+p_2}{p_1^2 p_2^2}$$

$$A_2 = \frac{1}{s^2(s+p_2)}\bigg|_{s=-p_1} = \frac{1}{p_1^2(p_2-p_1)}$$

$$A_3 = \frac{1}{s^2(s+p_1)}\bigg|_{s=-p_2} = -\frac{1}{p_2^2(p_2-p_1)}$$

Using (3.21), the desired z-domain transfer function is

$$G_{ZAS}(z) = (1-z^{-1})\mathcal{Z}\left\{K\left[\frac{1}{p_1 p_2 s^2} - \frac{p_1+p_2}{p_1^2 p_2^2 s} + \frac{1}{p_1^2(p_2-p_1)(s+p_1)} - \frac{1}{p_2^2(p_2-p_1)}\frac{1}{(s+p_2)}\right]\right\}$$

$$= K\left[\frac{1}{p_1 p_2(z-1)} - \frac{p_1+p_2}{p_1^2 p_2^2} + \frac{1}{p_1^2(p_2-p_1)} - \frac{1}{p_2^2(p_2-p_1)}\frac{1}{(z-1)(z-e^{-p_2 T})}\right]$$

Note that if the velocity of the motor is considered as output (i.e., $y(t) = \dot{\theta}(t)$), we have the transfer function

$$G(s) = \frac{Y(s)}{U(s)} = \frac{K_t}{(Js+b)(Ls+R)+K_tK_e}$$

and the calculations of Example 3.5 can be repeated to obtain the z-domain transfer function (see (3.22)).

In the preceding examples, we observe that if the analog system has a pole at p_s, then $G_{ZAS}(z)$ has a pole at $p_z = e^{p_sT}$. The division by s in (3.21) results in a pole at $z = 1$ that cancels, leaving the same poles as those obtained when sampling and z-transforming the impulse response of the analog subsystem. However, the zeros of the transfer function are different in the presence of a DAC.

3.6 SYSTEMS WITH TRANSPORT LAG

Many physical system models include a transport lag or delay in their transfer functions. These include chemical processes, automotive engines, sensors, digital systems, and so on. In this section, we obtain the z-transfer function $G_{ZAS}(z)$ of (3.21) for a system with transport delay.

The transfer function for systems with a transport delay is of the form

$$G(s) = G_a(s)e^{-T_ds} \tag{3.23}$$

where T_d is the transport delay. As in Section 2.7, the transport delay can be rewritten as

$$T_d = lT - mT, \quad 0 \le m < 1 \tag{3.24}$$

where m is a positive integer. For example, a time delay of 3.1 s with a sampling period T of 1 s corresponds to $l = 4$ and $m = 0.9$. A delay by an integer multiple of the sampling period does not affect the form of the impulse response of the system. Therefore, using the delay theorem and (3.20), the z-transfer function for the system of (3.23) can be rewritten as

$$G_{ZAS}(z) = (1 - z^{-1})\mathcal{Z}\left\{\mathcal{L}^{-1}\left[\frac{G_a(s)e^{-T(l-m)s}}{s}\right]_*\right\}$$

$$= z^{-l}(1 - z^{-1})\mathcal{Z}\left\{\mathcal{L}^{-1}\left[\frac{G_a(s)e^{mTs}}{s}\right]_*\right\} \tag{3.25}$$

From (3.25), we observe that the inverse Laplace transform of the function

$$G_s(s) = \frac{G_a(s)}{s} \tag{3.26}$$

is sampled and z-transformed to obtain the desired z-transfer function. We rewrite (3.26) in terms of $G_s(s)$ as

$$G_{ZAS}(z) = z^{-l}(1-z^{-1})\mathscr{Z}\left\{\mathscr{L}^{-1}\left[G_s(s)e^{mTs}\right]*\right\} \tag{3.27}$$

Using the time advance theorem of Laplace transforms gives

$$G_{ZAS}(z) = z^{-l}(1-z^{-1})\mathscr{Z}\left\{g_s^*(t+mT)\right\} \tag{3.28}$$

where the impulse-sampled waveform must be used to allow Laplace transformation. The remainder of the derivation does not require impulse sampling, and we can replace the impulse-sampled waveform with the corresponding sequence

$$G_{ZAS}(z) = z^{-l}(1-z^{-1})\mathscr{Z}\left\{g_s(kT+mT)\right\} \tag{3.29}$$

The preceding result hinges on our ability to obtain the effect of a time advance mT on a sampled waveform. In Section 2.7, we discussed the z-transform of a signal delayed by T and advanced by mT, which is known as the modified z-transform. To express (3.29) in terms of the modified z-transform, we divide the time delay lT into a delay $(l-1)T$ and a delay T and rewrite the transfer function as

$$G_{ZAS}(z) = z^{-(l-1)}(1-z^{-1})z^{-1}\mathscr{Z}\left\{g_s(kT+mT)\right\} \tag{3.30}$$

Finally, we express the z-transfer function in terms of the modified z-transform

$$G_{ZAS}(z) = \left(\frac{z-1}{z^l}\right)\mathscr{Z}_m\left\{g_s(kT)\right\} \tag{3.31}$$

We recall two important modified transforms that are given in Section 2.7:

$$\mathscr{Z}_m\left\{1(kT)\right\} = \frac{1}{z-1} \tag{3.32}$$

$$\mathscr{Z}_m\left\{e^{-pkT}\right\} = \frac{e^{-mpT}}{z-e^{-pT}} \tag{3.33}$$

Returning to our earlier numerical example, a delay of 3.1 sampling periods gives a delay of 3 sampling periods and the corresponding z^{-3} term and the modified z-transform of $g(kT)$ with $m = 0.9$.

EXAMPLE 3.7

If the sampling period is 0.1 s, determine the z-transfer function $G_{ZAS}(z)$ for the system

$$G(s) = \frac{3e^{-0.31s}}{s+3}$$

Solution

First write the delay in terms of the sampling period as $0.31 = 3.1 \times 0.1 = (4-0.9) \times 0.1$. Thus, $l-1 = 3$ and $m = 0.9$. Next, obtain the partial fraction expansion

$$G_s(s) = \frac{3}{s(s+3)} = \frac{1}{s} - \frac{1}{s+3}$$

This is the transform of the continuous-time function shown in Figure 3.13, which must be sampled, shifted, and z-transformed to obtain the desired transfer function. Using the modified z-transforms obtained in Section 2.7, the desired transfer function is

$$G_{ZAS}(z) = \left(\frac{z-1}{z^4}\right)\left\{\frac{1}{z-1} - \frac{e^{-0.3\times0.9}}{z-e^{-0.3}}\right\}$$

$$= z^{-4}\left\{\frac{z - 0.741 - 0.763(z-1)}{z - 0.741}\right\} = \frac{0.237z + 0.022}{z^4(z - 0.741)}$$

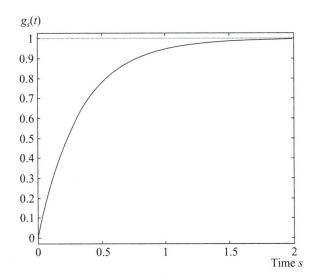

$g_s(t)$

FIGURE 3.13

Continuous time function $g_s(t)$.

3.7 THE CLOSED-LOOP TRANSFER FUNCTION

Using the results of Section 3.5, the digital control system of Figure 3.1 yields the closed-loop block diagram of Figure 3.14. The block diagram includes a comparator, a digital controller with transfer function $C(z)$, and the ADC-analog subsystem-DAC transfer function $G_{ZAS}(z)$. The controller and comparator are actually computer programs and replace the computer block in Figure 3.1. The block diagram is identical to those commonly encountered in s-domain analysis of analog systems

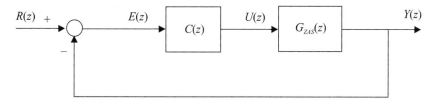

FIGURE 3.14

Block diagram of a single-loop digital control system.

with the variable s replaced by z. Hence, the closed-loop transfer function for the system is given by

$$G_{cl}(z) = \frac{C(z)G_{ZAS}(z)}{1 + C(z)G_{ZAS}(z)} \tag{3.34}$$

and the closed-loop characteristic equation is

$$1 + C(z)G_{ZAS}(z) = 0 \tag{3.35}$$

The roots of the equation are the closed-loop system poles, which can be selected for desired time response specifications as in s-domain design. Before we discuss this in some detail, we first examine alternative system configurations and their transfer functions.

When deriving closed-loop transfer functions of other configurations, the results of Section 3.4 must be considered carefully, as seen from the following example.

EXAMPLE 3.8

Find the Laplace transform of the analog and sampled output for the block diagram of Figure 3.15.

Solution
The analog variable $x(t)$ has the Laplace transform

$$X(s) = H(s)G(s)D(s)E(s)$$

which involves three multiplications in the s-domain. In the time domain, $x(t)$ is obtained after three convolutions.

From the block diagram

$$E(s) = R(s) - X^*(s)$$

Substituting in the $X(s)$ expression, sampling then gives

$$X(s) = H(s)G(s)D(s)[R(s) - X^*(s)]$$

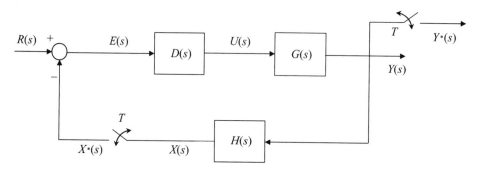

FIGURE 3.15

Block diagram of a system with sampling in the feedback path.

Thus, the impulse-sampled variable $x^*(t)$ has the Laplace transform

$$X^*(s) = (HGDR)^*(s) - (HGD)^*(s)X^*(s)$$

where, as in the first part of Example 3.1, several components are no longer separable. These terms are obtained as shown in Example 3.1 by inverse Laplace transforming, impulse sampling, and then Laplace transforming the impulse-sampled waveform.

Next, we solve for $X^*(s)$

$$X^*(s) = \frac{(HGDR)^*(s)}{1 + (HGD)^*(s)}$$

and then $E(s)$

$$E(s) = R(s) - \frac{(HGDR)^*(s)}{1 + (HGD)^*(s)}$$

With some experience, the last two expressions can be obtained from the block diagram directly. The combined terms are clearly the ones not separated by samplers in the block diagram.

From the block diagram the Laplace transform of the output is $Y(s) = G(s)D(s)E(s)$. Substituting for $E(s)$ gives

$$Y(s) = G(s)D(s)\left[R(s) - \frac{(HGDR)^*(s)}{1 + (HGD)^*(s)}\right]$$

Thus, the sampled output is

$$Y^*(s) = (GDR)^*(s) - (GD)^*(s)\frac{(HGDR)^*(s)}{1 + (HGD)^*(s)}$$

With the transformation $z = e^{st}$, we can rewrite the sampled output as

$$Y(z) = (GDR)(z) - (GD)(z)\frac{(HGDR)(z)}{1 + (HGD)(z)}$$

The last equation demonstrates how for some digital systems, no expression is available for the transfer function excluding the input. Nevertheless, the preceding system has a closed-loop characteristic equation similar to (3.35) given by $1 + (HGD)(z) = 0$.

This equation can be used in design as in cases where a closed-loop transfer function is defined.

3.8 ANALOG DISTURBANCES IN A DIGITAL SYSTEM

Disturbances are variables that are not included in the system model but affect its response. They can be deterministic, such as load torque in a position control system, or stochastic, such as sensor or actuator noise. However, almost all disturbances are analog and are inputs to the analog subsystem in a digital control loop. We use the results of Section 3.7 to obtain a transfer function with a disturbance input.

Consider the system with disturbance input shown in Figure 3.16. Because the system is linear, the reference input can be treated separately and is assumed to be zero.

The Laplace transform of the impulse-sampled output is

$$Y^*(s) = (GG_dD)^*(s) - (GG_{ZOH})^*(s)C^*(s)Y^*(s) \tag{3.36}$$

Solving for $Y^*(s)$, we obtain

$$Y^*(s) = \frac{(GG_dD)^*(s)}{1 + (GG_{ZOH})^*(s)C^*(s)} \tag{3.37}$$

The denominator involves the transfer function for the zero-order hold, analog subsystem, and sampler. We can therefore rewrite (3.37) using the notation of (3.21) as

$$Y^*(s) = \frac{(GG_dD)^*(s)}{1 + (G_{ZAS})^*(s)C^*(s)} \tag{3.38}$$

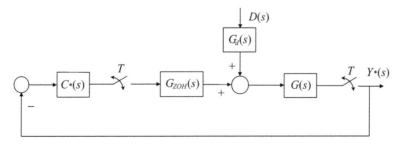

FIGURE 3.16

Block diagram of a digital system with an analog disturbance.

or in terms of z as

$$Y(z) = \frac{(GG_d D)(z)}{1 + G_{ZAS}(z)C(z)} \tag{3.39}$$

EXAMPLE 3.9

Consider the block diagram of Figure 3.16 with the transfer functions

$$G(s) = \frac{K_p}{s+1}, \quad G_d(s) = \frac{1}{s}, \quad C(z) = K_c$$

Find the steady-state response of the system to an impulse disturbance of strength A.

Solution
We first evaluate

$$G(s)G_d(s)D(s) = \frac{K_p A}{s(s+1)} = K_p A\left[\frac{1}{s} - \frac{1}{s+1}\right]$$

The z-transform of the corresponding impulse response sequence is

$$(GG_d D)(z) = K_p A\left[\frac{z}{z-1} - \frac{z}{z-e^{-T}}\right]$$

Using (3.21), we obtain the transfer function

$$G_{ZAS}(z) = K_p \frac{1-e^{-T}}{z-e^{-T}}$$

From (3.38), we obtain the sampled output

$$Y(z) = \frac{K_p A\left[\dfrac{z}{z-1} - \dfrac{z}{z-e^{-T}}\right]}{1 + K_c\left[K_p \dfrac{1-e^{-T}}{z-e^{-T}}\right]}$$

To obtain the steady-state response, we use the final value theorem
$$y(\infty) = (z-1)Y(z)\big|_{z=1}$$
$$= \frac{K_p A}{1 + K_c K_p}$$

Thus, as with analog systems, increasing the controller gain reduces the error due to the disturbance. Equivalently, an analog amplifier before the point of disturbance injection can increase the gain and reduce the output due to the disturbance and is less likely to saturate the DAC. Note that it is simpler to apply the final value theorem without simplification because terms not involving $(z - 1)$ drop out.

3.9 STEADY-STATE ERROR AND ERROR CONSTANTS

In this section, we consider the unity feedback block diagram shown in Figure 3.14 subject to standard inputs and determine the associated **tracking error** in each case. The standard inputs considered are the **sampled step**, the **sampled ramp**, and the **sampled parabolic**. As with analog systems, an error constant is associated with each input, and a type number can be defined for any system from which the nature of the error constant can be inferred. All results are obtained by direct application of the final value theorem.

From Figure 3.13, the tracking error is given by

$$E(z) = \frac{R(z)}{1 + G_{ZAS}(z)C(z)}$$
$$= \frac{R(z)}{1 + L(z)} \tag{3.40}$$

where $L(z)$ denotes the loop gain of the system.

Applying the final value theorem yields the steady-state error

$$e(\infty) = (1 - z^{-1})E(z)\big|_{z=1}$$
$$= \frac{(z-1)R(z)}{z(1+L(z))}\bigg|_{z=1} \tag{3.41}$$

The limit exists if all $(z - 1)$ terms in the denominator cancel. This depends on the reference input as well as on the loop gain.

To examine the effect of the loop gain on the limit, rewrite it in the form

$$L(z) = \frac{N(z)}{(z-1)^n D(z)}, \quad n \geq 0 \tag{3.42}$$

where $N(z)$ and $D(z)$ are numerator and denominator polynomials, respectively, with no unity roots. The following definition plays an important role in determining the steady-state error of unity feedback systems.

Definition 3.1: Type Number. The type number of the system is the number of unity poles in the system z-transfer function.   ∎

The loop gain of (3.42) has n poles at unity and is therefore type n. These poles play the same role as poles at the origin for an s-domain transfer function in determining the steady-state response of the system. Note that s-domain poles at zero play the same role as z-domain poles at e^0.

Substituting from (3.42) in the error expression (3.41) gives

$$e(\infty) = \left. \frac{(z-1)^{n+1} D(z) R(z)}{z(N(z)+(z-1)^n D(z))} \right|_{z=1}$$

$$= \left. \frac{(z-1)^{n+1} D(1) R(z)}{N(1)+(z-1)^n D(1)} \right|_{z=1} \tag{3.43}$$

Next, we examine the effect of the reference input on the steady-state error.

3.9.1 Sampled Step Input

The z-transform of a sampled unit step input is

$$R(z) = \frac{z}{z-1}$$

Substituting in (3.41) gives the steady-state error

$$e(\infty) = \left. \frac{1}{1+L(z)} \right|_{z=1} \tag{3.44}$$

The steady-state error can also be written as

$$e(\infty) = \frac{1}{1+K_p} \tag{3.45}$$

where K_p is the position error constant given by

$$K_p = L(1) \tag{3.46}$$

Examining (3.42) shows that K_p is finite for type 0 systems and infinite for systems of type 1 or higher. Therefore, the steady-state error for a sampled unit step input is

$$e(\infty) = \begin{cases} \dfrac{1}{1+L(1)}, & n = 0 \\ 0, & n \geq 1 \end{cases} \tag{3.47}$$

3.9.2 Sampled Ramp Input

The z-transform of a sampled unit ramp input is

$$R(z) = \frac{Tz}{(z-1)^2}$$

Substituting in (3.41) gives the steady-state error

$$e(\infty) = \left. \frac{T}{[z-1][1+L(z)]} \right|_{z=1}$$

$$= \frac{1}{K_v} \tag{3.48}$$

where K_v is the velocity error constant. The velocity error constant is thus given by

$$K_v = \frac{1}{T}(z-1)L(z)\Big|_{z=1} \tag{3.49}$$

From (3.49), the velocity error constant is zero for type 0 systems, finite for type 1 systems and infinite for type 2 or higher systems. The corresponding steady-state error is

$$e(\infty) = \begin{cases} \infty, & n = 0 \\ \dfrac{T}{(z-1)L(z)|_{z=1}}, & n = 1 \\ 0 & n \geq 2 \end{cases} \tag{3.50}$$

Similarly, it can be shown that for a sampled parabolic input, an acceleration error constant given by

$$K_a = \frac{1}{T^2}(z-1)^2 L(z)\Big|_{z=1} \tag{3.51}$$

can be defined, and the associated steady-state error is

$$e(\infty) = \begin{cases} \infty, & n \leq 1 \\ \dfrac{T^2}{(z-1)^2 L(z)|_{z=1}}, & n = 2 \\ 0, & n \geq 3 \end{cases} \tag{3.52}$$

EXAMPLE 3.10

Find the steady-state position error for the digital position control system with unity feedback and with the transfer functions

$$G_{ZAS}(z) = \frac{K(z+a)}{(z-1)(z-b)}, \quad C(z) = \frac{K_c(z-b)}{z-c}, \quad 0 < a, b, c < 1$$

1. For a sampled unit step input.
2. For a sampled unit ramp input.

Solution
The loop gain of the system is given by

$$L(z) = C(z)G_{ZAS}(z) = \frac{KK_c(z+a)}{(z-1)(z-c)}$$

The system is type 1. Therefore, it has zero steady-state error for a sampled step input and a finite steady-state error for a sampled ramp input given by

$$e(\infty) = \frac{T}{(z-1)L(z)|_{z=1}} = \frac{T}{KK_c}\left(\frac{1-c}{1+a}\right)$$

Clearly, the steady-state error is reduced by increasing the controller gain and is also affected by the choice of controller pole and zero.

EXAMPLE 3.11

Find the steady-state error for the analog system

$$G(s) = \frac{K}{s+a} \quad a > 0$$

1. For proportional analog control with a unit step input.
2. For proportional digital control with a sampled unit step input.

Solution

The transfer function of the system can be written as

$$G(s) = \frac{K/a}{s/a+1} \quad a > 0$$

Thus, the position error constant for analog control is K/a, and the steady-state error is

$$e(\infty) = \frac{1}{1+K_p} = \frac{a}{K+a}$$

For digital control, it can be shown that for sampling period T, the DAC-plant-ADC z-transfer function is

$$G_{ZAS}(z) = \frac{K}{a}\left(\frac{1-e^{-aT}}{z-e^{-aT}}\right)$$

Thus, the position error constant for digital control is

$$K_p = G_{ZAS}(z)|_{z=1} = K/a$$

and the associated steady-state error is the same as that of the analog system with proportional control. In general, it can be shown that the steady-state error for the same control strategy is identical for digital or analog implementation.

3.10 MATLAB COMMANDS

The transfer function for the ADC, analog subsystem, and DAC combination can be easily obtained using the MATLAB program. Assume that the sampling period is equal to 0.1 s and that the transfer function of the analog subsystem is *G*.

3.10.1 MATLAB

The MATLAB command to obtain a digital transfer function from an analog transfer function is

$$\gg g = \textbf{tf(num,den)}$$

$$\gg \textbf{gd} = \textbf{c2d(g, 0.1, 'method')}$$

where **num** is a vector containing the numerator coefficients of the analog transfer function in descending order, and **den** is a similarly defined vector of denominator coefficients. For example, the numerator polynomial $(2s^2 + 4s + 3)$ is entered as

$$\gg \textbf{num} = [2, 4, 3]$$

The term "method" specifies the method used to obtain the digital transfer function. For a system with a zero-order hold and sampler (DAC and ADC), we use

$$\gg \textbf{gd} = \textbf{c2d(g, 0.1, 'zoh')}$$

For a first-order hold, we use

$$\gg \textbf{gd} = \textbf{c2d(g, 0.1, 'foh')}$$

Other options of MATLAB commands are available but are not relevant to the material presented in this chapter.

For a system with a time delay, the discrete transfer function can be obtained using the commands

$$\textbf{gdelay} = \textbf{tf(num, den, 'inputdelay', Td)} \% \text{ Delay} = \text{Td}$$

$$\textbf{gdelay\_d} = \textbf{c2d(gdelay, 0.1, 'method')}$$

RESOURCES

Franklin, G. F., J. D. Powell, and M. L. Workman, *Digital Control of Dynamic Systems,* Addison-Wesley, 1998.

Jacquot, R. G., *Modern Digital Control Systems,* Marcel Dekker, 1981.

Katz, P., *Digital Control Using Microprocessors,* Prentice Hall International, 1981.

Kuo, B. C., *Digital Control Systems,* Saunders, 1992.

Ogata, K., *Discrete-Time Control Systems,* Prentice Hall, 1987.

PROBLEMS

3.1 Find the magnitude and phase at frequency $\omega = 1$ of a zero-order hold with sampling time $T = 0.1$ s.

3.2 The first-order hold uses the last two numbers in a sequence to generate its output. The output of both the zero-order hold and the first-order hold is given by

$$u(t) = u(kT) + a\frac{u(kT) - u[(k-1)T]}{T}(t - kT), kT \leq t \leq (k+1)T \quad k = 0, 1, 2, \ldots$$

with $a = 0,1$, respectively.

(a) For a discrete impulse input, obtain and sketch the impulse response for the preceding equation with $a = 0$ and $a = 1$.

(b) Write an impulse-sampled version of the preceding impulse response, and show that the transfer function is given by

$$G_H(s) = \frac{1}{s}(1 - e^{-sT})\left[1 - ae^{-sT} + \frac{a}{sT}(1 - e^{-sT})\right]$$

(c) Obtain the transfer functions for the zero-order hold and for the first-order hold from the preceding transfer function. Verify that the transfer function of the zero-order hold is the same as that obtained in Section 3.3.

3.3 Many chemical processes can be modelled by the following transfer function:

$$G(s) = \frac{K}{\tau s + 1}e^{-T_d s}$$

where K is the gain, τ is the time constant and T_d is the time delay. Obtain the transfer function $G_{ZAS}(z)$ for the system in terms of the system parameters. Assume that the time delay T_d is a multiple of the sampling time T.

3.4 Obtain the transfer function of a point mass (m) with force as input and displacement as output neglecting actuator dynamics; then find $G_{ZAS}(z)$ for the system.

3.5 For an internal combustion engine, the transfer function with injected fuel flow rate as input and fuel flow rate into the cylinder as output is given by[1]

$$G(s) = \frac{\varepsilon \tau s + 1}{\tau s + 1}$$

where τ is a time constant and ε is known as the fuel split parameter. Obtain the transfer function $G_{ZAS}(z)$ for the system in terms of the system parameters.

---

[1] J. Moskwa, *Automotive Engine Modeling and Real time Control*, MIT doctoral thesis, 1988.

3.6 Repeat Problem 3.5 including a delay of 25 ms in the transfer function with a sampling period of 10 ms.

3.7 Find the equivalent sampled impulse response sequence and the equivalent z-transfer function for the cascade of the two analog systems with sampled input

$$H_1(s) = \frac{1}{s+6} \quad H_2(s) = \frac{10}{s+1}$$

(a) If the systems are directly connected.
(b) If the systems are separated by a sampler.

3.8 Obtain expressions for the analog and sampled outputs from the block diagrams shown in Figure P3.8.

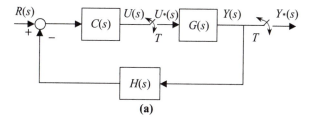

(a)

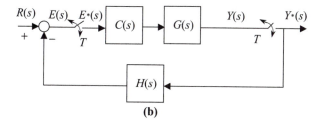

(b)

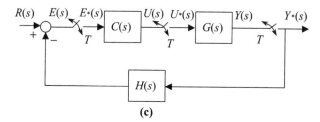

(c)

FIGURE P3.8

Block diagrams for systems with multiple samplers.

3.9 For the unity feedback system shown in Figure P3.9, we are given the analog subsystem

$$G(s) = \frac{s+8}{s+5}$$

The system is digitally controlled with a sampling period of 0.02 s. The controller transfer function was selected as

$$C(z) = \frac{0.35z}{z-1}$$

(a) Find the z-transfer function for the analog subsystem with DAC and ADC.
(b) Find the closed-loop transfer function and characteristic equation.
(c) Find the steady-state error for a sampled unit step and a sampled unit ramp. Comment on the effect of the controller on steady-state error.

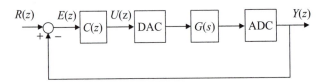

FIGURE P3.9

Block diagram for a closed-loop system with digital control.

3.10 Find the steady-state error for a unit step disturbance input for the systems shown in Figure P3.10 with a sampling period of 0.03 s and the transfer functions

$$G_d(s) = \frac{2}{s+1} \quad G(s) = \frac{4(s+2)}{s(s+3)} \quad C^*(s) = \frac{e^{sT} - 0.95}{e^{sT} - 1}$$

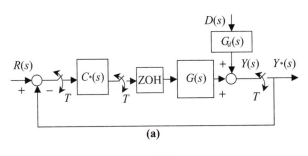

(a)

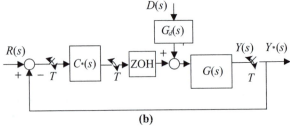

(b)

FIGURE P3.10

Block diagrams for systems with disturbance inputs.

3.11 For the following systems with unity feedback, find
   (a) The position error constant.
   (b) The velocity error constants.
   (c) The steady-state error for a unit step input.
   (d) The steady-state error for a unit ramp input.

   (i) $G(z) = \dfrac{0.4(z+0.2)}{(z-1)(z-0.1)}$

   (ii) $G(z) = \dfrac{0.5(z+0.2)}{(z-0.1)(z-0.8)}$

COMPUTER EXERCISES

3.12 For the analog system with a sampling period of 0.05 s

$$G(s) = \frac{10(s+2)}{s(s+5)}$$

   (a) Obtain the transfer function for the system with sampled input and output.
   (b) Obtain the transfer function for the system with DAC and ADC.
   (c) Obtain the unit step response of the system with sampled output and analog input.
   (d) Obtain the poles of the systems in (a) and (b), and the output of (c), and comment on the differences between them.

3.13 For the system of Problem 3.9
   (a) Obtain the transfer function for the analog subsystem with DAC and ADC.
   (b) Obtain the step response of the open-loop analog system and the closed-loop digital control system and comment on the effect of the controller on the time response.
   (c) Obtain the frequency response of the digital control system, and verify that 0.02 s is an acceptable choice of sampling period. Explain briefly why the sampling period is chosen based on the closed-loop rather than the open-loop dynamics.

3.14 Consider the internal combustion engine model of Problem 3.5. Assume that, for the operational conditions of interest, the time constant τ is approximately 1.2 s, whereas the parameter ε can vary in the range 0.4 to 0.6. The digital cascade controller

$$C(z) = \frac{0.02z}{z-1}$$

   was selected to improve the time response of the system with unity feedback. Simulate the digital control system with $\varepsilon = 0.4$, 0.5, and 0.6, and discuss the behavior of the controller in each case.

3.15 Simulate the continuous-discrete system discussed in Problem 3.9 and examine the behavior of both the continuous output and the sampled output. Repeat the simulation with a 10% error in the plant gain. Discuss the simulation results, and comment on the effect of the parameter error on disturbance rejection.

Stability of Digital Control Systems

Stability is a basic requirement for digital and analog control systems. Digital control is based on samples and is updated every sampling period, and there is a possibility that the system will become unstable between updates. This obviously makes stability analysis different in the digital case. We examine different definitions and tests of the stability of linear time-invariant (LTI) digital systems based on transfer function models. In particular, we consider input-output stability and internal stability. We provide several tests for stability: the Routh-Hurwitz criterion, the Jury criterion, and the Nyquist criterion. We also define the gain margin and phase margin for digital systems.

Objectives

After completing this chapter, the reader will be able to do the following:

1. Determine the input-output stability of a z-transfer function.
2. Determine the asymptotic stability of a z-transfer function.
3. Determine the internal stability of a digital feedback control system.
4. Determine the stability of a z-polynomial using the Routh-Hurwitz criterion.
5. Determine the stability of a z-polynomial using the Jury criterion.
6. Determine the stable range of a parameter for a z-polynomial.
7. Determine the closed-loop stability of a digital system using the Nyquist criterion.
8. Determine the gain margin and phase margin of a digital system.

4.1 DEFINITIONS OF STABILITY

The most commonly used definitions of stability are based on the magnitude of the system response in the steady state. If the steady-state response is unbounded, the system is said to be unstable. In this chapter, we discuss two stability definitions that concern the boundedness or exponential decay of the system output.

The first stability definition considers the system output due to its initial conditions. To apply it to transfer function models, we need the assumption that no pole-zero cancellation occurs in the transfer function. Reasons for this assumption are given later and are discussed further in the context of state-space models.

Definition 4.1: Asymptotic Stability. A system is said to be asymptotically stable if its response to any initial conditions decays to zero asymptotically in the steady state—that is, the response due to the initial conditions satisfies

$$\lim_{k \to \infty} y(k) = 0 \tag{4.1}$$

If the response due to the initial conditions remains bounded but does not decay to zero, the system is said to be **marginally stable**.   ∎

The second definition of stability concerns the forced response of the system for a bounded input. A bounded input satisfies the condition

$$|u(k)| < b_u, \quad k = 0, 1, 2, \ldots$$
$$0 < b_u < \infty \tag{4.2}$$

For example, a bounded sequence satisfying the constraint $|u(k)| < 3$ is shown in Figure 4.1.

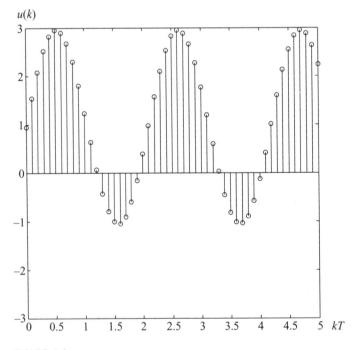

FIGURE 4.1

Bounded sequence with bound $b_u = 3$.

Definition 4.2: Bounded-Input–Bounded-Output Stability. A system is said to be bounded-input–bounded-output (BIBO) stable if its response to any bounded input remains bounded—that is, for any input satisfying (4.2), the output satisfies

$$|y(k)| < b_y, \quad k = 0, 1, 2, \ldots$$
$$0 < b_y < \infty \tag{4.3}$$

∎

4.2 STABLE z-DOMAIN POLE LOCATIONS

The examples provided in Chapter 2 show that the locations of the poles of the system z-transfer function determine the time response. The implications of this fact for system stability are now examined more closely.

Consider the sampled exponential and its z-transform

$$p^k, \quad k = 0, 1, 2, \ldots \xleftarrow{\;\mathcal{z}\;} \frac{z}{z - p} \tag{4.4}$$

where p is real or complex. Then the time sequence for large k is given by

$$|p|^k \to \begin{cases} 0, & |p| < 1 \\ 1, & |p| = 1 \\ \infty, & |p| > 1 \end{cases} \tag{4.5}$$

Any time sequence can be described by

$$f(k) = \sum_{i=1}^{n} A_i p_i^k, \quad k = 0, 1, 2, \ldots \xleftarrow{\;\mathcal{z}\;} F(z) = \sum_{i=1}^{n} A_i \frac{z}{z - p_i} \tag{4.6}$$

where A_i are partial fraction coefficients and p_i are z-domain poles. Hence, we conclude that the sequence is bounded if its poles lie in the closed unit disc (i.e., on or inside the unit circle) and decays exponentially if its poles lie in the open unit disc (i.e., inside the unit circle). This conclusion allows us to derive stability conditions based on the locations of the system poles. Note that the case of repeated poles on the unit circle corresponds to an unbounded time sequence (see, for example, the transform of the sampled ramp).

Although the preceding conclusion is valid for complex as well as real-time sequences, we will generally restrict our discussions to real-time sequences. For real-time sequences, the poles and partial fraction coefficients in (4.6) are either real or complex conjugate pairs.

4.3 STABILITY CONDITIONS

The analysis of Section 4.2 allows the derivation of conditions for asymptotic and BIBO stability based on transfer function models. It is shown that, in the absence of pole-zero cancellation, conditions for BIBO stability and asymptotic stability are identical.

4.3.1 Asymptotic Stability

The following theorem gives conditions for asymptotic stability.

Theorem 4.1: Asymptotic Stability. In the absence of pole-zero cancellation, an LTI digital system is asymptotically stable if its transfer function poles are in the open unit disc and marginally stable if the poles are in the closed unit disc with no repeated poles on the unit circle.

PROOF. Consider the LTI system governed by the constant coefficient difference equation

$$y(k+n) + a_{n-1}y(k+n-1) + \ldots + a_1 y(k+1) + a_0 y(k)$$
$$= b_m u(k+m) + b_{m-1}u(k+m-1) + \ldots + b_1 u(k+1) + b_0 u(k), \quad k = 0, 1, 2, \ldots$$

with initial conditions $y(0)$, $y(1)$, $y(2)$, . . . , $y(n-1)$. Using the z-transform of the output, we observe that the response of the system due to the initial conditions with the input zero is of the form

$$Y(z) = \frac{N(z)}{z^n + a_{n-1}z^{n-1} + \ldots + a_1 z + a_0}$$

where $N(z)$ is a polynomial dependent on the initial conditions. Because transfer function zeros arise from transforming the input terms, they have no influence on the response due to the initial conditions. The denominator of the output z-transform is the same as the denominator of the z-transfer function in the absence of pole-zero cancellation. Hence, the poles of the function $Y(z)$ are the poles of the system transfer function.

$Y(z)$ can be expanded as partial fractions of the form (4.6). Thus, the output due to the initial conditions is bounded for system poles in the closed unit disc with no repeated poles on the unit circle. It decays exponentially for system poles in the open unit disc (i.e., inside the unit circle). ∎

Theorem 4.1 applies even if pole-zero cancellation occurs, provided that the poles that cancel are stable. This follows from the fact that stability testing is essentially a search for unstable poles, with the system being declared stable if none are found. Invisible but stable poles would not lead us to a wrong conclusion. However, invisible but stable poles would. The next example shows how to determine asymptotic stability using Theorem 4.1.

EXAMPLE 4.1

Determine the asymptotic stability of the following systems:

(a) $H(z) = \dfrac{4(z-2)}{(z-2)(z-0.1)}$

(b) $H(z) = \dfrac{4(z-0.2)}{(z-0.2)(z-0.1)}$

(c) $H(z) = \dfrac{5(z-0.3)}{(z-0.2)(z-0.1)}$

(d) $H(z) = \dfrac{8(z-0.2)}{(z-0.1)(z-1)}$

Solution

Theorem 4.1 can only be used for transfer functions (a) and (b) if their poles and zeros are not canceled. Ignoring the zeros, which do not affect the response to the initial conditions, (a) has a pole outside the unit circle and the poles of (b) are inside the unit circle. Hence, (a) is unstable, whereas (b) is asymptotically stable.

Theorem 4.1 can be applied to the transfer functions (c) and (d). The poles of (c) are all inside the unit circle, and the system is therefore asymptotically stable. However, (d) has one pole on the unit circle and is only marginally stable.

4.3.2 BIBO Stability

BIBO stability concerns the response of a system to a bounded input. The response of the system to any input is given by the convolution summation

$$y(k) = \sum_{i=0}^{k} h(k-i)u(i), \quad k = 0, 1, 2, \ldots \tag{4.7}$$

where $h(k)$ is the impulse response sequence.

It may seem that a system should be BIBO stable if its impulse response is bounded. To show that this is generally false, let the impulse response of a linear system be bounded and strictly positive with lower bound b_{h1} and upper bound b_{h2}—that is,

$$0 < b_{h1} < h(k) < b_{h2} < \infty, \quad k = 0, 1, 2, \ldots \tag{4.8}$$

Then using the bound (4.8) in (4.7) gives the inequality

$$|y(k)| = \sum_{i=0}^{k} h(k-i)u(i) > b_{h1}\sum_{i=0}^{k} u(i), \quad k = 0, 1, 2, \ldots \tag{4.9}$$

which is unbounded as $k \to \infty$ for the bounded input $u(k) = 1, k = 0,1,2, \ldots$

The following theorem establishes necessary and sufficient conditions for BIBO stability of a discrete-time linear system.

Theorem 4.2: A discrete-time linear system is BIBO stable if and only if its impulse response sequence is absolutely summable—that is,

$$\sum_{i=0}^{\infty} |h(i)| < \infty \tag{4.10}$$

PROOF

1. Necessity (only if)

To prove necessity by contradiction, assume that the system is BIBO stable but does not satisfy (4.10). Then consider the input sequence given by

$$u(k-i) = \begin{cases} 1, & h(i) \geq 0 \\ -1, & h(i) < 0 \end{cases}$$

The corresponding output is

$$y(k) = \sum_{i=0}^{k} |h(i)|$$

which is unbounded as $k \to \infty$. This contradicts the assumption of BIBO stability.

2. Sufficiency (if)

To prove sufficiency, we assume that (4.10) is satisfied and then show that the system is BIBO stable. Using the bound (4.2) in the convolution summation (4.7) gives the inequality

$$|y(k)| \leq \sum_{i=0}^{k} |h(i)||u(k-i)| < b_u \sum_{i=0}^{k} |h(i)|, \quad k = 0, 1, 2, \ldots$$

which remains bounded as $k \to \infty$ if (4.10) is satisfied. ∎

Because the z-transform of the impulse response is the transfer function, BIBO stability can be related to pole locations as follows.

Theorem 4.3: A discrete-time linear system is BIBO stable if and only if the poles of its transfer function lie inside the unit circle.

PROOF. Applying (4.6) to the impulse response and transfer function shows that the impulse response is bounded if the poles of the transfer function are in the closed unit disc and decays exponentially if the poles are in the open unit disc. It has already been established that systems with a bounded impulse response that does not decay exponentially are not BIBO stable. Thus, it is necessary for BIBO stability that the system poles lie inside the unit circle.

To prove sufficiency, assume an exponentially decaying impulse response (i.e., poles inside the unit circle). Let A_r be the coefficient of largest magnitude and $|p_s| < 1$ be the system pole of largest magnitude in (4.6). Then the impulse response (assuming no repeated poles for simplicity) is bounded by

$$|h(k)| = \left| \sum_{i=1}^{n} A_i p_i^k \right| \leq \sum_{i=1}^{n} |A_i||p_i|^k \leq n|A_r||p_s|^k, \quad k = 0, 1, 2, \ldots$$

Hence, the impulse response decays exponentially at a rate determined by the largest system pole. Substituting the upper bound in (4.10) gives

$$\sum_{i=0}^{\infty} |h(i)| \leq n|A_r| \sum_{i=0}^{\infty} |p_s|^i = n|A_r| \frac{1}{1-|p_s|} < \infty$$

Thus, the condition is sufficient by Theorem 4.2. ∎

EXAMPLE 4.2

Investigate the BIBO stability of the class of systems with the impulse response

$$h(k) = \begin{cases} K, & 0 \le k \le m < \infty \\ 0, & elsewhere \end{cases}$$

where K is a finite constant.

Solution
The impulse response satisfies

$$\sum_{i=0}^{\infty} |h(i)| = \sum_{i=0}^{m} |h(i)| = (m+1)|K| < \infty$$

Using condition (4.10), the systems are all BIBO stable. This is the class of **finite impulse response** (FIR) systems (i.e., systems whose impulse response is nonzero over a finite interval). Thus, we conclude that all FIR systems are BIBO stable.

EXAMPLE 4.3

Investigate the BIBO stability of the systems discussed in Example 4.1.

Solution
After pole-zero cancellation, the transfer functions (a) and (b) have all poles inside the unit circle and are therefore BIBO stable. The transfer function (c) has all poles inside the unit circle and is stable; (d) has a pole on the unit circle and is not BIBO stable.

The preceding analysis and examples show that for LTI systems, with no pole-zero cancellation, BIBO and asymptotic stability are equivalent and can be investigated using the same tests. Hence, the term **stability** is used in the sequel to denote either BIBO or asymptotic stability with the assumption of no unstable pole-zero cancellation. Pole locations for a stable system (inside the unit circle) are shown in Figure 4.2.

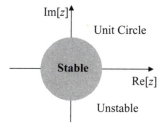

FIGURE 4.2

Stable pole locations in the z-plane. *Im*[z] denotes the imaginary part and *Re*[z] denotes the real part of z.

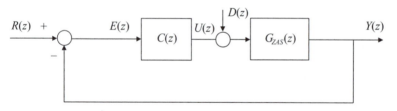

FIGURE 4.3

Digital control system with disturbance $D(z)$.

4.3.3 Internal Stability

So far, we have only considered stability as applied to an open-loop system. For closed-loop systems, these results are applicable to the closed-loop transfer function. However, the stability of the closed-loop transfer function is not always sufficient for proper system operation because some of the internal variables may be unbounded. In a feedback control system, it is essential that all the signals in the loop be bounded when bounded exogenous inputs are applied to the system.

Consider the unity feedback digital control scheme of Figure 4.3 where, for simplicity, a disturbances input is added to the controller output before the ADC. We consider that system as having two outputs, Y and U, and two inputs, R and D. Thus, the transfer functions associated with the system are given by

$$\begin{bmatrix} Y(z) \\ U(z) \end{bmatrix} = \begin{bmatrix} \dfrac{C(z)G_{ZAS}(z)}{1+C(z)G_{ZAS}(z)} & \dfrac{G_{ZAS}(z)}{1+C(z)G_{ZAS}(z)} \\ \dfrac{C(z)}{1+C(z)G_{ZAS}(z)} & -\dfrac{C(z)G_{ZAS}(z)}{1+C(z)G_{ZAS}(z)} \end{bmatrix} \begin{bmatrix} R(z) \\ D(z) \end{bmatrix} \tag{4.11}$$

Clearly, it is not sufficient to prove that the output of the controlled system Y is bounded for bounded reference input R because the controller output U can be unbounded. In addition, the system output must be bounded when a different input is applied to the system, namely, in the presence of a disturbance. This suggests the following definition of stability.

Definition 4.3: Internal Stability. If all the transfer functions that relate system inputs (R and D) to the possible system outputs (Y and U) are BIBO stable, then the system is said to be **internally stable**.   ∎

Because internal stability guarantees the stability of the transfer function from R to Y, among others, it is obvious that an internally stable system is also externally stable (i.e., the system output Y is bounded when the reference input R is bounded). However, external stability does not, in general, imply internal stability.

We now provide some results that allow us to test internal stability.

Theorem 4.4: The system shown in Figure 4.3 is internally stable if and only if all its closed-loop poles are in the open unit disc.

PROOF

1. Necessity (only if)

To prove necessity, we write $C(z)$ and $G_{ZAS}(z)$ as ratios of coprime polynomials (i.e., polynomials with no common factors):

$$C(z) = \frac{N_C(z)}{D_C(z)} \quad G_{ZAS}(z) = \frac{N_G(z)}{D_G(z)} \tag{4.12}$$

Substituting in (4.11), we rewrite it as

$$\begin{bmatrix} Y \\ U \end{bmatrix} = \frac{1}{D_C D_G + N_C N_G} \begin{bmatrix} N_C N_G & D_C N_G \\ N_C D_G & -N_C N_G \end{bmatrix} \begin{bmatrix} R \\ D \end{bmatrix} \tag{4.13}$$

where we have dropped the argument z for brevity. If the system is internally stable, then the four transfer functions in (4.11) have no poles on or outside the unit circle. Thus, we can conclude that polynomial $D_C D_G + N_C N_G$ has no zeros on or outside the unit circle because it cannot have a zero that is also a zero of the four numerators (which cancels, leaving four stable transfer functions).

2. Sufficiency (if)

Sufficiency is evident from (4.13). In fact, if the characteristic polynomial $D_C D_G + N_C N_G$ has no zeros on or outside the unit circle, then all the transfer functions are asymptotically stable and the system is internally stable. ∎

Theorem 4.5: The system of Figure 4.3 is internally stable if and only if the following two conditions hold:

1. The characteristic polynomial $1 + C(z)G_{ZAS}(z)$ has no zeros on or outside the unit circle.
2. The loop gain $C(z)G_{ZAS}(z)$ has no pole-zero cancellation on or outside the unit circle.

PROOF

1. Necessity (only if)

Condition 1 is clearly necessary by Theorem 4.4. To prove the necessity of condition 2, we first factor $C(z)$ and $G_{ZAS}(z)$ as in (4.12) to write the characteristic polynomial in the form $D_C D_G + N_C N_G$. We also have that $C(z) G_{ZAS}(z)$ is equal to $N_C N_G / D_C D_G$. Assume that condition 2 is violated and that there exists Z_0, $|Z_0| \geq 1$, which is a zero of $D_C D_G$ as well as a zero of $N_C N_G$. Then clearly Z_0 is also a zero of the characteristic polynomial $D_C D_G + N_C N_G$, and the system is unstable. This establishes the necessity of condition 2.

2. Sufficiency (if)

By Theorem 4.4, condition 1 implies internal stability unless unstable pole-zero cancellation occurs in the characteristic polynomial $1 + C(z)G_{ZAS}(z)$. We therefore have internal stability if condition 2 implies the absence of unstable pole-zero cancellation. If the loop gain $C(z)G_{ZAS}(z) = N_CN_G/D_CD_G$ has no unstable pole-zero cancellation, then $1 + C(z)G_{ZAS}(z) = [D_CD_G+N_CN_G]/D_CD_G$ does not have unstable pole-zero cancellation, and the system is internally sable. ∎

EXAMPLE 4.4

An isothermal chemical reactor where the product concentration is controlled by manipulating the feed flow rate is modeled by the following transfer function[1]:

$$G(s) = \frac{0.5848(-0.3549s + 1)}{0.1828s^2 + 0.8627 + 1}$$

Determine $G_{ZAS}(Z)$ with a sampling rate $T = 0.1$, and then verify that the closed-loop system with the feedback controller

$$C(z) = \frac{-10(z - 0.8149)(z - 0.7655)}{(z - 1)(z - 1.334)}$$

is not internally stable.

Solution
The discretized process transfer function is

$$G_{ZAS}(z)=(1-z^{-1})\mathcal{Z}\left\{\frac{G(s)}{s}\right\} = \frac{-0.075997(z - 1.334)}{(z - 0.8149)(z - 0.7655)}$$

The transfer function from the reference input to the system output is given by

$$\frac{Y(z)}{R(z)} = \frac{C(z)G_{ZAS}(z)}{1+C(z)G_{ZAS}(z)}$$

$$= \frac{1.2054(z - 0.8149)(z - 0.7655)}{(z - 0.8756)(z^2 - 1.27z + 0.6908)}$$

The system appears to be asymptotically stable with all its poles inside the unit circle. However, the system is not internally stable as seen by examining the transfer function

$$\frac{U(z)}{R(z)} = -12\frac{(z - 0.8149)(z - 0.7655)}{(z - 1.334)(z - 0.08798)}$$

which has a pole at 1.334 outside the unit circle. The control variable is unbounded even when the reference input is bounded. In fact, the system violates condition 2 of Theorem 4.5 because the pole at 1.334 cancels in the loop gain

$$C(z)G_{ZAS}(z) = \frac{-10(z - 0.8149)(z - 0.7655)}{(z - 1)(z - 1.334)} \times \frac{-0.075997(z - 1.334)}{(z - 0.8149)(z - 0.7655)}$$

---

[1]B. W. Bequette, *Process Control: Modeling, Design, and Simulation*, Prentice Hall, 2003.

4.4 STABILITY DETERMINATION

The simplest method for determining the stability of a discrete-time system given its z-transfer function is by finding the system poles. This can be accomplished using a suitable numerical algorithm based on Newton's method.

4.4.1 MATLAB

The roots of a polynomial are obtained using one of the MATLAB commands,

>> **roots(den)**

>> **zpk(g)**

where **den** is a vector of denominator polynomial coefficients. The command **zpk** factorizes the numerator and denominator of the transfer function **g** and displays it. The poles of the transfer function can be obtained with the command **pole** and then sorted with the command **dsort** in order of decreasing magnitude.

Alternatively, one may use the command **ddamp**, which yields the pole locations (eigenvalues), the damping ratio, and the undamped natural frequency. For example, given a sampling period of 0.1 s and the denominator polynomial with coefficients

>> **den = [1.0, 0.2, 0.0, 0.4]**

the command is

>> **ddamp(den, 0.1)**

The command yields the output.

Eigenvalue	Magnitude	Equiv. Damping	Equiv. Freq. (rad/sec)
0.2306 + 0.7428I	0.7778	0.1941	12.9441
0.2306 − 0.7428I	0.7778	0.1941	12.9441
−0.6612	0.6612	0.1306	31.6871

The MATLAB command

>> **T = feedback(g, gf, ±1)**

calculates the closed-loop transfer function **T** using the forward transfer function **g** and the feedback transfer function **gf**. For negative feedback, the third argument is **−1** or is omitted. For unity feedback, we replace the argument **gf** with **1**. We can solve for the poles of the closed-loop transfer function as before using **zpk** or **ddamp**.

4.4.2 Routh-Hurwitz Criterion

The Routh-Hurwitz criterion determines conditions for left half plane (LHP) polynomial roots and cannot be directly used to investigate the stability of discrete-time systems. The *bilinear* transformation

$$z = \frac{1+w}{1-w} \Leftrightarrow w = \frac{z-1}{z+1} \tag{4.14}$$

transforms the inside of the unit circle to the LHP. This allows the use of the Routh-Hurwitz criterion for the investigation of discrete-time system stability. For the general z-polynomial,

$$F(z) = a_n z^n + a_{n-1} z^{n-1} + \ldots + a_0 \xrightarrow{z=\frac{1+w}{1-w}} a_n \left(\frac{1+w}{1-w}\right)^n + a_{n-1} \left(\frac{1+w}{1-w}\right)^{n-1} + \ldots + a_0 \tag{4.15}$$

The Routh-Hurwitz approach becomes progressively more difficult as the order of the z-polynomial increases. But for low-order polynomials, it easily gives stability conditions. For high-order polynomials, a symbolic manipulation package can be used to perform the necessary algebraic manipulations. The Routh-Hurwitz approach is demonstrated in the following example.

EXAMPLE 4.5

Find stability conditions for

1. The first-order polynomial $a_1 z + a_0$, $a_1 > 0$
2. The second-order polynomial $a_2 z^2 + a_1 z + a_0$, $a_2 > 0$

Solution

1. The stability of the first-order polynomial can be easily determined by solving for its root. Hence, the stability condition is

$$\left|\frac{a_0}{a_1}\right| < 1 \tag{4.16}$$

2. The roots of the second-order polynomial are in general given by

$$z_{1,2} = \frac{-a_1 \pm \sqrt{a_1^2 - 4a_0 a_2}}{2a_2} \tag{4.17}$$

Thus, it is not easy to determine the stability of the second-order polynomial by solving for its roots. For a monic polynomial (coefficient of z^2 is unity), the constant term is equal to the product of the poles. Hence, for pole magnitudes less than unity, we obtain the necessary stability condition

$$\left|\frac{a_0}{a_2}\right| < 1 \tag{4.18}$$

or equivalently

$$-a_0 < a_2 \quad \text{and} \quad a_0 < a_2$$

This condition is also sufficient in the case of complex conjugate poles where the two poles are of equal magnitude. The condition is only necessary for real poles because the product of a number greater than unity and a number less than unity can be less than unity. For example, for poles at 0.01 and 10, the product of the two poles has magnitude 0.1, which satisfies (4.18), but the system is clearly unstable.

Substituting the bilinear transformation in the second-order polynomial gives

$$a_2\left(\frac{1+w}{1-w}\right)^2 + a_1\left(\frac{1+w}{1-w}\right) + a_0$$

which reduces to

$$(a_2 - a_1 + a_0)w^2 + 2(a_2 - a_0)w + (a_2 + a_1 + a_0)$$

By the Routh-Hurwitz criterion, it can be shown that the poles of the second-order w-polynomial remain in the LHP if and only if its coefficients are all positive. Hence, the stability conditions are given by

$$a_2 - a_1 + a_0 > 0$$

$$a_2 - a_0 > 0 \qquad\qquad (4.19)$$

$$a_2 + a_1 + a_0 > 0$$

Adding the first and third conditions gives

$$a_2 + a_0 > 0 \Rightarrow -a_0 < a_2$$

This condition, obtained earlier in (4.18), is therefore satisfied if the three conditions of (4.19) are satisfied. The reader can verify through numerical examples that if real roots satisfying conditions (4.19) are substituted in (4.17), we obtain roots between −1 and +1.

Without loss of generality, the coefficient a_2 can be assumed to be unity, and the stable parameter range can be depicted in the a_0 versus a_1 parameter plane as shown in Figure 4.4.

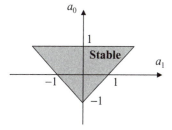

FIGURE 4.4

Stable parameter range for a second-order z-polynomial.

4.5 JURY TEST

It is possible to investigate the stability of z-domain polynomials directly using the **Jury test** for real coefficients or the **Schur-Cohn test** for complex coefficients. These tests involve determinant evaluations as in the Routh-Hurwitz test for s-domain polynomials but are more time consuming. The Jury test is given next.

Theorem 4.6: For the polynomial

$$F(z) = a_n z^n + a_{n-1} z^{n-1} + \ldots + a_1 z + a_0 = 0, \quad a_n > 0 \qquad (4.20)$$

the roots of the polynomial are inside the unit circle if and only if

$$
\begin{aligned}
&(1) \quad F(1) > 0 \\
&(2) \quad (-1)^n F(-1) > 0 \\
&(3) \quad |a_0| < a_n \\
&(4) \quad |b_0| > |b_{n-1}| \\
&(5) \quad |c_0| > |c_{n-2}| \\
\end{aligned}
\qquad (4.21)
$$

$$ \vdots $$

$$(n+1) \quad |r_0| > |r_2|$$

where the terms in the $n + 1$ conditions are calculated from Table 4.1. The entries of the table are calculated as follows

$$ b_k = \begin{vmatrix} a_0 & a_{n-k} \\ a_n & a_k \end{vmatrix}, \quad k = 0, 1, \ldots, n-1 $$

$$ c_k = \begin{vmatrix} b_0 & b_{n-k} \\ b_{n-1} & b_k \end{vmatrix}, \quad k = 0, 1, \ldots, n-2 $$

$$ (4.22) $$

$$ \vdots $$

$$ r_0 = \begin{vmatrix} s_0 & s_3 \\ s_3 & s_0 \end{vmatrix}, \quad r_1 = \begin{vmatrix} s_0 & s_2 \\ s_3 & s_1 \end{vmatrix}, \quad r_2 = \begin{vmatrix} s_0 & s_1 \\ s_3 & s_2 \end{vmatrix} \qquad ■ $$

Based on the Jury table and the Jury stability conditions, we make the following observations:

1. The first row of the Jury table is a listing of the coefficients of the polynomial $F(z)$ in order of increasing power of z.

2. The number of rows of the table $2 n - 3$ is always odd, and the coefficients of each even row are the same as the odd row directly above it with the order of the coefficients reversed.

Table 4.1 Jury Table

Row	z^0	z^1	z^2	...	z^{n-k}	...	z^{n-1}	z^n
1	a_0	a_1	a_2	...	a_{n-k}	...	a_{n-1}	a_n
2	a_n	a_{n-1}	a_{n-2}	...	a_k	...	a_1	a_0
3	b_0	b_1	b_2	...	b_{n-k}	...	b_{n-1}	
4	b_{n-1}	b_{n-2}	b_{n-3}	...	b_k	...	b_0	
5	c_0	c_1	c_2	...	...	c_{n-2}		
6	c_{n-2}	c_{n-3}	c_{n-4}	...	...	c_0		
.	.	.	.	...	...			
.	.	.	.	...	...			
.	.	.	.	...	...			
$2n-5$	s_0	s_1	s_2	s_3				
$2n-4$	s_3	s_2	s_1	s_0				
$2n-3$	r_0	r_1	r_2					

3. There are $n+1$ conditions in (4.21) that correspond to the $n+1$ coefficients of $F(z)$.

4. Conditions 3 through $2n-3$ of (4.21) are calculated using the coefficient of the first column of the Jury table together with the last coefficient of the preceding row. The middle coefficient of the last row is never used and need not be calculated.

5. Conditions 1 and 2 of (4.21) are calculated from $F(z)$ directly. If one of the first two conditions is violated, we conclude that $F(z)$ has roots on or outside the unit circle without the need to construct the Jury table or test the remaining conditions.

6. Condition 3 of (4.21), with $a_n = 1$, requires the constant term of the polynomial to be less than unity in magnitude. The constant term is simply the product of the roots and must be smaller than unity for all the roots to be inside the unit circle.

7. Conditions (4.21) reduce to conditions (4.18) and (4.19) for first and second-order systems respectively where the Jury table is simply one row.

8. For higher-order systems, applying the Jury test by hand is laborious, and it is preferable to test the stability of a polynomial $F(z)$ using a computer-aided design (CAD) package.

9. If the coefficients of the polynomial are functions of system parameters, the Jury test can be used to obtain the stable ranges of the system parameters.

EXAMPLE 4.6

Test the stability of the polynomial.

$$F(z) = z^5 + 2.6z^4 - 0.56z^3 - 2.05z^2 + 0.0775z + 0.35 = 0$$

We compute the entries of the Jury table using the coefficients of the polynomial (see Table 4.2).

The first two conditions require the evaluation of $F(z)$ at $z = \pm 1$.

1. $F(1) = 1 + 2.6 - 0.56 - 2.05 + 0.0775 + 0.35 = 1.4175 > 0$
2. $(-1)^5 F(-1) = (-1)(-1 + 2.6 + 0.56 - 2.05 - 0.0775 + 0.35) = -0.3825 < 0$

Conditions 3 through 6 can be checked quickly using the entries of the first column of the Jury table.

3. $|\,0.35\,| < 1$
4. $|-0.8775\,| > |\,0.8325\,|$
5. $|\,0.0770\,| < |\,0.5151\,|$
6. $|-0.2593\,| < |-0.3472\,|$

Conditions 2, 5, and 6 are violated, and the polynomial has roots on or outside the unit circle. In fact, the polynomial can be factored as

$$F(z) = (z - 0.7)(z - 0.5)(z + 0.5)(z + 0.8)(z + 2.5) = 0$$

and has a root at −2.5 outside the unit circle. Note that the number of conditions violated is not equal to the number of roots outside the unit circle and that condition 2 is sufficient to conclude the instability of $F(z)$.

Table 4.2 Jury Table for Example 4.6

Row	z^0	z^1	z^2	z^3	z^4	z^5
1	0.35	0.0775	−2.05	−0.56	2.6	1
2	1	2.6	−0.56	−2.05	0.0775	0.35
3	−0.8775	−2.5729	−0.1575	1.854	0.8325	
4	0.8325	1.854	−0.1575	−2.5729	−0.8775	
5	0.0770	0.7143	0.2693	0.5151		
6	0.5151	0.2693	0.7143	0.0770		
7	−0.2593	−0.0837	−0.3472			

EXAMPLE 4.7

Find the stable range of the gain K for the unity feedback digital cruise control system of Example 3.2 with the analog plant transfer function

$$G(s) = \frac{K}{s+3}$$

and with digital-to-analog converter (DAC) and analog-to-digital converter (ADC) if the sampling period is 0.02 s.

Solution

The transfer function for analog subsystem ADC and DAC is

$$G_{ZAS}(z) = (1 - z^{-1}) \mathcal{Z} \left\{ \mathcal{L}^{-1} \left[\frac{G(s)}{s} \right] \right\}$$

$$= (1 - z^{-1}) \mathcal{Z} \left\{ \mathcal{L}^{-1} \left[\frac{K}{s(s+3)} \right] \right\}$$

Using the partial fraction expansion

$$\frac{K}{s(s+3)} = \frac{K}{3} \left[\frac{1}{s} - \frac{1}{s+3} \right]$$

we obtain the transfer function

$$G_{ZAS}(z) = \frac{1.9412 \times 10^{-2}\, K}{z - 0.9418}$$

For unity feedback, the closed-loop characteristic equation is

$$1 + G_{ZAS}(z) = 0$$

which can be simplified to

$$z - 0.9418 + 1.9412 \times 10^{-2} \quad K = 0$$

The stability conditions are

$$0.9418 - 1.9412 \times 10^{-2} \quad K < 1$$

$$-0.9418 + 1.9412 \times 10^{-2} \quad K < 1$$

Thus, the stable range of K is

$$-3 < K < 100.03$$

EXAMPLE 4.8

Find the stable range of the gain K for the vehicle position control system (see Example 3.3) with the analog plant transfer function

$$G(s) = \frac{K}{s(s+10)}$$

and with DAC and ADC if the sampling period is 0.05 s.

Solution
The transfer function for analog subsystem, ADC, and DAC is

$$G_{ZAS}(z) = (1 - z^{-1})\mathcal{Z}\left\{\mathcal{L}^{-1}\left[\frac{G(s)}{s}\right]\right\}$$

$$= (1 - z^{-1})\mathcal{Z}\left\{\mathcal{L}^{-1}\left[\frac{K}{s^2(s+10)}\right]\right\}$$

Using the partial fraction expansion

$$\frac{K}{s^2(s+10)} = 0.1K\left[\frac{10}{s^2} - \frac{1}{s} - \frac{1}{s+10}\right]$$

we obtain the transfer function

$$G_{ZAS}(z) = \frac{1.0653 \times 10^{-2}\, K(z + 0.8467)}{(z-1)(z-0.6065)}$$

For unity feedback, the closed-loop characteristic equation is

$$1 + G_{ZAS}(z) = 0$$

which can be simplified to

$$(z-1)(z-0.6065) + 1.0653 \times 10^{-2}\, K(z + 0.8467)$$
$$= z^2 + (1.0653 \times 10^{-2}\, K - 1.6065)z + 0.6065 + 9.02 \times 10^{-3}\, K = 0$$

The stability conditions are

1. $F(1) = 1 + (1.0653 \times 10^{-2}\, K - 1.6065) + 0.6065 + 9.02 \times 10^{-3}\, K > 0 \Leftrightarrow K > 0$
2. $F(-1) = 1 - (1.0653 \times 10^{-2}\, K - 1.6065) + 0.6065 + 9.02 \times 10^{-3}\, K > 0 \Leftrightarrow$
   $K < 1967.582$
3. $|\,0.6065 + 0.0902\,K\,| < 1 \Leftrightarrow + (0.6065 + 0.0902\,K\,) < 1$
   $- (0.6065 + 0.0902\,K\,) < 1$
   $\Leftrightarrow -178.104 < K < 43.6199$

The three conditions yield the stable range

$$0 < K < 43.6199$$

4.6 NYQUIST CRITERION

The Nyquist criterion allows us to answer two questions:

1. Does the system have closed-loop poles outside the unit circle?
2. If the answer to the first question is yes, how many closed-loop poles are outside the unit circle?

We begin by considering the closed-loop characteristic polynomial

$$p_{cl}(z) = 1 + C(z)G(z) = 1 + L(z) = 0 \tag{4.23}$$

where $L(z)$ denotes the loop gain. We rewrite the characteristic polynomial in terms of the numerator of the loop gain N_L and its denominator D_L in the form

$$p_{cl}(z) = 1 + \frac{N_L(z)}{D_L(z)} = \frac{N_L(z) + D_L(z)}{D_L(z)} \tag{4.24}$$

We observe that the zeros of the rational function are the closed-loop poles, whereas its poles are the open-loop poles of the system. We assume that we are given the number of open-loop poles outside the unit circle, and we denote this number by P. The number of closed-loop poles outside the unit circle is denoted by Z and is unknown.

To determine Z, we use some simple results from complex analysis. We first give the following definition.

Definition 4.4: Contour. A contour is a closed directed simple (does not cross itself) curve. ∎

An example of a contour is shown in Figure 4.5. In the figure, shaded text denotes a vector. Recall that in the complex plane the vector connecting any point a to a point z is the vector $(z - a)$. We can calculate the net angle change for the term $(z - a)$ as the point z traverses the contour in the shown (clockwise) direction by determining the net number of rotations of the corresponding vector. From Figure 4.5, we observe that the net rotation is one full turn or 360° for the point a_1, which is inside the contour. The net rotation is zero for the point a_2, which is outside the contour. If the point in question corresponds to a zero, then the rotation gives a numerator angle; if it is a pole, we have a denominator angle. The net angle change for a rational function is the angle of the numerator minus

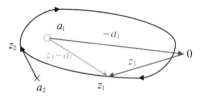

FIGURE 4.5

Closed contours (shaded letters denote vectors).

the angle of the denominator. So for Figure 4.5, we have one clockwise rotation because of a_1 and no rotation as a result of a_2 for a net angle change of one clockwise rotation. Angles are typically measured in the counterclockwise direction. We therefore count clockwise rotations as negative.

The preceding discussion shows how to determine the number of zeros of a rational function in a specific region; given the number of poles, we perform the following steps:

1. Select a closed contour surrounding the region.
2. Compute the net angle change for the function as we traverse the contour once.
3. The net angle change or number of rotations N is equal to the number of zeros inside the contour Z minus the angle of poles inside the contour P. Calculate the number of zeros as

$$Z = N + P \qquad (4.25)$$

To use this to determine closed-loop stability, we need to select a contour that encircles the outside of the unit circle. The contour is shown in Figure 4.6. The smaller circle is the unit circle, whereas the large circle is selected with a large radius so as to enclose all poles and zeros of the functions of interest.

The value of the loop gain on the unit circle is $L(e^{j\omega T})$, which is the frequency response of the discrete-time system for angles ωT in the interval $[-\pi, \pi]$. The values obtained for negative frequencies $L(e^{-j\omega T})$ are simply the complex conjugate of the values $L(e^{j\omega T})$ and need not be separately calculated. Because the order of the numerator is equal to that of the denominator or less, points on the large circle map to zero or to a single constant value. Because the value ε is infinitesimal, the values on the straight-line portions close to the real axis cancel.

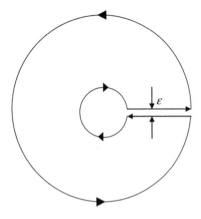

FIGURE 4.6

Contour for stability determination.

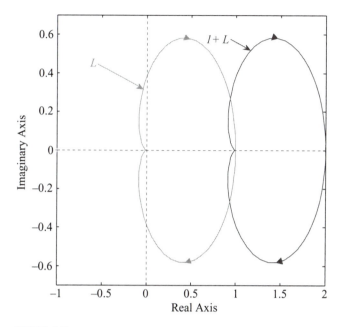

FIGURE 4.7

Nyquist plots of L and $1 + L$.

We can simplify the test by plotting $L(e^{j\omega T})$ as we traverse the contour and then counting its encirclements of the point $-1 + j\, 0$. As Figure 4.7 shows, this is equivalent to plotting $p_{cl}(z)$ and counting encirclements of the origin.

If the system has open-loop poles on the unit circle, the contour passes through poles and the test fails. To avoid this, we modify the contour to avoid these open-loop poles. The most common case is a pole at unity for which the modified contour is shown in Figure 4.8. The contour includes an additional circular arc of infinitesimal radius. Because the portions on the positive real axis cancel, the contour effectively reduces to the one shown in Figure 4.9. For m poles at unity, the loop gain is given by

$$L(z) = \frac{N_L(s)}{(z-1)^m D(s)} \tag{4.26}$$

where N_L and D have no unity roots. The value of the transfer function on the circular arc is approximately given by

$$L(z)]_{z \to 1 + \varepsilon e^{j\theta}} = \frac{K}{\varepsilon e^{j\theta m}}, \quad \theta \in \left[-\frac{\pi}{2}, \frac{\pi}{2} \right] \tag{4.27}$$

where K is equal to $N_L(1)/D(1)$ and $(z - 1) = \varepsilon e^{j\theta}$, with ε the radius of the small circular arc. Therefore, the small circle maps to a large circle and traversing the

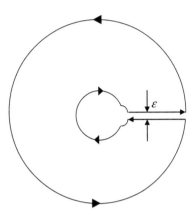

FIGURE 4.8

Modified contour for stability determination.

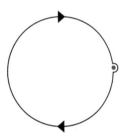

FIGURE 4.9

Simplification of the modified contour for stability determination with (1, 0) shown in gray.

small circle once causes a net denominator angle change of $-m\pi$ radians (clockwise) (i.e., m half circles). The net angle change for the quotient on traversing the small circular arc is thus $m\pi$ radians (counterclockwise). We conclude that for a type m system the Nyquist contour will include m large clockwise semicircles.

We now summarize the results obtained in the following theorem.

Theorem 4.7: Nyquist Criterion. Let the number of counterclockwise encirclements of the point (−1, 0) for a loop gain $L(z)$ when traversing the stability contour be N (i.e., −N for clockwise encirclements), where $L(z)$ has P open-loop poles inside the contour. Then the system has Z closed-loop poles outside the unit circle with Z given by

$$Z = (-N) + P \qquad (4.28)$$

∎

Corollary: An open-loop stable system is stable if and only if it does not encircle the point (−1, 0) (i.e., if $N = 0$).

Although counting encirclements appears complicated, it is actually quite simple using the following recipe:

1. Starting at a distant point, move toward the point (−1, 0).
2. Count all lines of the stability contour crossed. Count each line with an arrow pointing from your left to your right as negative and every line with an arrow pointing from your right to your left as positive.
3. The net number of lines counted is equal to the number of encirclements of the point (−1, 0).

The recipe is demonstrated in the following examples.

EXAMPLE 4.9

Consider a linearized model of a furnace:

$$G(s) = \frac{T_i(s)}{U(s)} = \frac{K g_{rw} g_{iw}}{s^2 + (2g_{iw} + g_{rw})s + g_{rw} g_{iw}}$$

During the heating phase, we have the model[2]

$$G(s) = \frac{1}{s^2 + 3s + 1}$$

Determine the closed-loop stability of the system with digital control and a sampling period of 0.01.

Solution

In Example 3.5, we determined the transfer function of the system to be

$$G_{ZAS}(z) = K \left[\frac{(p_1 e^{-p_2 T} - p_2 e^{-p_1 T} + p_2 - p_1)z + \left(\dfrac{p_1 e^{-p_1} - p_2 e^{-p_2} + p_2 e^{-(p_1 + p_2)} - }{p_1 e^{-(p_1 + p_2)}} \right)}{p_1 p_2 (p_2 - p_1)(z - e^{-p_1 T})(z - e^{-p_2 T})} \right]$$

Substituting the values of the parameters in the general expression gives the z-transfer function

$$G_{ZAS}(z) = 10^{-5} \frac{4.95z + 4.901}{z^2 - 1.97z + 0.9704}$$

The Nyquist plot of the system is shown in Figure 4.10. The plot shows that the Nyquist plot does not encircle the point (−1, 0). Because the open-loop transfer function has no

---

[2]T. Hagglund and A. Tengall, An automatic tuning procedure for unsymmetrical processes, *Proceedings of the European Control Conference*, Rome, 1995.

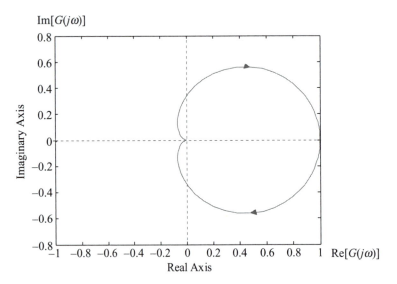

FIGURE 4.10

Nyquist plot of the furnace transfer function.

RHP poles, the system is closed-loop stable. Note that there is a free path to the point $(-1, 0)$ without crossing any lines of the plot because there are no encirclements.

4.6.1 Phase Margin and Gain Margin

In practice, the stability of a mathematical model is not sufficient to guarantee acceptable system performance or even to guarantee the stability of the physical system that the model represents. This is because of the approximate nature of mathematical models and our imperfect knowledge of the system parameters. We therefore need to determine how far the system is from instability. This degree of stability is known as **relative stability**. To keep our discussion simple, we restrict it to open-loop stable systems where zero encirclements guarantee stability. For open-loop stable systems that are nominally closed-loop stable, the distance from instability can be measured by the distance between the set of points of the Nyquist plot and the point $(-1, 0)$.

Typically, the distance between a set of points and the single point $(-1, 0)$ is defined as the minimum distance over the set of points. However, it is more convenient to define relative stability in terms of two distances: a magnitude distance and an angular distance. The two distances are given in the following definitions.

Definition 4.5: Gain Margin. The gain margin is the gain perturbation that makes the system marginally stable. ∎

Definition 4.6: Phase Margin. The phase margin is the negative phase perturbation that makes the system marginally stable.   ∎

The two stability margins become clearer by examining the block diagram of Figure 4.11. If the system has a multiplicative gain perturbation $\Delta G(z) = \Delta K$, then the gain margin is the magnitude of ΔK that makes the system on the verge of instability. If the system has a multiplicative gain perturbation $\Delta G(z) = e^{-j\Delta\theta}$, then the gain margin is the lag angle $\Delta\theta$ that makes the system on the verge of instability. Clearly, the perturbations corresponding to the gain margin and phase margin are limiting values, and satisfactory behavior would require smaller model perturbations.

The Nyquist plot of Figure 4.12 shows the gain margin and phase margin for a given polar plot (the positive frequency portion of the Nyquist plot). Recall that each point on the plot represents a complex number, which is represented by a

FIGURE 4.11

Model perturbation $\Delta G(s)$.

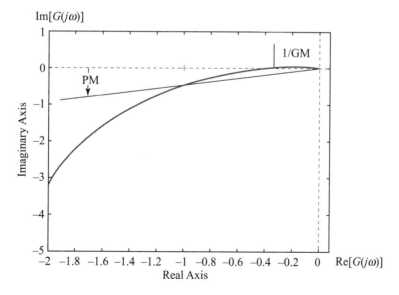

FIGURE 4.12

Nyquist plot with phase margin and gain margin.

vector from the origin. Scaling the plot with a gain ΔK results in scaled vectors without rotation. Thus, the vector on the negative real axis is the one that reaches the point $(-1, 0)$ if appropriately scaled, and the magnitude of that vector is the reciprocal of the gain margin. On the other hand, multiplication by $e^{-j\Delta\theta}$ rotates the plot clockwise without changing the magnitudes of the vectors, and it is the vector of magnitude unity that can reach the point $(-1, 0)$ if rotated by the phase margin.

For an unstable system, a counterclockwise rotation or a reduction in gain is needed to make the system on the verge of instability. The system will have a negative phase margin and a gain margin less than unity, which is also negative if it is expressed in decibels—that is, in units of $20 \log\{|G(j\omega)|\}$. The polar plot of a system with negative gain margin and phase margin is shown in Figure 4.13.

The gain margin can be obtained analytically by equating the imaginary part of the frequency response to zero and solving for the real part. The phase margin can be obtained by equating the magnitude of the frequency response to unity and solving for angle, and then adding 180°. However, because only approximate values are needed in practice, it is easier to use MATLAB to obtain both margins. In some cases, the intercept with the real axis can be obtained as the value where $z = -1$ provided that the system has no pole at -1 (i.e., the frequency response has no discontinuity at the folding frequency $\omega_s/2$).

It is sometimes convenient to plot the frequency response using the Bode plot, but the availability of frequency response plotting commands in MATLAB reduces the necessity for such plots. The MATLAB commands for obtaining frequency

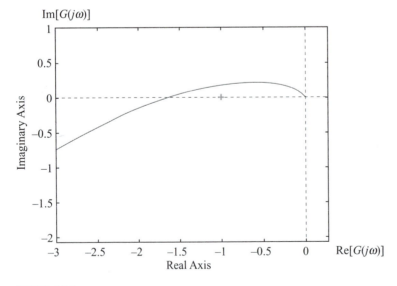

FIGURE 4.13

Nyquist plot with negative gain margin (dBs) and phase margin.

response plots (which work for both continuous-time and discrete-time systems) are

>> **nyquist(gd)**   **% Nyquist plot**

>> **bode(gd)**   **% Bode plot**

It is also possible to find the gain and phase margins with the command

>> **[gm,pm] = margin(gd)**

An alternative form of the command is

>> **margin(gd)**

The latter form shows the gain margin and phase margin on the Bode plot of the system. We can also obtain the phase margin and gain margin using the Nyquist plot by clicking on the plot and selecting

Characteristics

All stability margins

The concepts of the gain margin and phase margin and their evaluation using MATLAB are illustrated by the following example.

EXAMPLE 4.10

Determine the closed-loop stability of the digital control system for the furnace model of Example 3.4 with a discrete-time first-order actuator of the form

$$G_a(z) = \frac{0.9516}{z - 0.9048}$$

and a sampling period of 0.01. If an amplifier of gain $K = 5$ is added to the actuator, how does the value of the gain affect closed-loop stability?

Solution

We use MATLAB to obtain the z-transfer function of the plant and actuator:

$$G_{ZAS}(z) = 10^{-5} \frac{4.711z + 4.644}{z^3 - 2.875z^2 + 2.753z - 0.8781}$$

The Nyquist plot for the system, Figure 4.14, is obtained with no additional gain and then for a gain $K = 5$. We also show the plot in the vicinity of the point $(-1, 0)$ in Figure 4.15 from which we see that the system with $K = 5$ encircles the point twice clockwise.

We count the encirclements by starting away from the point $(-1, 0)$ and counting the lines crossed as we approach it. We cross the gray curve twice and at each crossing the arrow indicates that the line is moving from our right to our left (i.e., two clockwise encirclements). The system is unstable and the number of closed-loop poles outside the unit circle is given by

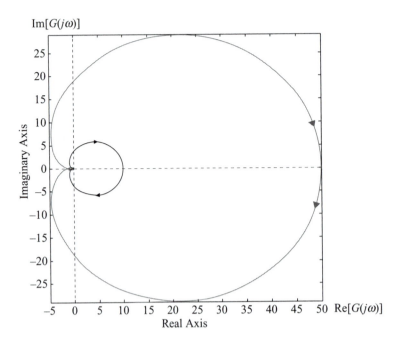

FIGURE 4.14

Nyquist plot for the furnace and actuator ($K = 1$, black, $K = 5$, gray).

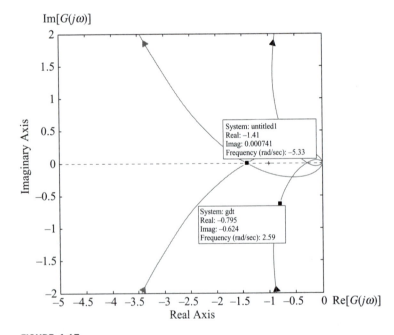

FIGURE 4.15

Nyquist plot for the furnace and actuator in the vicinity of the point $(-1, 0)$ ($K = 1$, black, $K = 5$, gray).

$$Z = (-N) + P$$
$$= 2 + 0$$

For the original gain of unity, the intercept with the real axis is at a magnitude of approximately 0.28 and can be increased by a factor of about 3.5 before the system becomes unstable.

At a magnitude of unity, the phase is about 38 degrees less negative than the instability value of $-180°$. We therefore have a gain margin of about 3.5 and a phase margin of about 38 degrees. Using MATLAB, we find approximately the same values for the margins

>> **[gm,pm] = margin(gtd)**

gm = 3.4817

pm = 37.5426

Thus, an additional gain of over 3 or an additional phase lag of over 37° can be tolerated without causing instability. However, such perturbations may cause a significant deterioration in the time response of the system. Perturbations in gain and phase may actually occur upon implementing the control, and the margins are needed for successful implementation. In fact, the phase margin of the system is rather low, and a controller may be needed to improve the response of the system.

To obtain the Bode plot showing the phase margin and gain margin of Figure 4.16, we use the following command.

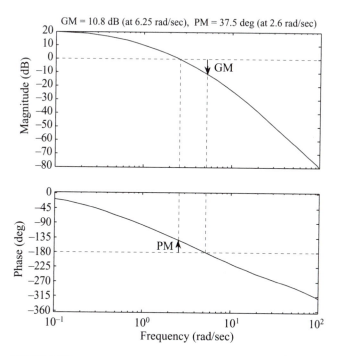

FIGURE 4.16

Phase margin and gain margin for the oven control system shown on the Bode plot.

$$>> \mathbf{margin\,(gtd)}$$

1. The phase margin for unity gain shown on the plot is as obtained with the first form of the command **margin**, but the gain margin is in dBs. The values are nevertheless identical as verified with the MATLAB command

$$>> \mathbf{20^*log10\,(gm)}$$

$$\mathbf{ans = 10.8359}$$

2. The gain margin and phase margin can also be obtained using the Nyquist command as shown in Figure 4.17.

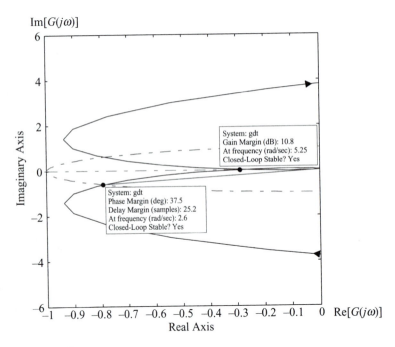

FIGURE 4.17

Phase margin and gain margin for the oven control system shown on the Nyquist plot.

EXAMPLE 4.11

Determine the closed-loop stability of the digital control system for the position control system with analog transfer function

$$G(s) = \frac{10}{s(s+1)}$$

and with a sampling period of 0.01. If the system is stable, determine the gain margin and the phase margin.

Solution

We first obtain the transfer function for the analog plant with ADC and DAC. The transfer function is given by

$$G_{ZAS}(z) = 4.983 \times 10^{-4} \frac{z + 0.9967}{(z-1)(z-0.99)}$$

Note that the transfer function has a pole at unity because the analog transfer function has a pole at the origin or is type I. Although such systems require the use of the modified Nyquist contour, this has no significant impacts on the steps required for stability testing using the Nyquist criterion. The Nyquist plot obtained using the MATLAB command **nyquist** is shown in Figure 4.18.

The plot does not include the large semicircle corresponding to the small semicircle on the modified contour of Figure 4.9. However, this does not prevent us from investigating stability. It is obvious that the contour does not encircle the point (−1, 0) because the point is to the left of the observer moving along the polar plot (lower half). In addition, we can reach the (−1, 0) point without crossing any of the lines of the Nyquist plot. The system is stable because the number of closed-loop poles outside the unit circle is given by

$$Z = (-N) + P$$
$$= 0 + 0 = 0$$

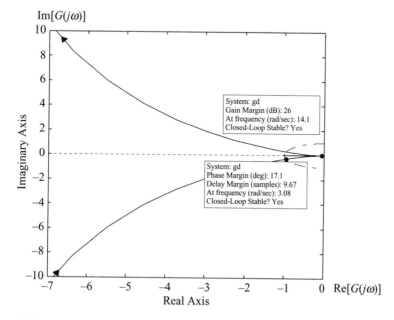

FIGURE 4.18

Nyquist plot for the position control system of Example 4.11.

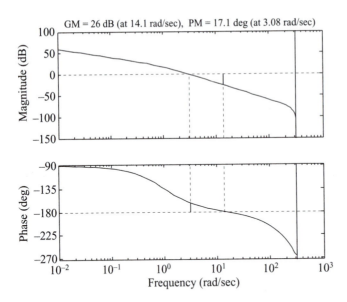

FIGURE 4.19

Bode diagram with phase margin and gain margin for the position control system of Example 4.11.

The gain margin is 17.1°, and the phase margin is 26 dB. The gain margin and phase margin can also be obtained using the margin command as shown in Figure 4.19.

RESOURCES

Franklin, G. F., J. D. Powell, and M. L. Workman, *Digital Control of Dynamic Systems,* Addison-Wesley, 1990.

Gupta, S. C., and L. Hasdorff, *Fundamentals of Automatic Control,* Wiley, 1970.

Jury, E. I., *Theory and Applications of the z-Transform Method,* Krieger, 1973.

Kuo, B. C., *Digital Control Systems,* Saunders, 1992.

Ogata, K., *Digital Control Engineering,* Prentice Hall, 1987.

Oppenheim, A. V., A. S. Willsky, and I. T. Young, *Signals and Systems,* Prentice Hall, 1983.

Ragazzini, J. R., and G. F. Franklin, *Sampled-Data Control Systems,* McGraw-Hill, 1958.

PROBLEMS

4.1 Determine the asymptotic stability and the BIBO stability of the following systems:

(a) $y(k+2)+0.8y(k+1)+0.07y(k) = 2u(k+1)+0.2u(k)$ $k = 0, 1, 2, \ldots$

(b) $y(k+2)-0.8y(k+1)+0.07y(k) = 2u(k+1)+0.2u(k)$ $k = 0, 1, 2, \ldots$

(c) $y(k+2)+0.1y(k+1)+0.9y(k) = 3.0u(k)$ $k = 0, 1, 2, \ldots$

4.2 Biochemical reactors are used in different processes such as waste treatment and alcohol fermentation. By considering the dilution rate as the manipulated variable and the biomass concentration as the measured output, the biochemical reactor can be modeled by the following transfer function in the vicinity of an unstable steady-state operating point[3]:

$$G(s) = \frac{5.8644}{-5.888s + 1}$$

Determine $G_{ZAS}(z)$ with a sampling rate $T = 0.1$, and then consider the feedback controller

$$C(z) = -\frac{z - 1.017}{z - 1}$$

Verify that the resulting feedback system is not internally stable.

4.3 Use the Routh-Hurwitz criterion to investigate the stability of the following systems:

(a) $G(z) = \dfrac{5(z - 2)}{(z - 0.1)(z - 0.8)}$

(b) $G(z) = \dfrac{10(z + 0.1)}{(z - 0.7)(z - 0.9)}$

4.4 Repeat Problem 4.3 using the Jury criterion.

4.5 Obtain the impulse response for the systems shown in Problem 4.3, and verify the results obtained using the Routh-Hurwitz criterion. Also determine the exponential rate of decay for each impulse response sequence.

4.6 Use the Routh-Hurwitz criterion to find the stable range of K for the closed-loop unity feedback systems with loop gain

(a) $G(z) = \dfrac{K(z - 1)}{(z - 0.1)(z - 0.8)}$

(b) $G(z) = \dfrac{K(z + 0.1)}{(z - 0.7)(z - 0.9)}$

4.7 Repeat Problem 4.6 using the Jury criterion.

4.8 Use the Jury criterion to determine the stability of the following polynomials:
(a) $z^5 + 0.2z^4 + z^2 + 0.3z - 0.1 = 0$
(b) $z^5 - 0.25z^4 + 0.1z^3 + 0.4z^2 + 0.3z - 0.1 = 0$

4.9 Determine the stable range of the parameter a for the closed-loop unity feedback systems with loop gain

(a) $G(z) = \dfrac{1.1(z - 1)}{(z - a)(z - 0.8)}$

(b) $G(z) = \dfrac{1.2(z + 0.1)}{(z - a)(z - 0.9)}$

---

[3]B. W. Bequette, *Process Control: Modeling, Design, and Simulation*, Prentice Hall, 2003.

4.10 For a gain of 0.5, derive the gain margin and phase margin of the systems shown in Problem 4.5 analytically. Let $T = 1$ with no loss of generality because the value of ωT in radians is all that is needed for the solution. Explain why the phase margin is not defined for the system shown in Problem 4.6(a). *Hint:* The gain margin is obtained by finding the point where the imaginary part of the frequency response is zero. The phase margin is obtained by finding the point where the magnitude of the frequency response is unity.

COMPUTER EXERCISES

4.11 Write a computer program to perform the Routh-Hurwitz test using a suitable CAD tool.

4.12 Write a computer program to perform the Jury test using a suitable CAD tool.

4.13 Write a computer program that uses the Jury test program in Exercise 4.12 to determine the stability of a system with an uncertain gain K in a given range $[K_{min}, K_{max}]$. Verify the answers obtained for Problem 4.6 using your program.

4.14 Show how the program written for Exercise 4.13 can be used to test the stability of a system with uncertain zero location. Use the program to test the effect of a ±20% variation in the location of the zero for the systems shown in Problem 4.6, with a fixed gain equal to half the critical value.

4.15 Show how the program written for Exercise 4.13 can be used to test the stability of a system with uncertain pole location. Use the program to test the effect of a ±20% variation in the location of the first pole for the systems shown in Problem 4.6, with a fixed gain equal to half the critical value.

4.16 Simulate the closed-loop systems shown in Problem 4.6 with a unit step input and (a) gain K equal to half the critical gain and (b) gain K equal to the critical gain. Discuss their stability using your simulation results.

4.17 For unity gain, obtain the Nyquist plots of the systems shown in Problem 4.6 using MATLAB and determine the following:
   (a) The intersection with the real axis using the Nyquist plot and then using the Bode plot
   (b) The stable range of positive gains K for the closed-loop unity feedback systems
   (c) The gain margin and phase margin for a gain $K = 0.5$

4.18 For twice the nominal gain, use MATLAB to obtain the Nyquist and Bode plots of the systems of the oven control system of Example 4.10 with a sampling period of 0.01 and determine the following:
   (a) The intersection with the real axis using the Nyquist plot and then using the Bode plot

(b) The stable range of additional positive gains K for the closed-loop unity feedback systems

(c) The gain margin and phase margin for twice the nominal gain

4.19 In many applications, there is a need for accurate position control at the nanometer scale. This is known as **nanopositioning** and is now feasible because of advances in nanotechnology. The following transfer function represents a single-axis nanopositioning system[4]:

$$G(s) = \frac{4.29 \times 10^{10}\left(s^2 + 631.2s + 9.4 \times 10^6\right)}{\left(s^2 + 178.2s + 6 \times 10^6\right)\left(s^2 + 412.3s + 16 \times 10^6\right)}$$
$$\frac{\left(s^2 + 638.8s + 45 \times 10^6\right)}{\left(s^2 + 209.7s + 56 \times 10^6\right)\left(s + 5818\right)}$$

(a) Obtain the DAC-analog system-ADC transfer function for a sampling period of 100 ms, and determine its stability using the Nyquist criterion.

(b) Obtain the DAC-analog system-ADC transfer function for a sampling period of 1 ms, and determine its stability using the Nyquist criterion.

(c) Plot the closed-loop step response of the system of (b), and explain the stability results of (a) and (b) based on your plot.

---

[4]A. Sebastian and S. M. Salapaka, Design methodologies of robust nano-positioning, *IEEE Trans. Control Systems Tech.*, *13*(6), 2005.

Analog Control System Design

Analog controllers can be implemented using analog components or approximated with digital controllers using standard analog-to-digital transformations. In addition, direct digital control system design in the z-domain is very similar to the s-domain design of analog systems. Thus, a review of classical control design is the first step toward understanding the design of digital control systems. This chapter reviews the design of analog controllers in the s-domain and prepares the reader for the digital controller design methods presented in Chapter 6. The reader is assumed to have had some exposure to the s-domain and its use in control system design.

Objectives

After completing this chapter, the reader will be able to do the following:

1. Obtain root locus plots for analog systems.
2. Characterize a system's step response based on its root locus plot.
3. Design proportional (P), proportional-derivative (PD), proportional-integral (PI), and proportional-integral-derivative (PID) controllers in the s-domain.
4. Tune PID controllers using the Ziegler-Nichols approach.

5.1 ROOT LOCUS

The **root locus** method provides a quick means of predicting the closed-loop behavior of a system based on its open-loop **poles** and **zeros**. The method is based on the properties of the closed-loop **characteristic equation**

$$1 + KL(s) = 0 \qquad (5.1)$$

where the gain K is a design parameter and $L(s)$ is the loop gain of the system. We assume a loop gain of the form

$$L(s) = \frac{\prod_{i=1}^{n_z}(s - z_i)}{\prod_{j=1}^{n_p}(s - p_j)} \tag{5.2}$$

where z_i, $i = 1, 2, \ldots, n_z$, are the open-loop system zeros and p_j, $j = 1, 2, \ldots, n_p$ are the open-loop system poles. It is required to determine the loci of the closed-loop poles of the system (root loci) as K varies between zero and infinity.[1] Because of the relationship between pole locations and the time response, this gives a preview of the closed-loop system behavior for different K.

The complex equality (5.1) is equivalent to the two real equalities

- *Magnitude condition $K|L(s)| = 1$*
- *Angle condition $\angle L(s) = \pm(2m + 1)180°$, $m = 0,1,2,\ldots$*

Using (5.1) or the preceding conditions, the following rules for sketching root loci can be derived:

1. The number of root locus branches is equal to the number of open-loop poles of $L(s)$.
2. The root locus branches start at the open-loop poles and end at the open-loop zeros or at infinity.
3. The real axis root loci have an odd number of poles plus zeros to their right.
4. The branches going to infinity asymptotically approach the straight lines defined by the angle

$$\theta_a = \pm\frac{(2m+1)180°}{n_p - n_z}, \quad m = 0, 1, 2, \ldots \tag{5.3}$$

and the intercept

$$\sigma_a = \frac{\sum_{i=1}^{n_p} p_i - \sum_{j=1}^{n_z} z_j}{n_p - n_z} \tag{5.4}$$

5. **Breakaway points** (points of departure from the real axis) correspond to local maxima of K, whereas **break-in points** (points of arrival at the real axis) correspond to local minima of K.
6. The **angle of departure** from a complex pole p_n is given by

$$180° - \sum_{i=1}^{n_p-1} \angle(p_n - p_i) + \sum_{j=1}^{n_z} \angle(p_n - z_j) \tag{5.5}$$

The **angle of arrival** at a complex zero is similarly defined.

---

[1] In rare cases, negative gain values are allowed, and the corresponding root loci are obtained. Negative root loci are not addressed in this text.

EXAMPLE 5.1

Sketch the root locus plots for the loop gains

1. $L(s) = \dfrac{1}{(s+1)(s+3)}$

2. $L(s) = \dfrac{1}{(s+1)(s+3)(s+5)}$

3. $L(s) = \dfrac{s+5}{(s+1)(s+3)}$

Comment on the effect of adding a pole or a zero to the loop gain.

Solution

The root loci for the three loop gains as obtained using MATLAB are shown in Figure 5.1. We now discuss how these plots can be sketched using root locus sketching rules.

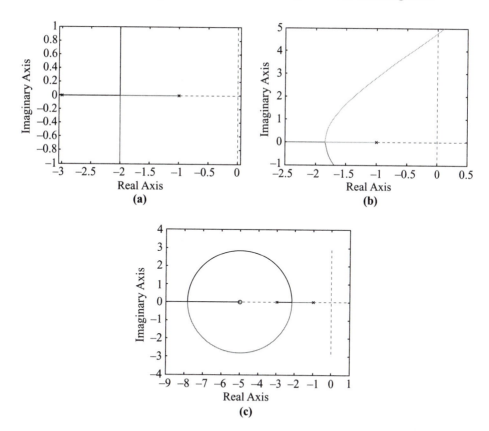

FIGURE 5.1

Root loci of second- and third-order systems. (a) Root locus of a second-order system. (b) Root locus of a third-order system. (c) Root locus of a second-order system with zero.

1. Using rule 1, the function has two root locus branches. By rule 2, the branches start at −1 and −3 and go to infinity. By rule 3, the real axis locus is between (−1) and (−3). Rule 4 gives the asymptote angles

$$\theta_a = \pm\frac{(2m+1)180°}{2}, \quad m = 0, 1, 2, \ldots$$

$$= \pm 90°, \pm 270°, \ldots$$

and the intercept

$$\sigma_a = \frac{-1-3}{2} = -2$$

To find the breakaway point using Rule 5, we express real K using the characteristic equation as

$$K = -(\sigma+1)(\sigma+3) = -(\sigma^2 + 4\sigma + 3)$$

We then differentiate with respect to σ and equate to zero for a maximum to obtain

$$-\frac{dK}{d\sigma} = 2\sigma + 4 = 0$$

Hence, the breakaway point is at $\sigma_b = -2$. This corresponds to a maximum of K because the second derivative is equal to −2 (negative). It can be easily shown that for any system with only two real axis poles, the breakaway point is midway between the two poles.

2. The root locus has three branches, with each branch starting at one of the open-loop poles (−1, −3, −5). The real axis loci are between −1 and −3 and to the left of −5. The branches all go to infinity, with one branch remaining on the negative real axis and the other two breaking away. The breakaway point is given by the maximum of the real gain K

$$K = -(\sigma+1)(\sigma+3)(\sigma+5)$$

Differentiating gives

$$-\frac{dK}{d\sigma} = (\sigma+1)(\sigma+3) + (\sigma+3)(\sigma+5) + (\sigma+1)(\sigma+5)$$

$$= 3\sigma^2 + 18\sigma + 23$$

$$= 0$$

which yields $\sigma_b = -1.845$ or -4.155. The first value is the actual breakaway point because it lies on the real axis locus between the poles and −1 and −3. The second value corresponds to a negative gain value and is therefore inadmissible. The gain at the breakaway point can be evaluated from the magnitude condition and is given by

$$K = -(-1.845+1)(-1.845+3)(-1.845+5) = 3.079$$

The asymptotes are defined by the angles

$$\theta_a = \pm\frac{(2m+1)180°}{3}, \quad m = 0, 1, 2, \ldots$$

$$= \pm 60°, \pm 180°, \ldots$$

and the intercept by

$$\sigma_a = \frac{-1-3-5}{3} = -3$$

The closed-loop characteristic equation corresponds to the **Routh table**

$$
\begin{array}{c|cc}
s^3 & 1 & 23 \\
s^2 & 9 & 15+K \\
s^1 & \dfrac{192-K}{9} & \\
s^0 & 15+K &
\end{array}
$$

Thus, at $K = 192$, a zero row results. This value defines the auxiliary equation

$$9s^2 + 207 = 0$$

Thus, the intersection with the $j\omega$-axis is $\pm j4.796$ rad/s.

3. The root locus has two branches as in (1), but now one of the branches ends at the zero. From the characteristic equation, the gain is given by

$$K = -\frac{(\sigma+1)(\sigma+3)}{\sigma+5}$$

Differentiating gives

$$
\begin{aligned}
\frac{dK}{d\sigma} &= -\frac{(\sigma+1+\sigma+3)(\sigma+5)-(\sigma+1)(\sigma+3)}{(\sigma+5)^2} \\
&= -\frac{\sigma^2 +10\sigma+17}{(\sigma+5)^2} \\
&= 0
\end{aligned}
$$

which yields $\sigma_b = -2.172$ or -7.828. The first value is the breakaway point because it lies between the poles, whereas the second value is to the left of the zero and corresponds to the break-in point. The second derivative

$$
\begin{aligned}
\frac{d^2K}{d\sigma^2} &= -\frac{(2\sigma+10)(\sigma+5)-2(\sigma^2+10\sigma+17)}{(\sigma+5)^3} \\
&= -16/(\sigma+5)^3
\end{aligned}
$$

is negative for the first value and positive for the second value. Hence, K has a maximum at the first value and a minimum at the second. It can be shown that the root locus is a circle centered at the zero with radius given by the geometric mean of the distances between the zero and the two real poles.

Clearly, adding a pole pushes the root locus branches toward the RHP, whereas adding a zero pulls them back into the left half plane. Thus, adding a zero allows the use of higher gain values without destabilizing the system. In practice, the allowable increase in gain is limited by the cost of the associated increase in control effort and by the possibility of driving the system outside the linear range of operation.

5.2 ROOT LOCUS USING MATLAB

While the above rules together with (5.1) allow the sketching of root loci for any loop gain of the form (5.2), it is often sufficient to use a subset of these rules to obtain the root loci. For higher-order or more complex situations, it is easier to use a CAD tool like MATLAB. These packages do not actually use root locus sketching rules. Instead they numerically solve for the roots of the characteristic equation as K is varied in a given range and then display the root loci.

The MATLAB command to obtain root locus plots is "rlocus". To obtain the root locus of the system

$$G(s) = \frac{s+5}{s^2 + 2s + 10}$$

using MATLAB enter

>> g = tf([1, 5], [1, 2, 10]);

>> rlocus(g);

To obtain specific points on the root locus and the corresponding data, we simply click or the root locus. Dragging the mouse allows us to change the referenced point to obtain more data.

5.3 DESIGN SPECIFICATIONS AND THE EFFECT OF GAIN VARIATION

The objective of control system design is to construct a system that has a desirable response to standard inputs. A desirable transient response is one that is sufficiently fast without excessive oscillations. A desirable steady-state response is one that follows the desired output with sufficient accuracy. In terms of the response to a unit step input, the transient response is characterized by the following criteria:

1. *Time constant τ.* Time required to reach about 63% of the final value.
2. *Rise time T_r.* Time to go from 10% to 90% of the final value.
3. *Percentage overshoot (PO).*

$$PO = \frac{Peak\ value - Final\ value}{Final\ value} \times 100\%$$

4. *Peak time T_p.* Time to first peak of an oscillatory response.
5. *Settling time T_s.* Time after which the oscillatory response remains within a specified percentage (usually 2 percent) of the final value.

Clearly, the percentage overshoot and the peak time are intended for use with an oscillatory response (i.e., for a system with at least one pair of complex conjugate poles). For a single complex conjugate pair, these criteria can be expressed in terms of the pole locations.

Consider the second-order system

$$L(s) = \frac{\omega_n^2}{s^2 + 2\zeta\omega_n s + \omega_n^2} \tag{5.6}$$

where ζ is the damping ratio and ω_n is the undamped natural frequency. Then criteria 3 through 5 are given by

$$PO = e^{-\frac{\pi\zeta}{\sqrt{1-\zeta^2}}} \times 100\% \tag{5.7}$$

$$T_p = \frac{\pi}{\omega_d} = \frac{\pi}{\omega_n\sqrt{1-\zeta^2}} \tag{5.8}$$

$$T_s = \frac{4}{\zeta\omega_n} \tag{5.9}$$

From (5.7) and (5.9), the damping ratio ζ is an indicator of the oscillatory nature of the response, with excessive oscillations occurring at low ζ values. Hence, ζ is used as a measure of the relative stability of the system. From (5.8), the time to first peak drops as the undamped natural frequency ω_n increases. Hence, ω_n is used as a measure of speed of response. For higher-order systems, these measures and equations (5.7) through (5.9) can provide approximate answers if the time response is dominated by a single pair of complex conjugate poles. This occurs if additional poles and zeros are far in the left half plane or almost cancel. For systems with zeros, the percentage overshoot is higher than predicted by (5.7) unless the zero is located far in the LHP or almost cancels with a pole. However, equations (5.7) through (5.9) are always used in design because of their simplicity.

Thus, the design process reduces to the selection of pole locations and the corresponding behavior in the time domain. The root locus summarizes information on the time response of a closed-loop system as dictated by the pole locations in a single plot. Together with the previously stated criteria, it provides a powerful design tool, as demonstrated by the next example.

EXAMPLE 5.2

Discuss the effect of gain variation on the time response of the position control system described in Example 3.3 with

$$L(s) = \frac{1}{s(s+p)}$$

Solution

The root locus of the system is similar to that of Example 5.1(1), and it is shown in Figure 5.2(a) for $p = 4$. As the gain K is increased, the closed-loop system poles become complex conjugate; then the damping ratio ζ decreases progressively. Thus, the relative stability of the system deteriorates for high gain values. However, large gain values are required to reduce the steady-state error of the system due to a unit ramp, which is given by

$$e(\infty)\% = \frac{100}{K_v} = \frac{100 p}{K}$$

In addition, increasing K increases the undamped natural frequency ω_n (i.e., the magnitude of the pole), and hence the speed of response of the system increases. Thus, the chosen gain must be a compromise value that is large enough for a low steady-state error and an acceptable speed of response, but small enough to avoid excessive oscillations. The time response for a gain of 10 is shown in Figure 5.2(b).

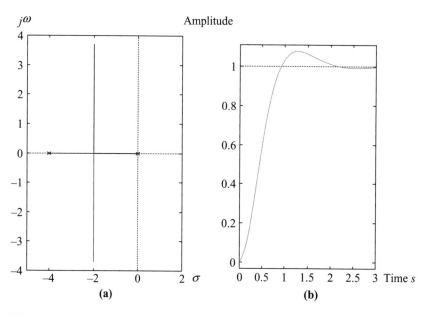

FIGURE 5.2

Use of the root locus in the design of a second-order system. (a) Root locus for $a = 4$. (b) Step response for $K = 10$.

The preceding example illustrates an important feature of design—namely, that it typically involves a compromise between conflicting requirements. The designer must always remember this when selecting design specifications so as to avoid over-optimizing some design criteria at the expense of others.

Note that the settling time of the system does not change in this case when the gain is increased. For a higher-order system or a system with a zero, this is usually not the case. Yet for simplicity the second-order equations (5.7) through (5.9) are still used in design. The designer must always be alert to errors that this simplification may cause. In practice, design is an iterative process where the approximate results from the second-order approximation are checked and, if necessary, the design is repeated until satisfactory results are obtained.

5.4 ROOT LOCUS DESIGN

Laplace transformation of a time function yields a function of the complex variable s that contains information about the transformed time function. We can therefore use the poles of the s-domain function to characterize the behavior of the time function without inverse transformation. Figure 5.3 shows pole locations in the s-domain and the associated time functions. Real poles are associated with an exponential time response that decays for LHP poles and increases for RHP poles. The magnitude of the pole determines the rate of exponential change. A pole at the origin is associated with a unit step. Complex conjugate poles are associated with an oscillatory response that decays exponentially for LHP poles and increases exponentially for RHP poles. The real part of the pole determines the rate of exponential change, and the imaginary part determines the frequency of oscillations. Imaginary axis poles are associated with sustained oscillations.

The objective of control system design in the s-domain is to indirectly select a desirable time response for the system through the selection of closed-loop pole locations. The simplest means of shifting the system poles is through the use of an amplifier or proportional controller. If this fails, then the pole locations can be more drastically altered by adding a dynamic controller with its own open-loop poles and zeros.

As Examples 5.1 and 5.2 illustrated, adding a zero to the system allows the improvement of its time response because it pulls the root locus into the LHP. Adding a pole at the origin increases the type number of the system and reduces its steady-state error but may adversely affect the transient response. If an improvement of both transient and steady-state performance is required, then it may be necessary to add two zeros as well as a pole at the origin. At times, more complex controllers may be needed to achieve the desired design objectives.

The controller could be added in the forward path, in the feedback path, or in an inner loop. A prefilter could also be added before the control loop to allow more freedom in design. Several controllers could be used simultaneously, if necessary, to meet all the design specifications. Examples of these control configurations are shown in Figure 5.4.

In this section, we review the design of analog controllers. We restrict the discussion to proportional (P), proportional-derivative (PD), proportional-integral

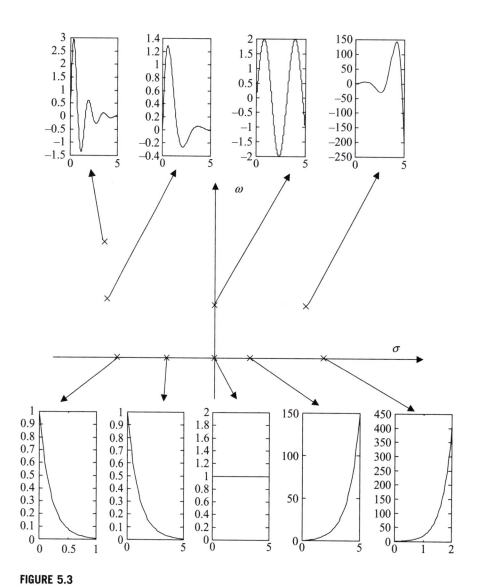

FIGURE 5.3

Pole locations and the associated time responses.

(PI), and proportional-integral-derivative (PID) control. Similar procedures can be developed for the design of lead, lag, and lag-lead controllers.

5.4.1 Proportional Control

Gain adjustment or proportional control allows the selection of closed-loop pole locations from among the poles given by the root locus plot of the system loop

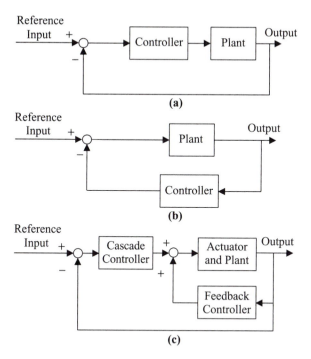

Reference
Input +

Controller   Plant   Output

−

(a)

Reference
Input +

Plant   Output

−

Controller

(b)

Reference
Input +

Cascade
Controller +

Actuator
and Plant   Output

−   +

Feedback
Controller

(c)

FIGURE 5.4

Control configurations. (a) Cascade compensation. (b) Feedback compensation. (c) Inner loop feedback compensation.

gain. For lower-order systems, it is possible to design proportional control systems analytically, but a sketch of the root locus is still helpful in the design process, as seen from the following example.

EXAMPLE 5.3

A position control system (see Example 3.3) with load angular position as output and motor armature voltage as input consists of an armature-controlled DC motor driven by a power amplifier together with a gear train. The overall transfer function of the system is

$$G(s) = \frac{K}{s(s + p)}$$

Design a proportional controller for the system to obtain

1. A specified damping ratio ζ
2. A specified undamped natural frequency ω_n

Solution

The root locus of the system was discussed in Example 5.2 and shown in Figure 5.2(a). The root locus remains in the LHP for all positive gain values. The closed-loop characteristic equation of the system is given by

$$s(s+p) + K = s^2 + 2\zeta\omega_n s + \omega_n^2 = 0$$

Equating coefficients gives

$$p = 2\zeta\omega_n \qquad K = \omega_n^2$$

which can be solved to yield

$$\omega_n = \sqrt{K} \qquad \zeta = \frac{p}{2\sqrt{K}}$$

Clearly, with one free parameter either ζ or ω_n can be selected, but not both. We now select a gain value that satisfies the design specifications.

1. If ζ is given and p is known, then the gain of the system and its undamped natural frequency are obtained from the equations

$$K = \left(\frac{p}{2\zeta}\right)^2 \qquad \omega_n = \frac{p}{2\zeta}$$

2. If ω_n is given and p is known, then the gain of the system and its damping ratio are obtained from the equations

$$K = \omega_n^2 \qquad \zeta = \frac{p}{2\omega_n}$$

The preceding example reveals some of the advantages and disadvantages of proportional control. The design is simple, and this simplicity carries over to higher-order systems if a CAD tool is used to assist in selecting the pole locations. Using cursor commands, the CAD tools allow the designer to select desirable pole locations from the root locus plot directly. The step response of the system can then be examined using the MATLAB command **step**. But the single free parameter available limits the designer's choice to one design criterion. If more than one aspect of the system time response must be improved, a dynamic controller is needed.

5.4.2 PD Control

As seen from Example 5.1, adding a zero to the loop gain improves the time response in the system. Adding a zero is accomplished using a cascade or feedback controller of the form

$$C(s) = K_p + K_d s = K_d(s + a)$$

$$a = K_p/K_d$$

(5.10)

This is known as a **proportional-derivative**, or **PD, controller**. The derivative term is only approximately realizable and is also undesirable because differentiating a noisy input results in large errors. However, if the derivative of the output is measured, an equivalent controller is obtained without differentiation. Thus, PD compensation is often feasible in practice.

The design of PD controllers depends on the specifications given for the closed-loop system and on whether a feedback or cascade controller is used. For a cascade controller, the system block diagram was shown in Figure 5.4(a) and the closed-loop transfer function is of the form

$$G_{cl}(s) = \frac{G(s)C(s)}{1 + G(s)C(s)}$$

$$= \frac{K_d(s+a)N(s)}{D(s) + K_d(s+a)N(s)}$$

(5.11)

where $N(s)$ and $D(s)$ are the numerator and denominator of the open-loop gain, respectively. Pole-zero cancellation occurs if the loop gain has a pole at $(-a)$. In the absence of pole-zero cancellation, the closed-loop system has a zero at $(-a)$, which may drastically alter the time response of the system. In general, the zero results in greater percentage overshoot than is predicted using (5.7).

Figure 5.5 shows feedback compensation including a preamplifier in cascade with the feedback loop and an amplifier in the forward path. We show that both amplifiers are often needed. The closed-loop transfer function is

$$G_{cl}(s) = \frac{K_p K_a G(s)}{1 + K_a G(s)C(s)}$$

$$= \frac{K_p K_a N(s)}{D(s) + K_a K_d(s+a)N(s)}$$

(5.12)

where K_a is the feedforward amplifier gain and K_p is the preamplifier gain. Note that although the loop gain is the same for both cascade and feedback compensation, the closed-loop system does not have a zero at $(-a)$ in the feedback case. If $D(s)$ has a pole at $(-a)$, the feedback-compensated system has a closed-loop pole at $(-a)$ that appears to cancel with a zero in the root locus when in reality it does not.

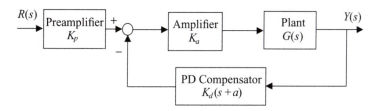

FIGURE 5.5

Block diagram of a PD feedback compensated system.

In both feedback and cascade compensation, two free design parameters are available and two design criteria can be selected. For example, if the settling time and percentage overshoot are specified, the steady-state error can only be checked after the design is completed and cannot be independently chosen.

EXAMPLE 5.4

Design a PD controller for the type 1 system described in Example 5.3 to meet the following specifications:

1. Specified ζ and ω_n
2. Specified ζ and steady-state error $e(\infty)\%$ due to a ramp input

Consider both cascade and feedback compensation, and compare them using a numerical example.

Solution
The root locus of the PD-compensated system is of the form of Figure 5.1(3). This shows that the system gain can be increased with no fear of instability. Even though this example is solved analytically, a root locus plot is needed to give the designer a feel for the variation in pole locations with gain.

With a PD controller the closed-loop characteristic equation is of the form

$$s^2 + ps + K(s+a) = s^2 + (p+K)s + Ka$$
$$= s^2 + 2\zeta\omega_n s + \omega_n^2$$

where $K = K_a$ for cascade compensation and $K = K_a K_a$ for feedback compensation.
Equating coefficients gives the equations

$$Ka = \omega_n^2 \qquad p + K = 2\zeta\omega_n$$

1. In this case, there is no difference between (K, a) in cascade and in feedback compensation. But the feedback case requires a preamplifier with the correct gain to yield zero steady-state error due to unit step. We examine the steady-state error in part 2. In either case, solving for K and a gives

$$K = 2\zeta\omega_n - p \qquad a = \frac{\omega_n^2}{2\zeta\omega_n - p}$$

2. For cascade compensation, the velocity error constant of the system is

$$K_v = \frac{Ka}{p} = \frac{100}{e(\infty)\%}$$

The undamped natural frequency is fixed at

$$\omega_n = \sqrt{Ka} = \sqrt{pK_v}$$

Solving for K and a gives

$$K = 2\zeta\sqrt{pK_v} - p \qquad a = \frac{pK_v}{2\zeta\sqrt{pK_v} - p}$$

For feedback compensation with preamplifier gain K_p and cascade amplifier gain K_a, as in Figure 5.5, the error is given by

$$R(s) - Y(s) = R(s)\left[1 - \frac{K_p K_a}{s^2 + (p+K)s + Ka}\right]$$

$$= R(s)\frac{s^2 + (p+K)s + Ka - K_p K_a}{s^2 + (p+K)s + Ka}$$

Using the final value theorem gives the steady-state error due to a unit ramp input as

$$e(\infty)\% = \underset{s \to 0}{\mathcal{L}im}\, s\left[\frac{1}{s^2}\right]\frac{s^2 + (p+K)s + Ka - K_p K_a}{s^2 + (p+K)s + Ka} \times 100\%$$

This error is infinite unless the amplifier gain is selected such that $K_p K_a = Ka$. The steady-state error is then given by

$$e(\infty)\% = \frac{p+K}{Ka} \times 100\%$$

The steady-state error $e(\infty)$ is simply the percentage error divided by 100. Hence, using the equations governing the closed-loop characteristic equation

$$Ka = \frac{p+K}{e(\infty)} = \frac{2\zeta\omega_n}{e(\infty)} = \omega_n^2$$

the undamped natural frequency is fixed at

$$\omega_n = \frac{2\zeta}{e(\infty)}$$

Then solving for K and a we obtain

$$K = \frac{4\zeta^2}{e(\infty)} - p$$

$$a = \frac{4\zeta^2}{e(\infty)(4\zeta^2 - pe(\infty))}$$

Note that, unlike cascade compensation, ω_n can be freely selected if the steady-state error is specified and ζ is free. To further compare cascade and feedback compensation, we consider the system with the pole $p = 4$, and require $\zeta = 0.7$ and $\omega_n = 10$ rad/s for part 1. These values give $K = K_a = 10$ and $a = 10$. In cascade compensation, (5.11) gives the closed-loop transfer function

$$G_{cl}(s) = \frac{10(s+10)}{s^2 + 14s + 100}$$

For feedback compensation, amplifier gains are selected such that the numerator is equal to 100 for unity steady-state output due to unit step input. For example, one may select

$$K_p = 10, \quad K_a = 10$$
$$K_d = 1, \quad a = 10$$

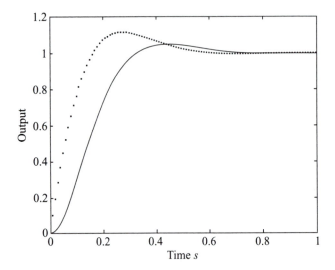

FIGURE 5.6

Step response of PD cascade (*dotted*) and feedback (*solid*) compensated systems with a given damping ratio and undamped natural frequency.

Substituting in (5.12) gives the closed-loop transfer function

$$G_{cl}(s) = \frac{100}{s^2 + 14s + 100}$$

The responses of the cascade- and feedback-compensated systems are shown together in Figure 5.6. The PO for the feedback case can be predicted exactly using (5.7) and is equal to about 4.6%. For cascade compensation, the PO is higher due to the presence of the zero. The zero is at a distance from the imaginary axis less than one and a half times the negative real part of the complex conjugate poles. Therefore, its effect is significant, and the PO increases to over 10% with a faster response.

For part 2 with $p = 4$, we specify $\zeta = 0.7$, and a steady-state error of 4%. Cascade compensation requires $K = 10$, $a = 10$. These are identical to the values of part (i) and correspond to an undamped natural frequency $\omega_n = 10$ rad/s.

For feedback compensation we obtain $K = 45$, $a = 27.222$. Using (5.12) gives the closed-loop transfer function

$$G_{cl}(s) = \frac{1225}{s^2 + 49s + 1225}$$

with $\omega_n = 35$ rad/s.

The responses of cascade- and feedback-compensated systems are shown in Figure 5.7. The PO for the feedback-compensated case is still 4.6% as calculated using (5.7). For cascade compensation, the PO is higher due to the presence of the zero that is close to the complex conjugate poles.

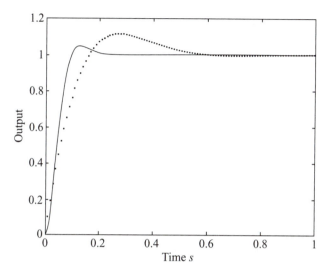

FIGURE 5.7

Step response of PD cascade (*dotted*) and feedback (*solid*) compensated systems with a given damping ratio and steady-state error.

Although the response of the feedback-compensated system is superior, several amplifiers are required for its implementation with high gains. The high gains may cause nonlinear behavior such as saturation in some situations.

Having demonstrated the differences between cascade and feedback compensation, we restrict our discussion in the sequel to cascade compensation. Similar procedures can be developed for feedback compensation.

Example 5.4 is easily solved analytically because the plant is only second order. For higher-order systems, the design is more complex, and a solution using CAD tools is preferable. We develop design procedures using CAD tools based on the classical graphical solution methods. These procedures combine the convenience of CAD tools and the insights that have made graphical tools successful. The procedures find a controller transfer function such that the angle of its product with the loop gain function at the desired pole location is an odd multiple of 180°. From the angle condition, this ensures that the desired location is on the root locus of the compensated system. The angle contribution required from the controller for a desired closed-loop pole location s_d is

$$\theta_C = \pm 180° - \angle L(s_d) \tag{5.13}$$

where $L(s)$ is the open-loop gain with numerator $N(s)$ and denominator $D(s)$. For a PD controller, the controller angle is simply the angle of the zero at the desired pole location. Applying the angle condition at the desired closed-loop location, it can be shown that the zero location is given by

$$a = \frac{\omega_d}{\tan(\theta_c)} + \zeta\omega_n \tag{5.14}$$

The proof of (5.13) and (5.14) is straightforward and is left as an exercise (see Problem 5.5).

In some special cases, a satisfactory design is obtained by cancellation, or near cancellation, of a system pole with the controller zero. The desired specifications are then satisfied by tuning the reduced transfer function's gain. In practice, exact pole-zero cancellation is impossible. However, with near cancellation the effect of the pole-zero pair on the time response is usually negligible.

A key to the use of powerful calculators and CAD tools in place of graphical methods is the ease with which transfer functions can be evaluated for any complex argument using direct calculation or cursor commands. The following CAD procedure exploits the MATLAB command **evalfr** to obtain the design parameters for a specified damping ratio and undamped natural frequency. The command **evalfr** evaluates a transfer function **g** for any complex argument **s** as follows:

>> **evalfr(g, s)**

The **angle** command gives the angle of any complex number. The complex value and its angle can also be evaluated using any hand calculator.

Procedure 5.1: Given ζ and ω_n

MATLAB or Calculator

1. Calculate the angle of the loop gain function evaluated at the desired location s_d, and subtract the angle from π using a hand calculator or the MATLAB commands **evalfr** and **angle**.
2. Calculate the zero location using equation (5.14) using **tan(theta)**, where **theta** is in radians.
3. Calculate the magnitude of the numerator of the new loop gain function, including the controller zero, using the command **abs** and the * pole-zero operator to multiply transfer functions; then calculate the gain using the magnitude condition.
4. Check the time response of the PD-compensated system, and modify the design to meet the desired specifications if necessary. Most currently available calculators cannot perform this step.

The following MATLAB function calculates the gain and zero location for PD control.

```
% L is the open loop gain
% zeta and wn specify the desired closed-loop pole
% scl is the closed-loop pole, theta is the controller angle at scl
% k (a) are the corresponding gain (zero location)

function [k, a, scl] = pdcon( zeta, wn, L)
scl = wn*exp( j*( pi-acos(zeta) ) ); % Find the desired closed-loop
                                     % pole location.
   theta = pi - angle( evalfr( L, scl) ) ; % Calculate the controller
                                           % angle.
  a = wn * sqrt(1-zeta^2)/ tan(theta) + zeta*wn; % Calculate the
                                                 % controller zero.
Lcomp = L*tf([1, a],1) ; % Include the controller zero.
k = 1/abs( evalfr( Lcomp, scl) ); % Calculate the gain that yields the
                                  % desired pole.
```

For a specified steady-state error, the system gain is fixed and the zero location is varied. Other design specifications require varying parameters other than the gain K. Root locus design with a free parameter other than the gain is performed using the following procedure.

Procedure 5.2: Given Steady-State Error and ζ

1. Obtain the error constant from the steady-state error, and determine a system parameter that remains free after the error constant is fixed for the system with PD control.

2. Rewrite the closed-loop characteristic equation of the PD-controlled system in the form

$$1 + K_f G_f(s) = 0 \qquad\qquad (5.15)$$

   where K_f is a gain dependent on the free system parameter and $G_f(s)$ is a function of s.

3. Obtain the value of the free parameter K_f corresponding to the desired closed-loop pole location. As in Procedure 5.1, K_f can be obtained by applying the magnitude condition using MATLAB or a calculator.

4. Calculate the free parameter from the gain K_f using the MATLAB command **rlocus**.

5. Check the time response of the PD-compensated system, and modify the design to meet the desired specifications if necessary.

EXAMPLE 5.5

Using a CAD package, design a PD controller for the type 1 position control system of Example 3.3 with transfer function

$$G(s) = \frac{1}{s(s+4)}$$

to meet the following specifications:
1. $\zeta = 0.7$ and $\omega_n = 10$ rad/s.
2. $\zeta = 0.7$ and 4% steady-state error due to a unit ramp input.

Solution

1. We solve the problem using Procedure 5.1 and the MATLAB function **pdcon**. Figure 5.8 shows the root locus of the compensated plot with the desired pole location at the intersection of the radial line for a damping ratio of 0.7 and the circular arc for an undamped natural frequency of 10. A compensator angle of 67.2 is obtained using (5.13) with a hand calculator or MATLAB. The MATLAB function **pdcon** gives

```
>> [k, a, scl] = pdcon( 0.7, 10,tf( 1, [1,4,0]) )
k =
    10.0000
a =
    10.0000
scl =
    -7.0000 + 7.1414i
```

Figure 5.9 shows the root locus of the compensated plot with the cursor at the desired pole location and a corresponding gain of 10. The results are approximately equal to those obtained analytically in Example 5.4.

2. The specified steady-state error gives

$$K_v = \frac{100}{e(\infty)\%} = \frac{100}{4\%} = 25 = \frac{Ka}{4} \Rightarrow Ka = 100$$

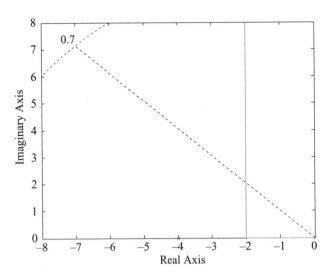

FIGURE 5.8

Root locus plot of uncompensated systems.

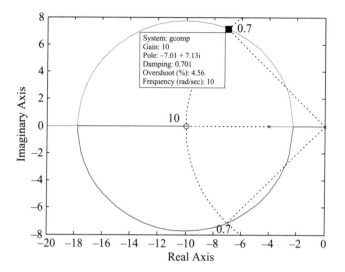

FIGURE 5.9

Root locus plot of PD-compensated systems.

The closed-loop characteristic equation of the PD-compensated system is given by

$$1 + K \frac{s+a}{s(s+4)} = 0$$

Let K vary with a so that their product Ka remains equal to 100; then Procedure 5.2 requires that the characteristic equation be rewritten as

$$1 + K \frac{s}{s^2 + 4s + 100} = 0$$

The corresponding root locus is a circle centered at the origin as shown in Figure 5.10 with the cursor at the location corresponding to the desired damping ratio. The desired location is at the intersection of the root locus with the $\zeta = 0.7$ radial line. The corresponding gain value is $K = 10$, which yields $a = 10$, that is, the same values as in Example 5.3. We obtain the value of K using the MATLAB commands

>> **g = tf([1, 0], [1, 4, 100]); rlocus(g)**

(Click on the root locus and drag the mouse until the desired gain is obtained.)

The time responses of the two designs are identical and were obtained earlier as the cascade-compensated responses of Figures 5.6 and 5.7, respectively.

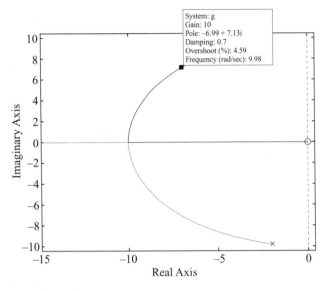

System: g
Gain: 10
Pole: –6.99 + 7.13i
Damping: 0.7
Overshoot (%): 4.59
Frequency (rad/sec): 9.98

FIGURE 5.10

Root locus plot of PD-compensated systems with **Ka** fixed.

5.4.3 PI Control

Increasing the type number of the system drastically improves its steady-state response. If an **integral controller** is added to the system, its type number is increased by one but its transient response deteriorates or the system becomes unstable. If a proportional control term is added to the integral control, the controller has a pole and a zero. The transfer function of the **proportional-integral (PI) controller** is

$$C(s) = K_p + \frac{K_i}{s} = K_p \frac{s+a}{s}$$
$$a = K_i/K_p$$
(5.16)

and is used in cascade compensation. An integral term in the feedback path is equivalent to a differentiator in the forward path and is therefore undesirable (see Problem 5.6).

PI design for a plant transfer function $G(s)$ can be viewed as PD design for the plant $G(s)/s$. Thus, Procedure 5.1 or 5.2 can be used for PI design. However, a better design is often possible by placing the controller zero close to the pole at the origin so that the controller pole and zero "almost cancel." An almost canceling pole-zero pair has a negligible effect on the time response. Thus, the PI controller results in a small deterioration in the transient response with a significant improvement in the steady-state error. The following procedure can be used for PI controller design.

Procedure 5.3

1. Design a proportional controller for the system to meet the transient response specifications (i.e., place the dominant closed-loop system poles at a desired location $s_{cl} = -\zeta\omega_n \pm j\omega_d$).

2. Add a PI controller with the zero location specified by

$$a = \frac{\omega_n}{\zeta + \sqrt{1-\zeta^2}/\tan(\phi)} \tag{5.17}$$

or

$$a = \frac{\zeta\omega_n}{10} \tag{5.18}$$

where ϕ is a small angle $(3 \rightarrow 5°)$.

3. Tune the gain of the system to move the closed-loop pole closer to s_{cl}.

4. Check the system time response.

If a PI controller is used, it is implicitly assumed that proportional control meets the transient response but not the steady-state error specifications. Failure of the first step in Procedure 5.3 indicates that a different controller (one that improves both transient and steady-state behavior) must be used.

To prove (5.17) and justify (5.18), we use the pole-zero diagram of Figure 5.11. The figure shows the angle contribution of the controller at the closed-loop pole location s_{cl}. The contribution of the open-loop gain $L(s)$ is not needed and is not shown.

PROOF. The controller angle at s_{cl} is

$$-\phi = \theta_z - \theta_p = (180° - \theta_p) - (180° - \theta_z)$$

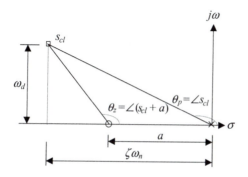

FIGURE 5.11

Pole-zero diagram of a PI controller.

From Figure 5.11, the tangents of the two angles in the preceding equation are

$$\tan(180° - \theta_p) = \frac{\omega_d}{\zeta\omega_n} = \frac{\sqrt{1-\zeta^2}}{\zeta}$$

$$\tan(180° - \theta_z) = \frac{\omega_d}{\zeta\omega_n - a} = \frac{\sqrt{1-\zeta^2}}{\zeta - x}$$

where x is the ratio (a/ω_n). Next we use the trigonometric identity

$$\tan(A - B) = \frac{\tan(A) - \tan(B)}{1 + \tan(A)\tan(B)}$$

to obtain

$$\tan(\phi) = \frac{\dfrac{\sqrt{1-\zeta^2}}{\zeta - x} - \dfrac{\sqrt{1-\zeta^2}}{\zeta}}{1 + \dfrac{1-\zeta^2}{\zeta(\zeta - x)}} = \frac{x\sqrt{1-\zeta^2}}{1 - \zeta x}$$

Solving for x, we have

$$x = \frac{1}{\zeta + \sqrt{1-\zeta^2}/\tan(\phi)}$$

Multiplying by ω_n gives (5.17).

If the controller zero is chosen using (5.8), then $x = \zeta/10$. Solving for ϕ we obtain

$$\phi = \tan^{-1}\left(\frac{\zeta\sqrt{1-\zeta^2}}{10 - \zeta^2}\right)$$

This yields the plot of Figure 5.12, which clearly shows an angle ϕ of 3° or less. ∎

The use of Procedure 5.1 or Procedure 5.3 to design PI controllers is demonstrated in the following example.

EXAMPLE 5.6

Design a controller for the position control system

$$G(s) = \frac{1}{s(s+10)}$$

to perfectly track a ramp input and have a dominant pair with a damping ratio of 0.7 and an undamped natural frequency of 4 rad/s.

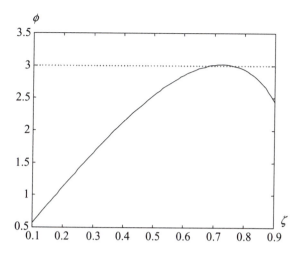

FIGURE 5.12

Plot of the controller angle ϕ at the underdamped closed-loop pole versus ζ.

Solution

Design 1

Apply Procedure 5.1 to the modified plant

$$G_i(s) = \frac{1}{s^2(s+10)}$$

This plant is unstable for all gains as seen from its root locus plot of Figure 5.13. The controller must provide an angle of about 111° at the desired closed-loop pole location. Substituting in (5.14) gives a zero at −1.732. Then moving the cursor to the desired pole location on the root locus of the compensated system (Figure 5.14) gives a gain of about 40.6.

The design can also be obtained analytically by writing the closed-loop characteristic polynomial as

$$s^3 + 10s^2 + Ks + Ka = (s+\alpha)(s^2 + 2\zeta\omega_n s + \omega_n^2)$$
$$= s^3 + (\alpha + 2\zeta\omega_n)s^2 + (2\zeta\omega_n\alpha + \omega_n^2)s + \alpha\omega_n^2$$

Then equating coefficients gives

$$\alpha = 10 - 2\zeta\omega_n = 10 - 2(0.7)(4) = 4.4$$
$$K = \omega_n(2\zeta\alpha + \omega_n) = 4 \times [2(0.7)(4.4) + 4] = 40.64$$
$$a = \frac{\alpha\omega_n^2}{K} = \frac{4.4 \times 4^2}{40.64} = 1.732$$

which are approximately the same as the values obtained earlier.

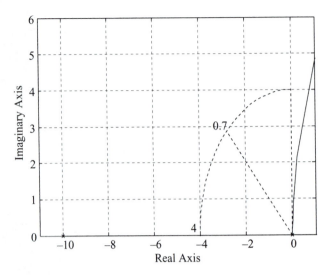

FIGURE 5.13

Root locus of a system with an integrator.

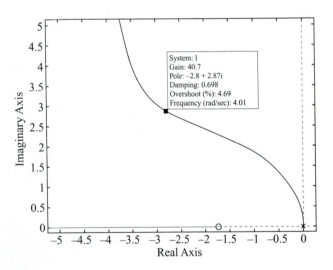

FIGURE 5.14

Root locus of a PI compensated system.

The MATLAB commands to obtain the zero location and the corresponding gain are

>> g = tf(1, [1, 10, 0, 0]); scl = 4*(−0.7 + j * sqrt(1 − 0.7^2))

scl = −2.8000 + 2.8566i

>> theta = pi − angle(polyval([1, 10, 0, 0], scl))

theta = 1.9285

>> a = imag(scl)/tan(theta) − real(scl)

a = 1.7323

>> k = 1/abs(evalfr(g * tf([1, a],1), scl))

k = 40.6400

The closed-loop transfer function for the preceding design (Design 1) is

$$G_{cl}(s) = \frac{40.64(s+1.732)}{s^3 + 10s^2 + 40.64s + 69.28}$$
$$= \frac{40.64(s+1.732)}{(s+4.4)(s^2 + 5.6s + 16)}$$

The system has a zero close to the closed-loop poles, which results in excessive overshoot in the time response of Figure 5.15 (together with the response for a later design).

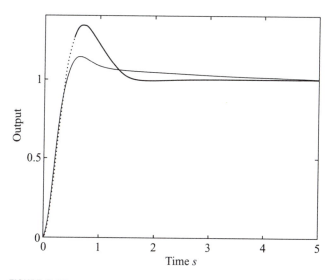

FIGURE 5.15

Step response of a PI compensated system: Design 1 (*dotted*), Design 2 (*solid*).

Design 2
Next, we apply Procedure 5.3 to the same problem. The proportional control gain for a damping ratio of 0.7 is approximately 51.02 and yields an undamped natural frequency of 7.143 rad/s. This is a faster design than required and is therefore acceptable. Then we use (5.18) and obtain the zero location

$$a = \frac{\omega_n}{\zeta + \sqrt{1-\zeta^2}/\tan(-\phi)}$$

$$= \frac{7.143}{0.7 + \sqrt{1-0.49}/\tan(3°)} \cong 0.5$$

If (5.18) is used, we have

$$a = \frac{\zeta\omega_n}{10} = \frac{7.143 \times 0.7}{10} \cong 0.5$$

That is, the same zero value is obtained.
    The closed-loop transfer function for this design is

$$G_{cl}(s) = \frac{50(s+0.5)}{s^3 + 10s^2 + 50s + 25}$$

$$= \frac{50(s+0.5)}{(s+0.559)(s^2 + 9.441s + 44.7225)}$$

where the gain value has been slightly reduced to bring the damping ratio closer to 0.7. This actually does give a damping ratio of about 0.7 and an undamped natural frequency of 6.6875 rad/s for the dominant closed-loop poles.
    The time response for this design (Design 2) is shown, together with that of Design 1, in Figure 5.15. The percentage overshoot for Design 2 is much smaller because its zero almost cancels with a closed-loop pole. Design 2 is clearly superior to Design 1.

5.4.4 PID Control

If both the transient and steady-state response of the system must be improved, then neither a PI nor a PD controller may meet the desired specifications. Adding a zero (PD) may improve the transient response but does not increase the type number of the system. Adding a pole at the origin increases the type number but may yield an unsatisfactory time response even if one zero is also added. With a **proportional-integral-derivative (PID) controller**, two zeros and a pole at the origin are added. This both increases the type number and allows satisfactory reshaping of the root locus.
    The transfer function of a PID controller is given by

$$C(s) = K_p + \frac{K_i}{s} + K_d s = K_d \frac{s^2 + 2\zeta\omega_n s + \omega_n^2}{s} \tag{5.19}$$

$$2\zeta\omega_n = K_p/K_d, \quad \omega_n^2 = K_i/K_d$$

where K_p, K_i, and K_d are the proportional, integral, and derivative gain, respectively.

The zeros of the controller can be real or complex conjugate, allowing the cancellation of real or complex conjugate LHP poles if necessary. In some cases, good design can be obtained by canceling the pole closest to the imaginary axis. The design then reduces to a PI controller design that can be completed by applying Procedure 5.3. Alternatively, one could apply Procedure 5.1 or 5.2 to the reduced transfer function with an added pole at the origin. A third approach to PID design is to follow Procedure 5.3 with the proportional control design step modified to PD design. The PD design is completed using Procedure 5.1 or 5.2 to meet the transient response specifications. PI control is then added to improve the steady-state response. The following examples illustrate these design procedures.

EXAMPLE 5.7

Design a PID controller for an armature-controlled DC motor with transfer function

$$G(s) = \frac{1}{s(s+1)(s+10)}$$

to obtain zero steady-state error due to ramp, a damping ratio of 0.7, and an undamped natural frequency of 4 rad/s.

Solution
Canceling the pole at −1 with a zero and adding an integrator yields the transfer function

$$G_i(s) = \frac{1}{s^2(s+10)}$$

This is identical to the transfer function obtained in Example 5.6. Hence, the overall PID controller is given by

$$C(s) = 50\frac{(s+1)(s+0.5)}{s}$$

This design is henceforth referred to as Design 1.

A second design (Design 2) is obtained by first selecting a PD controller to meet the transient response specifications. We seek an undamped natural frequency of 5 rad/s in anticipation of the effect of adding PI control. The PD controller is designed using the MATLAB commands (using the function **pdcon**)

>> [k, a, scl] = pdcon(0.7, 5, tf(1, [1, 11, 10, 0]))

k = 43.0000

a = 2.3256

scl = -3.5000 + 3.5707i

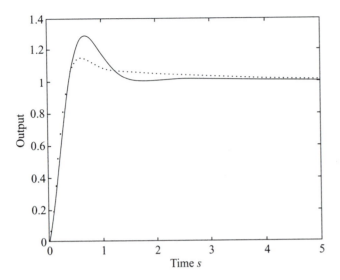

FIGURE 5.16

Time response for Design 1 (*dotted*) and Design 2 (*solid*).

The PI zero is obtained using the command

>> b = 5 /(0.7 + sqrt(1-.49)/tan(3*pi/ 180))

b = 0.3490

The gain is reduced to 40, and the controller transfer function for Design 2 is

$$C(s) = 40\frac{(s+0.349)(s+2.326)}{s}$$

The step responses for Designs 1 and 2 are shown in Figure 5.16. Clearly, Design 1 is superior because the zeros in Design 2 result in excessive overshoot. The plant transfer function favors pole cancellation in this case because the remaining real axis pole is far in the LHP. If the remaining pole is at −3, say, the second design procedure would give better results. The lesson to be learned is that there are no easy solutions in design. There are recipes with which the designer should experiment until satisfactory results are obtained.

EXAMPLE 5.8

Design a PID controller to obtain zero steady-state error due to step, a damping ratio of 0.7, and an undamped natural frequency of at least 4 rad/s for the transfer function

$$G(s) = \frac{1}{(s+10)(s^2 + 2s + 10)}$$

Solution

The system has a pair of complex conjugate poles that slow down its time response and a third pole that is far in the LHP. Canceling the complex conjugate poles with zeros and adding the integrator yields the transfer function

$$G(s) = \frac{1}{s(s+10)}$$

The root locus of the system is similar to Figure 5.2(a), and we can increase the gain without fear of instability. The closed-loop characteristic equation of the compensated system with gain K is

$$s^2 + 10s + K = 0$$

Equating coefficients as in Example 5.3, we observe that for a damping ratio of 0.7 the undamped natural frequency is

$$\omega_n = \frac{10}{2\zeta} = \frac{5}{0.7} = 7.143 \, \text{rad/s}$$

This meets the design specifications. The corresponding gain is 51.02, and the PID controller is given by

$$C(s) = 51.02 \frac{s^2 + 2s + 10}{s}$$

In practice, pole-zero cancellation may not occur, but near cancellation is sufficient to obtain a satisfactory time response, as shown in Figure 5.17.

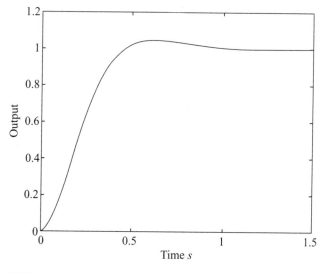

FIGURE 5.17

Step response of the PID-compensated system described in Example 5.8.

5.5 EMPIRICAL TUNING OF PID CONTROLLERS

In industrial applications, PID controllers are often tuned empirically. Typically, the controller parameters are selected based on a simple process model using a suitable tuning rule. This allows us to address (1) load disturbance rejection specifications (which are often a major concern in process control) and (2) the presence of a time delay in the process. We first write the PID controller transfer function (5.19) in the form

$$C(s) = K_p \left(1 + \frac{1}{T_i s} + T_d s \right) \tag{5.20}$$

$$T_i = K_p / K_i, \quad T_d = K_d / K_p$$

where T_i denotes the **integral time constant** and T_d denotes the **derivative time constant**. The three controller parameters K_p, T_i, and T_d have a clear physical meaning. Increasing K_p (i.e., increasing the proportional action) provides a faster but more oscillatory response. The same behavior results from increasing the integral action by decreasing the value of T_i. Finally, increasing the value of T_d leads to a slower but more stable response.

These considerations allow tuning the controller by a trial-and-error procedure. However, this can be time consuming, and the achieved performance depends on the skill of the designer. Fortunately, tuning procedures are available to simplify the PID design. Typically, the parameters of the process model are determined assuming a first-order-plus-dead-time model. That is,

$$G(s) = \frac{K}{\tau s + 1} e^{-Ls} \tag{5.21}$$

where K is the process gain, τ is the process (dominant) time constant, and L is the (apparent) dead time of the process. These parameters can be estimated based on the step response of the process through the **tangent method.**[2] The method consists of the following steps.

TANGENT METHOD

1. Obtain the step response of the process experimentally.
2. Draw a tangent to the step response at the inflection point as shown in Figure 5.18.
3. Compute the process gain as the ratio of the steady-state change in the process output y to the amplitude of the input step A.

---

[2]See Åström and Hägglund (2006) and Visioli (2006) for a detailed description and analysis of different methods for estimating the parameters of a first-order-plus-dead-time system.

4. Compute the apparent dead time L as the time interval between the application of the step input and the intersection of the tangent line with the time axis.
5. Determine the sum $\tau + L$ (from which the value of τ can be easily computed) as the time interval between the application of the step input and the intersection of the tangent line with the straight representing the final steady-state value of the process output. Alternatively, the value of $\tau + L$ can be determined as the time interval between the application of the step input and the time when the process output attains 63.2% of its final value. Note that in case the dynamics of the process can be perfectly described by a first-order-plus-dead-time model, the values of τ obtained in the two cases are identical.

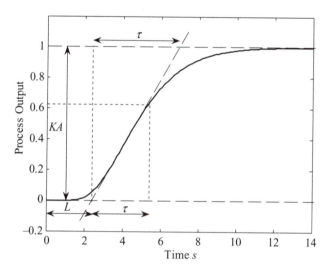

FIGURE 5.18

Application of the tangent method.

Given the process model parameters, several tuning rules are available for determining the PID controller parameter values, but different rules address different design specifications. The most popular tuning rules are those attributed to **Ziegler-Nichols**. Their aim is to provide satisfactory load disturbance rejection. Table 5.1 shows the Ziegler-Nichols rules for P, PI, and PID controllers. Although the rules are empirical, they are consistent with the physical meaning of the parameters. For example, consider the effect the derivative action in PID control governed by the third row of Table 5.1. The derivative action provides added damping to the system, which increases its relative stability. This allows us to

Table 5.1 Ziegler-Nichols Tuning Rules for a First-Order-Plus-Dead-Time Model of the Process

Controller Type	K_p	T_i	T_d
P	$\dfrac{\tau}{KL}$	—	—
PI	$0.9\dfrac{\tau}{KL}$	$3\,L$	—
PID	$1.2\dfrac{\tau}{KL}$	$2\,L$	$0.5\,L$

increase both the proportional and integral action while maintaining an acceptable time response. We demonstrate the Ziegler-Nichols procedure using the following example.

EXAMPLE 5.9

Consider the control system shown in Figure 5.19, where the process has the following transfer function:

$$G(s) = \frac{1}{(s+1)^4} e^{-0.2s}$$

Estimate a first-order-plus-dead-time model of the process, and design a PID controller by applying the Ziegler-Nichols tuning rules.

Solution
The process step response is shown in Figure 5.20. By applying the tangent method, a first-order-plus-dead-time model with gain $K = 1$, $L = 1.55$ and a delay $\tau = 3$ is estimated. The Ziegler-Nichols rules of Table 5.1 provide the following PID parameters: $K_p = 2.32$,

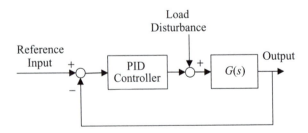

FIGURE 5.19
Block diagram of the process of Example 5.9.

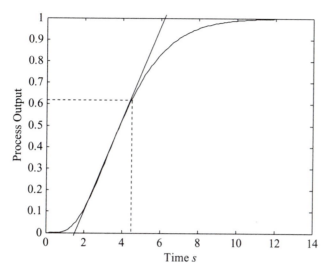

FIGURE 5.20

Application of the tangent method in Example 5.9.

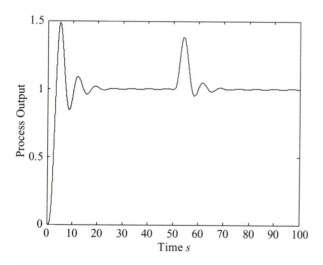

FIGURE 5.21

Process output with the PID controller tuned with the Ziegler-Nichols method.

$T_i = 3.1$, and $T_d = 0.775$. Figure 5.21 shows the response of the closed-loop system due a step input at $t = 0$ followed by a step load disturbance input at $t = 50$.

The system response is oscillatory, which is a typical feature of the Ziegler-Nichols method, whose main objective is to satisfy load disturbance response specifications. However, the effect of the disturbance on the step response is minimized.

RESOURCES

Åström, K. J., and T. Hägglund, *Advanced PID Controllers,* ISA Press, 2006.

D'Azzo, J. J., and C. H. Houpis, *Linear Control System Analysis and Design,* McGraw-Hill, 1988.

Kuo, B. C., *Automatic Control Systems,* Prentice Hall, 1995.

Nise, N. S., *Control Systems Engineering,* Benjamin Cummings, 1995.

O'Dwyer, A., *Handbook of PI and PID Tuning Rules,* Imperial College Press, 2006.

Ogata, K., *Modern Control Engineering,* Prentice Hall, 1990.

Van de Vegte, J., *Feedback Control Systems,* Prentice Hall, 1986.

Visioli, A., *Practical PID Control,* Springer, 2006.

PROBLEMS

5.1 Prove that for a system with two real poles and a real zero

$$L(s) = \frac{s+a}{(s+p_1)(s+p_2)}, \quad a < p_1 < p_2 \text{ or } p_1 < p_2 < a$$

the breakaway point is at a distance $\sqrt{(a-p_1)(a-p_2)}$ from the zero.

5.2 Use the result of Problem 5.1 to draw the root locus of the system

$$KL(s) = \frac{K(s+4)}{s(s+2)}$$

5.3 Sketch the root loci of the following systems:

(a) $KL(s) = \dfrac{K}{s(s+2)(s+5)}$

(b) $KL(s) = \dfrac{K(s+2)}{s(s+3)(s+5)}$

5.4 Consider the system in 5.3(b) with a required steady-state error of 20%, and an adjustable PI controller zero location. Show that the corresponding closed-loop characteristic equation is given by

$$1 + K\left(\frac{s+a}{s}\right)\frac{1}{(s+3)(s+5)} = 0$$

Next, rewrite the equation as

$$1 + K_f G_f(s) = 0$$

where $K_f = K$, Kz is constant, and $G_f(s)$ is a function of s, and examine the effect of shifting the zero on the closed-loop poles.

(a) Design the system for a dominant second-order pair with a damping ratio of 0.5. What is ω_n for this design?

(b) Obtain the time response using a CAD program. How does the time response compare with that of a second-order system with the same ω_n and ζ as the dominant pair? Give reasons for the differences.

(c) Discuss briefly the trade-off between error, speed of response, and relative stability in this problem.

5.5 Prove equations (5.13) and (5.14), and justify the design Procedures 5.1 and 5.2.

5.6 Show that a PI feedback controller is undesirable because it results in a differentiator in the forward path. Discuss the step response of the closed-loop system.

5.7 Design a controller for the transfer function

$$G(s) = \frac{1}{(s+1)(s+5)}$$

to obtain (a) zero steady-state error due to step, (b) a settling time of less than 2 s, and (c) an undamped natural frequency of 5 rad/s. Obtain the response due to a unit step, and find the percentage overshoot, the time to the first peak, and the steady-state error percentage due to a ramp input.

5.8 Repeat Problem 5.7 with a required settling time less than 0.5 s and an undamped natural frequency of 10 rad/s.

5.9 Consider the oven temperature control system of Example 3.5 with transfer function

$$G(s) = \frac{K}{s^2 + 3s + 10}$$

(a) Design a proportional controller for the system to obtain a percentage overshoot less than 5%.

(b) Design a controller for the system to reduce the steady-state error due to step to zero without significant deterioration in the transient response.

5.10 For the inertial system governed by the differential equation

$$\ddot{\theta} = \tau$$

design a feedback controller to stabilize the system and reduce the percentage overshoot below 10% with a settling time of less that 4 s.

COMPUTER EXERCISES

5.11 Consider the oven temperature control system described in Example 3.5 with transfer function

$$G(s) = \frac{K}{s^2 + 3s + 10}$$

(a) Obtain the step response of the system with a PD cascade controller with gain 80 and a zero at −5.

(b) Obtain the step response of the system with PD feedback controller with a zero at −5 and unity gain and forward gain of 80.

(c) Why is the root locus identical for both systems?

(d) Why are the time responses different although the systems have the same loop gains?

(e) Complete a comparison table using the responses of (a) and (b) including the percentage overshoot, the time to first peak, the settling time, and the steady-state error. Comment on the results, and explain the reason for the differences in the response.

5.12 Use Simulink to examine a practical implementation of the cascade controller described in Exercise 5.11. The compensator transfer function includes a pole because PD control is only approximately realizable. The controller transfer is of the form

$$C(s) = 80 \frac{0.2s + 1}{0.02s + 1}$$

(a) Simulate the system with a step reference input both with and without a saturation block with saturation limits ±5 between the controller and plant. Export the output to MATLAB for plotting (you can use a Scope block and select "Save data to workspace").

(b) Plot the output of the system with and without saturation together, and comment on the difference between the two step responses.

5.13 Consider the system

$$G(s) = \frac{1}{(s+1)^4}$$

and apply the Ziegler-Nichols procedure to design a PID controller. Obtain the response due to a unit step input as well as a unit step disturbance signal.

5.14 Write a computer program that implements the estimation of a first-order-plus-dead-time transfer function with the tangent method and then determines the PID parameters using the Ziegler-Nichols formula. Apply the program to the system

$$G(s) = \frac{1}{(s+1)^8}$$

and simulate the response of the control system when a set-point step change and a load disturbance step are applied. Discuss the choice of the time constant value based on the results.

5.15 Apply the script of Exercise 5.14 to the system

$$G(s) = \frac{1}{(s+1)^2}$$

and simulate the response of the control system when a set-point step change and a load disturbance step are applied. Compare the results obtained with those of Problem 5.14.

Digital Control System Design

To design a digital control system, we seek a *z*-domain transfer function or difference equation model of the controller that meets given design specifications. The controller model can be obtained from the model of an analog controller that meets the same design specifications. Alternatively, the digital controller can be designed in the *z*-domain using procedures that are almost identical to *s*-domain analog controller design. We discuss both approaches in this chapter. We begin by introducing the *z*-domain root locus.

Objectives

After completing this chapter, the reader will be able to do the following:

1. Sketch the *z*-domain root locus for a digital control system, or obtain it using MATLAB.
2. Obtain and tune a digital controller from an analog design.
3. Design a digital controller in the *z*-domain directly using the root locus approach.
4. Design a digital controller directly using frequency domain techniques.
5. Design a digital controller directly using the synthesis approach of Ragazzini.
6. Design a digital control system with finite settling time.

6.1 *z*-DOMAIN ROOT LOCUS

In Chapter 3, we showed that the closed-loop characteristic equation of a digital control system is of the form

$$1 + C(z) G_{ZAS}(z) = 0 \tag{6.1}$$

where $C(z)$ is the controller transfer function, and $G_{ZAS}(z)$ is the transfer function of the DAC, analog subsystem, and ADC combination. If the controller is assumed to include a constant gain multiplied by a rational z-transfer function, then (6.1) is equivalent to

$$1 + K L(z) = 0 \tag{6.2}$$

where $L(z)$ is the open-loop gain.

Equation (6.2) is identical in form to the s-domain characteristic equation (5.1) with the variable s replaced by z. Thus, all the rules derived for equation (5.1) are applicable to (6.2) and can be used to obtain z-domain root locus plots. The plots can also be obtained using the root locus plots of most computer-aided design (CAD) programs. Thus, we can use the MATLAB command **rlocus** for z-domain root loci.

EXAMPLE 6.1

Obtain the root locus plot and the critical gain for the first-order type 1 system with loop gain

$$L(z) = \frac{1}{z-1}$$

Solution

Using root locus rules gives the root locus plot of Figure 6.1, which can be obtained using the MATLAB command **rlocus**. The root locus lies entirely on the real axis between the open-loop pole and the open-loop zero. For a stable discrete system, real axis z-plane

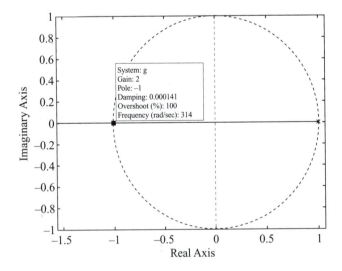

FIGURE 6.1

Root locus of a type 1 first-order system.

poles must lie between the point (−1, 0) and the point (1, 0). The critical gain for the system corresponds to the point (−1, 0). The closed-loop characteristic equation of the system is

$$z - 1 + K = 0$$

Substituting $z = -1$ gives the critical gain $K_{cr} = 2$ as shown on the root locus plot.

EXAMPLE 6.2

Obtain the root locus plot and the critical gain for the second-order type 1 system with loop gain

$$L(z) = \frac{1}{(z-1)(z-0.5)}$$

Solution

Using root locus rules gives the root locus plot of Figure 6.2, which has the same form as the root locus of Example 5.1(1) but is entirely in the right-hand plane (RHP). The breakaway point is midway between the two open-loop poles at $z_b = 0.75$. The critical gain now occurs at the intersection of the root locus with the unit circle. To obtain the critical gain value, first write the closed-loop characteristic equation

$$(z-1)(z-0.5) + K = z^2 - 1.5z + K + 0.5 = 0$$

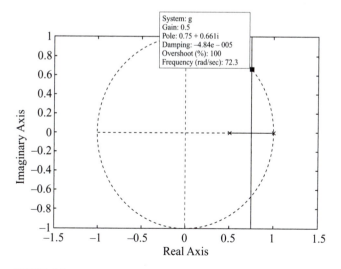

System: g
Gain: 0.5
Pole: 0.75 + 0.661i
Damping: −4.84e − 005
Overshoot (%): 100
Frequency (rad/sec): 72.3

FIGURE 6.2

Z-Root locus of a type 1 second-order system.

On the unit circle, the closed-loop poles are complex conjugate and of magnitude unity. Hence, the magnitude of the poles satisfies the equation

$$|z_{1,2}|^2 = K_{cr} + 0.5 = 1$$

where K_{cr} is the critical gain. The critical gain is equal to 0.5, which, from the closed-loop characteristic equation, corresponds to unit circle poles at

$$z_{1,2} = 0.75 \pm j0.661$$

6.2 z-DOMAIN DIGITAL CONTROL SYSTEM DESIGN

In Chapter 5, we were able to design analog control systems by selecting their poles and zeros in the s-domain to correspond to the desired time response. This approach was based on the relation between any time function and its s-domain poles and zeros. If the time function is sampled and the resulting sequence is z-transformed, the z-transform contains information about the transformed time sequence and the original time function. The poles of the z-domain function can therefore be used to characterize the sequence, and possibly the sampled continuous time function, without inverse z-transformation. However, this latter characterization is generally more difficult than characterization based on the s-domain functions described by Figure 5.3.

Figure 6.3 shows z-domain pole locations and the associated temporal sequences. As in the continuous case, positive real poles are associated with exponentials. Unlike the continuous case, the exponentials decay for poles inside the unit circle and increase for poles outside it. In addition, negative poles are associated with sequences of alternating sign. Poles on the unit circle are associ-

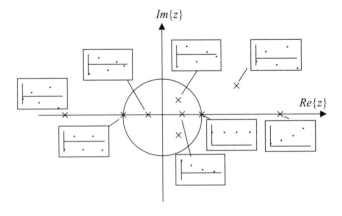

FIGURE 6.3

z-domain pole locations and the associated temporal sequences.

ated with a response of constant magnitude. For complex conjugate poles, the response is oscillatory with the rate of decay determined by the pole distance from the origin, and the frequency of oscillations determined by the magnitude of the pole angle. Complex conjugate poles on the unit circle are associated with sustained oscillations.

Because it is easier to characterize time function using s-domain poles, it may be helpful to reexamine Figure 5.3 and compare it to Figure 6.3. Comparing the figures suggests a relationship between s-domain and z-domain poles that greatly simplifies z-domain pole characterization. To obtain the desired relationship, we examine two key cases, both in the s-domain and in the z-domain. One case yields a real pole and the other a complex conjugate pair of poles. More complex time functions can be reduced to these cases by partial fraction expansion of the trans-forms. The two cases are summarized in Tables 6.1 and 6.2.

Using the two tables, it appears that if $F(s)$ has a pole at $-\alpha$, $F(z)$ has a pole at $e^{-\alpha T}$, and if $F(s)$ has poles at $-\zeta\omega_n + j\omega_d$, $F(z)$ has poles at $e^{(-\zeta\omega_n + j\omega_d)T}$. We therefore make the following observation.

Observation

If the Laplace transform $F(s)$ of a continuous-time function $f(t)$ has a pole p_s, then the z-transform $F(z)$ of its sampled counterpart $f(kT)$ has a pole at

$$p_z = e^{p_s T} \tag{6.3}$$

where T is the sampling period.

REMARKS

1. The preceding observation is valid for a unit step with its s-domain pole at the origin because the z-transform of a sampled step has a pole at $1 = e^0$.
2. There is no general mapping of s-domain zeros to z-domain zeros.
3. From (6.3), the z-domain poles in the complex conjugate case are given by

$$p_z = e^{\sigma T} e^{j\omega_d T}$$
$$= e^{\sigma T} e^{j(\omega_d T + k2\pi)}, \quad k = 0, 1, 2, \ldots$$

Thus, pole locations are a periodic function of the damped natural frequency ω_d with period $(2\pi/T)$ (i.e., the sampling angular frequency ω_s). The mapping of distinct s-domain poles to the same z-domain location is clearly undesirable in situations where a sampled waveform is used to represent its continuous coun-terpart. The strip of width ω_s over which no such ambiguity occurs (frequencies in the range $[(-\omega_s/2), \omega_s/2]$ rad/s is known as the *primary strip* (Figure 6.4). The

Table 6.1 Time Functions and Real Poles

Continuous	Laplace Transform	Sampled	z-Transform
$f(t) = \begin{cases} e^{-\alpha t}, & t \geq 0 \\ 0, & t < 0 \end{cases}$	$F(s) = \dfrac{1}{s + \alpha}$	$f(kT) = \begin{cases} e^{-\alpha kT}, & k \geq 0 \\ 0, & k < 0 \end{cases}$	$F(z) = \dfrac{z}{z - e^{-\alpha T}}$

Table 6.2 Time Functions and Complex Conjugate Poles

Continuous	Laplace Transform	Sampled	z-Transform
$f(t) = \begin{cases} e^{-\alpha t} \sin(\omega_d t), & t \geq 0 \\ 0, & t < 0 \end{cases}$	$F(s) = \dfrac{\omega_d}{(s + \alpha)^2 + \omega_d^2}$	$f(t) = \begin{cases} e^{-\alpha t} \sin(\omega_d t), & t \geq 0 \\ 0, & t < 0 \end{cases}$	$F(z) = \dfrac{\sin(\omega_d T) e^{-\alpha T} z}{z^2 - 2\cos(\omega_d T) e^{-\alpha T} z + e^{-2\alpha T}}$

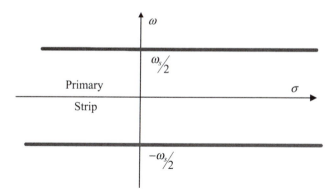

FIGURE 6.4

Primary strip in the *s*-plane.

width of this strip can clearly be increased by faster sampling, with the choice of suitable sampling rate dictated by the nature of the continuous time function. We observe that the minimum sampling frequency for good correlation between the analog and digital signals is twice the frequency ω_d as expected based on the sampling theorem of Section 2.9.

6.2.1 *z*-Domain Contours

Using the observation (6.3), one can use *s*-domain contours over which certain characteristics of the function poles are fixed to obtain the shapes of similar *z*-domain contours. In particular, the important case of a second-order underdamped system yields Table 6.3.

   The information in the table is shown in Figures 6.5 and 6.6 and can be used to predict the time responses of Figure 6.3. From Figure 6.5, we see that negative values of σ correspond to the inside of the unit circle in the *z*-plane, whereas positive values correspond to the outside of the unit circle. The unit circle is the $\sigma = 0$ contour. Both Table 6.3 and Figure 6.5 also show that large positive σ values correspond to circuits of large radii, whereas large negative σ values correspond to circles of small radii. In particular, a positive infinite σ corresponds to the point at ∞, and a negative infinite σ corresponds to the origin of the *z*-plane.

Table 6.3 Pole Contours in the *s*-Domain and the *z*-Domain.

Contour	s-Domain Poles	Contour	z-Domain Poles
$\alpha = constant$	Vertical line	$\lvert z \rvert = e^{\sigma T} = constant$	Circle
$\omega_d = constant$	Horizontal line	$\angle z = \omega_d T = constant$	Radial line

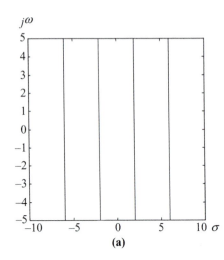
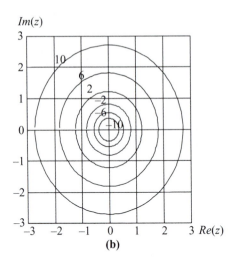

FIGURE 6.5

Constant σ contours in the s-plane and in the z-plane. (a) Constant σ contours in the s-plane. (b) Constant σ contours in the z-plane.

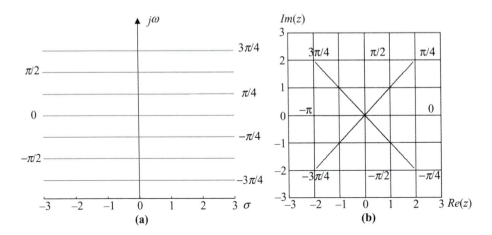

FIGURE 6.6

Constant ω_d contours in the s-plane and the z-plane. (a) Constant ω_d contours in the s-plane. (b) Constant ω_d contours in the z-plane.

From Figure 6.6, we see that larger ω_d values correspond to larger angles, with $\omega_d = \pm\omega_s/2$ corresponding to $\pm\pi$. As observed earlier using a different argument, a system with poles outside this range (outside the primary strip) does not have a one-to-one correspondence between s-domain and z-domain poles.

The z-domain characteristic polynomial for a second-order underdamped system is

$$\left(z - e^{(-\zeta\omega_n + j\omega_d)T}\right)\left(z - e^{(-\zeta\omega_n - j\omega_d)T}\right) = z^2 - 2\cos(\omega_d T)e^{-\zeta\omega_n T}z + e^{-2\zeta\omega_n T}$$

Hence, the poles of the system are given by

$$z_{1,2} = e^{-\zeta\omega_n T} \angle \pm \omega_d T \qquad (6.4)$$

This confirms that constant $\zeta\omega_n$ contours are circles, whereas constant ω_d contours are radial lines.

Constant ζ lines are logarithmic spirals that get smaller for larger values of ζ. The spirals are defined by the equation

$$|z| = e^{\frac{-\zeta\theta}{\sqrt{1-\zeta^2}}} = e^{\frac{-\zeta\left(\pi\theta^\circ/180^\circ\right)}{\sqrt{1-\zeta^2}}} \qquad (6.5)$$

where $|z|$ is the magnitude of the pole and θ is its angle. Constant ω_n contours are defined by the equation

$$|z| = e^{-\sqrt{(\omega_n T)^2 - \theta^2}} \qquad (6.6)$$

Figure 6.7 shows constant ζ contours and constant ω_n contours.

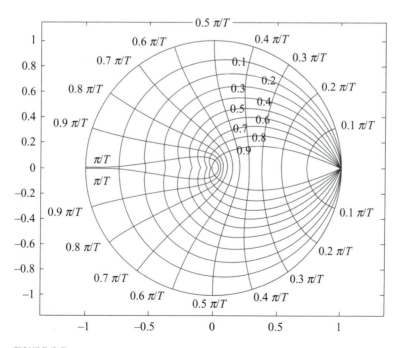

FIGURE 6.7

Constant ζ and ω_n contours in the z-plane.

To prove (6.5), rewrite the complex conjugate poles as

$$z_{1,2} = e^{-\zeta\omega_n T} \angle \pm \omega_d T = |z| \angle \pm \theta \tag{6.7}$$

or equivalently,

$$\theta = \omega_d T = \omega_n T \sqrt{1 - \zeta^2} \tag{6.8}$$

$$|z| = e^{-\zeta\omega_n T} \tag{6.9}$$

Eliminating $\omega_n T$ from (6.9) using (6.8), we obtain the spiral equation (6.5). The proof of (6.6) is similar and is left as an exercise (Problem 6.2).

The following observations can be made by examining the spiral equation (6.5):

1. For every ζ value, there are two spirals corresponding to negative and positive angles θ. The negative θ spiral is below the real axis and is the mirror image of the positive θ spiral.

2. For a given spiral, the magnitude of the pole drops logarithmically with its angle.

3. At the same angle θ, increasing the damping ratio gives smaller pole magnitudes. Hence, the spirals are smaller for larger ζ values.

4. All spirals start at $\theta = 0$, $|z| = 1$ but end at different points.

5. For a given damping ratio and angle θ, the pole magnitude can be obtained by substituting in (6.5). For a given damping ratio and pole magnitude, the pole angle can be obtained by substituting in the equation

$$\theta = \frac{\sqrt{1 - \zeta^2}}{\zeta} |\ln(|z|)| \tag{6.10}$$

The standard contours can all be plotted using a hand-held calculator, a protractor, and a compass. Their intersection with the root locus can then be used to obtain the pole locations for desired closed-loop characteristics. However, it is more convenient to obtain the root loci and contours using a CAD tool, especially for higher-order systems. The unit circle and constant ζ and ω_n contours can be added to root locus plots obtained with CAD packages to provide useful information on the significance of z-domain pole locations. In MATLAB, this is accomplished using the command

>> **zgrid(zeta, wn)**

where **zeta** is a vector of damping ratios and **wn** is a vector of the undamped natural frequencies for the contours.

Clearly, the significance of pole and zero locations in the z-domain is completely different from identical locations in the s-domain. For example, the stability boundary in the z-domain is the unit circle, not the imaginary axis. The characterization of the time response of a discrete-time system based on z-domain information is more complex than the analogous process for continuous-time systems based on the s-domain information discussed in Chapter 5. These are factors that slightly complicate z-domain design, although the associated difficulties are not insurmountable.

The specifications for z-domain design are similar to those for s-domain design. Typical design specifications are as follows:

Time constant. This is the time constant of exponential decay for the continuous envelope of the sampled waveform. The sampled signal is therefore not necessarily equal to a specified portion of the final value after one time constant. The time constant is defined as

$$\tau = \frac{1}{\zeta\omega_n} \tag{6.11}$$

Settling time. The settling time is defined as the period after which the envelope of the sampled waveform stays within a specified percentage (usually 2%) of the final value. It is a multiple of the time constant depending on the specified percentage. For a 2% specification, the settling time is given by

$$T_s = \frac{4}{\zeta\omega_n} \tag{6.12}$$

Frequency of oscillations ω_d. This frequency is equal to the angle of the dominant complex conjugate poles divided by the sampling period.

Other design criteria such as the percentage overshoot, the damping ratio, and the undamped natural frequency can also be defined analogously to the continuous case.

6.2.2 Proportional Control Design in the z-Domain

Proportional control involves the selection of a DC gain value that corresponds to a time response satisfying design specifications. As in s-domain design, a satisfactory time response is obtained by tuning the gain to select a dominant closed-loop pair in the appropriate region of the complex plane. Analytical design is possible for low-order systems but is more difficult than its analog counterpart. The following example illustrates the design of proportional digital controllers.

EXAMPLE 6.3

Design a proportional controller for the digital system described in Example 6.2 with a sampling period $T = 0.1$ s to obtain

1. A damped natural frequency of 5 rad/s
2. A time constant of 0.5 s
3. A damping ratio of 0.7

Solution

After some preliminary calculations, the design results can be easily obtained using the **rlocus** command of MATLAB. The following calculations, together with the information provided by a cursor command, allow us to determine the desired closed-loop pole locations:

1. The angle of the pole is $\omega_d T = 5 \times 0.1 = 0.5$ rad or 28.65°.
2. The reciprocal of the time constant is $\zeta \omega_n = 1/0.5 = 2$ rad/s. This yields a pole magnitude of $e^{-\zeta \omega_n T} = 0.82$.
3. The damping ratio given can be used directly to locate the desired pole.

Using MATLAB, we obtain the results shown in Table 6.4. The corresponding sampled step response plots obtained using the command **step** (MATLAB) are shown in Figure 6.8. As expected, the higher gain designs are associated with a low damping ratio and a more oscillatory response.

Table 6.4 results can also be obtained analytically using the characteristic equation for the complex conjugate poles of (6.4). The system's closed-loop characteristic equation is

$$z^2 - 1.5z + K + 0.5 = z^2 - 2\cos(\omega_d T)e^{-\zeta \omega_n T} z + e^{-2\zeta \omega_n T}$$

Table 6.4 Proportional Control Design Results

Design	Gain	ζ	ω_n rad/s
(a)	0.23	0.3	5.24
(b)	0.17	0.4	4.60
(c)	0.10	0.7	3.63

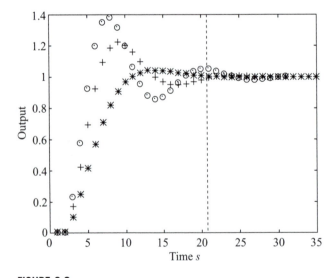

FIGURE 6.8

Time response for the designs of Table 6.4: (a) ⊙, (b) +, (c) *.

Equating coefficients gives the two equations

$$z^1: \quad 1.5 = 2\cos(\omega_d T)e^{-\zeta\omega_n T}$$

$$z^0: \quad K + 0.5 = e^{-2\zeta\omega_n T}$$

1. From the z^1 equation,

$$\zeta\omega_n = \frac{1}{T}\ln\left(\frac{1.5}{2\cos(\omega_d T)}\right) = 10\left|\ln\left(\frac{1.5}{2\cos(0.5)}\right)\right| = 1.571$$

In addition,

$$\omega_d^2 = \omega_n^2(1 - \zeta^2) = 25$$

Hence, we obtain the ratio

$$\frac{\omega_d^2}{(\zeta\omega_n)^2} = \frac{1 - \zeta^2}{\zeta^2} = \frac{25}{(1.571)^2}$$

This gives a damping ratio $\zeta = 0.3$ and an undamped natural frequency $\omega_n = 5.24$ **rad/s**. Finally, the z^0 equation gives a gain

$$K = e^{-2\zeta\omega_n T} - 0.5 = e^{-2\times1.571\times0.1} - 0.5 = 0.23$$

2. From (6.11) and the z^1 equation, we obtain

$$\zeta\omega_n = \frac{1}{\tau} = \frac{1}{0.5} = 2 \text{ rad/s}$$

$$\omega_d = \frac{1}{T}\cos^{-1}\left(\frac{1.5e^{\zeta\omega_n T}}{2}\right) = 10\cos^{-1}(0.75e^{0.2}) = 4.127 \text{ rad/s}$$

Solving for ζ gives

$$\frac{\omega_d^2}{(\zeta\omega_n)^2} = \frac{1 - \zeta^2}{\zeta^2} = \frac{(4.127)^2}{2^2}$$

which gives a damping ratio $\zeta = 0.436$ and an undamped natural frequency $\omega_n = 4.586$ **rad/s**. The gain for this design is

$$K = e^{-2\zeta\omega_n T} - 0.5 = e^{-2\times2\times0.1} - 0.5 = 0.17$$

3. For a damping ratio of 0.7, the z^1 equation obtained by equating coefficients remains nonlinear and is difficult to solve analytically. The equation now becomes

$$1.5 = 2\cos(0.0714\omega_n)e^{-0.07\omega_n}$$

The equation can be solved numerically by trial and error with a calculator to obtain the undamped natural frequency $\omega_n = 3.63$ **rad/s**. The gain for this design is

$$K = e^{-0.14\times3.63} - 0.5 = 0.10$$

This controller can also be designed graphically by drawing the root locus and a segment of the constant ζ spiral and finding their intersection. But the results obtained graphically are often very approximate, and the solution is difficult for all but a few simple root loci.

EXAMPLE 6.4

Consider the vehicle position control system of Example 3.3 with the transfer function

$$G(s) = \frac{1}{s(s+5)}$$

Design a proportional controller for the unity feedback digital control system with analog process and a sampling period $T = 0.04$ s to obtain

1. A steady-state error of 10% due to a ramp input
2. A damping ratio of 0.7

Solution

The analog transfer function together with a DAC and ADC has the z-transfer function

$$G_{ZAS}(z) = \frac{7.4923 \times 10^{-4}(z + 0.9355)}{(z-1)(z-0.8187)}$$

and the closed-loop characteristic equation is

$$1 + KG_{ZAS}(z) = z^2 - (1.8187 - 7.4923 \times 10^{-4} K)z + 0.8187 - 7.009 \times 10^{-4} K$$
$$= z^2 - 2\cos(\omega_d T)e^{-\zeta\omega_n T}z + e^{-2\zeta\omega_n T}$$

The equation involves three parameters ζ, ω_n, and K. As in Example 6.3, equating coefficients yields two equations that we can use to evaluate two unknowns. The third parameter must be obtained from a design specification.

1. The system is type 1, and the velocity error constant is

$$K_v = \frac{1}{T} \frac{z-1}{z} KG(z)\Big|_{z=1}$$

$$= K\frac{7.4923 \times 10^{-4}(1 + 0.9355)}{(0.04)(1 - 0.8187)}$$

$$= \frac{K}{5}$$

This is identical to the velocity error constant for the analog proportional control system. In both cases, a steady-state error due to ramp of 10% is achieved with

$$\frac{K}{5} = K_v = \frac{100}{e(\infty)\%}$$

$$= \frac{100}{10} = 10$$

Hence, the required gain is $K = 50$.

2. As in Example 6.3, the analytical solution for constant ζ is difficult. But the design results are easily obtained using the root locus cursor command of any CAD program. As shown in Figure 6.9, moving the cursor to the $\zeta = 0.7$ contour yields a gain value of approximately 11.7

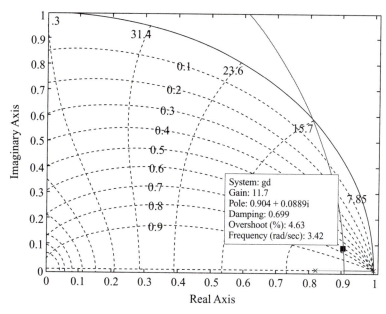

FIGURE 6.9

Root locus for the constant ζ design.

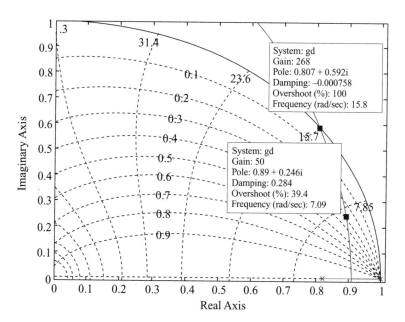

FIGURE 6.10

Root locus for $K = 50$.

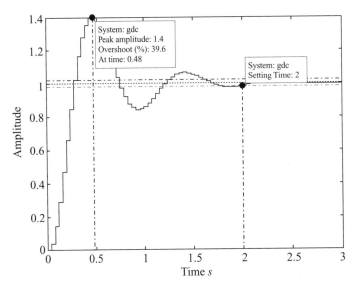

FIGURE 6.11

Time response for $K = 50$.

The root locus of Figure 6.10 shows that the critical gain for the system K_{cr} is approximately 268, and the system is therefore stable at the desired gain and can meet the design specifications for both (1) and (2). However, other design criteria, such as the damping ratio and the undamped natural frequency, should be checked. Their values can be obtained using a CAD cursor command or by equating characteristic equation coefficients as in Example 6.3. For the gain of 50 selected in (1), the root locus of Figure 6.10 and the cursor give a damping ratio of 0.28. This corresponds to the highly oscillatory response of Figure 6.11, which is likely to be unacceptable in practice. For the gain of 11.7 selected in (2), the steady-state error is 42.7% due to a unit ramp. It is therefore clear that to obtain the steady-state error specified together with an acceptable transient response, proportional control is inadequate.

6.3 DIGITAL IMPLEMENTATION OF ANALOG CONTROLLER DESIGN

This section introduces an indirect approach to digital controller design. The approach is based on designing an analog controller for the analog subsystem and then obtaining an equivalent digital controller and using it to digitally implement the desired control. The digital controller can be obtained using a number of recipes that are well known in the field of signal processing, where they are used in the design of digital filters. In fact, a controller can be viewed as a filter that attenuates some dynamics and accentuates others so as to obtain the desired time response. We limit our discussion of digital filters and the comparison of various

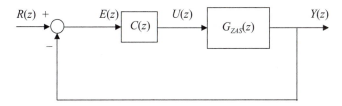

FIGURE 6.12

Block diagram of a single-loop digital control system.

recipes for obtaining them from analog filters to differencing methods and bilinear transformation. The system configuration we consider is shown in Figure 6.12. The system includes (1) a z-transfer function model of a DAC, analog subsystem, and ADC; and (2) a cascade controller. We begin with a general procedure to obtain a digital controller using analog design.

Procedure 6.1

1. Design a controller $C_a(s)$ for the analog subsystem to meet the desired design specifications.
2. Map the analog controller to a digital controller $C(z)$ using a suitable transformation.
3. Tune the gain of the transfer function $C(z)G_{ZAS}(z)$ using proportional z-domain design to meet the design specifications.
4. Check the sampled time response of the digital control system and repeat steps 1 to 3, if necessary, until the design specifications are met.

Step 2 of Procedure 6.1—that is, the transformation from an analog to a digital filter—must satisfy the following requirements:

1. A stable analog filter (poles in the left half plane (LHP) must transform to a stable digital filter.
2. The frequency response of the digital filter must closely resemble the frequency response of the analog filter in the frequency range $0 \rightarrow \omega_s/2$ where ω_s is the sampling frequency.

Most filter transformations satisfy these two requirements to varying degrees. However, this is not true of all analog-to-digital transformations, as illustrated by the following section.

6.3.1 Differencing Methods

An analog filter can be represented by a transfer function or differential equation. Numerical analysis provides standard approximations of the derivative so as to obtain the solution to a differential equation. The approximations reduce a differential equation to a difference equation and could thus be used to obtain the difference equation of a digital filter from the differential equation of an analog filter. We examine two approximations of the derivative: forward differencing and backward differencing.

Forward Differencing

The forward differencing approximation of the derivative is

$$\dot{y}(k) \cong \frac{1}{T}[y(k+1) - y(k)] \qquad (6.13)$$

The approximation of the second derivative can be obtained by applying (6.13) twice, that is,

$$\ddot{y}(k) \cong \frac{1}{T}[\dot{y}(k+1) - \dot{y}(k)]$$

$$\cong \frac{1}{T}\left\{\frac{1}{T}[y(k+2) - y(k+1)] - \frac{1}{T}[y(k+1) - y(k)]\right\} \qquad (6.14)$$

$$= \frac{1}{T^2}\{y(k+2) - 2y(k+1) + y(k)\}$$

Approximations of higher-order derivatives can be similarly obtained. Alternatively, one may consider the Laplace transform of the derivative and the z-transform of the difference in (6.13). This yields the mapping

$$sY(s) \rightarrow \frac{1}{T}[z-1]Y(z) \qquad (6.15)$$

Therefore, the direct transformation of an s-transfer function to a z-transfer function is possible using the substitution

$$s \rightarrow \frac{z-1}{T} \qquad (6.16)$$

EXAMPLE 6.5: FORWARD DIFFERENCE

Apply the forward difference approximation of the derivative to the second-order analog filter

$$C_a(s) = \frac{\omega_n^2}{s^2 + 2\zeta\omega_n s + \omega_n^2}$$

and examine the stability of the resulting digital filter for a stable analog filter.

Solution

The given filter is equivalent to the differential equation

$$\ddot{y}(t) + 2\zeta\omega_n \dot{y}(t) + \omega_n^2 y(t) = \omega_n^2 u(t)$$

where $y(t)$ is the filter output and $u(t)$ is the filter input. The approximation of the first derivative by (6.13) and the second derivative by (6.14) gives the difference equation

$$\frac{1}{T^2}\{y(k+2)-2y(k+1)+y(k)\}+2\zeta\omega_n\frac{1}{T}[y(k+1)-y(k)]+\omega_n^2 y(k)=\omega_n^2 u(k)$$

Multiplying by T^2 and rearranging terms, we obtain the digital filter

$$y(k+2)+2[\zeta\omega_n T-1]y(k+1)+\left[(\omega_n T)^2-2\zeta\omega_n T+1\right]y(k)=(\omega_n T)^2 u(k)$$

Equivalently, we obtain the transfer function of the filter using the simpler transformation (6.16)

$$C(z)=\frac{\omega_n^2}{s^2+2\zeta\omega_n s+\omega_n^2}\bigg|_{s=\frac{z-1}{T}}$$

$$=\frac{(\omega_n T)^2}{z^2+2[\zeta\omega_n T-1]z+\left[(\omega_n T)^2-2\zeta\omega_n T+1\right]}$$

For a stable analog filter, we have $\zeta>0$ and $\omega_n>0$ (positive denominator coefficients are sufficient for a second-order polynomial). However, the digital filter is unstable if the magnitude of the constant term in its denominator polynomial is greater than unity. This gives the instability condition

$$(\omega_n T)^2-2\zeta\omega_n T+1>1$$

$$\text{i.e.,}\quad \zeta<\omega_n T/2$$

For example, a sampling period of 0.2 s and an undamped natural frequency of 10 rad/s yield unstable filters for any underdamped analog filter.

Backward Differencing

The backward differencing approximation of the derivative is

$$\dot{y}(k)\cong\frac{1}{T}[y(k)-y(k-1)] \tag{6.17}$$

The approximation of the second derivative can be obtained by applying (6.17) twice, that is,

$$\ddot{y}(k)\cong\frac{1}{T}[\dot{y}(k)-\dot{y}(k-1)]$$

$$\cong\frac{1}{T}\left\{\frac{1}{T}[y(k)-y(k-1)]-\frac{1}{T}[y(k-1)-y(k-2)]\right\} \tag{6.18}$$

$$=\frac{1}{T^2}\{y(k)-2y(k-1)+y(k-2)\}$$

Approximations of higher-order derivatives can be similarly obtained. One may also consider the Laplace transform of the derivative and the z-transform of the difference in (6.17). This yields the substitution

$$s\to\frac{z-1}{zT} \tag{6.19}$$

EXAMPLE 6.6: BACKWARD DIFFERENCE

Apply the backward difference approximation of the derivative to the second-order analog filter

$$C_a(s) = \frac{\omega_n^2}{s^2 + 2\zeta\omega_n s + \omega_n^2}$$

and examine the stability of the resulting digital filter for a stable analog filter.

Solution

We obtain the transfer function of the filter using (6.19)

$$C(z) = \frac{\omega_n^2}{s^2 + 2\zeta\omega_n s + \omega_n^2}\Bigg|_{s=\frac{z-1}{zT}}$$

$$= \frac{(\omega_n Tz)^2}{\left[(\omega_n T)^2 + 2\zeta\omega_n T + 1\right]z^2 - 2[\zeta\omega_n T + 1]z + 1}$$

The stability conditions for the digital filter are (see Chapter 4)

$$\left[(\omega_n T)^2 + 2\zeta\omega_n T + 1\right] + 2[\zeta\omega_n T + 1] + 1 > 0$$

$$\left[(\omega_n T)^2 + 2\zeta\omega_n T + 1\right] - 1 > 0$$

$$\left[(\omega_n T)^2 + 2\zeta\omega_n T + 1\right] - 2[\zeta\omega_n T + 1] + 1 > 0$$

The conditions are all satisfied for $\zeta > 0$ and $\omega_n > 0$, that is, for all stable analog filters.

6.3.2 Bilinear Transformation

The relationship

$$s = c\frac{z-1}{z+1} \tag{6.20}$$

with a linear numerator and a linear denominator and a constant scale factor c is known as a bilinear transformation. The relationship can be obtained from the equality $z = e^{sT}$ using the first-order approximation

$$s = \frac{1}{T}\ln(z) \cong \frac{2}{T}\left[\frac{z-1}{z+1}\right] \tag{6.21}$$

where the constant $c = 2/T$. A digital filter $C(z)$ is obtained from an analog filter $C_a(s)$ by the substitution

$$C(z) = C_a(s)\Bigg|_{s=c\left[\frac{z-1}{z+1}\right]} \tag{6.22}$$

The resulting digital filter has the frequency response

$$C\left(e^{j\omega T}\right) = C_a(s)\Bigg|_{s=c\left[\frac{e^{j\omega T}-1}{e^{j\omega T}+1}\right]}$$

$$= C_a\left(c\left[\frac{e^{j\omega T/2} - e^{-j\omega T/2}}{e^{j\omega T/2} + e^{-j\omega T/2}}\right]\right)$$

Thus, the frequency responses of the digital and analog filters are related by

$$C\left(e^{j\omega T}\right) = C_a\left(jc\tan\left[\frac{\omega T}{2}\right]\right) \tag{6.23}$$

Evaluating the frequency response at the folding frequency $\omega_s/2$ gives

$$C\left(e^{j\omega_s T/2}\right) = C_a\left(jc\tan\left[\frac{\omega_s T}{4}\right]\right)$$

$$= C_a\left(jc\tan\left[\frac{2\pi}{4}\right]\right) = C_a(j\infty)$$

We observe that bilinear mapping squeezes the entire frequency response of the analog filter for a frequency range $0 \rightarrow \infty$ into the frequency range $0 \rightarrow \omega_s/2$. This implies the absence of aliasing (which makes the bilinear transformation a popular method for digital filter design) but also results in distortion or warping of the frequency response. The relationship between the frequency ω_a of the analog filter and the frequency ω of the digital filter for the case $c = 2/T$, namely,

$$\omega_a = \frac{2}{T}\tan\left(\frac{\omega T}{2}\right)$$

is plotted in Figure 6.13 for $T = 1$. Note that, in general, if the sampling time is sufficiently small so that $\omega \ll \pi/T$, then

$$\tan\left(\frac{\omega T}{2}\right) \cong \frac{\omega T}{2}$$

and therefore $\omega_a \cong \omega$, so that the effect of the warping is negligible.

In any case, the distortion of the frequency response can be corrected at a single frequency ω_0 using the *prewarping* equality

$$C\left(e^{j\omega_0 T}\right) = C_a\left(jc\tan\left[\frac{\omega_0 T}{2}\right]\right) = C_a(j\omega_0) \tag{6.24}$$

The equality holds provided that the constant c is chosen as

$$c = \frac{\omega_0}{\tan\left(\frac{\omega_0 T}{2}\right)} \tag{6.25}$$

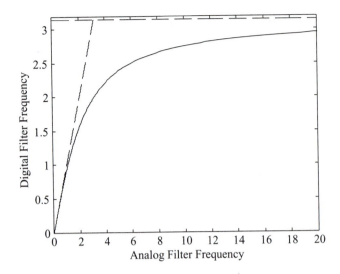

FIGURE 6.13

Relationship between analog filter frequencies and digital filter frequencies with bilinear transformation.

The choice of the prewarping frequency ω_0 depends on the mapped filter. In control applications, a suitable choice of ω_0 is the 3-dB frequency for a PI or PD controller and the upper 3-dB frequency for a PID controller. This is explored further in design examples.

In MATLAB, the bilinear transformation is accomplished using the following command:

$$\texttt{>> gd = c2d(g,tc,'tustin')}$$

where **g** is the analog system and **tc** is the sampling time. If prewarping is requested at a frequency **w**, then the command is

$$\texttt{>> gd = c2d(g,tc,'prewarp',w)}$$

EXAMPLE 6.7

Design a digital filter by applying the bilinear transformation to the analog filter

$$C_a(s) = \frac{1}{0.1s + 1} \tag{6.26}$$

with $T = 0.1$ s. Examine the warping effect and then apply prewarping at the 3-dB frequency.

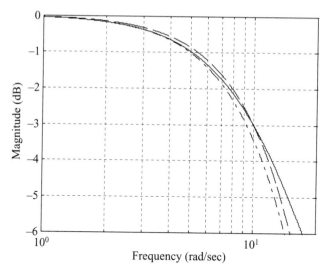

FIGURE 6.14

Bode plots of the analog filter (*solid*) and the digital filter obtained without (*dashed*) and with prewarping (*dash-dot*).

Solution

By applying the bilinear transformation (6.21) to (6.26), we obtain

$$C(z) = \frac{1}{0.1\dfrac{2}{0.1}\dfrac{z-1}{z+1}+1} = \frac{z+1}{3z-1}$$

The Bode plots of $C_a(s)$ (solid line) and $C(z)$ (dashed line) are shown in Figure 6.13, where the warping effect can be evaluated. We select the 3-dB frequency $\omega_0 = 10$ as a prewarping frequency and apply (6.25) to obtain

$$C(z) = \frac{1}{0.1\dfrac{10}{\tan\left(\dfrac{10\cdot 0.1}{2}\right)}\dfrac{z-1}{z+1}+1} \cong \frac{0.35z+0.35}{z-0.29}$$

The corresponding Bode plot is shown again in Figure 6.14 (dash-dot line). It coincides with the Bode plot of $C(s)$ at $\omega_0 = 10$. Note that for lower values of the sampling time, the three Bode plots tend to coincide.

Another advantage of bilinear transformation is that it maps points in the LHP to points inside the unit circle and thus guarantees the stability of a digital filter

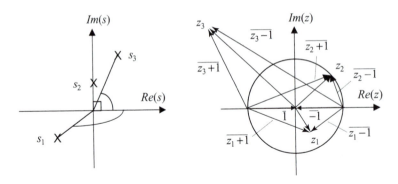

FIGURE 6.15

Angles associated with bilinear transformation.

for a stable analog filter. To verify this property, consider the three cases shown in Figure 6.15. They represent the mapping of a point in the LHP, a point in the RHP, and a point on the $j\omega$ axis. The angle of s after bilinear transformation is

$$\angle s = \angle(z-1) - \angle(z+1) \tag{6.27}$$

For a point inside the unit circle, the angle of s is of a magnitude greater than $90°$, which corresponds to points in the LHP. For a point on the unit circle, the angle is $\pm 90°$, which corresponds to points on the imaginary axis. And for points outside the unit circle, the magnitude of the angle is less than $90°$, which corresponds to points in the RHP.

Bilinear transformation of the analog PI controller gives the following digital PI controller:

$$C(z) = K \frac{(s+a)}{s}\bigg|_{s=c\left[\frac{z-1}{z+1}\right]}$$

$$= K\left(\frac{a+c}{c}\right)\frac{z + \left(\frac{a-c}{a+c}\right)}{z-1} \tag{6.28}$$

The digital PI controller increases the type of the system by one and can therefore be used to improve steady-state error. As in the analog case, it has a zero that reduces the deterioration of the transient response due to the increase in system type. The PI controller of (6.28) has a numerator order equal to its denominator order. Hence, the calculation of its output from its difference equation requires knowledge of the input at the current time. Assuming negligible computational time, the controller is approximately realizable.

Bilinear transformation of the analog PD controller gives the digital PD controller

$$C(z) = K(s+a)\Big|_{s=c\left[\frac{z-1}{z+1}\right]}$$

$$= K(a+c)\frac{z+\left(\dfrac{a-c}{a+c}\right)}{z+1} \tag{6.29}$$

This corresponds to a zero that can be used to improve the transient response and a pole at $z = -1$. A pole at $z = -1$ corresponds to an unbounded frequency response at the folding frequency, as $e^{j\omega_s T/2} = e^{j\pi} = -1$ and must therefore be eliminated. However, eliminating the undesirable pole would result in an unrealizable controller. An approximately realizable PD controller is obtained by replacing the pole at $z = -1$ with a pole at the origin to obtain

$$C(z) = K(a+c)\frac{z+\left(\dfrac{a-c}{a+c}\right)}{z} \tag{6.30}$$

A pole at the origin is associated with a term that decays as rapidly as possible so as to have the least effect on the controller dynamics. However, this variation from direct transformation results in additional distortion of the analog filter and complication of the digital controller design.

Bilinear transformation of the analog PID controller gives the digital PD controller

$$C(z) = K\frac{(s+a)(s+b)}{s}\Big|_{s=c\left[\frac{z-1}{z+1}\right]}$$

$$= K\frac{(a+c)(b+c)}{c}\frac{\left[z+\left(\dfrac{a-c}{a+c}\right)\right]\left[z+\left(\dfrac{b-c}{b+c}\right)\right]}{(z+1)(z-1)}$$

The controller has two zeros that can be used to improve the transient response and a pole at $z = 1$ to improve the steady-state error. As with PD control, the pole at $z = -1$ is replaced by a pole at the origin to yield a transfer function with a bounded frequency response at the folding frequency. The resulting transfer function is approximately realizable and is given by

$$C(z) = K\frac{(a+c)(b+c)}{c}\frac{\left[z+\left(\dfrac{a-c}{a+c}\right)\right]\left[z+\left(\dfrac{b-c}{b+c}\right)\right]}{z(z-1)} \tag{6.31}$$

Using Procedure 6.1 and equations (6.28) through (6.31), digital PI, PD, and PID controllers can be designed to yield satisfactory transient and steady-state performance.

EXAMPLE 6.8

Design a digital controller for the DC motor speed control system described in Example 3.6, where the (type 0) analog plant has the transfer function

$$G(s) = \frac{1}{(s+1)(s+10)}$$

to obtain (1) a zero steady-state error due to a unit step, (2) a damping ratio of 0.7, and (3) a settling time of about 1 s.

Solution

The design is completed following Procedure 6.1. First, an analog controller is designed for the given plant. For zero steady-state error due to unit step, the system type must be increased by one. A PI controller effects this increase, but the location of its zero must be chosen so as to obtain an acceptable transient response. The simplest possible design is obtained by pole-zero cancellation and is of the form

$$C_a(s) = K \frac{s+1}{s}$$

The corresponding loop gain is

$$C_a(s)G(s) = \frac{K}{s(s+10)}$$

Hence, the closed-loop characteristic equation of the system is

$$s(s+10) + K = s^2 + 2\zeta\omega_n s + \omega_n^2$$

Equating coefficients gives $\zeta\omega_n = 5$ **rad/s** and the settling time

$$T_s = \frac{4}{\zeta\omega_n} = \frac{4}{5} = 0.8\,\text{s}$$

as required. The damping ratio of the analog system can be set equal to 0.7 by appropriate choice of the gain K. The gain selected at this stage must often be tuned after filter transformation to obtain the same damping ratio for the digital controller. We solve for the undamped natural frequency

$$\omega_n = 10/(2\zeta) = 10/(2 \times 0.7) = 7.142\,\text{rad/s}$$

The corresponding analog gain is

$$K = \omega_n^2 = 51.02$$

We therefore have the analog filter

$$C_a(s) = 51.02 \frac{s+1}{s}$$

Next, we select a suitable sampling period for an undamped natural frequency of about 7.14 rad/s. We select $T = 0.02$ s $< 2\pi/(40\omega_d)$, which corresponds to a sampling frequency higher than 10 times the damped natural frequency (see Chapter 2). The model of the analog plant together with an ADC and sampler is

$$G_{ZAS}(z) = (1 - z^{-1}) \mathscr{Z} \left\{ \frac{G(s)}{s} \right\}$$

$$= 1.8604 \times 10^{-4} \frac{z + 0.9293}{(z - 0.8187)(z - 0.9802)}$$

Bilinear transformation of the PI controller, with gain K included as a free parameter, gives

$$C(z) = 1.01K \frac{z - 0.9802}{z - 1}$$

Because the analog controller was obtained using pole-zero cancellation, near pole-zero cancellation occurs when the digital controller $C(z)$ is multiplied by $G_{ZAS}(z)$. The gain can now be tuned for a damping ratio of 0.7 using a CAD package with the root locus of the loop gain $C(z)G_{ZAS}(z)$. From the root locus, shown in Figure 6.16, at $\zeta = 0.7$, the gain K is about 46.7, excluding the 1.01 gain of $C(z)$ (i.e., a net gain of 47.2). The undamped natural frequency is $\omega_n = 6.85$ **rad/s**. This yields the approximate settling time

$$T_s = \frac{4}{\zeta \omega_n} = \frac{4}{6.85 \times 0.7}$$

$$= 0.83 \, \text{s}$$

The settling time is acceptable but is slightly worse than the settling time for the analog controller. The step response of the closed-loop digital control system shown in Figure 6.17

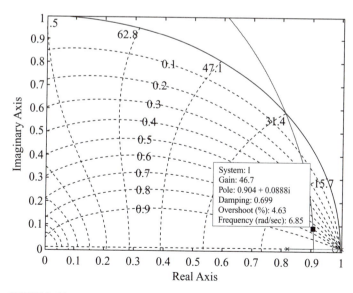

FIGURE 6.16

Root locus for PI design.

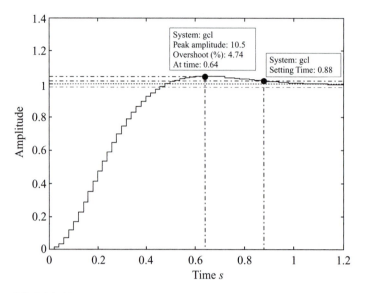

FIGURE 6.17

Step response for PI design with $K = 46.2$.

is also acceptable and confirms the estimated settling time. The net gain value of 47.2, which meets the design specifications, is significantly less than the gain value of 51.02 for the analog design. This demonstrates the need for tuning the controller gain after mapping the analog controller to a digital controller.

EXAMPLE 6.9

Design a digital controller for the DC motor position control system described in Example 3.6, where the (type 1) analog plant has the transfer function

$$G(s) = \frac{1}{s(s+1)(s+10)}$$

to obtain a settling time of about 1 second and a damping ratio of 0.7.

Solution
Using Procedure 6.1, we first observe that an analog PD controller is needed to improve the system transient response. Pole-zero cancellation yields the simple design

$$C_a(s) = K(s+1)$$

We can solve for the undamped natural frequency analytically, or we can use a CAD package to obtain the values $K = 51.02$ and $\omega_n = 7.143$ rad/s for $\zeta = 0.7$.

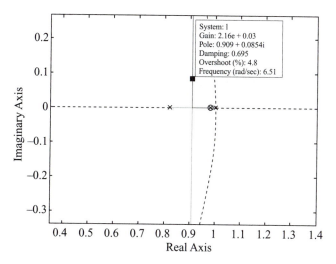

FIGURE 6.18

Root locus for PD design.

A sampling period of 0.02 s is appropriate because it is less than $2\pi/(40\omega_d)$. The plant with the ADC and DAC has the z-transfer function

$$G_{ZAS}(z) = (1 - z^{-1}) \mathscr{Z}\left\{\frac{G(s)}{s}\right\}$$

$$= 1.2629 \times 10^{-6} \frac{(z + 0.2534)(z + 3.535)}{(z - 1)(z - 0.8187)(z - 0.9802)}$$

Bilinear transformation of the PD controller gives

$$C(z) = K\frac{z - 0.9802}{z} = K(1 - 0.9802z^{-1})$$

The root locus of the system with PD control (Figure 6.18) gives a gain K of 2,160 and an undamped natural frequency of 6.79 rad/s at a damping ratio $\zeta = 0.7$. The settling time for this design is about

$$T_s = \frac{4}{\zeta\omega_n} = \frac{4}{0.7 \times 6.51} = 0.88\,s$$

which meets the design specifications.

Checking the step response with MATLAB gives Figure 6.19 with a settling time of 0.94 s, a peak time of 0.6 s, and 5% overshoot. The time response shows a slight deterioration from the characteristics of the analog system but meets all the design specifications. In some cases, the deterioration may necessitate repeatedly modifying the digital design or modifying the analog design and then mapping it to the z-domain until the resulting digital filter meets the desired specifications.

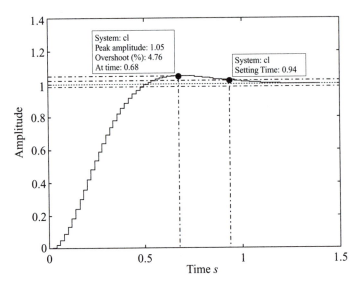

FIGURE 6.19

Time step response for PD design with $K = 2{,}160$.

Note that for a prewarping frequency $\omega_0 = 1$ **rad/s**, the 3-dB frequency of the PD controller, $\omega_0 T = 0.02$ **rad** and $\tan(\omega_0\ T/2) = \tan(0.01) \cong 0.01$. Hence, equation (6.25) is approximately valid and prewarping has a negligible effect on the design.

EXAMPLE 6.10

Design a digital controller for the speed control system of Example 3.6, where the analog plant has transfer function

$$G(s) = \frac{1}{(s+1)(s+3)}$$

to obtain a time constant of less than 0.3 second, a dominant pole damping ratio of at least 0.7, and zero steady-state error due to a step input.

Solution
The root locus of the analog system is shown in Figure 6.20. To obtain zero steady-state error due to a step input, the system type must be increased to one by adding an integrator in the forward path. However, adding an integrator results in significant deterioration of the time response or in instability. If the pole at −1 is canceled, the resulting system is stable but has $\zeta\omega_n = 1.5$—that is, a time constant of 2/3 s and not less than 0.3 s as specified. Using a PID controller provides an additional zero that can be used to stabilize the system and satisfy the remaining design requirements.

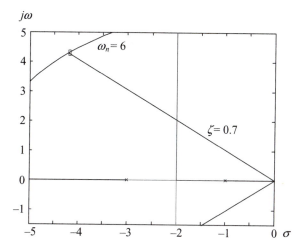

FIGURE 6.20

Root locus for the analog system for PD design.

For a time constant τ of 0.3 s, we have $\zeta\omega_n = 1/\tau \geq 3.33$ rad/s. A choice of $\zeta = 0.7$, ω_n of about 6 rad/s meets the design specifications. The design appears conservative, but we choose a larger undamped natural frequency than the minimum needed in anticipation of the deterioration due to adding PI control. We first design a PD controller to meet these specifications using MATLAB. We obtain the controller angle of about 52.4° using the angle condition. The corresponding zero location is

$$a = \frac{6\sqrt{1-(0.7)^2}}{\tan(52.4°)} + (0.7)(6) \cong 7.5$$

The root locus for the system with PD control (Figure 6.21) shows that the system with $\zeta = 0.7$ has ω_n of about 6 rad/s and meets the transient response specifications with a gain of 4.4 and $\zeta\omega_n = 4.2$. Following the PI design procedure, we place the second zero of the PID controller at one-tenth this distance from the $j\omega$ axis to obtain

$$C_a(s) = K\frac{(s+0.4)(s+7.5)}{s}$$

To complete the analog PID design, the gain must be tuned to ensure that $\zeta = 0.7$. Although this step is not needed, we determine the gain $K \approx 5.8$, and $\omega_n = 6.7$ rad/s (Figure 6.22) for later comparison to the actual gain value used in the digital design. The analog design meets the transient response specification with $\zeta\omega_n = 4.69 > 3.33$, and the dynamics allow us to choose a sampling period of 0.025 s ($\omega_s > 50\omega_d$).

The model of the analog plant with DAC and ADC is

$$G_{ZAS}(z) = (1 - z^{-1})\mathscr{Z}\left\{\frac{G(s)}{s}\right\}$$

$$= 1.170 \times 10^{-3}\frac{z + 0.936}{(z - 0.861)(z - 0.951)}$$

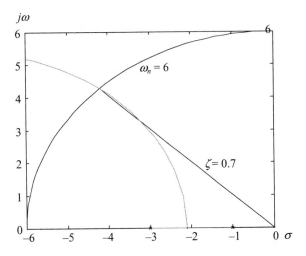

FIGURE 6.21

Root locus of a PD-controlled system.

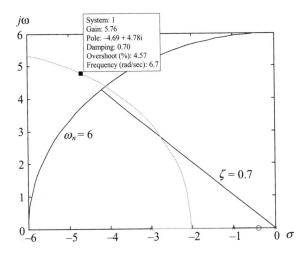

FIGURE 6.22

Root locus of an analog system with PID control.

Bilinear transformation and elimination of the pole at −1 yields the digital PID controller

$$C(z) = 47.975K \frac{(z - 0.684)(z - 0.980)}{z(z - 1)}$$

The root locus for the system with digital PID control is shown in Figure 6.23, and the system is seen to be **minimum phase** (i.e., its zeros are inside the unit circle).

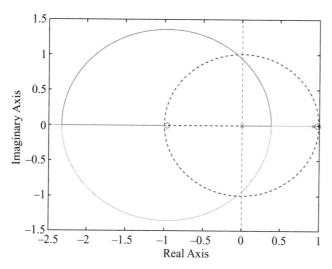

FIGURE 6.23

Root locus of a system with digital PID control.

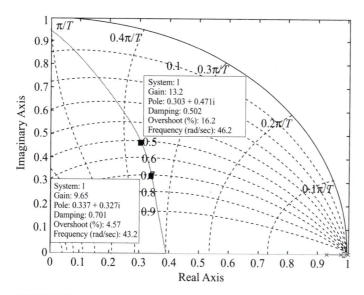

FIGURE 6.24

Detail of the root locus of a system with digital PID control.

For design purposes, we zoom in on the most significant portion of the root locus and obtain the plot of Figure 6.24. With $K = 9.62, 13.2$, the system has $\zeta = 0.7, 0.5$, and $\omega_n = 43.2, 46.2$ rad/s, respectively. Both designs have a sufficiently fast time constant, but the second damping ratio is less than the specified value of 0.7. The time response of the

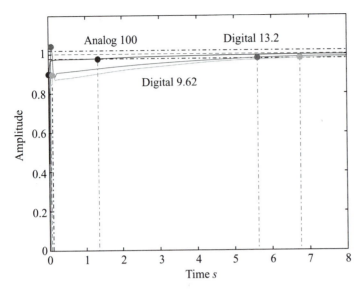

FIGURE 6.25

Time step response for the digital PID design with $K = 9.62$ (*light gray*), $K = 13.2$ (*dark gray*) and for analog design (*black*).

two digital systems and for analog control with $K = 100$ are shown in Figure 6.25. Lower gains give an unacceptably slow analog design. The time response for the high-gain digital design is very fast. However, it has an overshoot of over 4% but has a settling time of 5.63 s. The digital design for $\zeta = 0.7$ has a much slower time response than its analog counterpart.

It is possible to improve the design by trial and error, including redesign of the analog controller, but the design with $\zeta = 0.5$ may be acceptable. One must weigh the cost of redesign against that of relaxing the design specifications for the particular application at hand. The final design must be a compromise between speed of response and relative stability.

6.3.3 Empirical Digital PID Controller Tuning

As explained in Section 5.5, the parameters of a PID controller are often selected by means of tuning rules. This concept can also be exploited to design a digital PID controller. The reader can show (Problem 6.4) that bilinear transformation of the PID controller expression (5.20) yields

$$C(z) = K_p\left(1 + \frac{1}{T_i}\frac{T}{2}\frac{z+1}{z-1} + T_d\frac{2}{T}\frac{z-1}{z+1}\right) \qquad (6.32)$$

If parameters K_p, T_i, and T_d are obtained by means of a tuning rule as in the analog case, then the expression of the digital controller is obtained by substituting in (6.32). The transfer function of a zero-order hold can be approximated by truncating the series expansions as

$$G_{ZOH}(s) = \frac{1-e^{-sT}}{s} \cong \frac{1-1+Ts-(Ts)^2/2+\cdots}{Ts} = 1 - \frac{Ts}{2} + \cdots \cong e^{-\frac{T}{2}s}$$

Thus, the presence of the ZOH can be considered as an additional time delay equal to half of the sampling period. The tuning rules of Table 5.1 can then be applied to a system with a delay equal to the sum of the process time delay and a delay of $T/2$ due to the zero-order hold.

EXAMPLE 6.11

Design a digital PID controller (with sampling time $T = 0.1$) for the analog plant (see Example 5.9)

$$G(s) = \frac{1}{(s+1)^4} e^{-0.2s}$$

by applying the Ziegler-Nichols tuning rules of Table 5.1.

Solution
A first-order-plus-dead-time model of the plant was obtained in Example 5.9 by applying the tangent method. We obtained a process gain $K = 1$, a process dominant time constant $\tau = 3$, and an apparent time delay $L = 1.55$. The apparent time delay for digital control obtained by adding half of the value of the sampling period is $L = 1.6$. The application of the tuning rules shown earlier in Table 5.1 yields

$$K_p = 1.2\frac{\tau}{KL} = 2.25$$

$$T_i = 2L = 3.2$$

$$T_d = 0.5L = 0.8$$

Thus, the digital PID controller has the transfer function

$$C(z) = 2.25\left(1 + \frac{1}{3.2}\frac{0.1}{2}\frac{z+1}{z-1} + 0.8\frac{2}{0.1}\frac{z-1}{z+1}\right)$$
$$= \frac{38.29z^2 - 71.93z + 33.79}{z^2 - 1}$$

The response of the digital control system due to a unit step input applied to the set point at time $t = 0$ and to a unit step change in the control variable at time $t = 50$ is shown

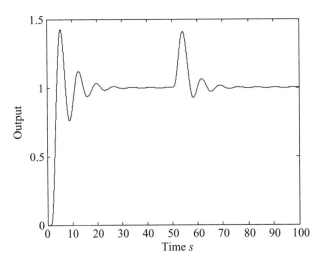

FIGURE 6.26

Process output with the digital PID controller tuned with the Ziegler-Nichols method.

in Figure 6.26. The response is similar to the result obtained with the analog PID controller (see Example 5.9).

6.4 DIRECT z-DOMAIN DIGITAL CONTROLLER DESIGN

Obtaining digital controllers from analog designs involves approximation that may result in significant controller distortion. In addition, the locations of the controller poles and zeros are often restricted to subsets of the unit circle. For example, bilinear transformation of the term $(s + a)$ gives $[z - (c - a)/(c + a)]$ as seen from (6.28) through (6.29). This yields only RHP zeros because a is almost always smaller than c. The plant poles are governed by $p_z = e^{p_s T}$, where p_s and p_z are the s-domain and z-domain poles, respectively, and can be canceled with RHP zeros. Nevertheless, the restrictions on the poles and zeros in (6.28) through (6.29) limit the designer's ability to reshape the system root locus.

Another complication in digital approximation of analog filters is the need to have a pole at 0 in place of the pole at -1, as obtained by direct digital transformation, to avoid an unbounded frequency response at the folding frequency. This may result in a significant difference between the digital and analog controllers and may complicate the design process considerably.

Alternatively, it is possible to directly design controllers in the less familiar z-plane. The controllers used are typically of the same form as those discussed in Section 6.3, but the poles of the controllers are no longer restricted as in

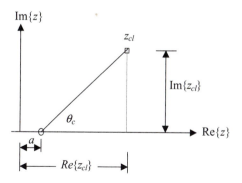

FIGURE 6.27

PD compensator zero.

s-domain-to-z-domain mapping. Thus, the poles can now be in either the LHP or the RHP as needed for design.

Because of the similarity of s-domain and z-domain root loci, Procedures 5.1, 5.2, and 5.3 are applicable with minor changes in the z-domain. However, equation (5.14) is no longer valid because the real and imaginary components of the complex conjugate poles are different in the z-domain. In addition, a digital controller with a zero and no poles is not realizable, and a pole must be added to the controller. To minimize the effect of the pole on the time response, it is placed at the origin. This is analogous to placing an s-plane pole far in the LHP to minimize its effect. We must now derive an expression similar to (5.14) for the digital PID controller.

Using the expression for the complex conjugate z-domain poles (6.3), we obtain (see Figure 6.27)

$$a = \text{Re}\{z_{cl}\} - \frac{\text{Im}\{z_{cl}\}}{\tan(\theta_a)}$$

$$= e^{-\zeta \omega_n T} \cos(\omega_d T) - \frac{e^{-\zeta \omega_n T} \sin(\omega_d T)}{\tan(\theta_a)}$$

$$(6.33)$$

where θ_a is the angle of the controller zero. θ_a is given by

$$\theta_a = \theta_c + \theta_p$$

$$= \theta_c + \theta_{zcl}$$

$$(6.34)$$

In (6.34), θ_{zcl} is the angle of the controller pole and θ_c is the controller angle contribution at the closed-loop pole location. The sign of the second term in (6.33) is negative, unlike (5.14), because the real part of a stable z-domain pole can be positive. In addition, a digital controller with a zero and no poles is not realizable and a pole, at the origin or other locations inside the unit circle, must first be added to the system before computing the controller angle. The computation of the zero location is simple using the following MATLAB function:

```
% Digital PD controller design with pole at origin and zero to be selected.
function [c,zcl]=dpdcon(zeta,wn,g,T)
% g (1) is the uncompensated (compensated) loop gain
% zeta and wn specify the desired closed-loop pole, T = sampling period
% zcl is the closed-loop pole, theta is the angle of the
% compensator zero at zcl.
% The corresponding gain is "k" and the zero is "a".
wdT = wn*T*sqrt(1-zeta^2);    % Pole angle: T*damped natural frequency
rzcl = exp(-zeta*wn*T)*cos(wdT);    % Real part of the closed-loop pole.
izcl = exp(-zeta*wn*T)*sin(wdT);    % Imaginary part of the closed-loop pole.
zcl = rzcl + j*izcl;    % Complex closed-loop pole.
% Find the angle of the compensator zero. Include the contribution
% of the pole at the origin.
theta = pi - angle(evalfr(g,zcl)) - angle(polyval([1,0],zcl));
a = rzcl-izcl/tan(theta);    % Calculate the zero location.
c = zpk([a],[0],1,T);    % Calculate the compensator transfer function.
L = c*g;    % Loop gain.
k = 1/abs( evalfr( 1, zcl)); % Calculate the gain.
c = k*c; %Include the correct gain.
```

Although equation (6.33) is useful in some situations, in many others PD or PID controllers can be more conveniently obtained by simply canceling the slow poles of the system with zeros. Design by pole-zero cancellation is the simplest possible and should be explored before more complex designs are attempted.

The main source of difficulty in z-domain design is the fact that the stable region is now the unit circle as opposed to the much larger left half of the s-plane. In addition, the selection of pole locations in the z-domain directly is less intuitive and is generally more difficult than s-domain pole selection. Pole selection and the entire design process are significantly simplified by the use of CAD tools and the availability of constant ζ contours, constant ω_n contours, and cursor commands.

No new theory is needed to introduce z-domain design, and we proceed directly to design examples. We repeat Examples 6.7, 6.8, and 6.9, using direct digital design to demonstrate its strengths and weaknesses compared to the indirect design approach.

EXAMPLE 6.12

Design a digital controller for the DC motor speed control system of Example 3.6, where the (type 0) analog plant has the transfer function

$$G(s) = \frac{1}{(s+1)(s+10)}$$

to obtain (1) zero steady-state error due to a unit step, (2) a damping ratio of 0.7, and (3) a settling time of about 1 s.

Solution

First, selecting $T = 0.1$ s, we obtain the *z*-transfer function

$$G_{ZAS}(z) = (1 - z^{-1})\mathcal{Z}\left\{\frac{G(s)}{s}\right\} = 3.55 \times 10^{-3}\frac{z + 0.694}{(z - 0.368)(z - 0.905)}$$

To perfectly track a step input, the system type must be increased by one and a PI controller is therefore used. The controller has a pole at $z = 1$ and a zero to be selected to meet the remaining design specifications. Directly canceling the pole at $z = 0.905$ gives a design that is almost identical to that of Example 6.7 and meets the design specifications.

EXAMPLE 6.13

Design a digital controller for the DC motor position control system described in Example 3.6, where the (type 1) analog plant has the transfer function

$$G(s) = \frac{1}{s(s+1)(s+10)}$$

for a settling time of less than 1 s and a damping ratio of 0.7.

Solution

For a sampling period of 0.01 s, the plant, ADC, and DAC have the *z*-transfer function

$$G_{ZAS}(z) = (1 - z^{-1})\mathcal{Z}\left\{\frac{G(s)}{s}\right\}$$

$$= 1.6217 \times 10^{-7}\frac{(z + 0.2606)(z + 3.632)}{(z - 1)(z - 0.9048)(z - 0.99)}$$

Using a digital PD controller improves the system transient response as in Example 6.8. Pole-zero cancellation yields the simple design

$$C(z) = K\frac{z - 0.99}{z}$$

which includes a pole at $z = 0$ to make the controller realizable. The design is almost identical to that of Example 6.8 and meets the desired transient response specifications with a gain of 4,580.

EXAMPLE 6.14

Design a digital controller for the DC motor speed control system described in Example 3.6, where the analog plant has the transfer function

$$G(s) = \frac{1}{(s+1)(s+3)}$$

for a time constant of less than 0.3 second, a dominant pole damping ratio of at least 0.7, and zero steady-state error due to a step input.

Solution

The plant is type 0 and is the same as in Example 6.9 with a sampling period $T = 0.005$ s.

$$G_{ZAS}(z) = (1 - z^{-1}) \mathcal{Z}\left\{\frac{G(s)}{s}\right\} = 1.2417 \times 10^{-5} \frac{z + 0.9934}{(z - 0.9851)(z - 0.995)}$$

For zero steady-state error due to step, the system type must be increased by one by adding a pole at $z = 1$. A controller zero can be used to cancel the pole at $z = 0.995$, leaving the loop gain

$$L(z) = 1.2417 \times 10^{-5} \frac{z + 0.9934}{(z - 1)(z - 0.9851)}$$

The system root locus of Figure 6.28 shows that the closed-loop poles are close to the unit circle at low gains, and the system is unstable at higher gains. Clearly, an additional zero is needed to meet the design specification, and a PID controller is required. To make the controller realizable, a pole must be added at $z = 0$. The simplest design is then to cancel the pole closest to but not on the unit circle, giving the loop gain

$$L(z) = 1.2417 \times 10^{-5} \frac{z + 0.9934}{z(z - 1)}$$

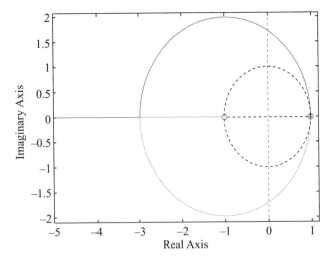

FIGURE 6.28

Root locus for digital PI control.

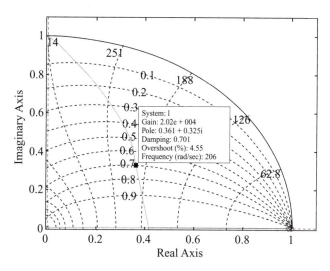

FIGURE 6.29

Root locus for digital PID control design by pole-zero cancellation.

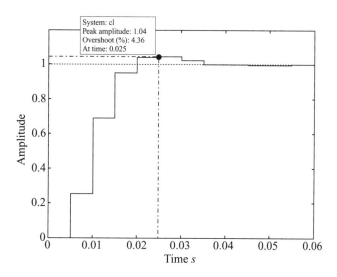

FIGURE 6.30

Time step response for digital PID control.

Adding the zero to the transfer function, we obtain the root locus of Figure 6.29 and select a gain of 20,200. The corresponding time response is shown in Figure 6.30. The time response shows less than 5% overshoot with a fast time response that meets all design specifications. The design is better than that of Example 6.9, where the digital controller was obtained via analog design.

Although it may be possible to improve the analog design to obtain better results than those of Example 6.9, this requires trial and error as well as considerable experience. By contrast, the digital design is obtained here directly in the z-domain without the need for trial and error. This demonstrates that direct design in the z-domain using CAD tools can be easier than indirect design.

6.5 FREQUENCY RESPONSE DESIGN

Frequency response design approaches, especially design based on Bode plots, are very popular in the design of analog control systems. They exploit the fact that the Bode plots of rational loop gains (i.e., ones with no delays) can be approximated with straight lines. Further, if the transfer function is minimum phase, the phase can be determined from the plot of the magnitude, and this allows us to simplify the design of a compensator that provides specified stability properties. These specifications are typically given in terms of the phase margin and the gain margin. As with other design approaches, the design is greatly simplified by the availability of CAD packages.

Unfortunately, discrete transfer functions are not rational functions in $j\omega$ because the frequency is introduced through the substitution $z = e^{j\omega T}$ in the z-transfer function (see Section 2.8). Hence, the simplification provided by methods based on Bode plots for analog systems is lost. A solution to this problem is to bilinearly transform the z-plane into a new plane, called the **w-plane**, where the corresponding transfer function is rational and where the Bode approximation is valid. For this purpose, we recall the bilinear transformation

$$w = c\frac{z-1}{z+1} \quad \text{where} \quad c = \frac{2}{T} \tag{6.35}$$

from Section 6.3.2, which maps points in the LHP into points inside the unit circle. To transform the inside of the unit circle to the LHP, we use the inverse bilinear transformation

$$z = \frac{1 + \dfrac{wT}{2}}{1 - \dfrac{wT}{2}} \tag{6.36}$$

This transforms the transfer function of a system from the z-plane to the w-plane. The w-plane is a complex plane whose imaginary part is denoted by v. To express the relationship between the frequency ω in the s-plane and the frequency v in the w-plane, we let $s = j\omega$ and therefore $z = e^{j\omega T}$. Substituting in (6.35) and using steps similar to those used to derive (6.23), we have

$$w = jv = j\frac{2}{T}\tan\frac{\omega T}{2} \tag{6.37}$$

From (6.37), as ω varies in the interval $[0, \pi/T]$, z moves on the unit circle and v goes from 0 to infinity. This implies that there is a distortion or warping of the frequency scale between v and ω (see Section 6.3.2). This distortion is significant especially at high frequencies. However, if $\omega << \omega_s/2 = \pi/T$, we have from (6.37) that $\omega \approx v$ and the distortion is negligible.

In addition to the problem of frequency distortion, the transformed transfer function $G(w)$ has two characteristics that can complicate the design: (1) The transfer function will always have a pole-zero deficit of zero (i.e., the same number of poles as zeros); (2) The bilinear transformation (6.36) can introduce RHP zeros and result in a nonminimum phase system.

Nonminimum phase systems limit the achievable performance. For example, because some root locus branches start at the open-loop pole and end at zeros, the presence of RHP zeros limits the stable range of gains K. Thus, attempting to reduce the steady-state error or to speed up the response by increasing the system gain can lead to instability. These limitations clearly make w-plane design challenging. Examples 6.15, 6.16, and 6.17 illustrate w-plane design and how to overcome its limitations.

We summarize the steps for controller design in the w-plane in the following procedure.

Procedure 6.2

1. Select a sampling period and obtain a transfer function $G_{ZAS}(z)$ of the discretized process.
2. Transform $G_{ZAS}(z)$ into $G(w)$ using (6.36).
3. Draw the Bode plot of $G(jv)$, and use analog frequency response methods to design a controller $C(w)$ that satisfies the frequency domain specifications.
4. Transform the controller back into the z-plane by means of (6.35), thus determining $C(z)$.
5. Verify that the performance obtained is satisfactory.

In controller design problems, the specifications are often given in terms of the step response of the closed-loop system, such as the settling time and the percentage overshoot. As shown in Chapter 5, the percentage overshoot specification yields the damping ratio ζ, which is then used with the settling time to obtain the undamped natural frequency ω_n. Thus, we need to obtain frequency response specifications from ζ and ω_n to follow Procedure 6.2.

The relationship between the time domain criteria and the frequency domain criteria is in general quite complicated. However, the relationship is much simpler if the closed-loop system can be approximated by the second-order underdamped transfer function

$$T(s) = \frac{\omega_n}{s^2 + 2\zeta\omega_n s + \omega_n^2} \tag{6.38}$$

The corresponding loop gain with unity feedback is given by

$$L(s) = \frac{\omega_n}{s^2 + 2\zeta\omega_n s} \tag{6.39}$$

We substitute $s = j\omega$ to obtain the corresponding frequency response and equate the square of its magnitude to unity

$$|L(j\omega)|^2 = \frac{1}{(\omega/\omega_n)^4 + 4\zeta^2(\omega/\omega_n)^2} = 1 \tag{6.40}$$

The magnitude of the loop gain, as well as its square, is unity at the gain crossover frequency. For the second-order underdamped case, we now have the relation

$$\omega_{gc} = \omega_n \left[\sqrt{4\zeta^4 + 1} - 2\zeta^2 \right]^{1/2} \tag{6.41}$$

Next, we consider the phase margin and derive

$$PM = 180° + \angle G(j\omega_{gc}) = \tan^{-1} \left(\frac{2\zeta}{\left[\sqrt{4\zeta^4 + 1} - 2\zeta^2 \right]^{1/2}} \right)$$

The last expression can be approximated by

$$PM \approx 100\zeta \tag{6.42}$$

Equations (6.41) and (6.42) provide the transformations we need to obtain frequency domain specifications from step response specifications. Together with Procedure 6.2, the equations allow us to design digital control systems using the w-plane approach. The following three examples illustrate w-plane design using Procedure 6.2.

EXAMPLE 6.15

Consider the cruise control system of Example 3.2, where the analog process is

$$G(s) = \frac{1}{s+1}$$

Transform the corresponding $G_{ZAS}(z)$ to the w-plane by considering both $T = 0.1$ and $T = 0.01$. Evaluate the role of the sampling period by analyzing the corresponding Bode plots.

Solution
When $T = 0.1$ we have

$$G_{ZAS}(z) = \frac{0.09516}{z - 0.9048}$$

and by applying (6.36), we obtain

$$G_1(w) = \frac{-0.05w + 1}{w + 1}.$$

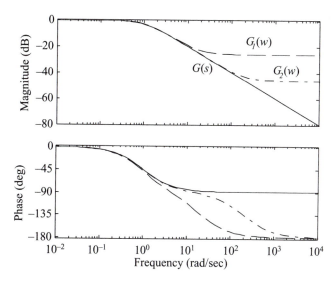

FIGURE 6.31

Bode plots for Example 6.15.

When $T = 0.01$ we have

$$G_{ZAS}(z) = \frac{0.00995}{z - 0.99}$$

and, again by applying (6.36), we obtain

$$G_2(w) = \frac{-0.005w + 1}{w + 1}$$

The Bode plots of $G(s)$, $G_1(w)$, and $G_2(w)$ are shown in Figure 6.31. For both sampling periods, the pole in the w-plane is in the same position as the pole in the s-plane. However, both $G_1(w)$ and $G_2(w)$ have a zero, whereas $G(s)$ does not. This results in a big difference between the frequency response of the analog system and that of the digital systems at the high frequencies. However, the influence of the zero on the system dynamics is clearly more significant when the sampling period is smaller. Note that for both sampling periods, distortion in the low frequency range is negligible. For both systems, the gain as w goes to zero is unity as is the DC gain of the analog system. This is true for the choice $c = 2/T$ in (6.35). Other possible choices are not considered because they do not yield DC gain equality.

EXAMPLE 6.16

Consider the DC motor speed control system of Example 3.6, where the (type 0) analog plant has the transfer function

$$G(s) = \frac{1}{(s+1)(s+10)}$$

Design a digital controller by using frequency response methods to obtain (1) zero steady-state error due to a unit step, (2) an overshoot less than 10%, (3) a settling time of about 1 s.

Solution

From the given specification, we have that the controller in the w-plane must contain a pole at the origin. For 10% overshoot, we calculate the damping ratio as

$$\zeta = \frac{|\ln(0.1)|}{\sqrt{|\ln(0.1)|^2 + \pi^2}} \approx 0.6$$

Using the rule of thumb that the phase margin is about 100 times the damping ratio of the closed-loop system, the required phase margin is about 60 degrees. For a settling time of 1 second, we calculate the undamped natural frequency

$$\omega_n = \frac{4}{\zeta T_s} \approx 6.7 \, \text{rad/s}$$

Using (6.41), we obtain the gain crossover $\omega_{gc} = 4.8$ rad/s.

A suitable sampling period for the selected dynamics is $T = 0.02$ s (see Example 6.8). The discretized process is then determined as

$$G_{ZAS}(z) = (1 - z^{-1}) \mathcal{Z}\left\{\frac{G(s)}{s}\right\} = 1.8604 \times 10^{-4} \frac{z + 0.9293}{(z - 0.8187)(z - 0.9802)}$$

Using (6.36), we obtain the w-plane transfer function

$$G(w) = \frac{-3.6519 \cdot 10^{-6}(w + 2729)(w - 100)}{(w + 9.967)(w + 1)}$$

Note the presence of the two additional zeros (with respect to $G(s)$), which, in any case do not significantly influence the system dynamics in the range of frequencies of interest for the design. The two poles are virtually in the same position as the poles of $G(s)$. The simplest design that meets the desired specifications is to insert a pole at the origin, to cancel the dominant pole at -1, and to increase the gain until the required gain crossover frequency is attained. Thus, the resulting controller transfer function is

$$C(w) = 54 \frac{w + 1}{w}$$

The Bode diagram of the loop transfer function $C(w)G(w)$, together with the Bode diagram of $G(w)$, is shown in Figure 6.32. The figure also shows the phase and gain margins. By transforming the controller back to the z-plane by means of (6.35), we obtain

$$C(z) = \frac{54.54z - 53.46}{z - 1}$$

The corresponding discretized closed-loop step response is plotted in Figure 6.33 and clearly meets the design specifications.

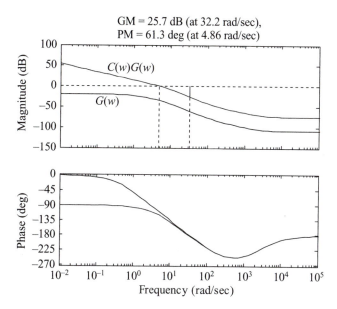

FIGURE 6.32

Bode plots of $C(w)G(w)$ and $G(w)$ for Example 6.16.

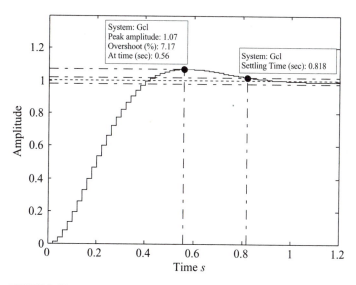

FIGURE 6.33

Discretized closed-loop step response for Example 6.16.

EXAMPLE 6.17

Consider the DC motor speed control system described in Example 3.6 with transfer function

$$G(s) = \frac{1}{(s+1)(s+3)}$$

Design a digital controller using frequency response methods to obtain (1) zero steady-state error due to a unit step, (2) an overshoot less than 10%, (3) a settling time of about 1 s. Use a sampling time of $T = 0.2$ s.

Solution

As in Example 6.15, (1) the given steady-state error specification requires a controller with a pole at the origin, (2) the percentage overshoot specification requires a phase margin of about 60 degrees, and (3) the settling time yields a gain crossover frequency of about 5 rad/s. The transfer function of the system with DAC and ADC is

$$G_{ZAS}(z) = (1 - z^{-1}) \mathscr{Z}\left\{\frac{G(s)}{s}\right\}$$

$$= 0.015437 \frac{z + 0.7661}{(z - 0.8187)(z - 0.5488)}$$

Transforming to the w-plane using (6.36), we obtain

$$G(w) = \frac{-12.819 \cdot 10^{-4}(w + 75.5)(w - 10)}{(w + 2.913)(w + 0.9967)}$$

Note again that the two poles are virtually in the same locations as the poles of $G(s)$. Here the RHP zero at $w = 10$ must be taken into account in the design because the required gain crossover frequency is about 5 rad/s. To achieve the required phase margin, both poles must be canceled with two controller zeros. For a realizable controller, we need at least two controller poles. In addition to the pole at the origin, we select a high-frequency controller pole so as not to impact the frequency response in the vicinity of the crossover frequency. Next, we adjust the system gain to achieve the desired specifications.

As a first attempt, we select a high-frequency pole in $w = -20$ and increase the gain to 78 for a gain crossover frequency of 4 rad/s with a phase margin of 60 degrees. The controller is therefore

$$C(w) = 78\frac{(w + 2.913)(w + 0.9967)}{w(w + 20)}$$

The corresponding open-loop Bode plot is shown in Figure 6.34 together with the Bode plot of $G(w)$. Transforming the controller back to the z-plane using (6.35), we obtain

$$C(z) = \frac{54.54z - 53.46}{z - 1}$$

The resulting discretized closed-loop step response, which satisfies the given requirements, is plotted in Figure 6.35.

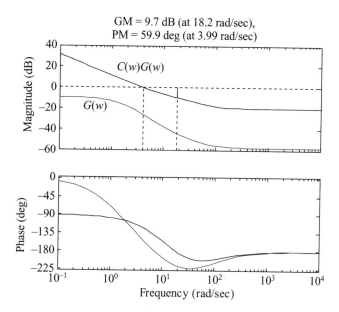

FIGURE 6.34

Bode plots of $C(w)G(w)$ and $G(w)$ for Example 6.17.

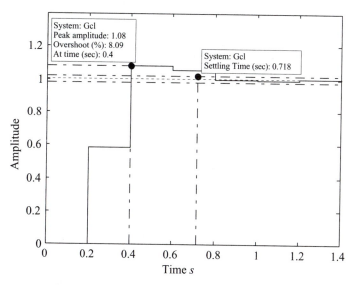

FIGURE 6.35

Discretized closed-loop step response for Example 6.17.

6.6 DIRECT CONTROL DESIGN

In some control applications, the desired transfer function of the closed-loop system is known from the design specification. For a particular system configuration, it is possible to calculate the controller transfer function for a given plant from the desired closed-loop transfer function. This approach to design is known as **synthesis**. Clearly, the resulting controller must be realizable for this approach to yield a useful design.

We consider the Figure 6.12 block diagram, with known closed-loop transfer function $G_{cl}(z)$. The controller transfer function $C(z)$ can be computed analytically, starting from the expression of the desired closed-loop transfer function $G_{cl}(z)$:

$$G_{cl}(z) = \frac{C(z)G_{ZAS}(z)}{1 + C(z)G_{ZAS}(z)}$$

We solve for the controller transfer function

$$C(z) = \frac{1}{G_{ZAS}(z)} \frac{G_{cl}(z)}{1 - G_{cl}(z)} \tag{6.43}$$

For the control system to be implementable, the controller must be causal and must ensure the asymptotic stability of the closed-loop control system. For a causal controller, the closed-loop transfer function $G_{cl}(z)$ must have the same pole-zero deficit as $G_{ZAS}(z)$. In other words, the delay in $G_{cl}(z)$ must be at least as long as the delay in $G_{ZAS}(z)$. We see this by examining (6.43) and observing that the second term on the right side of the equality has as many poles as zeros.

Recall from Chapter 4 that if unstable pole-zero cancellation occurs, the system is input-output stable but not asymptotically stable. This is because the response due to the initial conditions is unaffected by the zeros and is affected by the unstable poles, even if they cancel with a zero. Hence, one must be careful when designing closed-loop control systems to avoid unstable pole-zero cancellations. This implies that the set of zeros of $G_{cl}(z)$ must include all the zeros of $G_{ZAS}(z)$ that are outside the unit circle. Suppose that the process has un unstable pole $z = \bar{z}$, $|\bar{z}| > 1$, namely:

$$G_{ZAS}(z) = \frac{G_1(z)}{z - \bar{z}}$$

In view of equation (6.38), we avoid unstable pole-zero cancellation by requiring

$$1 - G_{cl}(z) = \frac{1}{1 + C(z)\dfrac{G_1(z)}{z - \bar{z}}} = \frac{z - \bar{z}}{z - \bar{z} + C(z)G_1(z)}$$

In other words, $z = \bar{z}$ must be a zero of $1 - G_{cl}(z)$.

An additional condition can be imposed to address state accuracy requirements. In particular, if zero steady-state error is required, the condition must be

$$G_{cl}(1) = 1$$

This condition is easily obtained by applying the final value theorem (see Section 2.8.1). The proof is left as an exercise for the reader.

Summarizing, the conditions required for the choice of $G_{cl}(z)$ are as follows:

- $G_{cl}(z)$ must have the same pole-zero deficit as $G_{ZAS}(z)$ (causality).
- $G_{cl}(z)$ must contain as zeros all the zeros of $G_{ZAS}(z)$ that are outside the unit circle (stability).
- The zeros of $1 - G_{cl}(z)$ must include all the poles of $G_{ZAS}(z)$ that are outside the unit circle (stability).
- $G_{cl}(1) = 1$ (zero steady-state error).

The choice of a suitable closed-loop transfer function is clearly the main obstacle in the application of the direct design method. The correct choice of closed-loop poles and zeros to meet the design requirements is difficult for higher-order systems. In practice, there are additional constraints on the control variable because of actuator limitations. Further, the performance of the control system relies heavily on an accurate process model.

To partially address the first problem, the poles of the continuous-time closed-loop system can be specified based on desired properties of the closed-loop system. For example, the Bessel pole locations shown in Tables 6.5 and 6.6 can be exploited to obtain a closed-loop system step response with a given settling time or a given bandwidth. The following ad hoc procedure often yields the desired transfer function.

Procedure 6.3

1. Select the desired settling time T_s (desired bandwidth ω_d).
2. Select the nth-order pole location from the table and divide it by T_s (multiply it by ω_d).
3. Obtain $G_{cl}(z)$ by converting the s-plane pole location to the z-plane pole location using the transformation $z = e^{sT}$.
4. Verify that $G_{cl}(z)$ meets the conditions for causality, stability, and steady-state error. If not, modify $G_{cl}(z)$ until the conditions are met.

Table 6.5 Bessel Pole Location for $T_s = 1$ s

Order	Poles		
1	-4.6200		
2	$-4.0530 \pm j\, 2.3400$		
3	$-3.9668 \pm j\, 3.7845$	-5.0093	
4	$-4.0156 \pm j\, 5.0723$	$-5.5281 \pm j\, 1.6553$	
5	$-4.1104 \pm j\, 6.3142$	$-5.9268 \pm j\, 3.0813$	-6.4480
6	$-4.2169 \pm j\, 7.5300$	$-6.2613 \pm j\, 4.4018$	$-7.1205 \pm j\, 1.4540$

Table 6.6 Bessel Pole Location for $\omega_d = 1$ rad/s

Order	Poles		
1	-1.0000		
2	$-0.8660 \pm j\,0.5000$		
3	$-0.7455 \pm j\,0.7112$	-0.9420	
4	$-0.6573 \pm j\,0.8302$	$-0.9047 \pm j\,0.2711$	
5	$-0.5906 \pm j\,0.9072$	$-0.8516 \pm j\,0.4427$	-0.9246
6	$-0.5385 \pm j\,0.9617$	$-0.7998 \pm j\,0.5622$	$-0.9093 \pm j\,0.1856$

EXAMPLE 6.18

Design a digital controller for the DC motor speed control system described in Example 3.6, where the (type 0) analog plant has the transfer function

$$G(s) = \frac{1}{(s+1)(s+10)}$$

to obtain (1) zero steady-state error due to a unit step and (2) a settling time of about 4 s. The sampling time is chosen as $T = 0.02$ s.

Solution

As in Example 6.8, the discretized process transfer function is

$$G_{ZAS}(z) = (1 - z^{-1})\mathcal{Z}\left\{\frac{G(s)}{s}\right\} = 1.8604 \times 10^{-4}\,\frac{z + 0.9293}{(z - 0.8187)(z - 0.9802)}$$

Note that there are no poles and zeros outside the unit circle and there are no zeros at infinity.

Then, according to Table 6.5, the desired continuous-time closed-loop transfer function is chosen as second-order with poles at $(-4.0530 \pm j2.3400)/4$ and with gain equal to one so that the requirements on the settling time and on the steady-state error are satisfied, namely:

$$G_d(s) = \frac{1.369}{s^2 + 2.027s + 1.369}$$

Then the desired closed-loop transfer function is obtained using $z = e^{sT}$, namely:

$$G_d(z) = 0.2683 \cdot 10^{-3}\,\frac{z + 1}{z^2 - 1.9597z + 0.9603}$$

By applying (6.38) we have

$$C(z) = \frac{1.422(z - 0.8187)(z - 0.9802)(z + 1)}{(z - 1)(z + 0.9293)(z - 0.96)}$$

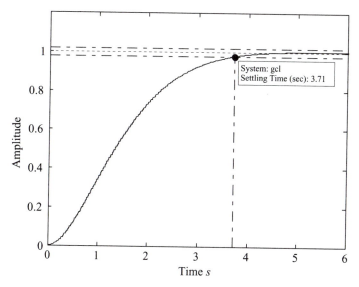

FIGURE 6.36

Step response for the direct design method described in Example 6.18.

We observe that stable pole-zero cancellation has occurred and that an integral term is present, as expected, in the controller because the zero steady-state error condition has been addressed. The digital closed-loop step response is shown in Figure 6.36.

EXAMPLE 6.19

Design a digital controller for the type 0 analog plant

$$G(s) = \frac{-0.16738(s - 9.307)(s + 6.933)}{(s^2 + 0.3311s + 9)}$$

to obtain (1) zero steady-state error due to a unit step, (2) a damping ratio of 0.7, and (3) a settling time of about 1 s. The sampling time is chosen as $T = 0.01$ s.

Solution

The discretized process transfer function is

$$G_{ZAS}(z) = (1 - z^{-1})\mathcal{Z}\left\{\frac{G(s)}{s}\right\} = \frac{-0.16738(z - 1.095)(z - 0.9319)}{z^2 - 1.996z + 0.9967}$$

Note that there is a zero outside the unit circle and this has to be included in the desired closed-loop transfer function, which is therefore selected as

$$G_d(z) = \frac{K(z-1.095)}{z^2 - 1.81z + 0.8269}$$

The closed-loop poles have been selected as in Example 6.8 and the value $K = -0.17789$ results from solving the equation $G_{cl}(1) = 1$. By applying (6.38), we have

$$C(z) = \frac{1.0628(z^2 - 1.81z + 0.8269)}{(z-1)(z-0.9319)(z-0.6321)}$$

The stable pole-zero cancellation is evident again, as well as the presence of the pole at $z = 1$. The digital closed-loop step response is shown in Figure 6.37. The presence of the undershoot is due to the unstable zero that cannot be modified by the direct design method. The achieved settling time is less than the required value, as the presence of the zero has not been considered when selecting the closed-loop poles to meet requirements (2) and (3).

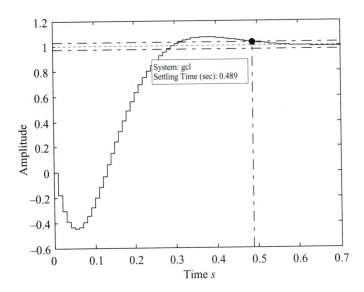

FIGURE 6.37

Step response for the direct design method described Example 6.19.

EXAMPLE 6.20

Design a digital controller for the type 0 analog plant

$$G(s) = \frac{1}{10s + 1} e^{-5s}$$

to obtain (1) zero steady-state error that results from a unit step, (2) a settling time of about 10 s with no overshoot. The sampling time is chosen as $T = 1$ s.

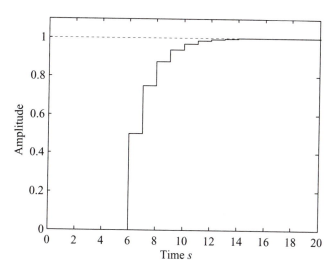

FIGURE 6.38

Step response for the direct design method of Example 6.20.

Solution

The discretized process transfer function is

$$G_{ZAS}(z) = (1 - z^{-1}) \mathscr{Z}\left\{\frac{G(s)}{s}\right\} = \frac{0.09516}{z - 0.9048} z^{-5}$$

To meet the causality requirements, a delay of five sampling periods must be included in the desired closed-loop transfer function. Then the settling time of 10 s (including the time delay) is achieved by considering a closed-loop transfer function with a pole in $z = 0.5$, namely, by selecting

$$G_{cl}(z) = \frac{K}{z - 0.5} z^{-5}$$

Setting $G_{cl}(1) = 1$ yields $K = 0.5$, then applying (6.38) we have

$$C(z) = \frac{5.2543(z - 0.9048) z^5}{z^6 - 0.5z^5 - 0.5}$$

The resulting closed-loop step response is shown in Figure 6.38.

6.7 FINITE SETTLING TIME DESIGN

Continuous-time systems can only reach the desired output asymptotically after an infinite time period. By contrast, digital control systems can be designed to

settle at the reference output after a finite time period and follow it exactly there-
after. By following the direct control design method described in Section 6.6, if
all the poles and zeros of the discrete-time process are inside the unit circle, an
attractive choice can be to select

$$G_{cl}(z) = z^{-k}$$

where k must be greater than or equal to the intrinsic delay of the discretized
process—namely, the difference between the degree of the denominator and
the degree of the numerator of the discrete process transfer function.

Disregarding the time delay, the definition implies that a unit step is tracked
perfectly starting at the first sampling point. From (6.43) we have the **deadbeat
controller**

$$C(z) = \frac{1}{G_{ZAS}(z)}\left[\frac{z^{-k}}{1 - z^{-k}}\right]. \tag{6.44}$$

In this case, the only design parameter is the sampling period T, and the overall
control system design is very simple. However, finite settling time designs may
exhibit undesirable intersample behavior (oscillations) because the control is
unchanged between two consecutive sampling points. Further, the control variable can easily assume values that may cause saturation of the DAC or exceed the
limits of the actuator in a physical system, resulting in unacceptable system behavior. The behavior of finite settling time designs such as the deadbeat controller
must therefore be carefully checked before implementation.

EXAMPLE 6.21

Design a deadbeat controller for the DC motor speed control system described in Example
3.6, where the (type 0) analog plant has transfer function

$$G(s) = \frac{1}{(s+1)(s+10)}$$

and the sampling time is initially chosen as $T = 0.02$ s. Redesign the controller with
$T = 0.1$ s.

Solution
Because the discretized process transfer function is

$$G_{ZAS}(z) = (1 - z^{-1})\mathcal{Z}\left\{\frac{G(s)}{s}\right\}$$

$$= 1.8604 \times 10^{-4} \frac{z + 0.9293}{(z - 0.8187)(z - 0.9802)}$$

we have no poles and zeros outside or on the unit circle and a deadbeat controller can
therefore be designed by setting

$$G_{cl}(z) = z^{-1}$$

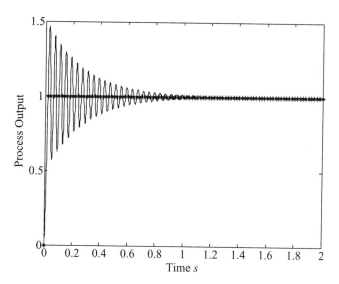

FIGURE 6.39

Sampled and analog step response for the deadbeat control of Example 6.21 ($T = 0.02$ s).

(Note that the difference between the order of the denominator and the order of the numerator is one.) By applying (6.39), we have

$$C(z) = \frac{5375.0533(z - 0.9802)(z - 0.8187)}{(z - 1)(z + 0.9293)}$$

The resulting sampled and analog closed-loop step response is shown in Figure 6.39, whereas the corresponding control variable is shown in Figure 6.40. It appears that, as expected, the sampled process output attains its steady-state value after just one sample—namely, at time $t = T = 0.02$ s, but between samples the output oscillates wildly and the control variable assumes very high values. In other words, the oscillatory behavior of the control variable causes an unacceptable intersample oscillation.

This can be ascertained analytically by considering the block diagram shown in Figure 6.41. Recall that the transfer function of the zero-order hold is

$$G_{ZOH}(s) = \frac{1 - e^{-sT}}{s}$$

Using simple block diagram manipulation, we obtain

$$E^*(s) = \frac{R^*(s)}{1 + (1 - e^{-sT})C^*(s)\left(\frac{G(s)}{s}\right)^*}$$

$$= \frac{R^*(s)}{1 + C^*(s)G^*_{ZAS}(s)}$$

(6.45)

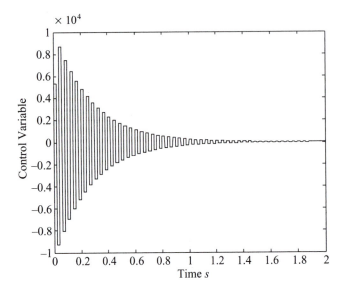

FIGURE 6.40

Control variable for the deadbeat control of Example 6.21 ($T = 0.02$ s).

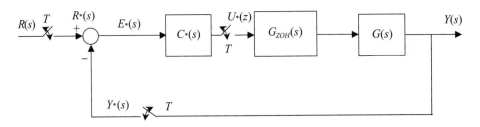

FIGURE 6.41

Block diagram for finite settling time design with analog output.

Thus, the output is

$$Y(s) = \left(\frac{1-e^{-sT}}{s}\right)G(s)C^*(s)\left[\frac{R^*(s)}{1+C^*(s)G_{ZAS}^*(s)}\right] \qquad (6.46)$$

Equations (6.45) and (6.46) are quite general and apply to any system described by the block diagram shown in Figure 6.41. To obtain the specific output pertaining to our example, we substitute $z = e^{sT}$ in the relevant expressions to obtain

$$G_{ZAS}^*(s) = 1.8604 \times 10^{-4} e^{-sT} \frac{1 + 0.9293 e^{-sT}}{(1 - 0.8187 e^{-sT})(1 - 0.9802 e^{-sT})}$$

$$C^*(s) = 5375.0533 \frac{(1 - 0.9802 e^{-sT})(1 - 0.8187 e^{-sT})}{(1 - e^{-sT})(1 + 0.9293 e^{-sT})}$$

$$R^*(s) = \frac{1}{1 - e^{-sT}}$$

Then substituting in (6.41) and simplifying gives

$$Y(s) = 5375.0533 \frac{(1 - 0.9802 e^{-sT})(1 - 0.8187 e^{-sT})}{s(s+1)(s+10)(1 + 0.9293 e^{-sT})}$$

Expanding the denominator using the identity

$$(1 + 0.9293 e^{-sT})^{-1} = 1 + (-0.9293 e^{-sT}) + (-0.9293 e^{-sT})^2 + (-0.9293 e^{-sT})^3 + \ldots$$
$$= 1 - 0.9293 e^{-sT} + 0.8636 e^{-2sT} - 0.8025 e^{-3sT} + \ldots$$

we obtain

$$Y(s) = \frac{5375.0533}{s(s+1)(s+10)}(1 - 2.782 e^{-sT} + 3.3378 e^{-2sT}$$
$$- 2.2993 e^{-3sT} + 0.6930 e^{-4sT} + \ldots)$$

Finally, we inverse Laplace transform to obtain the analog output

$$y(t) = 5375.0533 \left(\frac{1}{10} - \frac{1}{9} e^{-t} + \frac{1}{90} e^{-10t} \right) 1(t) +$$
$$5375.0533(-0.2728 + 0.3031 e^{-(t-T)} - 0.0303 e^{-10(t-T)}) 1(t-T) +$$
$$5375.0533(0.3338 - 0.3709 e^{-(t-2T)} + 0.0371 e^{-10(t-2T)}) 1(t-2T) +$$
$$5375.0533(-0.2299 + 0.2555 e^{-(t-3T)} - 0.0255 e^{-10(t-3T)}) 1(t-3T) +$$
$$5375.0533(0.0693 - 0.0770 e^{-(t-4T)} + 0.0077 e^{-10(t-4T)}) 1(t-4T) + \ldots$$

where $1(t)$ is the unit step function. It is easy to evaluate that, at the sampling points, we have $y(0.02) = y(0.04) = y(0.06) = y(0.08) = \ldots = 1$, but between samples the output oscillates wildly as shown in Figure 6.39. To reduce intersample oscillations, we set $T = 0.1$ s and obtain the transfer function

$$G_{ZAS}(z) = (1 - z^{-1}) \mathcal{Z} \left\{ \frac{G(s)}{s} \right\} = 35.501 \times 10^{-4} \frac{z + 0.6945}{(z - 0.9048)(z - 0.3679)}$$

For $G_c(z) = z^{-1}$ we obtain

$$C(z) = \frac{281.6855(z - 0.9048)(z - 0.3679)}{(z - 1)(z + 0.6945)}$$

   The resulting sampled and analog closed-loop step responses are shown in Figure 6.42, whereas the corresponding control variable is shown in Figure 6.43. The importance of the sampling period selection in the deadbeat controller design is clear. Note that the oscillations and the amplitude of the control signal are both reduced; but it is also evident that the intersample oscillations cannot be avoided completely with this approach.

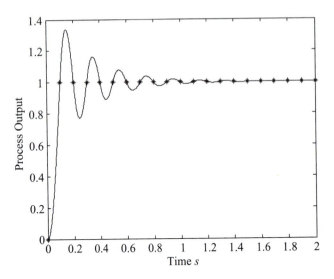

FIGURE 6.42

Sampled and analog step response for the deadbeat control of Example 6.21 ($T = 0.1$ s).

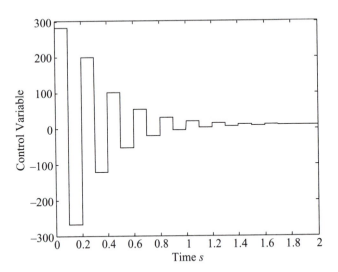

FIGURE 6.43

Control variable for the deadbeat control of Example 6.21 ($T = 0.1$ s).

To avoid intersample oscillations, we maintain the control variable constant after n samples, where n is the degree of the denominator of the discretized process. This is done at the expense of achieving the minimum settling time (by considering the sampled process output) as in the previous examples. By considering the control scheme shown in Figure 6.12, we have

$$U(z) = \frac{Y(z)}{G_{ZAS}(z)} = \frac{Y(z)}{R(z)}\frac{R(z)}{G_{ZAS}(z)} = G_{cl}(z)\frac{R(z)}{G_{ZAS}(z)} \tag{6.47}$$

If we solve this equation with the constraints $G_{cl}(1) = 1$, we obtain the expression of $U(z)$ or, alternatively, we obtain the expression of $G_{cl}(z)$, which yields to the expression of the controller $C(z)$ by applying (6.43). Obviously, the overall design requires the specification a priori of the reference signal to be tracked. In the following examples, a step reference signal is assumed.

EXAMPLE 6.22

Design a ripple-free deadbeat controller for the type 1 vehicle positioning system described in Example 3.3 with transfer function

$$G(s) = \frac{1}{s(s+1)}$$

The sampling time is chosen as $T = 0.1$.

Solution
The discretized process transfer function is

$$G_{ZAS}(z) = (1 - z^{-1})\mathcal{Z}\left\{\frac{G(s)}{s}\right\}$$

$$= \frac{0.0048374(1 + 0.9672z^{-1})z^{-1}}{(1 - z^{-1})(1 - 0.9048z^{-1})}$$

By considering that the z-transform of the step reference signal is

$$R(z) = \frac{1}{1 - z^{-1}}$$

we have

$$U(z) = G_{cl}(z)\frac{R(z)}{G_{ZAS}(z)}$$

$$= G_{cl}(z)\cdot\frac{206.7218(1 - 0.9048z^{-1})}{z^{-1}(1 + 0.9672z^{-1})} \tag{6.48}$$

Because the process is of type 1, we have to require that the control variable be zero after two samples (note that $n = 2$)—namely, that

$$U(z) = a_0 + a_1 z^{-1}$$

A solution of equation (6.48) is therefore

$$G_d(z) = K \cdot z^{-1}(1 + 0.9672 z^{-1})$$

and

$$U(z) = K \cdot 206.7218 (1 - 0.9048 z^{-1})$$

where the value $K = 0.5083$ is found by imposing $G_d(1) = 1$. Thus, by applying (6.38) we have

$$C(z) = \frac{105.1z - 95.08}{z + 0.4917}$$

The resulting sampled and analog closed-loop step response is shown in Figure 6.44, whereas the corresponding control variable is shown in Figure 6.45. Note that the control variable is constant after the second sample and that there is no intersample ripple.

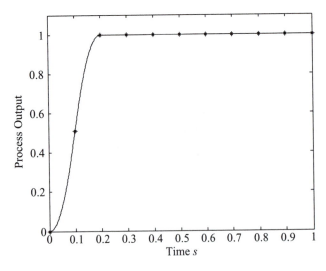

FIGURE 6.44

Sampled and analog step response for the deadbeat control of Example 6.22.

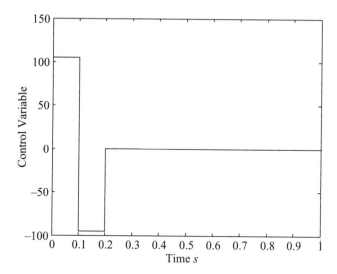

FIGURE 6.45

Control variable for the deadbeat control of Example 6.22.

EXAMPLE 6.23

Design a ripple-free deadbeat controller for the DC motor speed control system described in Example 3.6, where the (type 0) analog plant has the transfer function

$$G(s) = \frac{1}{(s+1)(s+10)}$$

The sampling period is chosen as $T = 0.1$ s.

Solution

For a sampling period of 0.01, the discretized process transfer function is

$$G_{ZAS}(z) = (1 - z^{-1})\mathcal{Z}\left\{\frac{G(s)}{s}\right\}$$

$$= \frac{0.0035501(1 + 0.6945z^{-1})z^{-1}}{(1 - 0.9048z^{-1})(1 - 0.3679z^{-1})}$$

For a sampled unit step input,

$$R(z) = \frac{1}{1 - z^{-1}}$$

we have the control input

$$U(z) = G_{cl}(z)\frac{R(z)}{G_{ZAS}(z)}$$

$$= G_{cl}(z) \cdot \frac{281.6855(1 - 0.9048z^{-1})(1 - 0.3679z^{-1})}{z^{-1}(1 - z^{-1})(1 + 0.6945z^{-1})}$$

By taking into account the delay of one sampling period in $G_{ZAS}(z)$, the control input condition is satisfied with the closed-loop transfer function

$$G_{cl}(z) = K \cdot z^{-1}(1 + 0.6945z^{-1})$$

where the value $K = 0.5901$ is found by imposing $G_{cl}(1) = 1$. Thus, by applying (6.38) we have

$$C(z) = \frac{166.2352(z - 0.9048)(z - 0.3679)}{(z - 1)(z + 0.4099)}$$

The resulting sampled and analog closed-loop step responses are shown in Figure 6.46, whereas the corresponding control variable is shown in Figure 6.47. Note also that in this case there is no intersample ripple.

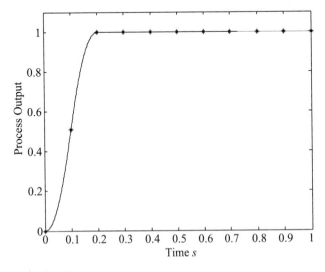

FIGURE 6.46

Sampled and analog step response for the deadbeat control of Example 6.23.

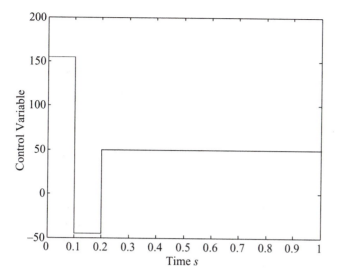

FIGURE 6.47

Control variable for the deadbeat control of Example 6.23.

RESOURCES

Jacquot, R. G., *Modern Digital Control Systems,* Marcel Dekker, 1981.
Kuo, B. C., *Automatic Control Systems,* Prentice Hall, 1991.
Kuo, B. C., *Digital Control Systems,* Saunders, 1992.
Ogata, K., *Digital Control Engineering,* Prentice Hall, 1987.
Oppenheim, A. V., and R. W. Schafer, *Digital Signal Processing,* Prentice Hall, 1975.
Ragazzini, J. R., and G. F. Franklin, *Sampled-Data Control systems,* McGraw-Hill, 1958.

PROBLEMS

6.1 Sketch the z-domain root locus, and find the critical gain for the following systems:

(a) $G(z) = \dfrac{K}{z - 0.4}$

(b) $G(z) = \dfrac{K}{(z + 0.9)(z - 0.9)}$

(c) $G(z) = \dfrac{Kz}{(z - 0.2)(z - 1)}$

(d) $G(z) = \dfrac{K(z + 0.9)}{(z - 0.2)(z - 0.8)}$

6.2 Prove that expression (6.6) describes a constant ω_n contour in the z-plane.

6.3 Hold equivalence is a digital filter design approach that approximates an analog filter using

$$C(z) = \left(\frac{z-1}{z} \right) \mathcal{Z} \left\{ \mathcal{L}^{-1} \left[\frac{C_a(s)}{s} \right]_* \right\}$$

(a) Obtain the hold-equivalent digital filter for the PD, PI, and PID controllers. Modify the results as necessary to obtain a realizable filter with finite frequency response at the folding frequency.

(b) Why are the filters obtained using hold equivalence always stable?

6.4 Show that the bilinear transformation of the PID controller expression (5.20) yields expression (6.32).

6.5 Design proportional controllers for the systems of Problem 6.1 to meet the following specifications where possible. If the design specification cannot be met, explain why and suggest a more appropriate controller.

(a) A damping ratio of 0.7

(b) A steady-state error of 10% due to a unit step

(c) A steady-state error of 10% due to a unit ramp

6.6 Design digital controllers to meet the desired specifications for the systems described in Problems 5.4, 5.7, and 5.8 by bilinearly transforming the analog designs.

6.7 Design a digital filter by applying the bilinear transformation to the analog (Butterworth) filter

$$C_a(s) = \frac{1}{s^2 + \sqrt{2}s + 1}$$

with $T = 0.1$ s. Then apply prewarping at the 3-dB frequency.

6.8 Design a digital PID controller (with $T = 0.1$) for the plant

$$G(s) = \frac{1}{10s + 1} e^{-5s}$$

by applying the Ziegler-Nichols tuning rules presented in Table 5.1.

6.9 Design digital controllers to meet the desired specifications for the systems described in Problems 5.4, 5.7, and 5.8 in the z-domain directly.

6.10 In Example 4.9, we examined the closed-loop stability of the furnace temperature digital control system with proportional control and a sampling period of 0.01 s. We obtained the z-transfer function

$$G_{ZAS}(z) = 10^{-5} \frac{4.95z + 4.901}{z^2 - 1.97z + 0.9704}$$

Design a controller for the system to obtain zero steady-state error due to a step input without significant deterioration in the transient response.

6.11 Consider the DC motor position control system described in Example 3.6, where the (type 1) analog plant has the transfer function

$$G(s) = \frac{1}{s(s+1)(s+10)}$$

and design a digital controller by using frequency response methods to obtain a settling time about 1 and an overshoot less than 5%.

6.12 Use direct control design for the system described in Problem 5.7 (with $T = 0.1$) to design a controller for the transfer function

$$G(s) = \frac{1}{(s+1)(s+5)}$$

to obtain (1) zero steady-state error due to step, (2) a settling time of less than 2 s, and (3) an undamped natural frequency of 5 rad/s. Obtain the discretized and the analog output. Then apply the designed controller to the system

$$G(s) = \frac{1}{(s+1)(s+5)(0.1s+1)}$$

and obtain the discretized and the analog output to verify the robustness of the control system.

6.13 Design a deadbeat controller for the system of Problem 5.7 to obtain perfect tracking of a unit step in minimum finite time. Obtain the analog output for the system, and compare your design to that obtained in Problem 5.7. Then apply the controller to the process

$$G(s) = \frac{1}{(s+1)(s+5)(0.1s+1)}$$

to verify the robustness of the control system.

6.14 Find a solution for Problem 6.13 that avoids intersample ripple.

COMPUTER EXERCISES

6.15 Write a MATLAB function to plot a constant damped natural frequency contour in the z-plane.

6.16 Write a MATLAB function to plot a time-constant contour in the z-plane.

6.17 Write a computer program that estimates a first-order-plus-dead-time transfer function with the tangent method and determines the digital PID parameters according to the Ziegler-Nichols formula. Apply the program to the system

$$G(s) = \frac{1}{(s+1)^8}$$

and simulate the response of the digital control system (with $T = 0.1$) when a set point step change and a load disturbance step are applied. Compare the results with those of Exercise 5.14.

6.18 To examine the effect of the sampling period on the relative stability and transient response of a digital control system, consider the system

$$G(s) = \frac{1}{(s+1)(s+5)}$$

(a) Obtain the transfer function of the system, the root locus, and the critical gain for $T = 0.01$ s, 0.05 s, 0.1 s.
(b) Obtain the step response for each system at a gain of 2.
(c) Discuss the effect of the sampling period on the transient response and relative stability of the system based on your results from (a) and (b).

State–Space Representation

In this chapter, we discuss an alternative system representation in terms of the system **state variables**, known as the **state–space** representation or **realization**. We examine the properties, advantages, and disadvantages of this representation. We also show how to obtain an input-output representation from a state-space representation. Obtaining a state–space representation from an input-output representation is further discussed in Chapter 8.

The term "realization" arises from the fact that this representation provides the basis for implementating digital or analog filters. In addition, state–space realizations can be used to develop powerful controller design methodologies. Thus, state–space analysis is an important tool in the arsenal of today's control system designer.

Objectives

After completing this chapter, the reader will be able to do the following:

1. Obtain a state–space model from the system transfer function or differential equation.
2. Determine a linearized model of a nonlinear system.
3. Determine the solution of linear (continuous-time and discrete-time) state–space equations.
4. Determine an input-output representation starting from a state–space representation.
5. Determine an equivalent state–space representation of a system by changing the basis vectors.

7.1 STATE VARIABLES

Linear continuous-time single-input-single-output (SISO) systems are typically described by the input–output differential equation

$$\frac{d^n y}{dt^n} + a_{n-1}\frac{d^{n-1} y}{dt^{n-1}} + \ldots + a_1 \frac{dy}{dt} + a_0 y = c_n \frac{d^n u}{dt^n}$$

$$+ c_{n-1}\frac{d^{n-1} u}{dt^{n-1}} + \ldots + c_1 \frac{du}{dt} + c_0 u \qquad (7.1)$$

where y is the system output, u is the system input, and a_i, $i = 0, 1, \ldots, n - 1$, c_j, $j = 0, 1, \ldots, n$ are constants. The description is valid for time-varying systems if the coefficients a_i and c_j are explicit functions of time. For a **multi-input-multi-output** (MIMO) system, the representation is in terms of l input-output differential equations of the form (7.1) where l is the number of outputs. The representation can also be used for nonlinear systems if (7.1) is allowed to include nonlinear terms.

The solution of the differential equation (7.1) requires knowledge of the system input $u(t)$ for the period of interest as well as a set of constant initial conditions

$$y(t_0), dy(t_0)/dt, \ldots, d^{n-1}y(t_0)/dt^{n-1}$$

where the notation signifies that the derivatives are evaluated at the initial time t_0. The set of initial conditions is minimal in the sense that incomplete knowledge of this set would prevent the complete solution of (7.1). On the other hand, additional initial conditions are not needed to obtain the solution. The initial conditions provide a summary of the history of the system up to the initial time. This leads to the following definition.

Definition 7.1: System State. The state of a system is the minimal set of numbers $\{x_i(t_0), i = 1, 2, \ldots, n\}$ needed together with the input $u(t)$, with t in the interval $[t_0, t_f)$ to uniquely determine the behavior of the system in the interval $[t_0, t_f]$. The number n is known as the order of the system.   ∎

As t increases, the state of the system evolves and each of the numbers $x_i(t)$ becomes a time variable. These variables are known as the **state variables**. In vector notation, the set of state variables form the **state vector**

$$\mathbf{x}(t) = [x_1 \quad x_2 \quad \ldots \quad x_n]^T = \begin{bmatrix} x_1 \\ x_2 \\ \vdots \\ x_n \end{bmatrix} \qquad (7.2)$$

The preceding equation follows standard notation in system theory where a column vector is bolded and a row vector is indicated by transposing a column.

State–space is an n-dimensional vector space where $\{x_i(t), i = 1, 2, \ldots, n\}$ represent the coordinate axes. So for a second-order system, the state space is two-dimensional and is known as the **state plane**. For the special case where the state variables are proportional to the derivatives of the output, the state plane is called the **phase plane** and the state variables are called the **phase variables**.

Curves in state space are known as the **state trajectories** and a plot of state trajectories in the plane is the **state portrait** (or **phase portrait** for the phase plane).

EXAMPLE 7.1

Consider the equation of motion of a point mass m driven by a force f

$$m\ddot{y} = f$$

where y is the displacement of the point mass. The solution of the differential equation is given by

$$y(t) = \frac{1}{m}\left\{y(t_0) + \dot{y}(t_0)t + \int_{t_0}^{t}\int_{t_0}^{t} f(\tau)d\tau\right\}$$

Clearly, a complete solution can only be obtained, given the force, if the two initial conditions $\{y(t_0), \dot{y}(t_0)\}$ are known. Hence, these constants define the state of the system at time t_0, and the system is second order. The state variables are

$$x_1(t) = y(t)$$
$$x_2(t) = \dot{y}(t)$$

and the state vector is

$$\mathbf{x}(t) = [x_1 \quad x_2]^T = \begin{bmatrix} x_1 \\ x_2 \end{bmatrix}$$

The state variables are phase variables in this case, because the second is the derivative of the first.

The state variables are governed by the two first-order differential equations

$$\dot{x}_1 = x_2$$
$$\dot{x}_2 = u/m$$

where $u = f$. The first of the two equations follows from the definitions of the two state variables. The second is obtained from the equation of motion for the point mass. The two differential equations together with the algebraic expression

$$y = x_1$$

are equivalent to the second-order differential equation because solving the first-order differential equations and then substituting in the algebraic expression yields the output y. For a force satisfying the state feedback law

$$\frac{u}{m} = -9x_1 - 3x_2$$

we have a second-order underdamped system with the solution depending only on the initial conditions. The solutions for different initial conditions can be obtained by repeatedly using the MATLAB commands **lsim** or **initial**. Each of these solutions yields position and velocity data for a phase trajectory, which is a plot of velocity versus position. A set of these trajec-

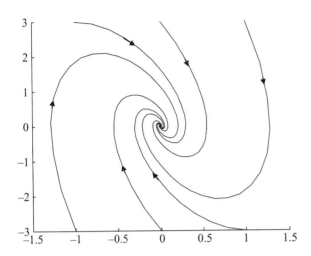

FIGURE 7.1

Phase portrait for a point mass.

tories corresponding to different initial states gives the phase portrait of Figure 7.1. The time variable does not appear explicitly in the phase portrait and is an implicit parameter. Arrows indicate the direction of increasing time.

Note that the choice of state variables is not unique. For example, one could use the displacement y and the sum of displacement and velocity as state variables (see Example 7.18). This choice has no physical meaning but nevertheless satisfies the definition of state variables. The freedom of choice is a general characteristic of state equations and is not restricted to this example. It allows us to represent a system so as to reveal its characteristics more clearly and is exploited in later sections.

7.2 STATE–SPACE REPRESENTATION

In Example 7.1, two first-order equations governing the state variables were obtained from the second-order input-output differential equation and the definitions of the state variables. These equations are known as **state equations**. In general, there are n state equations for an nth-order system. State equations can be obtained for state variables of systems described by input-output differential equations, with the form of the equations depending on the nature of the system. For example, the equations are time varying for time-varying systems and nonlinear for nonlinear systems. State equations for linear time-invariant systems can also be obtained from their transfer functions.

The algebraic equation expressing the output in terms of the state variables is called the **output equation**. For multi-output systems, a separate output equation is needed to define each output. The state and output equations together provide a complete representation for the system described by the differential equation, which is known as the **state–space representation**. For linear systems, it is often more convenient to write the state equations as a single matrix equation referred to as the **state equation**. Similarly, the output equations can be combined in a single output equation in matrix form. The matrix form of the state-space representation is demonstrated in the following example.

EXAMPLE 7.2

The state–space equations for the system of Example 7.1 in matrix form are

$$\begin{bmatrix} \dot{x}_1 \\ \dot{x}_2 \end{bmatrix} = \begin{bmatrix} 0 & 1 \\ 0 & 0 \end{bmatrix} \begin{bmatrix} x_1 \\ x_2 \end{bmatrix} + \begin{bmatrix} 0 \\ \dfrac{1}{m} \end{bmatrix} u$$

$$y = \begin{bmatrix} 1 & 0 \end{bmatrix} \begin{bmatrix} x_1 \\ x_2 \end{bmatrix}$$

The general form of the state-space equations for linear systems is

$$\dot{\mathbf{x}}(t) = A\mathbf{x}(t) + B\mathbf{u}(t)$$
$$\mathbf{y}(t) = C\mathbf{x}(t) + D\mathbf{u}(t) \tag{7.3}$$

where $\mathbf{x}(t)$ is an $n \times 1$ real vector, $\mathbf{u}(t)$ is an $m \times 1$ real vector, and $\mathbf{y}(t)$ is an $l \times 1$ real vector. The matrices in the equations are

$$A = n \times n \ \textbf{state matrix}$$
$$B = n \times m \ \textbf{input} \text{ or } \textbf{control matrix}$$
$$C = l \times n \ \textbf{output matrix}$$
$$D = l \times m \ \textbf{direct transmission matrix}$$

7.2.1 State–Space Representation in MATLAB

MATLAB has a special state-space representation obtained with the command **ss**. However, some state commands only operate on one or more of the matrices (A, B, C, D). To enter a matrix

$$A = \begin{bmatrix} 1 & 1 \\ -5 & -4 \end{bmatrix}$$

use the command

>> A = [0, 1; –5, –4]

If B, C, and D are similarly entered, we obtain the state–space quadruple **p** with the command

$$\text{>> p = ss(A, B, C, D)}$$

We can also specify names for the input (torque, say) and output (position) using the **set** command

$$\text{>> set(p,'inputn','torque','outputn','position')}$$

7.2.2 Linear versus Nonlinear State–Space Equations

The orders of the matrices are dictated by the dimensions of the vectors and the rules of vector-matrix multiplication. For example, in the single-input (SI) case, B is a column matrix, and in the single-output (SO) case, both C and D are row matrices. For the SISO case, D is a scalar. The entries of the matrices are constant for time-invariant systems and functions of time for time-varying systems.

EXAMPLE 7.3

The following are examples of state–space equations for linear systems:

1. A third-order 2-input–2-output (MIMO) linear time-invariant system:

$$
\begin{bmatrix} \dot{x}_1 \\ \dot{x}_2 \\ \dot{x}_3 \end{bmatrix} = \begin{bmatrix} 1.1 & 0.3 & -1.5 \\ 0.1 & 3.5 & 2.2 \\ 0.4 & 2.4 & -1.1 \end{bmatrix} \begin{bmatrix} x_1 \\ x_2 \\ x_3 \end{bmatrix} + \begin{bmatrix} 0.1 & 0 \\ 0 & 1.1 \\ 1.0 & 1.0 \end{bmatrix} \begin{bmatrix} u_1 \\ u_2 \end{bmatrix}
$$

$$
\begin{bmatrix} y_1 \\ y_2 \end{bmatrix} = \begin{bmatrix} 1 & 2 & 0 \\ 0 & 0 & 1 \end{bmatrix} \begin{bmatrix} x_1 \\ x_2 \\ x_3 \end{bmatrix} + \begin{bmatrix} 1 & 2 \\ 0 & 1 \end{bmatrix} \begin{bmatrix} u_1 \\ u_2 \end{bmatrix}
$$

2. A second-order 2-output–single-input (SIMO) linear time-varying system:

$$
\begin{bmatrix} \dot{x}_1 \\ \dot{x}_2 \end{bmatrix} = \begin{bmatrix} \sin(t) & \cos(t) \\ 1 & e^{-2t} \end{bmatrix} \begin{bmatrix} x_1 \\ x_2 \end{bmatrix} + \begin{bmatrix} 0 \\ 1 \end{bmatrix} u
$$

$$
\begin{bmatrix} y_1 \\ y_2 \end{bmatrix} = \begin{bmatrix} 1 & 0 \\ 0 & 1 \end{bmatrix} \begin{bmatrix} x_1 \\ x_2 \end{bmatrix}
$$

Here, the direct transmission matrix D is zero and the input and output matrices are constant. But the system is time varying because the state matrix has some entries that are functions of time.

It is important to remember that the form (7.3) is only valid for linear state equations. Nonlinear state equations involve nonlinear functions and cannot be written in terms of the matrix quadruple (A, B, C, D).

EXAMPLE 7.4

Obtain a state–space representation for the s-degree-of-freedom (s-D.O.F.) robotic manipulator from the equation of motion

$$M(\mathbf{q})\ddot{\mathbf{q}} + V(\mathbf{q}, \dot{\mathbf{q}})\dot{\mathbf{q}} + \mathbf{g}(\mathbf{q}) = \tau$$

where

$\mathbf{q}$ = vector of generalized coordinates
$M(\mathbf{q}) = s \times s$ positive definite inertia matrix
$V(\mathbf{q}, \dot{\mathbf{q}}) = s \times s$ matrix of velocity related terms
$\mathbf{g}(\mathbf{q}) = s \times 1$ vector of gravitational terms
τ = vector of generalized forces

The output of the manipulator is the position vector $\mathbf{q}$.

Solution

The system is of order $2s$, as $2s$ initial conditions are required to completely determine the solution. The most natural choice of state variables is the vector

$$\mathbf{x} = \text{col}\{\mathbf{x}_1, \mathbf{x}_2\} = \text{col}\{\mathbf{q}, \dot{\mathbf{q}}\}$$

where col{.} denotes a column vector. The associated state equations are

$$\begin{bmatrix} \dot{\mathbf{x}}_1 \\ \dot{\mathbf{x}}_2 \end{bmatrix} = \begin{bmatrix} \mathbf{x}_2 \\ -M^{-1}(\mathbf{x}_1)\{V(\mathbf{x}_1, \mathbf{x}_2)\mathbf{x}_2 + \mathbf{g}(\mathbf{x}_1)\} \end{bmatrix} + \begin{bmatrix} 0 \\ M^{-1}(\mathbf{x}_1) \end{bmatrix} \mathbf{u}$$

with the generalized force now denoted by the symbol $\mathbf{u}$
  The output equation is

$$\mathbf{y} = \mathbf{x}_1$$

This equation is linear and can be written in the standard form

$$\mathbf{y} = [I_s \mid 0_{s\times s}] \begin{bmatrix} \mathbf{x}_1 \\ \mathbf{x}_2 \end{bmatrix}$$

EXAMPLE 7.5

Write the state–space equations for the 2-D.O.F. **anthropomorphic manipulator** of Figure 7.2. The equations of motion of the manipulator are as in Example 7.4 with the definitions

$$M(\mathbf{\theta}) = \begin{bmatrix} (m_1 + m_2)l_1^2 + m_2 l_2^2 + 2m_2 l_1 l_2 \cos(\theta_2) & m_2 l_2^2 + m_2 l_1 l_2 \cos(\theta_2) \\ m_2 l_2^2 + m_2 l_1 l_2 \cos(\theta_2) & m_2 l_2^2 \end{bmatrix}$$

$$V(\mathbf{\theta}, \dot{\mathbf{\theta}})\dot{\mathbf{\theta}} = \begin{bmatrix} -m_2 l_1 l_2 \sin(\theta_2)\dot{\theta}_2(2\dot{\theta}_1 + \dot{\theta}_2) \\ m_2 l_1 l_2 \sin(\theta_2)\dot{\theta}_1^2 \end{bmatrix}$$

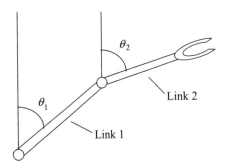

FIGURE 7.2

A 2-D.O.F. anthropomorphic manipulator.

$$\mathbf{g}(\boldsymbol{\theta}) = \begin{bmatrix} (m_1 + m_2)\, gl_1 \sin(\theta_1) + m_2 gl_2 \sin(\theta_1 + \theta_2) \\ m_2 gl_2 \sin(\theta_1 + \theta_2) \end{bmatrix}$$

where m_i, $i = 1,\ 2$ are the masses of the two links, l_i, $i = 1,\ 2$ are their lengths, and g is the acceleration due to gravity; $(\boldsymbol{\theta}, \dot{\boldsymbol{\theta}})$ are the vectors of angular positions and angular velocities, respectively.

Solution

The state equations can be written using the results of Example 7.4 as

$$\begin{bmatrix} \dot{\mathbf{x}}_1 \\ \dot{\mathbf{x}}_2 \end{bmatrix} = \begin{bmatrix} \overbrace{\mathbf{x}_2}^{} \\ -L \left\{ \begin{bmatrix} -m_2 l_1 l_2 \sin(\theta_2)\dot{\theta}_2 \left(2\dot{\theta}_1 + \dot{\theta}_2\right) \\ m_2 l_1 l_2 \sin(\theta_2)\dot{\theta}_1^2 \end{bmatrix} + \begin{bmatrix} (m_1 + m_2)\, gl_1 \sin(\theta_1) + m_2 gl_2 \sin(\theta_1 + \theta_2) \\ m_2 gl_2 \sin(\theta_1 + \theta_2) \end{bmatrix} \right\} \end{bmatrix} + \begin{bmatrix} 0 \\ \dfrac{}{L} \end{bmatrix} \mathbf{u}$$

where

$$L = \frac{1}{\det(D)} \begin{bmatrix} m_2 l_2^2 & -(m_2 l_2^2 + m_2 l_1 l_2 \cos(\theta_2)) \\ -(m_2 l_2^2 + m_2 l_1 l_2 \cos(\theta_2)) & (m_1 + m_2)\, l_1^2 + m_2 l_2^2 + 2m_2 l_1 l_2 \cos(\theta_2) \end{bmatrix}$$

$$\mathbf{x} = \text{col}\{x_1, x_2\} = \text{col}\{\boldsymbol{\theta}, \dot{\boldsymbol{\theta}}\}.$$

The general form of nonlinear state-space equations is

$$\dot{\mathbf{x}} = \mathbf{f}(\mathbf{x}, \mathbf{u})$$

$$\mathbf{y} = \mathbf{g}(\mathbf{x}, \mathbf{u})$$

(7.4)

where $\mathbf{f}(.)$ $(n \times 1)$ and $\mathbf{g}(.)$ $(l \times 1)$ are vectors of functions satisfying mathematical conditions that guarantee the existence and uniqueness of solution. But a form that is often encountered in practice and includes the equations of robotic manipulators is

$$\dot{\mathbf{x}} = \mathbf{f}(\mathbf{x}) + B(\mathbf{x})\mathbf{u}$$
$$\mathbf{y} = \mathbf{g}(\mathbf{x}) + D(\mathbf{x})\mathbf{u} \tag{7.5}$$

The state equation (7.5) is said to be affine in the control because the RHS is affine (includes a constant vector) for constant $\mathbf{x}$.

7.3 LINEARIZATION OF NONLINEAR STATE EQUATIONS

Nonlinear state equations of the form (7.4) or (7.5) can be approximated by linear state equations of the form (7.3) for small ranges of the control and state variables. The linear equations are based on the **first-order approximation**

$$f(x) = f(x_0) + \frac{df}{dx}\bigg|_{x_0} \Delta x + O(\Delta^2 x) \tag{7.6}$$

where x_0 is a constant and $\Delta x = x - x_0$ is a perturbation from the constant. The error associated with the approximation is of order $\Delta^2 x$ and is therefore acceptable for small perturbations. For a function of n variables, (7.6) can be modified to

$$f(\mathbf{x}) = f(\mathbf{x}_0) + \frac{\partial f}{\partial x_1}\bigg|_{\mathbf{x}_0} \Delta x_1 + \ldots + \frac{\partial f}{\partial x_n}\bigg|_{\mathbf{x}_0} \Delta x_n + O\left(\|\Delta \mathbf{x}\|^2\right) \tag{7.7}$$

where $\mathbf{x}_0$ is the constant vector

$$\mathbf{x}_0 = \begin{bmatrix} x_{10} & x_{20} & \ldots & x_{n0} \end{bmatrix}^T$$

and $\Delta\mathbf{x}$ denotes the perturbation vector

$$\Delta\mathbf{x} = \begin{bmatrix} x_1 - x_{10} & x_2 - x_{20} & \ldots & x_n - x_{n0} \end{bmatrix}^T$$

The term $\|\Delta\mathbf{x}\|^2$ denotes the sum of squares of the entries of the vector (i.e., its 2-norm), which is a measure of the length or "size" of the perturbation vector.[1] The error term dependent on this perturbation is assumed to be small and is neglected in the sequel.

For nonlinear state–space equations of the form (7.5), let the ith entry of the vector $\mathbf{f}$ be f_i. Then applying (7.7) to f_i yields the approximation

$$f_i(\mathbf{x}, \mathbf{u}) = f_i(\mathbf{x}_0, \mathbf{u}_0) + \frac{\partial f_i}{\partial x_1}\bigg|_{(\mathbf{x}_0, \mathbf{u}_0)} \Delta x_1 + \ldots + \frac{\partial f_i}{\partial x_n}\bigg|_{(\mathbf{x}_0, \mathbf{u}_0)} \Delta x_n +$$
$$\frac{\partial f_i}{\partial u_1}\bigg|_{(\mathbf{x}_0, \mathbf{u}_0)} \Delta u_1 + \ldots + \frac{\partial f_i}{\partial u_m}\bigg|_{(\mathbf{x}_0, \mathbf{u}_0)} \Delta u_m \tag{7.8}$$

---

[1]Other norms can be used, but the 2-norm is used most commonly.

which can be rewritten as

$$
f_i(\mathbf{x}, \mathbf{u}) - f_i(\mathbf{x}_0, \mathbf{u}_0) = \left[\left.\frac{\partial f_i}{\partial x_1}\right|_{(\mathbf{x}_0, \mathbf{u}_0)} \quad \cdots \quad \left.\frac{\partial f_i}{\partial x_n}\right|_{(\mathbf{x}_0, \mathbf{u}_0)} \right] \begin{bmatrix} \Delta x_1 \\ \Delta x_2 \\ \vdots \\ \Delta x_{n-1} \\ \Delta x_n \end{bmatrix} +
$$

$$
\left[\left.\frac{\partial f_i}{\partial u_1}\right|_{(\mathbf{x}_0, \mathbf{u}_0)} \quad \cdots \quad \left.\frac{\partial f_i}{\partial u_m}\right|_{(\mathbf{x}_0, \mathbf{u}_0)} \right] \begin{bmatrix} \Delta u_1 \\ \Delta u_2 \\ \vdots \\ \Delta u_{m-1} \\ \Delta u_m \end{bmatrix}
\tag{7.9}
$$

In most situations where we seek a linearized model, the nominal state is an **equilibrium point**. This term refers to an initial state where the system remains unless perturbed. In other words, it is a system where the state's rate of change, as expressed by the RHS of the state equation, must be zero (see Definition 8.1). Thus, if the perturbation is about an equilibrium point, then the derivative of the state vector is zero at the nominal state; that is, $f_i(\mathbf{x}_0, \mathbf{u}_0)$ is a zero vector.

The ith entry g_i of the vector $\mathbf{g}$ can be similarly expanded to yield the perturbation in the ith output

$$
\Delta y_i = g_i(\mathbf{x}, \mathbf{u}) - g_i(\mathbf{x}_0, \mathbf{u}_0)
$$

$$
= \left[\left.\frac{\partial g_i}{\partial x_1}\right|_{(\mathbf{x}_0, \mathbf{u}_0)} \quad \cdots \quad \left.\frac{\partial g_i}{\partial x_n}\right|_{(\mathbf{x}_0, \mathbf{u}_0)} \right] \begin{bmatrix} \Delta x_1 \\ \Delta x_2 \\ \vdots \\ \Delta x_{n-1} \\ \Delta x_n \end{bmatrix}
$$

$$
+ \left[\left.\frac{\partial g_i}{\partial u_1}\right|_{(\mathbf{x}_0, \mathbf{u}_0)} \quad \cdots \quad \left.\frac{\partial g_i}{\partial u_m}\right|_{(\mathbf{x}_0, \mathbf{u}_0)} \right] \begin{bmatrix} \Delta u_1 \\ \Delta u_2 \\ \vdots \\ \Delta u_{m-1} \\ \Delta u_m \end{bmatrix}
\tag{7.10}
$$

We also note that the derivative of the perturbation vector is

$$
\Delta\dot{\mathbf{x}} = \frac{d\Delta\mathbf{x}}{dt} = \frac{d(\mathbf{x} - \mathbf{x}_0)}{dt} = \dot{\mathbf{x}}
\tag{7.11}
$$

because the nominal state $\mathbf{x}_0$ is constant.

We now substitute the approximations (7.9) and (7.10) in the state and output equations, respectively, to obtain the linearized equations

$$\Delta\dot{\mathbf{x}} = \begin{bmatrix} \left.\dfrac{\partial f_1}{\partial x_1}\right|_{(\mathbf{x}_0,\mathbf{u}_0)} & \cdots & \left.\dfrac{\partial f_1}{\partial x_n}\right|_{(\mathbf{x}_0,\mathbf{u}_0)} \\ \cdots & \cdots & \cdots \\ \left.\dfrac{\partial f_n}{\partial x_1}\right|_{(\mathbf{x}_0,\mathbf{u}_0)} & \cdots & \left.\dfrac{\partial f_n}{\partial x_n}\right|_{(\mathbf{x}_0,\mathbf{u}_0)} \end{bmatrix} \begin{bmatrix} \Delta x_1 \\ \Delta x_2 \\ \vdots \\ \Delta x_{n-1} \\ \Delta x_n \end{bmatrix} + \begin{bmatrix} \left.\dfrac{\partial f_1}{\partial u_1}\right|_{(\mathbf{x}_0,\mathbf{u}_0)} & \cdots & \left.\dfrac{\partial f_1}{\partial u_m}\right|_{(\mathbf{x}_0,\mathbf{u}_0)} \\ \cdots & \cdots & \cdots \\ \left.\dfrac{\partial f_n}{\partial u_1}\right|_{(\mathbf{x}_0,\mathbf{u}_0)} & \cdots & \left.\dfrac{\partial f_n}{\partial u_m}\right|_{(\mathbf{x}_0,\mathbf{u}_0)} \end{bmatrix} \begin{bmatrix} \Delta u_1 \\ \Delta u_2 \\ \vdots \\ \Delta u_{m-1} \\ \Delta u_m \end{bmatrix}$$

$$\Delta\mathbf{y} = \begin{bmatrix} \left.\dfrac{\partial g_1}{\partial x_1}\right|_{(\mathbf{x}_0,\mathbf{u}_0)} & \cdots & \left.\dfrac{\partial g_1}{\partial x_n}\right|_{(\mathbf{x}_0,\mathbf{u}_0)} \\ \cdots & \cdots & \cdots \\ \left.\dfrac{\partial g_n}{\partial x_1}\right|_{(\mathbf{x}_0,\mathbf{u}_0)} & \cdots & \left.\dfrac{\partial g_n}{\partial x_n}\right|_{(\mathbf{x}_0,\mathbf{u}_0)} \end{bmatrix} \cdot \begin{bmatrix} \Delta x_1 \\ \Delta x_2 \\ \vdots \\ \Delta x_{n-1} \\ \Delta x_n \end{bmatrix} + \begin{bmatrix} \left.\dfrac{\partial g_1}{\partial u_1}\right|_{(\mathbf{x}_0,\mathbf{u}_0)} & \cdots & \left.\dfrac{\partial g_1}{\partial u_m}\right|_{(\mathbf{x}_0,\mathbf{u}_0)} \\ \cdots & \cdots & \cdots \\ \left.\dfrac{\partial g_n}{\partial u_1}\right|_{(\mathbf{x}_0,\mathbf{u}_0)} & \cdots & \left.\dfrac{\partial g_n}{\partial u_m}\right|_{(\mathbf{x}_0,\mathbf{u}_0)} \end{bmatrix} \begin{bmatrix} \Delta u_1 \\ \Delta u_2 \\ \vdots \\ \Delta u_{m-1} \\ \Delta u_m \end{bmatrix}$$

$$(7.12)$$

Dropping the Δ's reduces (7.12) to (7.13), with the matrices of the linear state equations defined as the **Jacobians**

$$A = \begin{bmatrix} \left.\dfrac{\partial f_1}{\partial x_1}\right|_{(\mathbf{x}_0,\mathbf{u}_0)} & \cdots & \left.\dfrac{\partial f_1}{\partial x_n}\right|_{(\mathbf{x}_0,\mathbf{u}_0)} \\ \cdots & \cdots & \cdots \\ \left.\dfrac{\partial f_n}{\partial x_1}\right|_{(\mathbf{x}_0,\mathbf{u}_0)} & \cdots & \left.\dfrac{\partial f_n}{\partial x_n}\right|_{(\mathbf{x}_0,\mathbf{u}_0)} \end{bmatrix} \qquad B = \begin{bmatrix} \left.\dfrac{\partial f_1}{\partial u_1}\right|_{(\mathbf{x}_0,\mathbf{u}_0)} & \cdots & \left.\dfrac{\partial f_1}{\partial u_m}\right|_{(\mathbf{x}_0,\mathbf{u}_0)} \\ \cdots & \cdots & \cdots \\ \left.\dfrac{\partial f_n}{\partial u_1}\right|_{(\mathbf{x}_0,\mathbf{u}_0)} & \cdots & \left.\dfrac{\partial f_n}{\partial u_m}\right|_{(\mathbf{x}_0,\mathbf{u}_0)} \end{bmatrix}$$

$$(7.13)$$

$$C = \begin{bmatrix} \left.\dfrac{\partial g_1}{\partial x_1}\right|_{(\mathbf{x}_0,\mathbf{u}_0)} & \cdots & \left.\dfrac{\partial g_1}{\partial x_n}\right|_{(\mathbf{x}_0,\mathbf{u}_0)} \\ \cdots & \cdots & \cdots \\ \left.\dfrac{\partial g_n}{\partial x_1}\right|_{(\mathbf{x}_0,\mathbf{u}_0)} & \cdots & \left.\dfrac{\partial g_n}{\partial x_n}\right|_{(\mathbf{x}_0,\mathbf{u}_0)} \end{bmatrix} \qquad D = \begin{bmatrix} \left.\dfrac{\partial g_1}{\partial u_1}\right|_{(\mathbf{x}_0,\mathbf{u}_0)} & \cdots & \left.\dfrac{\partial g_1}{\partial u_m}\right|_{(\mathbf{x}_0,\mathbf{u}_0)} \\ \cdots & \cdots & \cdots \\ \left.\dfrac{\partial g_n}{\partial u_1}\right|_{(\mathbf{x}_0,\mathbf{u}_0)} & \cdots & \left.\dfrac{\partial g_n}{\partial u_m}\right|_{(\mathbf{x}_0,\mathbf{u}_0)} \end{bmatrix}$$

EXAMPLE 7.6

Consider the equation of motion of the nonlinear spring-mass-damper system given by

$$m\ddot{y} + b(y)\dot{y} + k(y) = f$$

where y is the displacement, f is the applied force, m is a mass of 1 kg, $b(y)$ is a nonlinear damper constant, and $k(y)$ is a nonlinear spring force. Find the equilibrium position corresponding to a force f_0 in terms of the spring force; then linearize the equation of motion about this equilibrium.

Solution

The equilibrium of the system with a force f_0 is obtained by setting all the time derivatives equal to zero and solving for y to obtain

$$y_0 = k^{-1}(f_0)$$

where $k^{-1}(\cdot)$ denotes the inverse function. The equilibrium is therefore at zero velocity and the position y_0.

The nonlinear state equation for the system with state vector $\mathbf{x} = [x_1, x_2]^T = [y, \dot{y}]^T$ is

$$\begin{bmatrix} \dot{x}_1 \\ \dot{x}_2 \end{bmatrix} = \begin{bmatrix} x_2 \\ -k(x_1) - b(x_1)x_2 \end{bmatrix} + \begin{bmatrix} 0 \\ 1 \end{bmatrix} u$$

where $u = f$. Then linearizing about the equilibrium, we obtain

$$\begin{bmatrix} \dot{x}_1 \\ \dot{x}_2 \end{bmatrix} = \begin{bmatrix} 0 & 1 \\ -\dfrac{dk(x_1)}{dx_1}\bigg|_{y_0} & -b(x_1)|_{y_0} \end{bmatrix} \begin{bmatrix} x_1 \\ x_2 \end{bmatrix} + \begin{bmatrix} 0 \\ 1 \end{bmatrix} u$$

Clearly, the entries of the state matrix are constants whose values depend on the equilibrium position. In addition, terms that are originally linear do not change because of linearization.

7.4 THE SOLUTION OF LINEAR STATE–SPACE EQUATIONS

The state–space equations (7.3) are linear and can therefore be Laplace-transformed to obtain their solution. Clearly, once the state equation is solved for the state vector **x**, substitution in the output equation easily yields the output vector **y**. So we begin by examining the Laplace transform of the state equation.

The state equation involves the derivative $\dot{\mathbf{x}}$ of the state vector **x**. Because Laplace transformation is simply multiplication by a scalar followed by integration, the Laplace transform of this derivative is the vector of Laplace transforms of its entries. More specifically,

$$\mathcal{L}\{\dot{\mathbf{x}}(t)\} = [sX_i(s) - x_i(0)] = s[X_i(s)] - [x_i(0)]$$
$$= s\mathbf{X}(s) - \mathbf{x}(0) \tag{7.14}$$

Using a similar argument, the Laplace transform of the product $A\mathbf{x}$ is

$$\mathcal{L}\{A\mathbf{x}(t)\} = \mathcal{L}\left\{ \left[\sum_{j=1}^{n} a_{ij} x_j \right] \right\} = \left[\sum_{j=1}^{n} a_{ij} \mathcal{L}\{x_j\} \right] = \left[\sum_{j=1}^{n} a_{ij} X_j(s) \right] \tag{7.15}$$
$$= A\mathbf{X}(s)$$

Hence, the state equation

$$\dot{\mathbf{x}}(t) = A\mathbf{x}(t) + B\mathbf{u}(t), \quad \mathbf{x}(0)$$

has the Laplace transform

$$sX(s) - x(0) = AX(s) + BU(s) \tag{7.16}$$

Rearranging terms, we obtain

$$[sI_n - A]X(s) = x(0) + BU(s) \tag{7.17}$$

Then premultiplying by the inverse of $[sI_n - A]$ gives

$$X(s) = [sI_n - A]^{-1}[x(0) + BU(s)] \tag{7.18}$$

We now need to inverse Laplace-transform to obtain the solution of the state equation. So we first examine the inverse known as the **resolvent matrix**

$$[sI_n - A]^{-1} = \frac{1}{s}\left[I_n - \frac{1}{s}A\right]^{-1} \tag{7.19}$$

This can be expanded as

$$\frac{1}{s}\left[I_n - \frac{1}{s}A\right]^{-1} = \frac{1}{s}\left\{I_n + \frac{1}{s}A + \frac{1}{s^2}A^2 + \ldots + \frac{1}{s^i}A^i + \ldots\right\} \tag{7.20}$$

Then inverse Laplace-transforming yields the series

$$\mathscr{L}^{-1}\{[sI_n - A]^{-1}\} = I_n + At + \frac{(At)^2}{2!} + \ldots + \frac{(At)^i}{i!} + \ldots \tag{7.21}$$

This summation is a matrix version of the exponential function

$$e^{at} = 1 + at + \frac{(at)^2}{2!} + \ldots + \frac{(at)^i}{i!} + \ldots \tag{7.22}$$

It is therefore known as the **matrix exponential** and is written as

$$e^{At} = \sum_{i=0}^{\infty} \frac{(At)^i}{i!} = \mathscr{L}^{-1}\{[sI_n - A]^{-1}\} \tag{7.23}$$

Returning to (7.18), we see that the first term can now be easily inverse-transformed using (7.23). The second term requires the use of the convolution property of Laplace transforms, which states that multiplication of Laplace transforms is equivalent to convolution of their inverses. Hence, the solution of the state equation is given by

$$x(t) = e^{At}x(0) + \int_0^t e^{A(t-\tau)}Bu(\tau)\,d\tau \tag{7.24}$$

The solution for nonzero initial time is obtained by simply shifting the time variable to get

$$\mathbf{x}(t) = e^{A(t-t_0)}\mathbf{x}(t_0) + \int_{t_0}^{t} e^{A(t-\tau)}\mathbf{B}\mathbf{u}(\tau)d\tau \qquad (7.25)$$

The solution includes two terms. The first is due to the initial conditions with zero input and is known as the **zero-input response**. The second term is due to the input with zero initial conditions and is known as the **zero-state response**. By superposition, the total response of the system is the sum of the zero-state response and the zero-input response.

The zero-input response involves the change of the system state from the initial vector $\mathbf{x}(0)$ to the vector $\mathbf{x}(t)$ through multiplication by the matrix exponential. Hence, the matrix exponential is also called the **state-transition matrix**. This name is also given to a matrix that serves a similar function in the case of time-varying systems and depends on the initial as well as the final time and not just the difference between them. However, the matrix exponential form of the state-transition matrix is only valid for **linear time-invariant systems**.

To obtain the output of the system, we substitute (7.24) into the output equation

$$\mathbf{y}(t) = C\mathbf{x}(t) + D\mathbf{u}(t)$$

This gives the time response

$$\mathbf{y}(t) = C\left\{e^{A(t-t_0)}\mathbf{x}(t_0) + \int_{t_0}^{t} e^{A(t-\tau)}\mathbf{B}\mathbf{u}(\tau)d\tau\right\} + D\mathbf{u}(t) \qquad (7.26)$$

EXAMPLE 7.7

The state equations of an armature-controlled DC motor are given by

$$\begin{bmatrix} \dot{x}_1 \\ \dot{x}_2 \\ \dot{x}_3 \end{bmatrix} = \begin{bmatrix} 0 & 1 & 0 \\ 0 & 0 & 1 \\ 0 & -10 & -11 \end{bmatrix} \begin{bmatrix} x_1 \\ x_2 \\ x_3 \end{bmatrix} + \begin{bmatrix} 0 \\ 0 \\ 10 \end{bmatrix} u$$

where x_1 is the angular position, x_2 is the angular velocity, and x_3 is the armature current. Find the following:

1. The state-transition matrix
2. The response due to an initial current of 10 mA with zero angular position and zero angular velocity
3. The response due to a unit step input
4. The response due to the initial condition in part 2 together with the input in part 3.

Solution
1. The state-transition matrix is the matrix exponential given by the inverse Laplace transform of the matrix

$$[sI_3 - A]^{-1} = \begin{bmatrix} s & -1 & 0 \\ 0 & s & -1 \\ 0 & 10 & s+11 \end{bmatrix}^{-1}$$

$$= \frac{\begin{bmatrix} (s+1)(s+10) & s+11 & 1 \\ 0 & s(s+11) & s \\ 0 & -10s & s^2 \end{bmatrix}}{s(s+1)(s+10)}$$

$$= \begin{bmatrix} \dfrac{1}{s} & \dfrac{s+11}{s(s+1)(s+10)} & \dfrac{1}{s(s+1)(s+10)} \\ 0 & \dfrac{s+11}{(s+1)(s+10)} & \dfrac{1}{(s+1)(s+10)} \\ 0 & \dfrac{-10}{(s+1)(s+10)} & \dfrac{s}{(s+1)(s+10)} \end{bmatrix}$$

$$= \begin{bmatrix} \dfrac{1}{s} & \dfrac{1}{90}\left(\dfrac{99}{s} - \dfrac{100}{s+1} + \dfrac{1}{s+10}\right) & \dfrac{1}{90}\left(\dfrac{9}{s} - \dfrac{10}{s+1} + \dfrac{1}{s+10}\right) \\ 0 & \dfrac{1}{9}\left(\dfrac{10}{s+1} - \dfrac{1}{s+10}\right) & \dfrac{1}{9}\left(\dfrac{1}{s+1} - \dfrac{1}{s+10}\right) \\ 0 & -\dfrac{10}{9}\left(\dfrac{1}{s+1} - \dfrac{1}{s+10}\right) & \dfrac{1}{9}\left(\dfrac{10}{s+10} - \dfrac{1}{s+1}\right) \end{bmatrix}$$

The preceding operations involve writing s in the diagonal entries of a matrix, subtracting entries of the matrix A, then inverting the resulting matrix. The inversion is feasible in this example but becomes progressively more difficult as the order of the system increases. The inverse matrix is obtained by dividing the adjoint matrix by the determinant because numerical matrix inversion algorithms cannot be used in the presence of the complex variable s.

Next, we inverse Laplace-transform to obtain the state-transition matrix

$$e^{At} = \begin{bmatrix} 1 & \dfrac{1}{90}(99 - 100e^{-t} + e^{-10t}) & \dfrac{1}{90}(9 - 10e^{-t} + e^{-10t}) \\ 0 & \dfrac{1}{9}(10e^{-t} - e^{-10t}) & \dfrac{1}{9}(e^{-t} - e^{-10t}) \\ 0 & -\dfrac{10}{9}(e^{-t} - e^{-10t}) & \dfrac{1}{9}(10e^{-10t} - e^{-t}) \end{bmatrix}$$

The state-transition matrix can be decomposed as

$$e^{At} = \begin{bmatrix} 10 & 11 & 1 \\ 0 & 0 & 0 \\ 0 & 0 & 0 \end{bmatrix}\dfrac{e^0}{10} + \begin{bmatrix} 0 & -10 & -1 \\ 0 & 10 & 1 \\ 0 & -10 & -1 \end{bmatrix}\dfrac{e^{-t}}{9} + \begin{bmatrix} 0 & 1 & 1 \\ 0 & -10 & -10 \\ 0 & 100 & 100 \end{bmatrix}\dfrac{e^{-10t}}{90}$$

This last form reveals that a matrix exponential is nothing more than a matrix-weighted sum of scalar exponentials. The scalar exponentials involve the eigenvalues of the state matrix $\{0, -1, -10\}$ and are known as the **modes** of the system. The matrix weights have rank 1. This general property can be used to check the validity of the matrix exponential.

2. For an initial current of 10 mA and zero initial angular position and velocity, the initial state is

$$\mathbf{x}(0) = [0 \quad 0 \quad 0.01]^T$$

The zero-input response is

$$\mathbf{x}_{ZI}(t) = e^{At}\mathbf{x}(0)$$

$$= \left\{ \begin{bmatrix} 10 & 11 & 1 \\ 0 & 0 & 0 \\ 0 & 0 & 0 \end{bmatrix} \frac{e^0}{10} + \begin{bmatrix} 0 & -10 & -1 \\ 0 & 10 & 1 \\ 0 & -10 & -1 \end{bmatrix} \frac{e^{-t}}{9} + \begin{bmatrix} 0 & 1 & 1 \\ 0 & -10 & -10 \\ 0 & 100 & 100 \end{bmatrix} \frac{e^{-10t}}{90} \right\} \begin{bmatrix} 0 \\ 0 \\ 1 \\ 100 \end{bmatrix}$$

$$= \begin{bmatrix} 1 \\ 0 \\ 0 \end{bmatrix} \frac{e^0}{1000} + \begin{bmatrix} -1 \\ 1 \\ -1 \end{bmatrix} \frac{e^{-t}}{900} + \begin{bmatrix} 1 \\ -10 \\ 100 \end{bmatrix} \frac{e^{-10t}}{9000}$$

By virtue of the decomposition of the state-transition matrix in part 1, the result is obtained using multiplication by constant matrices rather than ones with exponential entries.

3. The response due to a step input is easier evaluated starting in the *s*-domain to avoid the convolution integral. To simplify the matrix operations, the resolvent matrix is decomposed as

$$[sI_3 - A]^{-1} = \begin{bmatrix} 10 & 11 & 1 \\ 0 & 0 & 0 \\ 0 & 0 & 0 \end{bmatrix} \frac{1}{10s} + \begin{bmatrix} 0 & -10 & -1 \\ 0 & 10 & 1 \\ 0 & -10 & -1 \end{bmatrix} \frac{1}{9(s+1)} + \begin{bmatrix} 0 & 1 & 1 \\ 0 & -10 & -10 \\ 0 & 100 & 100 \end{bmatrix} \frac{1}{90(s+10)}$$

The Laplace transform of the zero-state response is

$$\mathbf{X}_{ZS}(s) = [sI_3 - A]^{-1}BU(s)$$

$$= \left\{ \begin{bmatrix} 10 & 11 & 1 \\ 0 & 0 & 0 \\ 0 & 0 & 0 \end{bmatrix} \frac{1}{10s} + \begin{bmatrix} 0 & -10 & -1 \\ 0 & 10 & 1 \\ 0 & -10 & -1 \end{bmatrix} \frac{1}{9(s+1)} \right.$$

$$\left. + \begin{bmatrix} 0 & 1 & 1 \\ 0 & -10 & -10 \\ 0 & 100 & 100 \end{bmatrix} \frac{1}{90(s+10)} \right\} \begin{bmatrix} 0 \\ 0 \\ 10 \end{bmatrix} \frac{1}{s}$$

$$= \begin{bmatrix} 1 \\ 0 \\ 0 \end{bmatrix} \frac{1}{s^2} + \begin{bmatrix} -1 \\ 1 \\ -1 \end{bmatrix} \frac{(10/9)}{s(s+1)} + \begin{bmatrix} 1 \\ -10 \\ 100 \end{bmatrix} \frac{(1/9)}{s(s+10)}$$

$$= \begin{bmatrix} 1 \\ 0 \\ 0 \end{bmatrix} \frac{1}{s^2} + \begin{bmatrix} -1 \\ 1 \\ -1 \end{bmatrix} (10/9) \left[\frac{1}{s} - \frac{1}{s+1} \right] + \begin{bmatrix} 1 \\ -10 \\ 100 \end{bmatrix} (1/90) \left[\frac{1}{s} - \frac{1}{s+10} \right]$$

Inverse Laplace-transforming, we obtain the solution

$$
\mathbf{x}_{ZS}(t) = \begin{bmatrix} 1 \\ 0 \\ 0 \end{bmatrix} t + \begin{bmatrix} -1 \\ 1 \\ -1 \end{bmatrix} \left(\tfrac{10}{9}\right)[1 - e^{-t}] + \begin{bmatrix} 1 \\ -10 \\ 100 \end{bmatrix} \left(\tfrac{1}{90}\right)[1 - e^{-10t}]
$$

$$
= \begin{bmatrix} -11 \\ 10 \\ 0 \end{bmatrix} \left(\tfrac{1}{10}\right) + \begin{bmatrix} 1 \\ 0 \\ 0 \end{bmatrix} t + \begin{bmatrix} 1 \\ -1 \\ 1 \end{bmatrix} \left(\tfrac{10}{9}\right)e^{-t} + \begin{bmatrix} -1 \\ 10 \\ -100 \end{bmatrix} \left(\tfrac{1}{90}\right)e^{-10t}
$$

4. The complete solution due to the initial conditions of part 2 and the unit step input of part 3 is simply the sum of the two responses obtained earlier. Hence,

$$
\mathbf{x}(t) = \mathbf{x}_{ZI}(t) + \mathbf{x}_{ZS}(t)
$$

$$
= \begin{bmatrix} 1 \\ 0 \\ 0 \end{bmatrix} \frac{1}{1000} + \begin{bmatrix} -1 \\ 1 \\ -1 \end{bmatrix} \frac{e^{-t}}{900} + \begin{bmatrix} 1 \\ -10 \\ 100 \end{bmatrix} \frac{e^{-10t}}{9000} + \begin{bmatrix} -11 \\ 10 \\ 0 \end{bmatrix} \left(\tfrac{1}{10}\right) + \begin{bmatrix} 1 \\ 0 \\ 0 \end{bmatrix} t +
$$

$$
\begin{bmatrix} 1 \\ -1 \\ 1 \end{bmatrix} \left(\tfrac{10}{9}\right)e^{-t} + \begin{bmatrix} -1 \\ 10 \\ -100 \end{bmatrix} \left(\tfrac{1}{90}\right)e^{-10t}
$$

$$
= \begin{bmatrix} -1.099 \\ 1 \\ 0 \end{bmatrix} + \begin{bmatrix} 1 \\ -1 \\ 1 \end{bmatrix} 1.11e^{-t} + \begin{bmatrix} -1 \\ 10 \\ -100 \end{bmatrix} 1.1 \times 10^{-2} e^{-10t} + \begin{bmatrix} 1 \\ 0 \\ 0 \end{bmatrix} t
$$

7.4.1 The Leverrier Algorithm

The calculation of the resolvent matrix $[sI - A]^{-1}$ is clearly the bottleneck in the solution of state–space equations by Laplace transformation. The **Leverrier algorithm** is a convenient method to perform this calculation, which can be programmed on the current generation of hand-held calculators that are capable of performing matrix arithmetic. The derivation of the algorithm is left as an exercise (see Problem 7.8).

We first write the resolvent matrix as

$$
[sI_n - A]^{-1} = \frac{\mathrm{adj}[sI_n - A]}{\det[sI_n - A]} \tag{7.27}
$$

where adj[·] denotes the adjoint matrix, and det[·] denotes the determinant. Then we observe that, because its entries are determinants of matrices of order $n - 1$, the highest power of s in the adjoint matrix is $n - 1$. Terms of the same powers of s can be separated with matrices of their coefficients, and the resolvent matrix can be expanded as

$$
[sI_n - A]^{-1} = \frac{P_0 + P_1 s + \ldots + P_{n-1} s^{n-1}}{a_0 + a_1 s + \ldots + a_{n-1} s^{n-1} + s^n} \tag{7.28}
$$

where P_i, $i = 1, 2, \ldots, n - 1$ are $n \times n$ constant matrices and a_i, $i = 1, 2, \ldots,$ $n - 1$ are constant coefficients. The coefficients and matrices are calculated as follows.

Leverrier Algorithm

1. *Initialization:* $k = n - 1$

$$P_{n-1} = I_n \quad a_{n-1} = -\text{tr}\{A\} = -\sum_{i=1}^{n} a_{ii}$$

where tr{·} denotes the trace of the matrix.

2. *Backward iteration:* $k = n - 2, \ldots, 0$

$$P_k = P_{k+1}A + a_{k+1}I_n \quad a_k = -\frac{1}{n-k}\text{tr}\{P_k A\}$$

3. *Check:*

$$[0] = P_0 A + a_0 I_n$$

The algorithm requires matrix multiplication, matrix scalar multiplication, matrix addition, and trace evaluation. These are relatively simple operations available in many hand-held calculators with the exception of the trace operation. However, the trace operation can be easily programmed using a single repetition loop. The initialization of the algorithm is simple, and the backward iteration starts with the formulas

$$P_{n-2} = A + a_{n-1}I_n \quad a_{n-2} = -\frac{1}{2}\text{tr}\{P_{n-2}A\}$$

The Leverrier algorithm yields a form of the resolvent matrix that cannot be inverse Laplace-transformed directly to obtain the matrix exponential. It is first necessary to expand the following s-domain functions into the partial fractions:

$$\frac{s^{n-1}}{\prod_{i=1}^{n}(s - \lambda_i)} = \sum_{i=1}^{n}\frac{q_{i,n-1}}{s - \lambda_i}, \frac{s^{n-2}}{\prod_{i=1}^{n}(s - \lambda_i)}$$

$$= \sum_{i=1}^{n}\frac{q_{i,n-2}}{s - \lambda_i}, \ldots, \frac{1}{\prod_{i=1}^{n}(s - \lambda_i)} = \sum_{i=1}^{n}\frac{q_{i,0}}{s - \lambda_i}$$

(7.29)

where λ_i, $i = 1, \ldots, n$ are the eigenvalues of the state matrix A defined by

$$\det[sI_n - A] = s^n + a_{n-1}s^{n-1} + \ldots + a_1 s + a_0 = (s - \lambda_1)(s - \lambda_2)\ldots(s - \lambda_n) \quad (7.30)$$

To simplify the analysis, we assume that the eigenvalues in (7.30) are distinct (i.e., $\lambda_i \neq \lambda_j$, $i \neq j$). The repeated eigenvalue case can be handled similarly but requires higher-order terms in the partial fraction expansion.

Substituting in (7.28) and combining similar terms gives

$$[sI_n - A]^{-1} = P_{n-1} \sum_{i=1}^{n} \frac{q_{i,n-1}}{s - \lambda_i} + \ldots + P_1 \sum_{i=1}^{n} \frac{q_{i,1}}{s - \lambda_{i\,1}} + P_0 \sum_{i=1}^{n} \frac{q_{i,0}}{s - \lambda_i}$$

$$= \frac{1}{s - \lambda_1} \sum_{i=0}^{n-1} q_{1,j} P_j + \frac{1}{s - \lambda_2} \sum_{i=0}^{n-1} q_{2,j} P_j + \ldots + \frac{1}{s - \lambda_n} \sum_{i=0}^{n-1} q_{n,j} P_j$$

$$= \sum_{i=1}^{n} \frac{1}{s - \lambda_i} \left[\sum_{i=0}^{n-1} q_{i,j} P_j \right]$$

Thus, we can write the resolvent matrix in the simple form

$$[sI_n - A]^{-1} = \sum_{i=1}^{n} \frac{1}{s - \lambda_i} Z_i \tag{7.31}$$

where the matrices Z_i, $I = 1, 2, \ldots, n$ are given by

$$Z_i = \left[\sum_{j=0}^{n-1} q_{i,j} P_j \right] \tag{7.32}$$

Finally, we inverse the Laplace transform to obtain the matrix exponential

$$e^{At} = \sum_{i=1}^{n} Z_i e^{\lambda_i t} \tag{7.33}$$

This is the general form of the expansion used to simplify the computation in Example 7.7 and is the form we use throughout this text.

EXAMPLE 7.8

Use the Leverrier algorithm to compute the matrix exponential for the state matrix described in Example 7.7:

$$A = \begin{bmatrix} 0 & 1 & 0 \\ 0 & 0 & 1 \\ 0 & -10 & -11 \end{bmatrix}$$

Solution
1. *Initialization:*

$$P_2 = I_3 \quad a_2 = -\mathrm{tr}\{A\} = 11$$

2. *Backward iteration:* $k = 1, 0$
   (a) $k = 1$

$$P_1 = A + a_2 I_3 = A + 11 I_3 = \begin{bmatrix} 11 & 1 & 0 \\ 0 & 11 & 1 \\ 0 & -10 & 0 \end{bmatrix}$$

$$a_1 = -\frac{1}{2} \text{tr}\{P_1 A\} = -\frac{1}{2} \text{tr}\left\{\begin{bmatrix} 11 & 1 & 0 \\ 0 & 11 & 1 \\ 0 & -10 & 0 \end{bmatrix}\begin{bmatrix} 0 & 1 & 0 \\ 0 & 0 & 1 \\ 0 & -10 & -11 \end{bmatrix}\right\}$$

$$= -\frac{1}{2} \text{tr}\left\{\begin{bmatrix} 0 & 11 & 1 \\ 0 & -10 & 0 \\ 0 & 0 & -10 \end{bmatrix}\right\} = 10$$

(b) $k = 0$

$$P_0 = P_1 A + a_1 I_3 = \begin{bmatrix} 11 & 1 & 0 \\ 0 & 11 & 1 \\ 0 & -10 & 0 \end{bmatrix}\begin{bmatrix} 0 & 1 & 0 \\ 0 & 0 & 1 \\ 0 & -10 & -11 \end{bmatrix} + \begin{bmatrix} 10 & 0 & 0 \\ 0 & 10 & 0 \\ 0 & 0 & 10 \end{bmatrix} = \begin{bmatrix} 10 & 11 & 1 \\ 0 & 0 & 0 \\ 0 & 0 & 0 \end{bmatrix}$$

$$a_0 = -\frac{1}{3} \text{tr}\{P_0 A\} = -\frac{1}{3} \text{tr}\left\{\begin{bmatrix} 10 & 11 & 1 \\ 0 & 0 & 0 \\ 0 & 0 & 0 \end{bmatrix}\begin{bmatrix} 0 & 1 & 0 \\ 0 & 0 & 1 \\ 0 & -10 & -11 \end{bmatrix}\right\} = -\frac{1}{3} \text{tr}\left\{\begin{bmatrix} 0 & 0 & 0 \\ 0 & 0 & 0 \\ 0 & 0 & 0 \end{bmatrix}\right\} = 0$$

3. *Check:*

$$[0] = P_0 A + a_0 I_n = \begin{bmatrix} 0 & 1 & 0 \\ 0 & 0 & 1 \\ 0 & -10 & -11 \end{bmatrix}\begin{bmatrix} 10 & 11 & 1 \\ 0 & 0 & 0 \\ 0 & 0 & 0 \end{bmatrix} = [0]$$

Thus, we have

$$[sI_n - A]^{-1} = \frac{\begin{bmatrix} 10 & 11 & 1 \\ 0 & 0 & 0 \\ 0 & 0 & 0 \end{bmatrix} + \begin{bmatrix} 11 & 1 & 0 \\ 0 & 11 & 1 \\ 0 & -10 & 0 \end{bmatrix}s + \begin{bmatrix} 1 & 0 & 0 \\ 0 & 1 & 0 \\ 0 & 0 & 1 \end{bmatrix}s^2}{10s + 11s^2 + s^3}$$

The characteristic polynomial of the system is

$$s^3 + 11s^2 + 10s = s(s+1)(s+10)$$

and the system eigenvalues are $\{0, -1, -10\}$.

Next, we obtain the partial fraction expansions

$$\frac{s^2}{\prod_{i=1}^{3}(s - \lambda_i)} = \sum_{i=1}^{3} \frac{q_{i,2}}{s - \lambda_i} = \frac{0}{s} + \frac{\left(-\frac{1}{9}\right)}{s+1} + \frac{\left(\frac{10}{9}\right)}{s+10}$$

$$\frac{s}{\prod_{i=1}^{3}(s - \lambda_i)} = \sum_{i=1}^{3} \frac{q_{i,1}}{s - \lambda_i} = \frac{0}{s} + \frac{\left(\frac{1}{9}\right)}{s+1} + \frac{\left(-\frac{1}{9}\right)}{s+10}$$

$$\frac{1}{\prod\limits_{i=1}^{3}(s-\lambda_i)} = \sum_{i=1}^{3}\frac{q_{i,0}}{s-\lambda_i} = \frac{\left(\tfrac{1}{10}\right)}{s} + \frac{\left(-\tfrac{1}{9}\right)}{s+1} + \frac{\left(\tfrac{1}{90}\right)}{s+10}$$

where some of the coefficients are zero due to cancellation. This allows us to evaluate the matrices

$$Z_1 = \left(\tfrac{1}{10}\right)\begin{bmatrix} 10 & 11 & 1 \\ 0 & 0 & 0 \\ 0 & 0 & 0 \end{bmatrix}$$

$$Z_2 = \left(-\tfrac{1}{9}\right)\begin{bmatrix} 10 & 11 & 1 \\ 0 & 0 & 0 \\ 0 & 0 & 0 \end{bmatrix} + \left(\tfrac{1}{9}\right)\begin{bmatrix} 11 & 1 & 0 \\ 0 & 11 & 1 \\ 0 & -10 & 0 \end{bmatrix} + \left(-\tfrac{1}{9}\right)\begin{bmatrix} 1 & 0 & 0 \\ 0 & 1 & 0 \\ 0 & 0 & 1 \end{bmatrix} = \left(\tfrac{1}{9}\right)\begin{bmatrix} 0 & -10 & -1 \\ 0 & 10 & 1 \\ 0 & -10 & -1 \end{bmatrix}$$

$$Z_3 = \left(\tfrac{1}{90}\right)\begin{bmatrix} 10 & 11 & 1 \\ 0 & 0 & 0 \\ 0 & 0 & 0 \end{bmatrix} + \left(-\tfrac{1}{9}\right)\begin{bmatrix} 11 & 1 & 0 \\ 0 & 11 & 1 \\ 0 & -10 & 0 \end{bmatrix} + \left(\tfrac{10}{9}\right)\begin{bmatrix} 1 & 0 & 0 \\ 0 & 1 & 0 \\ 0 & 0 & 1 \end{bmatrix}$$

$$= \left(\tfrac{1}{90}\right)\begin{bmatrix} 0 & 1 & 1 \\ 0 & -10 & -10 \\ 0 & 100 & 100 \end{bmatrix}$$

Therefore, the state-transition matrix is

$$e^{At} = \begin{bmatrix} 10 & 11 & 1 \\ 0 & 0 & 0 \\ 0 & 0 & 0 \end{bmatrix}\frac{e^0}{10} + \begin{bmatrix} 0 & -10 & -1 \\ 0 & 10 & 1 \\ 0 & -10 & -1 \end{bmatrix}\frac{e^{-t}}{9} + \begin{bmatrix} 0 & 1 & 1 \\ 0 & -10 & -10 \\ 0 & 100 & 100 \end{bmatrix}\frac{e^{-10t}}{90}$$

Thus, we obtained the answer of Example 7.7 with far fewer partial fraction expansions but with some additional operations with constant matrices. Because these operations are easily performed using a calculator, this is a small price to pay for the resulting simplification.

7.4.2 Sylvester's Expansion

The matrices Z_i, $i = 1, 2, \ldots, n$, obtained in Section 7.4.1 using the Leverrier algorithm, are known as the **constituent matrices** of A. The constituent matrices can also be calculated using Sylvester's formula as follows:

$$Z_i = \frac{\prod\limits_{\substack{j=1 \\ j\neq i}}^{n}[A - \lambda_j I_n]}{\prod\limits_{\substack{j=1 \\ j\neq i}}^{n}[\lambda_i - \lambda_j]} \tag{7.34}$$

where λ_i, $i = 1, 2, \ldots, n$ are the eigenvalues of the matrix A.

Numerical computation of the matrix exponential using (7.34) can be problematic. For example, if two eigenvalues are almost equal, the scalar denominator in the equation is small, resulting in large computational errors. In fact, the numerical computation of the matrix exponential is not as simple as our presentation may suggest, with all known computational procedures failing in special cases. These issues are discussed in more detail in a well-known paper by Moler and Van Loan (1978).

EXAMPLE 7.9

Calculate constituent matrices of the matrix **A** given in Examples 7.7 and 7.8, using Sylvester's formula.

Solution

$$Z_1 = \frac{\prod_{\substack{j=1 \\ j\neq i}}^{3}[A - \lambda_j I_n]}{\prod_{\substack{j=1 \\ j\neq i}}^{3}[0 - \lambda_j]} = \frac{(A + I_n)(A + 10I_n)}{(1)(10)} = \left(\frac{1}{10}\right)\begin{bmatrix} 10 & 11 & 1 \\ 0 & 0 & 0 \\ 0 & 0 & 0 \end{bmatrix}$$

$$Z_2 = \frac{\prod_{\substack{j=1 \\ j\neq i}}^{3}[A - \lambda_j I_n]}{\prod_{\substack{j=1 \\ j\neq i}}^{3}[-1 - \lambda_j]} = \frac{A(A + 10I_n)}{(-1)(-1 + 10)} = \left(\frac{1}{9}\right)\begin{bmatrix} 0 & -10 & -1 \\ 0 & 10 & 1 \\ 0 & -10 & -1 \end{bmatrix}$$

$$Z_3 = \frac{\prod_{\substack{j=1 \\ j\neq i}}^{3}[A - \lambda_j I_n]}{\prod_{\substack{j=1 \\ j\neq i}}^{3}[-10 - \lambda_j]} = \frac{A(A + I_n)}{(-10)(-10 + 1)} = \left(\frac{1}{90}\right)\begin{bmatrix} 0 & 1 & 1 \\ 0 & -10 & -10 \\ 0 & 100 & 100 \end{bmatrix}$$

The constituent matrices can be used to define any analytic function of a matrix (i.e., a function possessing a Taylor series) and not just the matrix exponential. The following identity is true for any analytic function f(λ):

$$f(A) = \sum_{i=1}^{n} Z_i f(\lambda_i) \tag{7.35}$$

This identity allows us to calculate the functions e^{At} and A^t, among others. Using the matrices Z_i described in Example 7.9 and (7.35), we obtain the state-transition matrix described in Example 7.7.

7.4.3 The State-Transition Matrix for a Diagonal State Matrix

For a state matrix Λ in the form

$$\Lambda = \text{diag}\{\lambda_1, \lambda_2, \ldots, \lambda_n\} \tag{7.36}$$

The resolvent matrix is

$$[sI_n - \Lambda]^{-1} = \text{diag}\left\{\frac{1}{s-\lambda_1}, \frac{1}{s-\lambda_2}, \ldots, \frac{1}{s-\lambda_n}\right\} \tag{7.37}$$

The corresponding state-transition matrix is

$$
\begin{aligned}
e^{\Lambda t} &= \mathscr{L}\left\{[sI_n - \Lambda]^{-1}\right\} \\
&= \text{diag}\{e^{\lambda_1 t}, e^{\lambda_2 t}, \ldots, e^{\lambda_n t}\} \\
&= \text{diag}\{1, 0, \ldots, 0\}e^{\lambda_1 t} + \text{diag}\{0, 1, \ldots, 0\}e^{\lambda_2 t} + \ldots + \text{diag}\{0, 0, \ldots, n\}e^{\lambda_n t}
\end{aligned}
\tag{7.38}
$$

Thus, the ith constituent matrix for the diagonal form is a diagonal matrix with unity entry (i, i) and all other entries equal to zero.

EXAMPLE 7.10

Calculate the state-transition matrix if the state matrix Λ is

$$\Lambda = \text{diag}\{-1, -5, -6, -20\}$$

Solution

Using (7.38), we obtain the state-transition matrix

$$e^{\Lambda t} = \text{diag}\{e^{-t}, e^{-5t}, e^{-6t}, e^{-20t}\}$$

Assuming distinct eigenvalues, the state matrix can in general be written in the form

$$A = V\Lambda V^{-1} = V\Lambda W \tag{7.39}$$

where

$$V = [\mathbf{v}_1 \mid \mathbf{v}_2 \mid \cdots \mid \mathbf{v}_n] \quad W = \begin{bmatrix} \mathbf{w}_1^T \\ \hline \mathbf{w}_2^T \\ \hline \vdots \\ \hline \mathbf{w}_n^T \end{bmatrix}$$

$\mathbf{v}_i$, $\mathbf{w}_i$, $= i = 1, 2, \ldots, n$ are the right and left eigenvectors of the matrix A, respectively. The fact that W is the inverse of V implies that their product is the identity matrix—that is,

$$WV = [\mathbf{w}_i^T \mathbf{v}_j] = I_n$$

Equating matrix entries gives

$$\mathbf{w}_i^T \mathbf{v}_j = \begin{cases} 1, & i = j \\ 0, & i \neq j \end{cases} \tag{7.40}$$

The right eigenvectors are the usual eigenvectors of the matrix, whereas the left eigenvectors can be shown to be the eigenvectors of the matrix transpose.

The matrix A raised to any integer power i can be written in the form

$$A^i = (V \Lambda W)(V \Lambda W)\ldots(V \Lambda W) = V \Lambda^i W \tag{7.41}$$

Substituting from (7.41) in the matrix exponential series expansion gives

$$e^{At} = \sum_{i=1}^{n} \frac{(At)^i}{i!}$$

$$= \sum_{i=1}^{n} \frac{(V \Lambda W t)^i}{i!}$$

$$= \sum_{i=1}^{n} \frac{V(\Lambda t)^i W}{i!} = V \sum_{i=1}^{n} \frac{(\Lambda t)^i}{i!} W$$

That is,

$$e^{At} = V e^{\Lambda t} W \tag{7.42}$$

Thus, we have an expression for the matrix exponential of A in terms of the matrix exponential for the diagonal matrix Λ. The expression provides another method to calculate the state-transition matrix using the **eigenstructure** (eigenvalues and eigenvectors) of the state matrix. The drawback of the method is that the eigenstructure calculation is computationally costly.

EXAMPLE 7.11

Obtain the constituent matrices of the state matrix using (7.42); then obtain the state-transition matrix for the system:

$$\dot{\mathbf{x}}(t) = \begin{bmatrix} 0 & 1 \\ -2 & -3 \end{bmatrix} \mathbf{x}(t)$$

Solution
The state matrix is in companion form, and its characteristic equation can be directly written with the coefficient obtained by negating the last row of the matrix

$$\lambda^2 + 3\lambda + 2 = (\lambda + 1)(\lambda + 2) = 0$$

The system eigenvalues are therefore {−1, −2}. The matrix of eigenvectors is the Van der Monde matrix

$$V = \begin{bmatrix} 1 & 1 \\ -1 & -2 \end{bmatrix} \quad W = V^{-1} = \frac{\begin{bmatrix} -2 & -1 \\ 1 & 1 \end{bmatrix}}{-2+1} = \begin{bmatrix} 2 & 1 \\ -1 & -1 \end{bmatrix}$$

The matrix exponential is

$$e^{At} = Ve^{\Lambda t}W = \begin{bmatrix} 1 & | & 1 \\ -1 & | & -2 \end{bmatrix} \begin{bmatrix} e^{-t} & 0 \\ \hline 0 & e^{-2t} \end{bmatrix} \begin{bmatrix} 2 & 1 \\ -1 & -1 \end{bmatrix}$$

$$= \left\{ \begin{bmatrix} 1 \\ -1 \end{bmatrix} [e^{-t} \mid 0] + \begin{bmatrix} 1 \\ -2 \end{bmatrix} [0 \mid e^{-2t}] \right\} \begin{bmatrix} 2 & 1 \\ \hline -1 & -1 \end{bmatrix}$$

$$= \begin{bmatrix} 1 \\ -1 \end{bmatrix} [2 \quad 1] e^{-t} + \begin{bmatrix} 1 \\ -2 \end{bmatrix} [-1 \quad -1] e^{-2t}$$

$$= \begin{bmatrix} 2 & 1 \\ -2 & -1 \end{bmatrix} e^{-t} + \begin{bmatrix} -1 & -1 \\ 2 & 2 \end{bmatrix} e^{-2t}$$

As one may infer from the partitioned matrices used in Example 7.11, expression (7.42) can be used to write the state-transition matrix in terms of the constituent matrices. We first obtain the product

$$e^{\Lambda t}W = \begin{bmatrix} e^{\lambda_1 t} & 0 & \cdots & 0 \\ 0 & e^{\lambda_2 t} & \cdots & 0 \\ \vdots & \vdots & \ddots & \vdots \\ 0 & 0 & \cdots & e^{\lambda_n t} \end{bmatrix} \begin{bmatrix} \mathbf{w}_1^T \\ \mathbf{w}_2^T \\ \vdots \\ \mathbf{w}_n^T \end{bmatrix}$$

$$= \begin{bmatrix} \mathbf{w}_1^T \\ 0_{1\times n} \\ \vdots \\ 0_{1\times n} \end{bmatrix} e^{\lambda_1 t} + \begin{bmatrix} 0_{1\times n} \\ \mathbf{w}_2^T \\ \vdots \\ 0_{1\times n} \end{bmatrix} e^{\lambda_2 t} + \ldots + \begin{bmatrix} 0_{1\times n} \\ 0_{1\times n} \\ \vdots \\ \mathbf{w}_n^T \end{bmatrix} e^{\lambda_1 t}$$

Then we premultiply by the partition matrix V to obtain

$$Ve^{\Lambda t}W = [\mathbf{v}_1 \mid \mathbf{v}_2 \mid \cdots \mid \mathbf{v}_n] \left\{ \begin{bmatrix} \mathbf{w}_1^T \\ 0_{1\times n} \\ \vdots \\ 0_{1\times n} \end{bmatrix} e^{\lambda_1 t} + \begin{bmatrix} 0_{1\times n} \\ \mathbf{w}_2^T \\ \vdots \\ 0_{1\times n} \end{bmatrix} e^{\lambda_2 t} + \ldots + \begin{bmatrix} 0_{1\times n} \\ 0_{1\times n} \\ \vdots \\ \mathbf{w}_n^T \end{bmatrix} e^{\lambda_n t} \right\} \quad (7.43)$$

Substituting from (7.43) into (7.42) yields

$$e^{At} = \sum_{i=1}^{n} Z_i e^{\lambda_i t} = \sum_{i=1}^{n} \mathbf{v}_i \mathbf{w}_i^T e^{\lambda_i t} \quad (7.44)$$

Hence, the ith constituent matrix of A is given by the product of the ith right and left eigenvectors. The following properties of the constituent matrix can be proved using (7.44). The proofs are left as an exercise.

Properties of Constituent Matrices

1. Constituent matrices have rank 1.
2. The product of two constituent matrices is

$$Z_i Z_j = \begin{cases} Z_i, & i = j \\ 0, & i \neq j \end{cases}$$

Raising Z_i to any power gives the matrix Z_i. Z_i is said to be *idempotent*.
3. The sum of the n constituent matrices of an $n \times n$ matrix is equal to the identity matrix

$$\sum_{i=1}^{n} Z_i = I_n$$

EXAMPLE 7.12

Obtain the constituent matrices of the state matrix of Example 7.11 using (7.44), and verify that they satisfy Properties 1 through 3. Then obtain the state-transition matrix for the system.

Solution
The constituent matrices are

$$Z_1 = \begin{bmatrix} 1 \\ -1 \end{bmatrix} \begin{bmatrix} 2 & 1 \end{bmatrix} = \begin{bmatrix} 2 & 1 \\ -2 & -1 \end{bmatrix}$$

$$Z_2 = \begin{bmatrix} 1 \\ -2 \end{bmatrix} \begin{bmatrix} -1 & -1 \end{bmatrix} = \begin{bmatrix} -1 & -1 \\ 2 & 2 \end{bmatrix}$$

Both matrices have a second column equal to the first and clearly have rank 1. The product of the first and second matrices is

$$Z_1 Z_2 = \begin{bmatrix} 2 & 1 \\ -2 & -1 \end{bmatrix} \begin{bmatrix} -1 & -1 \\ 2 & 2 \end{bmatrix}$$

$$= \begin{bmatrix} 0 & 0 \\ 0 & 0 \end{bmatrix} = \begin{bmatrix} -1 & -1 \\ 2 & 2 \end{bmatrix} \begin{bmatrix} 2 & 1 \\ -2 & -1 \end{bmatrix} = Z_2 Z_1$$

The squares of the matrices are

$$Z_1 Z_1 = \begin{bmatrix} 2 & 1 \\ -2 & -1 \end{bmatrix} \begin{bmatrix} 2 & 1 \\ -2 & -1 \end{bmatrix} = \begin{bmatrix} 2 & 1 \\ -2 & -1 \end{bmatrix}$$

$$Z_2 Z_2 = \begin{bmatrix} -1 & -1 \\ 2 & 2 \end{bmatrix} \begin{bmatrix} -1 & -1 \\ 2 & 2 \end{bmatrix} = \begin{bmatrix} -1 & -1 \\ 2 & 2 \end{bmatrix}$$

The state-transition matrix is

$$e^{At} = \sum_{i=1}^{2} Z_i e^{\lambda_i t} = \begin{bmatrix} 2 & 1 \\ -2 & -1 \end{bmatrix} e^{-t} + \begin{bmatrix} -1 & -1 \\ 2 & 2 \end{bmatrix} e^{-2t}$$

EXAMPLE 7.13

Obtain the state-transition matrix for the system with state matrix

$$A = \begin{bmatrix} 0 & 1 & 3 \\ 3 & 5 & 6 \\ 5 & 6 & 7 \end{bmatrix}$$

Solution

Using the MATLAB command **eig**, we obtain the matrices

$$V = \begin{bmatrix} 0.8283 & -0.2159 & 0.5013 \\ 0.1173 & -0.6195 & -0.7931 \\ -0.5479 & -0.7547 & 0.3459 \end{bmatrix} \quad W = \begin{bmatrix} 0.8355 & 0.3121 & -0.4952 \\ -0.4050 & -0.5768 & -0.7357 \\ 0.4399 & -0.7641 & 0.5014 \end{bmatrix}$$

and the eigenvalues {−1.8429, 13.3554, 0.4875}. Then multiplying each column of V by the corresponding row of W, we obtain the constituent matrices

$$Z_1 = \begin{bmatrix} 0.6920 & 0.2585 & -0.4102 \\ 0.0980 & 0.0366 & -0.0581 \\ -0.4578 & -0.1710 & 0.2713 \end{bmatrix} \quad Z_2 = \begin{bmatrix} 0.0874 & 0.1245 & 0.1588 \\ 0.2509 & 0.3573 & 0.4558 \\ 0.3057 & 0.4353 & 0.5552 \end{bmatrix}$$

$$Z_3 = \begin{bmatrix} 0.2205 & -0.3831 & 0.2513 \\ -0.3489 & 0.6061 & -0.3977 \\ -0.1521 & -0.2643 & 0.1734 \end{bmatrix}$$

The state-transition matrix is

$$e^{At} = \begin{bmatrix} 0.6920 & 0.2585 & -0.4102 \\ 0.0980 & 0.0366 & -0.0581 \\ -0.4578 & -0.1710 & 0.2713 \end{bmatrix} e^{-1.8429t} + \begin{bmatrix} 0.0874 & 0.1245 & 0.1588 \\ 0.2509 & 0.3573 & 0.4558 \\ 0.3057 & 0.4353 & 0.5552 \end{bmatrix} e^{13.3554t}$$

$$+ \begin{bmatrix} 0.2205 & -0.3831 & 0.2513 \\ -0.3489 & 0.6061 & -0.3977 \\ -0.1521 & -0.2643 & 0.1734 \end{bmatrix} e^{0.4875t}$$

7.5 THE TRANSFER FUNCTION MATRIX

The transfer function matrix of a system can be derived from its state and output equations. We begin by Laplace transforming the state equation (7.3) with zero initial conditions to obtain

$$\mathbf{X}(s) = [sI_n - A]^{-1} B \mathbf{U}(s) \tag{7.45}$$

Then we Laplace transform the output equation and substitute from (7.45) to get

$$\mathbf{Y}(s) = C[sI_n - A]^{-1} B \mathbf{U}(s) + D\mathbf{U}(s)$$

The last equation can be rewritten in the form

$$\mathbf{Y}(s) = H(s)\mathbf{U}(s) \xleftarrow{\mathscr{L}} \mathbf{y}(t) = H(t) * \mathbf{u}(t)$$

$$H(s) = C[sI_n - A]^{-1} B + D \xleftarrow{\mathscr{L}} H(t) = Ce^{At} B + D\delta(t) \tag{7.46}$$

where $H(s)$ is the transfer function matrix and $H(t)$ is the impulse response matrix. The equations emphasize the fact that the transfer function and the impulse response are Laplace transform pairs.

The preceding equation cannot be simplified further in the MIMO case because division by a vector $\mathbf{U}(s)$ is not defined. In the SI case, the transfer function can be expressed as a vector of ratios of outputs to inputs. This reduces to a single scalar ratio in the SISO case. In general, the ijth entry of the transfer function matrix denotes

$$h_{ij}(s) = \frac{Y_i(s)}{U_j(s)}\Bigg|_{\substack{U_l=0,\, l \neq j \\ \text{zero initial conditions}}} \tag{7.47}$$

Equation (7.46) can be rewritten in terms of the constituent matrices of A as

$$H(s) = C\left[\sum_{i=1}^{n} Z_i \frac{1}{s - \lambda_i}\right] B + D$$

Thus,

$$H(s) = \sum_{i=1}^{n} CZ_i B \frac{1}{s - \lambda_i} + D \xleftarrow{\mathscr{L}} H(t) = \sum_{i=1}^{n} CZ_i Be^{\lambda_i} + D\delta(t) \tag{7.48}$$

This shows that the poles of the transfer function are the eigenvalues of the state matrix A. In some cases, however, one or both of the matrix products CZ_i, $Z_i B$ are zero, and the eigenvalues do not appear in the reduced transfer function.

The evaluation of (7.46) requires the computation of the resolvent matrix and can be performed using the Leverrier algorithm or Sylvester's formula. Nevertheless, this entails considerable effort. Therefore, we should only use (7.46) to evaluate the transfer function if the state–space equations are given and the input-output differential equations are not. It is usually simpler to obtain the transfer function by Laplace-transforming the input-output differential equation.

EXAMPLE 7.14

Calculate the transfer function of the system of Example 7.7 with the angular position as output.

Solution

$$H(s) = \begin{bmatrix} 1 & 0 & 0 \end{bmatrix} \left\{ \begin{bmatrix} 10 & 11 & 1 \\ 0 & 0 & 0 \\ 0 & 0 & 0 \end{bmatrix} \frac{1}{10s} + \begin{bmatrix} 0 & -10 & -1 \\ 0 & 10 & 1 \\ 0 & -10 & -1 \end{bmatrix} \frac{1}{9(s+1)} + \begin{bmatrix} 0 & 1 & 1 \\ 0 & -10 & -10 \\ 0 & 100 & 100 \end{bmatrix} \frac{1}{90(s+10)} \right\} \begin{bmatrix} 0 \\ 0 \\ 10 \end{bmatrix}$$

$$= \frac{1}{s} - \frac{10}{9(s+1)} + \frac{1}{9(s+10)}$$

$$= \frac{10}{s(s+1)(s+10)}$$

7.5.1 MATLAB Commands

MATLAB obtains the transfer function for the matrices (A, B, C, D) with the commands

$$\text{>> g = tf(ss(A, B, C, D))}$$

For example, the matrices

$$A = \begin{bmatrix} 0 & 1 & 0 \\ 0 & 1 & 1 \\ -3 & -4 & -2 \end{bmatrix} \quad B = \begin{bmatrix} 0 & 0 \\ 1 & 0 \\ 0 & 1 \end{bmatrix}$$

$$C = \begin{bmatrix} 0 & 1 & 0 \\ 0 & 1 & 1 \end{bmatrix} \quad D = \begin{bmatrix} 0 & 1 \\ 0 & 1 \end{bmatrix}$$

are entered as

    >> A = [zeros(2, 1), [1, 0; 1, 1]; −3, −4, −2]; B = [zeros(1, 2); eye(2)]

    >> C = [zeros(2, 1), [1, 0; 1, 1]]; D = [zeros(2, 1), ones(2, 1)

Then the transfer function for the first input is obtained with the transformation command

$$\text{>> g = tf(ss(A, B, C, D))}$$

Transfer function from input 1 to output

```
          s^2 + 2s
#1:  -------------------
      s^3 + s^2 + 2s + 3
```

```
          s^2 - 2s - 3
#2:  -------------------
      s^3 + s^2 + 2s + 3
```

Transfer function from input 2 to output

```
      s^3 + s^2 + 3s + 3
#1:  -------------------
      s^3 + s^2 + 2s + 3
```

```
      s^3 + 2s^2 + 2s + 3
#2:  -------------------
      s^3 + s^2 + 2s + 3
```

These correspond to the transfer function matrix

$$H_1(s) = \frac{\begin{bmatrix} s^2 + 2s \\ s^2 - 2s - 3 \end{bmatrix}}{s^3 + s^2 + 2s + 3}$$

The transfer function column corresponding to the second input can be similarly obtained.

The **tf** command can also be used to obtain the resolvent matrix by setting $B = C = I_n$ with D zero. The command takes the form

>> **Resolvent = tf(ss(A, eye(n), eye(n), 0))**

7.6 DISCRETE-TIME STATE–SPACE EQUATIONS

Given an analog system with piecewise constant inputs over a given sampling period, the system state variables at the end of each period can be related by a difference equation. The difference equation is obtained by examining the solution of the analog state derived in Section 7.4, over a sampling period T. The solution is given by (7.25), which is repeated here for convenience:

$$\mathbf{x}(t_f) = e^{A(t_f - t_0)}\mathbf{x}(t_0) + \int_{t_0}^{t_f} e^{A(t_f - \tau)}B\mathbf{u}(\tau)d\tau \qquad (7.49)$$

Let the initial time $t_0 = kT$, and the final time $t_f = (k + 1)T$. Then the solution (7.49) reduces to

$$\mathbf{x}(k+1) = e^{AT}\mathbf{x}(k) + \int_{kT}^{(k+1)T} e^{A((k+1)T - \tau)}B\mathbf{u}(\tau)d\tau \qquad (7.50)$$

where $\mathbf{x}(k)$ denotes the vector at time kT. For a piecewise constant input

$$\mathbf{u}(t) = \mathbf{u}(k), \quad kT \le t < (k+1)T$$

the input can be moved outside the integral. The integrand can then be simplified by changing the variable of integration to

$$\lambda = (k+1)T - \tau$$

$$d\lambda = -d\tau$$

The integral now becomes

$$\left\{ \int_{T}^{0} e^{A\lambda} B(-d\lambda) \right\} \mathbf{u}(k) = \left\{ \int_{0}^{T} e^{A\lambda} Bd\lambda \right\} \mathbf{u}(k)$$

Substituting in (7.50), we obtain the discrete-time state equation

$$\mathbf{x}(k+1) = A_d \mathbf{x}(k) + B_d \mathbf{u}(k) \tag{7.51}$$

where

$$A_d = e^{AT} \tag{7.52}$$

$$B_d = \int_{0}^{T} e^{A\lambda} Bd\lambda \tag{7.53}$$

A_d is the discrete-time state matrix and B_d is the discrete input matrix, and they are clearly of the same orders as their continuous counterparts. The discrete-time state matrix is the state-transition matrix for the analog system evaluated at the sampling period T.

Equations (7.52) and (7.53) can be simplified further using properties of the matrix exponential. For invertible state matrix A, the integral of the matrix exponential is

$$\int e^{At} dt = A^{-1}[e^{At} - I_n] = [e^{At} - I_n]A^{-1} \tag{7.54}$$

This allows us to write B_d in the form

$$B_d = A^{-1}[e^{AT} - I_n]B = [e^{AT} - I_n]A^{-1}B \tag{7.55}$$

Using the expansion of the matrix exponential (7.33), we rewrite (7.52) and (7.53) as

$$A_d = \sum_{i=1}^{n} Z_i e^{\lambda_i T} \tag{7.56}$$

$$B_d = \int_{0}^{T} \left(\sum_{i=1}^{n} Z_i e^{\lambda_i \tau} \right) Bd\tau$$

$$= \sum_{i=1}^{n} Z_i B \int_{0}^{T} e^{\lambda_i \tau} d\tau \tag{7.57}$$

The integrands in (7.57) are scalar functions, and the integral can be easily evaluated. Because we assume distinct eigenvalues, only one eigenvalue can be zero. Hence, we obtain the following expression for B_d:

$$B_d = \begin{cases} \sum_{i=1}^{n} Z_i B \left[\dfrac{1-e^{\lambda_i T}}{-\lambda_i} \right] & \lambda_i \neq 0 \\[3ex] Z_1 B T + \sum_{i=2}^{n} Z_i B \left[\dfrac{1-e^{\lambda_i T}}{-\lambda_i} \right] & \lambda_1 = 0 \end{cases}$$

(7.58)

The output equation evaluated at time kT is

$$y(k) = Cx(k) + Du(k)$$

(7.59)

The discrete-time state–space representation is given by (7.51) and (7.59).

Equation (7.51) is approximately valid for a general input vector $u(t)$ provided that the sampling period T is sufficiently short. The equation can therefore be used to obtain the solution of the state equation in the general case.

EXAMPLE 7.15

Obtain the discrete-time state equations for the system of Example 7.7:

$$\begin{bmatrix} \dot{x}_1 \\ \dot{x}_2 \\ \dot{x}_3 \end{bmatrix} = \begin{bmatrix} 0 & 1 & 0 \\ 0 & 0 & 1 \\ 0 & -10 & -11 \end{bmatrix} \begin{bmatrix} x_1 \\ x_2 \\ x_3 \end{bmatrix} + \begin{bmatrix} 0 \\ 0 \\ 10 \end{bmatrix} u$$

for a sampling period $T = 0.01$ s.

Solution

From Example 7.3, the state-transition matrix of the system is

$$e^{At} = \begin{bmatrix} 10 & 11 & 1 \\ 0 & 0 & 0 \\ 0 & 0 & 0 \end{bmatrix} \frac{e^0}{10} + \begin{bmatrix} 0 & -10 & -1 \\ 0 & 10 & 1 \\ 0 & -10 & -1 \end{bmatrix} \frac{e^{-t}}{9} + \begin{bmatrix} 0 & 1 & 1 \\ 0 & -10 & -10 \\ 0 & 100 & 100 \end{bmatrix} \frac{e^{-10t}}{90}$$

Thus, the discrete-time state matrix is

$$A_d = e^{0.01 \times A} = \begin{bmatrix} 10 & 11 & 1 \\ 0 & 0 & 0 \\ 0 & 0 & 0 \end{bmatrix} \frac{e^0}{10} + \begin{bmatrix} 0 & -10 & -1 \\ 0 & 10 & 1 \\ 0 & -10 & -1 \end{bmatrix} \frac{e^{-0.01}}{9} + \begin{bmatrix} 0 & 1 & 1 \\ 0 & -10 & -10 \\ 0 & 100 & 100 \end{bmatrix} \frac{e^{-10 \times 0.01}}{90}$$

This simplifies to

$$A_d = \begin{bmatrix} 1.0 & 0.1 & 0.0 \\ 0.0 & 0.9995 & 0.0095 \\ 0.0 & -0.0947 & 0.8954 \end{bmatrix}$$

The discrete-time input matrix is

$$B_d = Z_1 B(0.01) + Z_2 B \left(1 - e^{-0.01} \right) + Z_3 B \left(1 - e^{-10 \times 0.01} \right) / 10$$

$$= \begin{bmatrix} 0.01 \\ 0 \\ 0 \end{bmatrix} + \begin{bmatrix} -1 \\ 1 \\ -1 \end{bmatrix} (10/9)(1 - e^{-0.01}) + \begin{bmatrix} 1 \\ -10 \\ 100 \end{bmatrix} (1/90)(1 - e^{-10 \times 0.01})$$

This simplifies to

$$B_d = \begin{bmatrix} 1.622 \times 10^{-6} \\ 4.821 \times 10^{-4} \\ 9.468 \times 10^{-2} \end{bmatrix}$$

7.6.1 MATLAB Commands for Discrete-Time State–Space Equations

The MATLAB command to obtain the discrete state-space quadruple **pd** from the continuous quadruple **p** with sampling period $T = 0.1$ is

>> pd = c2d(p, .1)

Alternatively, the matrices are obtained using (7.52) and (7.55) and the MATLAB commands

>> Ad = expm(A * 0.05)

>> Bd = A\ (Ad-eye(3))* B

7.7 SOLUTION OF DISCRETE-TIME STATE–SPACE EQUATIONS

We now seek an expression for the state at time k in terms of the initial condition vector $\mathbf{x}(0)$ and the input sequence $\mathbf{u}(k)$, $k = 0, 1, 2,\ldots, k - 1$. We begin by examining the discrete-time state equation (7.51)

$$\mathbf{x}(k+1) = A_d\mathbf{x}(k) + B_d\mathbf{u}(k)$$

At $k = 0, 1$, we have

$$\mathbf{x}(1) = A_d\mathbf{x}(0) + B_d\mathbf{u}(0)$$
$$\mathbf{x}(2) = A_d\mathbf{x}(1) + B_d\mathbf{u}(1) \tag{7.60}$$

Substituting from the first into the second equation in (7.60) gives

$$\begin{aligned} \mathbf{x}(2) &= A_d[A_d\mathbf{x}(0) + B_d\mathbf{u}(0)] + B_d\mathbf{u}(1) \\ &= A_d^2\mathbf{x}(0) + A_d B_d\mathbf{u}(0) + B_d\mathbf{u}(1) \end{aligned} \tag{7.61}$$

We then rewrite (7.61) as

$$\mathbf{x}(2) = A_d^2\mathbf{x}(0) + \sum_{i=0}^{2-1} A_d^{2-i-1}B_d\mathbf{u}(i) \tag{7.62}$$

and observe that the expression generalizes to

$$\mathbf{x}(k) = A_d^k\mathbf{x}(0) + \sum_{i=0}^{k-1} A_d^{k-i-1}B_d\mathbf{u}(i) \tag{7.63}$$

This expression is in fact the general solution. Left as an exercise are details of the proof by induction where (7.63) is assumed to hold and it is shown that a similar form holds for $\mathbf{x}(k + 1)$.

The expression (7.63) is the solution of the discrete-time state equation. The matrix A_d^k is known as the **state-transition matrix** for the discrete-time system, and it plays a role analogous to its continuous counterpart. A state-transition matrix can be defined for time-varying discrete-time systems, but it is not a matrix power and it is dependent on both time k and initial time k_0.

The solution (7.63) includes two terms as in the continuous-time case. The first is the **zero-input response** due to nonzero initial conditions and zero input. The second is the **zero-state response** due to nonzero input and zero initial conditions. Because the system is linear, each term can be computed separately and then added to obtain the total response for a forced system with nonzero initial conditions.

Substituting from (7.63) in the discrete-time output equation (7.59) gives the output

$$\mathbf{y}(k) = C\left\{ A_d^k \mathbf{x}(0) + \sum_{i=0}^{k-1} A_d^{k-i-1} B_d \mathbf{u}(i) \right\} + D\mathbf{u}(k) \qquad (7.64)$$

7.7.1 z-Transform Solution of Discrete-Time State Equations

Equation (7.63) can be obtained by z-transforming the discrete-time state equation (7.51). The z-transform is given by

$$z\mathbf{X}(z) - z\mathbf{x}(0) = A_d \mathbf{X}(z) + B_d \mathbf{U}(z) \qquad (7.65)$$

Hence, $\mathbf{X}(z)$ is given by

$$\mathbf{X}(z) = [zI_n - A_d]^{-1} [z\mathbf{x}(0) + B_d \mathbf{U}(z)] \qquad (7.66)$$

We therefore need to evaluate the inverse z-transform of the matrix $[zI_n - A_d]^{-1}z$. This can be accomplished by expanding the matrix in the form of the series

$$[zI_n - A_d]^{-1}z = \left[I_n - \frac{1}{z} A_d \right]^{-1} \qquad (7.67)$$
$$= I_n + A_d z^{-1} + A_d^2 z^{-2} + \ldots + A_d^i z^{-i} + \ldots$$

The inverse z-transform of the series is

$$\mathscr{Z}\left\{ [zI_n - A_d]^{-1}z \right\} = \left\{ I_n, A_d, A_d^2, \ldots, A_d^i, \ldots \right\} \qquad (7.68)$$

Hence, we have the z-transform pair

$$[zI_n - A_d]^{-1}z \xleftrightarrow{\;\mathscr{Z}\;} \left\{ A_d^k \right\}_{k=0}^{\infty} \qquad (7.69)$$

This result is analogous to the scalar transform pair

$$\frac{z}{z - a_d} \xleftrightarrow{\ z\ } \{a_d^k\}_{k=0}^{\infty}$$

The inverse matrix in (7.69) can be evaluated using the Leverrier algorithm of Section 7.4.1 to obtain the expression

$$[zI_n - A_d]^{-1}z = \frac{P_0 z + P_1 z^2 + \ldots + P_{n-1}z^n}{a_0 + a_1 z + \ldots + a_{n-1}z^{n-1} + z^n} \tag{7.70}$$

Then, after denominator factorization and partial fraction expansion, we obtain

$$[zI_n - A_d]^{-1}z = \sum_{i=1}^{n} \frac{z}{z - \lambda_i} Z_i \tag{7.71}$$

where $\lambda_i,\ i = 1, 2, \ldots, n$ are the eigenvalues of the discrete state matrix A_d. Finally, we inverse z-transform to obtain the discrete-time state-transition matrix

$$A_d^k = \sum_{i=1}^{n} Z_i \lambda_i^k \tag{7.72}$$

Writing (7.72) for $k = 1$ and using (7.52), we have

$$A_d = \sum_{i=1}^{n} Z_i \lambda_i = \sum_{i=1}^{n} Z_i(A)e^{\lambda_i(A)T} \tag{7.73}$$

where the parentheses indicate terms pertaining to the continuous-time state matrix A. Because the equality must hold for any sampling period T and any matrix A, we have the two equalities

$$Z_i = Z_i(A)$$
$$\lambda_i = e^{\lambda_i(A)T} \tag{7.74}$$

In other words, the constituent matrices of the discrete-time state matrix are simply those of the continuous-time state matrix A, and its eigenvalues are exponential functions of the continuous-time characteristic values times the sampling period. This allows us to write the discrete-time state-transition matrix as

$$A_d^k = \sum_{i=1}^{n} Z_i e^{\lambda_i(A)kT} \tag{7.75}$$

To complete the solution of the discrete-time state equation, we examine the zero-state response rewritten as

$$\mathbf{X}_{zs}(z) = \{[zI_n - A_d]^{-1}z\} z^{-1} B_d \mathbf{U}(z) \tag{7.76}$$

The term in braces in (7.76) has a known inverse transform, and multiplication by z^{-1} is equivalent to delaying its inverse transform by one sampling period. The remaining terms also have a known inverse transform. Using the convolution theorem, the inverse of the product is the convolution summation

$$\mathbf{x}_{zs}(k) = \sum_{i=0}^{k-1} A_d^{k-i-1} B_d \mathbf{u}(i) \tag{7.77}$$

This completes the solution using z-transformation.

Using (7.75), the zero-state response can be written as

$$\mathbf{x}_{zs}(k) = \sum_{i=0}^{k-1} \left[\sum_{j=1}^{n} Z_j e^{\lambda_j(A)(k-i-1)T} \right] B_d \mathbf{u}(i) \tag{7.78}$$

Then interchanging the order of summation gives

$$\mathbf{x}_{zs}(k) = \sum_{j=1}^{n} Z_j B_d e^{\lambda_j(A)(k-1)T} \left[\sum_{i=0}^{k-1} e^{-\lambda_j(A)iT} \mathbf{u}(i) \right] \tag{7.79}$$

This expression is useful in some special cases where the summation over i can be obtained in closed form.

Occasionally, the initial conditions are given at nonzero time. The complete solution (7.63) can be shifted to obtain the response

$$\mathbf{x}(k) = A_d^{k-k_0} \mathbf{x}(k_0) + \sum_{i=k_0}^{k-1} A_d^{k-i-1} B_d \mathbf{u}(i) \tag{7.80}$$

where $\mathbf{x}(k_0)$ is the state at time k_0.

EXAMPLE 7.16

Consider the state equation

$$\begin{bmatrix} \dot{x}_1 \\ \dot{x}_2 \end{bmatrix} = \begin{bmatrix} 0 & 1 \\ -2 & -3 \end{bmatrix} \begin{bmatrix} x_1 \\ x_2 \end{bmatrix} + \begin{bmatrix} 0 \\ 1 \end{bmatrix} u$$

1. Solve the state equation for a unit step input and the initial condition vector $\mathbf{x}(0) = [1\ 0]^T$.
2. Use the solution to obtain the discrete-time state equations for a sampling period of 0.1s.
3. Solve the discrete-time state equations with the same initial conditions and input as in part 1, and verify that the solution is the same as that shown in part 1 evaluated at multiples of the sampling period T.

Solution
1. We begin by finding the state-transition matrix for the given system. The resolvent matrix is

$$\Phi(s) = \begin{bmatrix} s & -1 \\ 2 & s+3 \end{bmatrix}^{-1} = \frac{\begin{bmatrix} s+3 & 1 \\ -2 & s \end{bmatrix}}{s^2 + 3s + 2} = \frac{\begin{bmatrix} 1 & 0 \\ 0 & 1 \end{bmatrix} s + \begin{bmatrix} 3 & 1 \\ -2 & 0 \end{bmatrix}}{(s+1)(s+2)}$$

To inverse Laplace transform, we need the partial fraction expansions

$$\frac{1}{(s+1)(s+2)} = \frac{1}{(s+1)} + \frac{-1}{(s+2)}$$

$$\frac{s}{(s+1)(s+2)} = \frac{-1}{(s+1)} + \frac{2}{(s+2)}$$

Thus, we reduce the resolvent matrix to

$$\Phi(s) = \frac{-\begin{bmatrix} 1 & 0 \\ 0 & 1 \end{bmatrix} + \begin{bmatrix} 3 & 1 \\ -2 & 0 \end{bmatrix}}{(s+1)} + \frac{2\begin{bmatrix} 1 & 0 \\ 0 & 1 \end{bmatrix} - \begin{bmatrix} 3 & 1 \\ -2 & 0 \end{bmatrix}}{(s+2)}$$

$$= \frac{\begin{bmatrix} 2 & 1 \\ -2 & -1 \end{bmatrix}}{(s+1)} + \frac{\begin{bmatrix} -1 & -1 \\ 2 & 2 \end{bmatrix}}{(s+2)}$$

The matrices in the preceding expansion are both rank 1 because they are the constituent matrices of the state matrix A. The expansion can be easily inverse Laplace transformed to obtain the state-transition matrix

$$\phi(t) = \begin{bmatrix} 2 & 1 \\ -2 & -1 \end{bmatrix} e^{-t} + \begin{bmatrix} -1 & -1 \\ 2 & 2 \end{bmatrix} e^{-2t}$$

The zero-input response of the system is

$$\mathbf{x}_{zi}(t) = \left\{ \begin{bmatrix} 2 & 1 \\ -2 & -1 \end{bmatrix} e^{-t} + \begin{bmatrix} -1 & -1 \\ 2 & 2 \end{bmatrix} e^{-2t} \right\} \begin{bmatrix} 1 \\ 0 \end{bmatrix}$$

$$= \begin{bmatrix} 2 \\ -2 \end{bmatrix} e^{-t} + \begin{bmatrix} -1 \\ 2 \end{bmatrix} e^{-2t}$$

For a step input, the zero-state response is

$$\mathbf{x}_{zs}(t) = \left\{ \begin{bmatrix} 2 & 1 \\ -2 & -1 \end{bmatrix} e^{-t} + \begin{bmatrix} -1 & -1 \\ 2 & 2 \end{bmatrix} e^{-2t} \right\} \begin{bmatrix} 0 \\ 1 \end{bmatrix} * 1(t)$$

$$= \begin{bmatrix} 1 \\ -1 \end{bmatrix} e^{-t} * 1(t) + \begin{bmatrix} -1 \\ 2 \end{bmatrix} e^{-2t} * 1(t)$$

where $*$ denotes convolution. Because convolution of an exponential and a step is a relatively simple operation, we do not need Laplace transformation. We use the identity

$$e^{-\alpha t} * 1(t) = \int_0^t e^{-\alpha \tau} d\tau = \frac{1 - e^{-\alpha t}}{\alpha}$$

to obtain

$$\mathbf{x}_{zs}(t) = \begin{bmatrix} 1 \\ -1 \end{bmatrix}(1 - e^{-t}) + \begin{bmatrix} -1 \\ 2 \end{bmatrix} \frac{1 - e^{-2t}}{2} = \begin{bmatrix} 1/2 \\ 0 \end{bmatrix} - \begin{bmatrix} 1 \\ -1 \end{bmatrix} e^{-t} - \begin{bmatrix} -1 \\ 2 \end{bmatrix} \frac{e^{-2t}}{2}$$

The total system response is the sum of the zero-input and zero-state responses

$$\mathbf{x}(t) = \mathbf{x}_{zi}(t) + \mathbf{x}_{zs}(t)$$

$$= \begin{bmatrix} 2 \\ -2 \end{bmatrix} e^{-t} + \begin{bmatrix} -1 \\ 2 \end{bmatrix} e^{-2t} + \begin{bmatrix} 1/2 \\ 0 \end{bmatrix} - \begin{bmatrix} 1 \\ -1 \end{bmatrix} e^{-t} - \begin{bmatrix} -1 \\ 2 \end{bmatrix} \frac{e^{-2t}}{2}$$

$$= \begin{bmatrix} 1/2 \\ 0 \end{bmatrix} + \begin{bmatrix} 1 \\ -1 \end{bmatrix} e^{-t} + \begin{bmatrix} -1 \\ 2 \end{bmatrix} \frac{e^{-2t}}{2}$$

2. To obtain the discrete-time state equations, we use the state-transition matrix obtained in step 1 with t replaced by the sampling period. For a sampling period of 0.1, we have

$$A_d = \Phi(0.1) = \begin{bmatrix} 2 & 1 \\ -2 & -1 \end{bmatrix} e^{-0.1} + \begin{bmatrix} -1 & -1 \\ 2 & 2 \end{bmatrix} e^{-2\times 0.1} = \begin{bmatrix} 0.9909 & 0.0861 \\ -0.1722 & 0.7326 \end{bmatrix}$$

The discrete-time input matrix B_d can be evaluated as shown earlier using (7.58). One may also observe that the continuous-time system response to a step input of the duration of one sampling period is the same as the response of a system due to a piecewise constant input discussed in Section 7.6. If the input is of unit amplitude, B_d can be obtained from the zero-state response of part 1 with t replaced by the sampling period $T = 0.1$. B_d is given by

$$B_d = \left\{ \begin{bmatrix} 2 & 1 \\ -2 & -1 \end{bmatrix} (1 - e^{-0.1}) + \begin{bmatrix} -1 & -1 \\ 2 & 2 \end{bmatrix} \frac{1 - e^{-2\times 0.1}}{2} \right\} \begin{bmatrix} 0 \\ 1 \end{bmatrix}$$

$$= \begin{bmatrix} 1/2 \\ 0 \end{bmatrix} - \begin{bmatrix} 1 \\ -1 \end{bmatrix} e^{-0.1} - \begin{bmatrix} -1 \\ 2 \end{bmatrix} \frac{e^{-0.2}}{2} = \begin{bmatrix} 0.0045 \\ 0.0861 \end{bmatrix}$$

3. The solution of the discrete-time state equations involves the discrete-time state-transition matrix

$$A_d^k = \Phi(0.1k) = \begin{bmatrix} 2 & 1 \\ -2 & -1 \end{bmatrix} e^{-0.1k} + \begin{bmatrix} -1 & -1 \\ 2 & 2 \end{bmatrix} e^{-0.2k}$$

The zero-input response is the product

$$\mathbf{x}_{zi}(k) = \Phi(0.1k)\mathbf{x}(0)$$

$$= \left\{ \begin{bmatrix} 2 & 1 \\ -2 & -1 \end{bmatrix} e^{-0.1k} + \begin{bmatrix} -1 & -1 \\ 2 & 2 \end{bmatrix} e^{-0.2k} \right\} \begin{bmatrix} 1 \\ 0 \end{bmatrix} = \begin{bmatrix} 2 \\ -2 \end{bmatrix} e^{-0.1k} + \begin{bmatrix} -1 \\ 2 \end{bmatrix} e^{-0.2k}$$

Comparing this result to the zero-input response of the continuous-time system reveals that the two are identical at all sampling points $k = 0, 1, 2,\ldots$
   Next, we z-transform the discrete-time state-transition matrix to obtain

$$\Phi(z) = \begin{bmatrix} 2 & 1 \\ -2 & -1 \end{bmatrix} \frac{z}{z - e^{-0.1}} + \begin{bmatrix} -1 & -1 \\ 2 & 2 \end{bmatrix} \frac{z}{z - e^{-0.2}}$$

Hence, the z-transform of the zero-state response for a unit step input is

$$\mathbf{X}_{zs}(z) = \Phi(z)z^{-1}B_d U(z)$$

$$= \left\{ \begin{bmatrix} 2 & 1 \\ -2 & -1 \end{bmatrix} \frac{z}{z - e^{-0.1}} + \begin{bmatrix} -1 & -1 \\ 2 & 2 \end{bmatrix} \frac{z}{z - e^{-0.2}} \right\} \begin{bmatrix} 0.0045 \\ 0.0861 \end{bmatrix} \frac{z^{-1}z}{z - 1}$$

$$= 9.5163 \times 10^{-2} \begin{bmatrix} 1 \\ -1 \end{bmatrix} \frac{z}{(z - e^{-0.1})(z - 1)} + 9.0635 \times 10^{-2} \begin{bmatrix} -1 \\ 2 \end{bmatrix} \frac{z}{(z - e^{-0.2})(z - 1)}$$

We now need the partial fraction expansions

$$\frac{z}{(z - e^{-0.1})(z - 1)} = \frac{1}{1 - e^{-0.1}} \left[\frac{z}{z - 1} + \frac{(-1)z}{z - e^{-0.1}} \right] = 10.5083 \left[\frac{z}{z - 1} + \frac{(-1)z}{z - 0.9048} \right]$$

$$\frac{z}{(z - e^{-0.2})(z - 1)} = \frac{1}{1 - e^{-0.2}} \left[\frac{z}{z - 1} + \frac{(-1)z}{z - e^{-0.2}} \right] = 5.5167 \left[\frac{z}{z - 1} + \frac{(-1)z}{z - 0.8187} \right]$$

Substituting in the zero-state response expression yields

$$\mathbf{X}_{zs}(z) = \begin{bmatrix} 0.5 \\ 0 \end{bmatrix} - \begin{bmatrix} 1 \\ -1 \end{bmatrix} \frac{z}{z - e^{-0.1}} - 0.5 \begin{bmatrix} -1 \\ 2 \end{bmatrix} \frac{z}{z - e^{-0.2}}$$

Then inverse-transforming gives the response

$$\mathbf{x}_{zs}(k) = \begin{bmatrix} 0.5 \\ 0 \end{bmatrix} - \begin{bmatrix} 1 \\ -1 \end{bmatrix} e^{-0.1k} - 0.5 \begin{bmatrix} -1 \\ 2 \end{bmatrix} e^{-0.2k}$$

This result is identical to the zero-state response for the continuous system at time $t = 0.1\, k$, $k = 0, 1, 2, \ldots$ The zero-state response can also be obtained using (7.79) and the well-known summation

$$\sum_{i=0}^{k-1} a^i = \frac{1 - a^k}{1 - a}, \quad |a| < 1$$

Substituting the summation in (7.79) with constant input and then simplifying the result gives

$$\mathbf{x}_{zs}(k) = \sum_{j=1}^{n} Z_j B_d e^{\lambda_j(A))k - 1\langle T} \left[\frac{1 - e^{-\lambda_j(A)kT}}{1 - e^{-\lambda_j(A)T}} \right]$$

$$= \sum_{j=1}^{n} Z_j B_d \left[\frac{1 - e^{\lambda_j(A)kT}}{1 - e^{\lambda_j(A)T}} \right]$$

Substituting numerical values in the preceding expression yields the zero-state response obtained earlier.

7.8 *Z*-TRANSFER FUNCTION FROM STATE–SPACE EQUATIONS

The *z*-transfer function can be obtained from the discrete-time state–space representation by *z*-transforming the output equation (7.59) to obtain

$$\mathbf{Y}(z) = C\mathbf{X}(z) + D\mathbf{U}(z) \tag{7.81}$$

The transfer function is derived under zero initial conditions. We therefore substitute the *z*-transform of the zero-state response (7.76) for $\mathbf{X}(z)$ in (7.81) to obtain

$$\mathbf{Y}(z) = C\left\{[zI_n - A_d]^{-1}\right\}B_d\mathbf{U}(z) + D\mathbf{U}(z) \tag{7.82}$$

Thus, the *z*-transfer function matrix is defined by

$$\mathbf{Y}(z) = G(z)\mathbf{U}(z) \tag{7.83}$$

where *G(z)* is the matrix

$$G(z) = C\left\{[zI_n - A_d]^{-1}\right\}B_d + D \xrightarrow{\;z\;} G(k) = \begin{cases} CA_d^{k-1}B_d, & k \geq 1 \\ D, & k = 0 \end{cases} \tag{7.84}$$

and *G(k)* is the impulse response matrix. The transfer function matrix and the impulse response matrix are *z*-transform pairs.

Substituting from (7.71) into (7.84) gives the alternative expression

$$G(z) = \sum_{i=1}^{n} CZ_iB_d \frac{1}{z - \lambda_i} + D \xrightarrow{\;z\;} G(k) = \begin{cases} \sum_{i=1}^{n} CZ_iB_d\lambda_i^{k-1}, & k \geq 1 \\ D, & k = 0 \end{cases} \tag{7.85}$$

Thus, the poles of the system are the eigenvalues of the discrete-time state matrix A_d. From (7.74), these are exponential functions of the eigenvalues $\lambda_i(A)$ of the continuous-time state matrix A. For a stable matrix A, the eigenvalues $\lambda_i(A)$ have negative real parts and the eigenvalues λ_i have magnitude less than unity. This implies that the discretization of Section 7.6 yields a stable discrete-time system for a stable continuous-time system.

Another important consequence of (7.85) is that the product CZ_iB can vanish and eliminate certain eigenvalues from the transfer function. This occurs if the product CZ_i is zero, the product Z_iB is zero, or both. If such cancellation occurs, the system is said to have an **output-decoupling zero** at λ_i, an **input-decoupling zero** at λ_i, or an **input-output-decoupling zero** at λ_i, respectively. The poles of the reduced transfer function are then a subset of the eigenvalues of the state matrix A_d. A state–space realization that leads to pole-zero cancellation is said to be **reducible** or **nonminimal**. If no cancellation occurs, the realization is said to be **irreducible** or **minimal**.

Clearly, in case of an output-decoupling zero at λ_i, the forced system response does not include the mode λ_i^k. In case of an input-decoupling zero, the mode is decoupled from or unaffected by the input. In case of an input-output-decoupling zero, the mode is decoupled from both the input and the output. These properties are related to the concepts of controllability and observability discussed in Chapter 8.

EXAMPLE 7.17

Obtain the z-transfer function for the position control system described in Example 7.16.

1. With x_1 as output
2. With $x_1 + x_2$ as output

Solution

From Example 7.16, we have

$$\Phi(z) = \begin{bmatrix} 2 & 1 \\ -2 & -1 \end{bmatrix} \frac{z}{z - e^{-0.1}} + \begin{bmatrix} -1 & -1 \\ 2 & 2 \end{bmatrix} \frac{z}{z - e^{-0.2}}$$

$$B_d = \begin{bmatrix} 1/2 \\ 0 \end{bmatrix} - \begin{bmatrix} 1 \\ -1 \end{bmatrix} e^{-0.1} - \begin{bmatrix} -1 \\ 2 \end{bmatrix} \frac{e^{-0.2}}{2} = \begin{bmatrix} 0.0045 \\ 0.0861 \end{bmatrix}$$

1. The output matrix C and the direct transmission matrix D for output x_1 are

$$C = \begin{bmatrix} 1 & 0 \end{bmatrix} \quad D = 0$$

Hence, the transfer function of the system is

$$G(z) = \begin{bmatrix} 1 & 0 \end{bmatrix} \left\{ \begin{bmatrix} 2 & 1 \\ -2 & -1 \end{bmatrix} \frac{1}{z - e^{-0.1}} + \begin{bmatrix} -1 & -1 \\ 2 & 2 \end{bmatrix} \frac{1}{z - e^{-0.2}} \right\} \begin{bmatrix} 0.0045 \\ 0.0861 \end{bmatrix}$$

$$= \frac{9.5163 \times 10^{-2}}{z - e^{-0.1}} - \frac{9.0635 \times 10^{-2}}{z - e^{-0.2}}$$

2. With $x_1 + x_2$ as output, $C = [1, 1]$ and $D = 0$. The transfer function is

$$G(z) = \begin{bmatrix} 1 & 1 \end{bmatrix} \left\{ \begin{bmatrix} 2 & 1 \\ -2 & -1 \end{bmatrix} \frac{1}{z - e^{-0.1}} + \begin{bmatrix} -1 & -1 \\ 2 & 2 \end{bmatrix} \frac{1}{z - e^{-0.2}} \right\} \begin{bmatrix} 0.0045 \\ 0.0861 \end{bmatrix}$$

$$= \frac{0}{z - e^{-0.1}} + \frac{9.0635 \times 10^{-2}}{z - e^{-0.2}}$$

The system has an output-decoupling zero at $e^{-0.1}$ because the product CZ_1 is zero. The response of the system to any input does not include the decoupling term. For example, the step response is the inverse of the transform

$$Y(z) = \frac{9.0635 \times 10^{-2} z}{(z - e^{-0.2})(z - 1)}$$

$$= \frac{9.0635 \times 10^{-2}}{1 - e^{-0.2}} \left[\frac{z}{z - 1} + \frac{(-1)z}{z - e^{-0.2}} \right] = 0.5 \left[\frac{z}{z - 1} + \frac{(-1)z}{z - 0.8187} \right]$$

That is, the step response is

$$y(k) = 0.5\left[1 - e^{-0.2k}\right]$$

7.8.1 z-Transfer Function in MATLAB

The expressions for obtaining z-domain and s-domain transfer functions differ only in that z in the former is replaced by s in the latter. The same MATLAB command is used to obtain s-domain and z-domain transfer functions. The transfer function for the matrices (A_d, B_d, C, D) is obtained with the commands

>> p = ss(Ad, Bd, C, D, T)

>> gd = tf(p)

where **T** is the sampling period. The poles and zeros of a transfer function are obtained with the command

>> zpk(gd)

For the system described in Example 7.17 (2), we obtain

```
Zero/pole/gain:
0.090635 (z-0.9048)
--------------------  .
(z-0.9048) (z-0.8187)

Sampling time: 0.1
```

The command reveals that the system has a zero at 0.9048 and poles at (0.9048, 0.8187) with a gain of 0.09035. With pole-zero cancellation, the transfer function is the same as that shown in Example 7.17 (2).

7.9 SIMILARITY TRANSFORMATION

Any given linear system has an infinite number of valid state–space representations. Each representation corresponds to a different set of basis vectors in state space. Some representations are preferred because they reveal certain system properties, whereas others may be convenient for specific design tasks. This section considers transformation from one representation to another.

Given a state vector $\mathbf{x}(k)$ with state–space representation of the form (7.3), we define a new state vector $\mathbf{z}(k)$

$$\mathbf{x}(k) = T_r \mathbf{z}(k) \Leftrightarrow \mathbf{z}(k) = T_r^{-1} \mathbf{x}(k) \tag{7.86}$$

where the transformation matrix T_r is assumed invertible. Substituting for $\mathbf{x}(k)$ from (7.86) in the state equation (7.51) and the output equation (7.59) gives

$$T_r \mathbf{z}(k+1) = A_d T_r \mathbf{z}(k) + B_d \mathbf{u}(k)$$
$$\mathbf{y}(k) = C T_r \mathbf{z}(k) + D \mathbf{u}(k) \tag{7.87}$$

Premultipying the state equation by T_r^{-1} gives

$$\mathbf{z}(k+1) = T_r^{-1}A_dT_r\mathbf{z}(k) + T_r^{-1}B_d\mathbf{u}(k) \tag{7.88}$$

Hence, we have the state–space quadruple for the state vector $\mathbf{z}(k)$

$$(\mathbf{A}, \mathbf{B}, \mathbf{C}, \mathbf{D}) = \left(T_r^{-1}A_dT_r, T_r^{-1}B_d, CT_r, D\right) \tag{7.89}$$

Clearly, the quadruple for the state vector $\mathbf{x}(k)$ can be obtained from the quadruple of $\mathbf{z}(k)$ using the inverse of the transformation T_r (i.e., the matrix T_r^{-1}).

For the continuous-time system of (7.3), a state vector $\mathbf{z}(t)$ can be defined as

$$\mathbf{x}(t) = T_r\mathbf{z}(t) \Leftrightarrow \mathbf{z}(t) = T_r^{-1}\mathbf{x}(t) \tag{7.90}$$

and substitution in (7.3) yields (7.89). Thus, a discussion of similarity transformation is identical for continuous-time and discrete-time systems. We therefore drop the subscript d in the sequel.

EXAMPLE 7.18

Consider the point mass m driven by a force f of Example 7.1, and determine the two state–space equations when $m = 1$ and

1. The displacement and the velocity are the state variables.
2. The displacement and the sum of displacement and velocity are the state variables.

Solution
From first principle, as shown in Example 7.1, we have that the equation of motion is

$$m\ddot{y} = f$$

and in case 1, by considering x_1 as the displacement and x_2 as the velocity, we can write

$$\dot{x}_1(t) = x_2(t)$$
$$\dot{x}_2(t) = u(t)$$
$$y(t) = x_1(t)$$

Thus, we obtain

$$A = \begin{bmatrix} 1 & 0 \\ 0 & 0 \end{bmatrix} \quad B = \begin{bmatrix} 0 \\ 1 \end{bmatrix} \quad C = [1 \quad 0] \quad D = 0$$

To solve case 2, we consider the new state variables z_1 and z_2, and we write

$$z_1(t) = x_1(t)$$
$$z_2(t) = x_1(t) + x_2(t)$$

That is,

$$T_r^{-1} = \begin{bmatrix} 1 & 0 \\ 1 & 1 \end{bmatrix}$$

By exploiting (7.89), we finally have

$$\mathbf{A} = T_r^{-1} A_d T_r = \begin{bmatrix} -1 & 1 \\ -1 & 1 \end{bmatrix} \quad \mathbf{B} = T_r^{-1} B_d = \begin{bmatrix} 0 \\ 1 \end{bmatrix}$$

$$\mathbf{C} = [1 \quad 0] \qquad\qquad\qquad \mathbf{D} = 0$$

To obtain the transformation to diagonal form, we recall the expression

$$A = V \Lambda V^{-1} \Leftrightarrow \Lambda = V^{-1} A V \qquad (7.91)$$

where V is the modal matrix of eigenvectors of A and $\Lambda = \text{diag}\{\lambda_1, \lambda_2, \ldots, \lambda_n\}$ is the matrix of eigenvalues of A. Thus, for $\mathbf{A} = \Lambda$ in (7.89), we use the modal matrix of A as the transformation matrix. The form thus obtained is not necessarily the same as a diagonal form obtained in Section 8.5 from the transfer function using partial fraction expansion. Even though all diagonal forms share the same state matrix, their input and output matrices may be different.

EXAMPLE 7.19

Obtain the diagonal form for the state–space equations

$$\begin{bmatrix} x_1(k+1) \\ x_2(k+1) \\ x_3(k+1) \end{bmatrix} = \begin{bmatrix} 0 & 1 & 0 \\ 0 & 0 & 1 \\ 0 & -0.04 & -0.5 \end{bmatrix} \begin{bmatrix} x_1(k) \\ x_2(k) \\ x_3(k) \end{bmatrix} + \begin{bmatrix} 0 \\ 0 \\ 1 \end{bmatrix} u(k)$$

$$y(k) = [1 \quad 0 \quad 0] \begin{bmatrix} x_1(k) \\ x_2(k) \\ x_3(k) \end{bmatrix}$$

Solution

The **eig** command of MATLAB yields the eigenvalues and the modal matrix

$$\Lambda = \text{diag}\{0, -0.1, -0.4\} \quad V = \begin{bmatrix} 1 & -0.995 & 0.9184 \\ 0 & 0.0995 & -0.36741 \\ 0 & -0.00995 & 0.1469 \end{bmatrix}$$

The state matrix is in companion form and the modal matrix is also known to be the **Van der Monde** matrix:

$$V = \begin{bmatrix} 1 & 1 & 1 \\ \lambda_1 & \lambda_2 & \lambda_3 \\ \lambda_1^2 & \lambda_2^2 & \lambda_3^2 \end{bmatrix} = \begin{bmatrix} 1 & 1 & 1 \\ 0 & -0.1 & -0.4 \\ 0 & 0.01 & 0.16 \end{bmatrix}$$

The MATLAB command for similarity transformation is **ss2ss**. It requires the inverse T_r^{-1} of the similarity transformation matrix T_r. Two MATLAB commands transform to diagonal form. The first requires the modal matrix of eigenvectors V and the system to be transformed

$$\text{>> s\_diag = ss2ss(system, inv(v))}$$

The second uses similarity transformation but does not require the transformation matrix

$$\text{>> s\_diag = canon(system, 'modal')}$$

The two commands yield

$$A_t = \text{diag}\{0, -0.1, -0.4\}, \qquad B_t = [25 \quad 33.5012 \quad 9.0738]^T$$
$$C_t = [1.0000 \quad -0.9950 \quad 0.9184] \quad D_t = 0$$

For complex conjugate eigenvalues, the command **canon** yields a real realization but its state matrix is not in diagonal form, whereas **ss2ss** will yield a diagonal but complex matrix.

7.9.1 Invariance of Transfer Functions and Characteristic Equations

Similar systems can be viewed as different representations of the same systems. This is justified by the following theorem.

Theorem 7.1: Similar systems have identical transfer functions and characteristic polynomials.

PROOF. Consider the characteristic polynomials of similar realizations (A, B, C, D) and (A_1, B_1, C_1, D):

$$\begin{aligned}
\det(zI_n - A_1) &= \det(zI_n - T_r^{-1}AT_r) \\
&= \det[T_r^{-1}(zI_n - A)T_r] \\
&= \det[T_r^{-1}]\det(zI_n - A)\det[T_r] = \det(zI_n - A)
\end{aligned}$$

where we used the identity $\det[T_r^{-1}] \times \det[T_r] = 1$.
    The transfer function matrix is

$$\begin{aligned}
G_1(s) &= C_1[zI_n - A_1]B_1 + D_1 \\
&= CT_r[zI_n - T_r^{-1}AT_r]T_r^{-1}B + D \\
&= C[T_r(zI_n - T_r^{-1}AT_r)T_r^{-1}]B + D \\
&= C[zI_n - A]B + D = G(s)
\end{aligned}$$

where we used the identity $(A\ B\ C)^{-1} = C^{-1}\ B^{-1}\ A^{-1}$.   ∎

Clearly, not all systems with the same transfer function are similar, due to the possibility of pole-zero cancellation. Systems that give rise to the same transfer function are said to be **equivalent**.

EXAMPLE 7.20

Show that the following system is equivalent to the system shown in Example 7.17(2).

$$x(k+1) = 0.8187x(k) + 9.0635 \times 10^{-2} u(k)$$

$$y(k) = x(k)$$

Solution

The transfer function of the system is

$$G(z) = \frac{9.0635 \times 10^{-2}}{z - 0.8187}$$

which is identical to the reduced transfer function of Example 7.17(2).

RESOURCES

D'Azzo, J. J., and C. H. Houpis, *Linear Control System Analysis and Design*, McGraw-Hill, 1988.

Belanger, P. R., *Control Engineering: A Modern Approach*, Saunders, 1995.

Brogan, W. L., *Modern Control Theory*, Prentice Hall, 1985.

Chen, C. T., *Linear System Theory and Design*, HRW, 1984.

Friedland, B., *Control System Design: an Introduction to State-Space Methods*, McGraw-Hill, 1986.

Gupta, S. C., and L. Hasdorff, *Fundamentals of Automatic Control*, Wiley, 1970.

Hou, S.-H., A simple proof of the Leverrier-Faddeev characteristic polynomial algorithm, *SIAM Review*, 40(3):706–709, 1998.

Kailath, T., *Linear Systems*, Prentice Hall, 1980.

Moler, C. B., and C. F. Van Loan, Nineteen dubious ways to calculate the exponential of a matrix, *SIAM Review*, 20:801–836, 1978.

Sinha, N. K., *Control Systems*, HRW, 1988.

PROBLEMS

7.1 Classify the state–space equations regarding linearity and time variance:

(a)
$$\begin{bmatrix} \dot{x}_1 \\ \dot{x}_2 \end{bmatrix} = \begin{bmatrix} \sin(t) & 1 \\ 0 & -2 \end{bmatrix} \begin{bmatrix} x_1 \\ x_2 \end{bmatrix} + \begin{bmatrix} 1 \\ 2 \end{bmatrix} u$$

$$y = \begin{bmatrix} 1 & 1 \end{bmatrix} \begin{bmatrix} x_1 \\ x_2 \end{bmatrix}$$

(b) $\begin{bmatrix} \dot{x}_1 \\ \dot{x}_2 \\ \dot{x}_3 \end{bmatrix} = \begin{bmatrix} 1 & 1 & 0 \\ 0 & 2 & 0 \\ 1 & 5 & 7 \end{bmatrix} \begin{bmatrix} x_1 \\ x_2 \\ x_3 \end{bmatrix} + \begin{bmatrix} 1 \\ 2 \\ 0 \end{bmatrix} u$

$y = \begin{bmatrix} 1 & 1 & 2 \end{bmatrix} \begin{bmatrix} x_1 \\ x_2 \\ x_3 \end{bmatrix}$

(c) $\dot{x} = -2x^2 + 7x + xu$
$y = 3x$

(d) $\dot{x} = -7x + u$
$y = 3x^2$

7.2 The equations of motion of a 2-D.O.F manipulator are

$$M\ddot{\theta} + D(\dot{\theta}) + \mathbf{g}(\theta) = \begin{bmatrix} \mathbf{T} \\ f \end{bmatrix}$$

$$M = \begin{bmatrix} m_{11} & m_{12} \\ m_{12} & m_{22} \end{bmatrix} \quad \mathbf{d}(\dot{\theta}) = \begin{bmatrix} 0 \\ D_2\dot{\theta}_2 \end{bmatrix} \quad \mathbf{g}(\theta) = \begin{bmatrix} g_1(\theta) \\ g_2(\theta) \end{bmatrix}$$

$$M_a = M^{-1} = \begin{bmatrix} m_{a11} & m_{a12} \\ m_{a12} & m_{a22} \end{bmatrix}$$

where $\theta = [\theta_1, \theta_2]^T$ is a vector of joint angles. The entries of the positive definite inertia matrix M depend on the robot coordinates θ. D_2 is a damping constant. The terms g_i, $i = 1, 2$, are gravity related terms that also depend on the coordinates. The right hand side is a vector of generalized forces.

a) Obtain a state-space representation for the manipulator then linearize it in the vicinity of a general operating point $(\mathbf{x}_0, \mathbf{u}_0)$.
b) Obtain the linearized model in the vicinity of zero coordinates, velocities and inputs.
c) Show that, if the entries of the state matrix are polynomials, the answer of (b) can be obtained from the (a) by letting all nonlinear terms go to zero.

7.3 Obtain the matrix exponentials for the state matrices using four different approaches:

(a) $A = \text{diag}\{-3, -5, -7\}$

(b) $A = \begin{bmatrix} 0 & 0 & 1 \\ 0 & -1 & 0 \\ -6 & 0 & 0 \end{bmatrix}$

(c) $A = \begin{bmatrix} 0 & 1 & 1 \\ 0 & 0 & 1 \\ 0 & -6 & -5 \end{bmatrix}$

(d) A is a block diagonal matrix with the matrices of (b) and (c) on its diagonal.

7.4 Obtain the zero-input responses of the systems of Problem 7.3 due to the initial condition vectors:
   (a), (b), (c) $[1, 1, 0]^T$ and $[1, 0, 0]^T$
   (d) $[1, 1, 0, 1, 0, 0]^T$

7.5 Determine the discrete-time state equations for the systems of Problem 7.3(a), (b), and (c), with $\mathbf{b} = [0, 0, 1]^T$ in terms of the sampling period T.

7.6 Prove that the (right) eigenvectors of the matrix A^T are the left eigenvectors of A and that A^T's eigenvalues are the eigenvalues of A using

$$A = V \Lambda V^{-1} = V \Lambda W$$

7.7 Prove the properties of the constituent matrices given in Section 7.4.3 using (7.44).

7.8 (a) Derive the expressions for the terms of the adjoint matrix used in the Leverrier algorithm. *Hint:* Multiply both sides of (7.28) by the matrix $[sI - A]$ and equate coefficients.
   (b) Derive the expressions for the coefficients of the characteristic equations used in the Leverrier algorithm. *Hint:* Laplace-transform the derivative expression for the matrix exponential to obtain

$$s \mathscr{L}\{e^{At}\} - I = \mathscr{L}\{e^{At}\} A$$

Take the trace, then use the identity

$$\operatorname{tr}\left([sI_n - A]^{-1}\right) = \frac{a_1 + 2a_2 s + \ldots + (n-1)a_{n-1}s^{n-2} + ns^{n-1}}{a_0 + a_1 s + \ldots + a_{n-1}s^{n-1} + s^n}$$

7.9 The biological component of a fishery system is assumed to be governed by the population dynamics equation

$$\frac{dx(t)}{dt} = rx(t)(1 - x(t)/K) - h(t)$$

where r is the intrinsic growth rate per unit time, K is the environment carrying capacity, $x(t)$ is the stock biomass, and $h(t)$ is the harvest rate in weight[2]:
   (a) Determine the harvest rate for a sustainable fish population $x_0 < K$.
   (b) Linearize the system in the vicinity of the fish population x_0.
   (c) Obtain a discrete-time model for the linearized model with a fixed average yearly harvest rate $h(k)$ in the *kth* year.
   (d) Obtain a condition for the stability of the fish population from your discrete-time model, and comment on the significance of the condition.

7.10 The following differential equations represent a simplified model of an overhead crane[3]:

---

[2]C. W. Clark, *Mathematical Bioeconomics: The Optimal Management of Renewable Resources,* Wiley, 1990.
[3]A. Piazzi and A. Visioli, Optimal dynamic-inversion-based control of an overhead crane, *IEE Proceedings: Control Theory and Applications,* 149(5):405–411, 2002.

$$(m_L + m_C)\ddot{x}_1(t) + m_L l(\ddot{x}_3(t)\cos x_3(t) - \dot{x}_3^2(t)\sin x_3(t)) = u$$

$$m_L \ddot{x}_1(t)\cos x_3(t) + m_L l\ddot{x}_3(t) = -m_L g \sin x_3(t)$$

where m_C is the mass of the trolley, m_L is the mass of the hook/load, l is the rope length, g is the gravity acceleration, u is the force applied to the trolley, x_1 is the position of the trolley, and x_3 is the rope angle. Consider the position of the load $y = x_1 + l \sin x_3$ as the output.

(a) Determine a linearized state–space model of the system about the equilibrium point $\mathbf{x} = 0$ with state variables x_1, x_3, the first derivative of x_1, and the first derivative of x_3.

(b) Determine a second state–space model when the sum of the trolley position and of the rope angle is substituted for the rope angle as a third state variable.

7.11 Consider the discretized armature-controlled DC motor system obtained in Example 7.15. Obtain the diagonal form for the system (note that the angular position of the motor is measured).

7.12 A system whose state and output responses are always nonnegative for any nonnegative initial conditions and any nonnegative input is called a **positive system**.[4] Positive systems arise in many applications where the system variable can never be negative, including chemical processes, biological systems, economics, among others. Show that the single-input-single-output discrete-time system $(A, \mathbf{b}, \mathbf{c}^T)$ is positive if and only if all the entries of the state, input, and output matrix are positive.

7.13 To monitor river pollution, we need to model the concentration of biodegradable matter contained in the water in terms of biochemical oxygen demand for its degradation. We also need to model the dissolved oxygen deficit defined as the difference between the highest concentration of dissolved oxygen and the actual concentration in mg/l. If the two variables of interest are the state variables x_1 and x_2, respectively, then an appropriate model is given by

$$\dot{\mathbf{x}} = \begin{bmatrix} -k_1 & 0 \\ k_1 & -k_2 \end{bmatrix} \mathbf{x}$$

where k_1 is a biodegradation constant and k_2 is a reaeration constant, and both are positive. Assume that the two positive constants are unequal. Obtain a discrete-time model for the system with sampling period T, and show that the system is positive.

7.14 Autonomous underwater vehicles (AUVs) are robotic submarines that can be used for a variety of studies of the underwater environment. The vertical and

---

[4]L. Farina and S. Rinaldi, *Positive Linear Systems: Theory & Applications*, Wiley-Interscience, 2000.

horizontal dynamics of the vehicle must be controlled to remotely operate the AUV. The INFANTE (Figure P7.14) is a research AUV operated by the Instituto Superior Tecnico of Lisbon, Portugal.[5] The variables of interest in horizontal motion are the sway speed and the yaw angle. A linearized model of the horizontal plane motion of the vehicle is given by

$$\begin{bmatrix} \dot{x}_1 \\ \dot{x}_2 \\ \dot{x}_3 \end{bmatrix} = \begin{bmatrix} -0.14 & -0.69 & 0.0 \\ -0.19 & -0.048 & 0.0 \\ 0.0 & 1.0 & 0.0 \end{bmatrix} \begin{bmatrix} x_1 \\ x_2 \\ x_3 \end{bmatrix} + \begin{bmatrix} 0.056 \\ -0.23 \\ 0.0 \end{bmatrix} u$$

$$\begin{bmatrix} y_1 \\ y_2 \end{bmatrix} = \begin{bmatrix} 1 & 0 & 0 \\ 0 & 1 & 0 \end{bmatrix} \begin{bmatrix} x_1 \\ x_2 \\ x_3 \end{bmatrix}$$

where x_1 is the sway speed, x_2 is the yaw angle, x_3 is the yaw rate, and u is the rudder deflection. Obtain the discrete state–space model for the system with a sampling period of 50 ms.

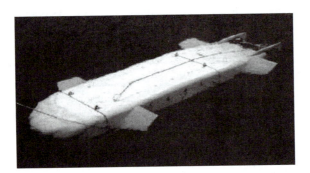

FIGURE P7.14

The INFANTE AUV. (*Source:* From Silvestrea and Pascoa, 2004; used with permission.)

COMPUTER EXERCISES

7.15 Write a computer program to simulate the systems described in Problem 7.1 for various initial conditions with zero input, and discuss your results referring to the solutions of Example 7.4. Obtain plots of the phase trajectories for any second-order system.

---

[5]C. Silvestrea and A. Pascoa, Control of the INFANTE AUV using gain scheduled static output feedback, *Control Engineering Practice*, 12:1501-1509, 2004.

7.16 Write a program to obtain the state-transition matrix using the Leverrier algorithm.

7.17 Simulate the systems of Problem 7.3(a–c) with the initial conditions of 7.4, and obtain state-trajectory plots with one state variable fixed for each system.

7.18 Repeat Problem 7.5 using a computer-aided design (CAD) package for two acceptable choices of the sampling period and compare the resulting systems.

7.19 Simulate the river pollution system of Problem 7.13 for the normalized parameter values of $k_1 = 1$, $k_2 = 2$, with a sampling period $T = 0.01$ s for the initial conditions $\mathbf{x}^T(0) = [1, 0]$, $[0, 1]$, $[1, 1]$, and plot all the results together.

7.20 Repeat Problem 7.14 using a CAD package.

Properties of State–Space Models

In Chapter 7, we described state–space models of linear discrete time and how these models can be obtained from transfer functions or input–output differential equations. We also obtained the solutions to continuous time and discrete-time state equations. Now, we examine some properties of these models that play an important role in system analysis and design.

We examine controllability, which determines the effectiveness of state feedback control; observability, which determines the possibility of state estimation from the output measurements; and stability. These three properties are independent, so that a system can be unstable but controllable, uncontrollable but stable, and so on. However, systems whose uncontrollable dynamics are stable are stabilizable, and systems whose unobservable dynamics are stable are called detectable. Stabilizability and detectability are more likely to be encountered in practice than controllability and observability.

Finally, we show how state–space representations of a system in several canonical forms can be obtained from its input-output representation.

To simplify our notation, the subscript d used in Chapter 7 with discrete-time state and input matrices is dropped if the discussion is restricted to discrete-time systems. In sections involving both continuous-time and discrete-time systems, the subscript d is retained. We begin with a discussion of stability.

Objectives

After completing this chapter, the reader will be able to do the following:

1. Determine the equilibrium point of a discrete-time system.
2. Determine the asymptotic stability of a discrete-time system.
3. Determine the input-output stability of a discrete-time system.
4. Determine the poles and zeros of multivariable systems.
5. Determine controllability and stabilizability.
6. Determine observability and detectability.
7. Obtain canonical state–space representations from an input-output representation of a system.

8.1 STABILITY OF STATE–SPACE REALIZATIONS

The concepts of **asymptotic stability** and **bounded-input-bounded-output (BIBO) stability** of transfer functions are discussed in Chapter 4. Here, we give more complete coverage using state–space models. We first discuss the asymptotic stability of state-space realizations. Then we discuss the BIBO stability of their input-output responses.

8.1.1 Asymptotic Stability

The natural response of a linear system because of its initial conditions may

1. Converge to the origin asymptotically.
2. Remain in a bounded region in the vicinity of the origin.
3. Grow unbounded.

In the first case, the system is said to be **asymptotically stable**; in the second, the system is **marginally stable**; and in the third, it is **unstable**. Clearly, physical variables are never actually unbounded even though linear models suggest this. Once a physical variable leaves a bounded range of values, the linear model ceases to be valid and the system must be described by a nonlinear model. Critical to the understanding of stability of both linear and nonlinear systems is the concept of an **equilibrium state**.

Definition 8.1: Equilibrium. An equilibrium point or state is an initial state from which the system never departs unless perturbed.   ∎

For the state equation

$$\mathbf{x}(k+1) = \mathbf{f}[\mathbf{x}(k)] \tag{8.1}$$

all equilibrium states $\mathbf{x}_e$ satisfy the condition

$$\begin{aligned} \mathbf{x}(k+1) &= \mathbf{f}[\mathbf{x}(k)] \\ &= \mathbf{f}[\mathbf{x}_e] = \mathbf{x}_e \end{aligned} \tag{8.2}$$

For linear state equations, equilibrium points satisfy the condition

$$\begin{aligned} \mathbf{x}(k+1) &= A\mathbf{x}(k) \\ &= A\mathbf{x}_e = \mathbf{x}_e \Leftrightarrow [A - I_n]\mathbf{x}_e = \mathbf{0} \end{aligned} \tag{8.3}$$

For an invertible matrix $A - I_n$, (8.3) has the unique solution $\mathbf{0}$ and the linear system has one equilibrium state at the origin. Later we show that invertibility of $A - I_n$ is a necessary condition for asymptotic stability.

Unlike linear systems, nonlinear models may have several equilibrium states. Leaving one equilibrium state, around which the linear model is valid, may drive the system to another equilibrium state where another linear model is required.

Nonlinear systems may also exhibit other more complex phenomena that are not discussed in this chapter but are discussed in Chapter 11.

The equilibrium of a system with constant input **u** can be derived from Definition 8.1 by first substituting the constant input value in the state equation to obtain the form of equation (8.2).

EXAMPLE 8.1

Find the equilibrium points of the following systems:

1. $x(k + 1) = x(k) [x(k) - 0.5]$
2. $x(k + 1) = 2x(k)$
3. $\begin{bmatrix} x_1(k+1) \\ x_2(k+1) \end{bmatrix} = \begin{bmatrix} 0.1 & 0 \\ 1 & 0.9 \end{bmatrix} \begin{bmatrix} x_1(k) \\ x_2(k) \end{bmatrix}$

Solution

1. At equilibrium, we have

$$x_e = x_e[x_e - 0.5]$$

We rewrite the equilibrium condition as

$$x_e[x_e - 1.5] = 0$$

Hence, the system has the two equilibrium states

$$x_e = 0 \quad \text{and} \quad x_e = 1.5$$

2. The equilibrium condition is

$$x_e = 2x_e$$

The system has one equilibrium point at $x_e = 0$.

3. The equilibrium condition is

$$\begin{bmatrix} x_{1e}(k) \\ x_{2e}(k) \end{bmatrix} = \begin{bmatrix} 0.1 & 0 \\ 1 & 0.9 \end{bmatrix} \begin{bmatrix} x_{1e}(k) \\ x_{2e}(k) \end{bmatrix} \Leftrightarrow \begin{bmatrix} 0.1-1 & 0 \\ 1 & 0.9-1 \end{bmatrix} \begin{bmatrix} x_{1e}(k) \\ x_{2e}(k) \end{bmatrix} = \begin{bmatrix} 0 \\ 0 \end{bmatrix}$$

The system has a unique equilibrium state at $x_e = [x_{1e}, x_{2e}]^T = [0, 0]^T$.

Although the systems of Example 8.1 all have equilibrium points, convergence to an equilibrium point is not guaranteed. This additional property defines asymptotic stability for linear systems assuming that the necessary condition of a unique equilibrium at the origin is satisfied.

Definition 8.2: Asymptotic Stability. A linear system is said to be asymptotically stable if all its trajectories converge to the origin; that is, for any initial state $\mathbf{x}(k_0)$, $\mathbf{x}(k) \to 0$ as $k \to \infty$.

Theorem 8.1: A discrete-time linear system is asymptotically **(Schur) stable** if and only if all the eigenvalues of its state matrix are inside the unit circle.

PROOF. To prove the theorem, we examine the zero-input response in (7.63) with the state-transition matrix given by (7.72). Substitution yields

$$\mathbf{x}_{zi}(k) = A^{k-k_0}\mathbf{x}(k_0)$$

$$= \sum_{i=1}^{n} Z_i \mathbf{x}(k_0) \lambda_i^{k-k_0} \qquad (8.4)$$

Sufficiency

The response decays to zero if the eigenvalues are all inside the unit circle. Hence, the condition is sufficient.

Necessity

To prove necessity, we assume that the system is stable but that one of its eigenvalues λ_j is outside the unit circle. Let the initial condition vector $\mathbf{x}(k_0) = \mathbf{v}_j$, the jth eigenvector of A. Then (see Section 7.4.3) the system response is

$$\mathbf{x}_{zi}(k) = \sum_{i=1}^{n} Z_i \mathbf{x}(k_0) \lambda_i^{k-k_0}$$

$$= \sum_{i=1}^{n} \mathbf{v}_i \mathbf{w}_i^T \mathbf{v}_j \lambda_i^{k-k_0} = \mathbf{v}_j \lambda_j^{k-k_0}$$

which is clearly unbounded as $k \rightarrow \infty$. Hence, the condition is also necessary by contradiction. ∎

REMARK

There are some nonzero initial states for which the product $Z_j \mathbf{x}(k_0)$ is zero because the matrix Z_j is rank 1 ($\mathbf{x}(k_0) = \mathbf{v}_i$, $i \neq j$). So the response to those initial states may converge to zero even if the corresponding λ_j has magnitude greater than unity. However, the solution grows unbounded for an initial state in the one direction for which the product is nonzero. Therefore, not *all* trajectories converge to the origin for $|\lambda_j| > 1$. Hence, the condition of Theorem 8.1 is necessary.

We now discuss the necessity of an invertible matrix $A - I_n$ for asymptotic stability. The matrix can be decomposed as

$$A - I_n = V[\Lambda - I_n]V^{-1} = V \operatorname{diag}\{\lambda_i - 1\}V^{-1}$$

An asymptotically stable matrix A has no unity eigenvalues, and $A - I_n$ is invertible.

EXAMPLE 8.2

Determine the stability of the systems of Example 8.1(2) and 8.1(3) using basic principles, and verify your results using Theorem 8.1.

Solution

8.1(2): Consider any initial state $x(0)$. Then the response of the system is

$$x(1) = 2\,x(0)$$
$$x(2) = 2\,x(1) = 2 \times 2\,x(0) = 4\,x(0)$$
$$x(3) = 2\,x(2) = 2 \times 4\,x(0) = 8\,x(0)$$
$$\vdots$$
$$x(k) = 2^k\,x(0)$$

Clearly, the response is unbounded as $k \to \infty$. Thus, the system is unstable. The system has one eigenvalue at $2 > 1$, which violates the stability condition of Theorem 8.1.

8.1(3): The state equation is

$$\begin{bmatrix} x_1(k+1) \\ x_2(k+1) \end{bmatrix} = \begin{bmatrix} 0.1 & 0 \\ 1 & 0.9 \end{bmatrix} \begin{bmatrix} x_1(k) \\ x_2(k) \end{bmatrix}$$

Using Sylvester's expansion (7.34), we obtain

$$\begin{bmatrix} 0.1 & 0 \\ 1 & 0.9 \end{bmatrix} = \begin{bmatrix} 1 & 0 \\ -1/0.8 & 0 \end{bmatrix} 0.1 + \begin{bmatrix} 0 & 0 \\ 1/0.8 & 1 \end{bmatrix} 0.9$$

The response of the system due to an arbitrary initial state $x(0)$ is

$$\begin{bmatrix} x_1(k) \\ x_2(k) \end{bmatrix} = \left\{ \begin{bmatrix} 1 & 0 \\ -1/0.8 & 0 \end{bmatrix} (0.1)^k + \begin{bmatrix} 0 & 0 \\ 1/0.8 & 1 \end{bmatrix} (0.9)^k \right\} \begin{bmatrix} x_1(0) \\ x_2(0) \end{bmatrix}$$

This decays to zero as $k \to \infty$. Hence, the system is asymptotically stable. Both system eigenvalues (0.1 and 0.9) are inside the unit circle, and the system satisfies the conditions of Theorem 8.1.

Note that the response due to any initial state with $x_1(0) = 0$ does not include the first eigenvalue. Had this eigenvalue been unstable, the response due to this special class of initial conditions would remain bounded. However, the response due to other initial conditions would be unbounded, and the system would be unstable.

In Chapter 4, asymptotic stability was studied in terms of the poles of the system with the additional condition of no pole-zero cancellation. The condition for asymptotic stability was identical to the condition imposed on the eigenvalues of a stable state matrix in Theorem 8.1. This is because the eigenvalues of the state matrix A and the poles of the transfer function are identical in the absence of pole-zero cancellation. Next, we examine a stability definition based on the input-output response.

8.1.2 BIBO Stability

For an input–output system description, the system output must remain bounded for any bounded-input function. To test the boundedness of an n-dimensional output vector, the norm of the vector (a measure of the vector length or size) must be used (see Appendix III). A vector $\mathbf{x}$ is bounded if it satisfies the condition

$$\|\mathbf{x}\| < b_x < \infty \qquad (8.5)$$

for some finite constant b_x where $\|\mathbf{x}\|$ denotes any vector norm.

For real or complex $n \times 1$ vectors, all norms are equivalent in the sense that if a vector has a bounded norm $\|\mathbf{x}\|_a$, then any other norm $\|\mathbf{x}\|_b$, is also bounded. This is true because for any two norms $\|\mathbf{x}\|_a$, and $\|\mathbf{x}\|_b$, there exist finite positive constants k_1 and k_2 such that

$$k_1\|\mathbf{x}\|_a \le \|\mathbf{x}\|_b \le \|\mathbf{x}\|_a k_2 \qquad (8.6)$$

Because a change of norm merely results in a scaling of the constant b_x, boundedness is independent of which norm is actually used in (8.5). Stability of the input-output behavior of a system is defined as follows.

Definition 8.3: Bounded-Input-Bounded-Output Stability. A system is BIBO stable if its output is bounded for any bounded input. That is,

$$\|\mathbf{u}(k)\| < b_u < \infty \Rightarrow \|\mathbf{y}(k)\| < b_y < \infty \qquad (8.7)$$

∎

To obtain a necessary and sufficient condition for BIBO stability, we need the concept of the norm of a matrix (see Appendix III). Matrix norms can be defined based on a set of axioms or properties. Alternatively, we define the norm of a matrix in terms of the vector norm as

$$\|A\| = \max_{\mathbf{x}} \frac{\|A\mathbf{x}\|}{\|\mathbf{x}\|}$$
$$= \max_{\|\mathbf{x}\|=1} \|A\mathbf{x}\| \qquad (8.8)$$

Multiplying by the norm of $\mathbf{x}$, we obtain the inequality

$$\|A\mathbf{x}\| \le \|A\| \cdot \|\mathbf{x}\| \qquad (8.9)$$

which applies to any matrix norm induced from a vector norm. Next, we give a condition for BIBO stability in terms of the norm of the impulse response matrix.

Theorem 8.2: A system is BIBO stable if and only if the norm of its impulse response matrix is **absolutely summable**. That is,

$$\sum_{k=0}^{\infty} \|G(k)\| < \infty \qquad (8.10)$$

PROOF.

Sufficiency

The output of a system with impulse response matrix $G(k)$ and input $\mathbf{u}(k)$ is

$$\mathbf{y}(i) = \sum_{k=0}^{i} G(k)\mathbf{u}(i-k)$$

The norm of $\mathbf{y}(i)$ satisfies

$$\|\mathbf{y}(i)\| = \left\| \sum_{k=0}^{i} G(k)\mathbf{u}(i-k) \right\|$$

$$\leq \sum_{k=0}^{i} \|G(k)\mathbf{u}(i-k)\| \leq \sum_{k=0}^{i} \|G(k)\|\|\mathbf{u}(i-k)\|$$

For a bounded input $\mathbf{u}(k)$

$$\|\mathbf{u}(k)\| < b_u < \infty$$

Hence, the output is bounded by

$$\|\mathbf{y}(i)\| \leq b_u \sum_{k=0}^{i} \|G(k)\| \leq b_u \sum_{k=0}^{\infty} \|G(k)\|$$

which is finite if condition (8.10) is satisfied.

Necessity

The proof is by contradiction. We assume that the system is BIBO stable but that (8.10) is violated. We write the output $\mathbf{y}(k)$ in the form

$$\mathbf{y}(k) = G(k)\mathbf{u}$$

$$= [G(0) | G(1) | \cdots | G(k)] \begin{bmatrix} \mathbf{u}(k) \\ \hline \mathbf{u}(k-1) \\ \hline \vdots \\ \hline \mathbf{u}(0) \end{bmatrix}$$

Using the definition of a matrix norm, the norm of the output vector can be written as

$$\|\mathbf{y}(k)\| = \max_{s} |y_s(k)|$$

$$= \max_{s} |\mathbf{g}_s^T(k)\mathbf{u}|$$

where $\mathbf{g}_s^T(k)$ is the s row of the matrix $G(k)$. Select the vectors $\mathbf{u}(i)$ to have the rth entry with unity magnitude and with sign opposite to that of $g_{sr}(i)$, where $g_{sr}(i)$ denotes the srth entry of the impulse response matrix $G(i)$. This gives the output norm

$$\|\mathbf{y}(k)\| = \max_{s} \sum_{i=0}^{k} \sum_{r=0}^{m} |g_{sr}(i)|$$

For BIBO stability, this sum remains finite. But for the violation of (8.10),

$$\sum_{i=0}^{k} \|G(k)\| = \sum_{i=0}^{k}\sum_{s=1}^{l}\sum_{r=1}^{m} |g_{sr}(i)| \to \infty \quad \text{as} \quad k \to \infty$$

The definition of the norm as the sum of absolute values of the entries of a matrix is valid and can be shown to (1) satisfy all norm properties and (2) remain finite if and only if other norms for the matrix are finite. Because the matrix has a finite number of entries, (8.10) is violated only if the sum for fixed (s, r) is unbounded (i.e., if the output is unbounded as $k \to \infty$). This contradicts the assumption of BIBO stability.   ∎

EXAMPLE 8.3

Determine the BIBO stability of the system with difference equations

$$y_1(k+2)+0.1y_2(k+1) = u(k)$$
$$y_2(k+2)+0.9y_1(k+1) = u(k)$$

Solution

To find the impulse response matrix, we first obtain the transfer function matrix

$$\begin{bmatrix} Y_1(z) \\ Y_2(z) \end{bmatrix} = \begin{bmatrix} z^2 & 0.1z \\ 0.9z & z^2 \end{bmatrix}^{-1} U(z)$$

$$= \frac{\begin{bmatrix} z^2 & -0.1z \\ -0.9z & z^2 \end{bmatrix}}{z^2(z^2-0.09)} U(z) = \frac{\begin{bmatrix} z & -0.1 \\ -0.9 & z \end{bmatrix}}{z(z-0.3)(z+0.3)} U(z)$$

Inverse z-transforming the matrix gives the impulse response

$$G(k) = \begin{bmatrix} 5.556\{(-0.3)^k+(0.3)^k\} & 1.111\delta(k-1)+1.852\{(-0.3)^k-(0.3)^k\} \\ 10\delta(k-1)+16.667\{(-0.3)^k-(0.3)^k\} & 5.556\{(-0.3)^k+(0.3)^k\} \end{bmatrix}$$

The entries of the impulse response matrix are all absolutely summable because

$$\sum_{k=0}^{\infty}(0.3)^k = \frac{1}{1-0.3} = (0.7)^{-1} < \infty$$

Hence, the impulse response satisfies condition (8.10), and the system is BIBO stable.

Although it is possible to test the impulse response for BIBO stability, it is much easier to develop tests based on the transfer function. The following theorem relates BIBO stability to asymptotic stability and sets the stage for such a test.

Theorem 8.3: If a discrete-time linear system is asymptotically stable, then it is BIBO stable. Furthermore, in the absence of unstable pole-zero cancellation, the system is asymptotically stable if it is BIBO stable.

PROOF. Substituting from (8.4) into (8.10) gives

$$\sum_{k=0}^{\infty}\|G(k)\| = \sum_{k=0}^{\infty}\left\|\sum_{j=1}^{n}CZ_jB\lambda_j^k\right\|$$

$$\leq \sum_{k=0}^{\infty}\sum_{j=1}^{n}\|CZ_jB\||\lambda_j|^k$$

For an asymptotically stable system, all poles are inside the unit circle. Therefore,

$$\sum_{k=0}^{\infty}\|G(k)\| \leq \sum_{j=1}^{n}\|CZ_jB\|\frac{1}{1-|\lambda_j|}$$

which is finite provided that the entries of the matrices are finite. Thus, whenever the product CZ_jB is nonzero for all j, BIBO and asymptotic stability are equivalent. However, the system is BIBO stable but not asymptotically stable if the product is zero for some λ_j with magnitude greater than unity. In other words, a system with unstable input-decoupling, output-decoupling, or input–output-decoupling zeros and all other modes stable is BIBO stable but not asymptotically stable. In the absence of unstable pole-zero cancellation, BIBO stability and asymptotic stability are equivalent.   ∎

Using Theorem 8.3, we can test BIBO stability by examining the poles of the transfer function matrix. Recall that, in the absence of pole-zero cancellation, the eigenvalues of the state matrix are the system poles. If the poles are all inside the unit circle, we conclude that the system is BIBO stable. However, we cannot conclude asymptotic stability except in the absence of unstable pole-zero cancellation. We reexamine this issue later in this chapter.

EXAMPLE 8.4

Test the BIBO stability of the transfer function of the system of Example 8.3.

Solution
The transfer function is

$$G(z) = \frac{\begin{bmatrix} z & -0.1 \\ -0.9 & z \end{bmatrix}}{z(z-0.3)(z+0.3)}$$

The system has poles at the origin, 0.3, and –0.3, all of which are inside the unit circle. Hence, the system is BIBO stable.

8.2 CONTROLLABILITY AND STABILIZABILITY

When obtaining z-transfer functions from discrete-time state–space equations, we discover that it is possible to have modes that do not affect the input–output relationship. This occurs as a result of pole-zero cancellation in the transfer function. We now examine this phenomenon more closely to assess its effect on control systems.

The time response of a system is given by a sum of the system modes weighted by eigenvectors. It is therefore important that each system mode be influenced by the input so as to be able to select a time response that meets design specifications. Systems possessing this property are said to be **completely controllable** or simply **controllable**. Modes that cannot be influenced by the control are called **uncontrollable modes**. A more precise definition of controllability is given next.

Definition 8.4: Controllability. A system is said to be controllable if for any initial state $\mathbf{x}(k_0)$ there exists a control sequence $\mathbf{u}(k)$, $k = k_0, k_0 + 1, \ldots, k_f - 1$, such that an arbitrary final state $\mathbf{x}(k_f)$ can be reached in finite k_f. ∎

Controllability can be given a geometric interpretation if we consider a second-order system. Figure 8.1 shows the decomposition of the state plane into a controllable subspace and an uncontrollable subspace. The controllable subspace represents states that are reachable using the control input, whereas the uncontrollable subspace represents states that are not reachable. Some authors prefer to use the term **reachability** instead of "controllability." Reachable systems are ones where every point in state space is reachable from the origin.[1] This definition is equivalent to our definition of controllability.

The following theorem establishes that the previous controllability definition is equivalent to the ability of the system input to influence all its modes.

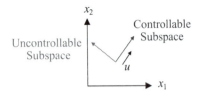

FIGURE 8.1

Controllable and uncontrollable subspaces.

---

[1]Traditionally, controllable systems are defined as systems where the origin can be reached from every point in the state space. This is equivalent to our definition for continuous-time systems but not for discrete-time systems. For simplicity, we adopt the more practically relevant Definition 8.4.

Theorem 8.4: Controllability Condition. A linear time-invariant system is completely controllable if and only if the products $\mathbf{w}_i^T B_d$, $i = 1, 2, \ldots, n$ are all nonzero where $\mathbf{w}_i^T$ is the ith left eigenvector of the state matrix. Furthermore, modes for which the product $\mathbf{w}_i^T B_d$ is zero are uncontrollable.

PROOF.

Necessity and Uncontrollable Modes

Definition 8.4 with finite final time k_f guarantees that all system modes are influenced by the control. To see this, we examine the zero-input response

$$\mathbf{x}_{zi}(k) = \sum_{i=1}^{n} Z_i \mathbf{x}(k_0) \lambda_i^k$$

For any eigenvalue λ_i inside the unit circle the corresponding mode decays to zero exponentially regardless of the influence of the input. However, to go to the zero state in finite time, it is necessary that the input influence all modes. For a zero product $Z_i B_d$, the ith mode is not influenced by the control and can only go to zero asymptotically. Therefore, the ith mode is uncontrollable for zero $Z_i B_d$. In Section 7.4.3, we showed that Z_i is a matrix of rank 1 given by the product of the ith right eigenvector $\mathbf{v}_i$ of the state matrix, and its ith left eigenvector is $\mathbf{w}_i$. Hence, the product $Z_i B_d$ is given by

$$Z_i B_d = \mathbf{v}_i \mathbf{w}_i^T B_d = \begin{bmatrix} v_{i1} \\ v_{i2} \\ \vdots \\ v_{in} \end{bmatrix} \mathbf{w}_i^T B_d = \begin{bmatrix} v_{i1} \mathbf{w}_i^T \\ v_{i2} \mathbf{w}_i^T \\ \vdots \\ v_{in} \mathbf{w}_i^T \end{bmatrix} B_d$$

where $v_{ij}, j = 1, \ldots, n$ are the entries of the ith eigenvector of the state matrix. Therefore, the product $Z_i B_d$ is zero if and only if the product $\mathbf{w}_i^T B_d$ is zero.

Sufficiency

We examine the total response for any initial and terminal state. From (7.80), we have

$$\mathbf{x} = \mathbf{x}(k) - A_d^k \mathbf{x}(k_0) = \sum_{i=k_0}^{k-1} A_d^{k-i-1} B_d \mathbf{u}(i) \tag{8.11}$$

Using the Cayley-Hamilton theorem, it can be shown that for $k > n$, no "new" terms are added to the preceding summation. By that we mean that the additional terms will be linearly dependent and are therefore superfluous. Thus, if the vector $\mathbf{x}$ cannot be obtained in n sampling periods by proper control sequence selection, it cannot be obtained over a longer duration. We therefore assume that $k = n$ and use the expansion of (7.78) to obtain

$$\mathbf{x} = \sum_{i=0}^{n-1} \left[\sum_{j=1}^{n} Z_j \lambda_j^{n-i-1} \right] B_d \mathbf{u}(i) \tag{8.12}$$

The outer summation in (8.12) can be written in matrix form to obtain

$$\mathbf{x} = \left[\sum_{j=1}^{n} Z_j B_d \lambda_j^{n-1} \Bigg| \sum_{j=1}^{n} Z_j B_d \lambda_j^{n-2} \Bigg| \ldots \Bigg| \sum_{j=1}^{n} Z_j B_d \lambda_j \Bigg| \sum_{j=1}^{n} Z_j B_d \right] \begin{bmatrix} \mathbf{u}(0) \\ \hline \mathbf{u}(1) \\ \hline \vdots \\ \hline \mathbf{u}(n-2) \\ \hline \mathbf{u}(n-1) \end{bmatrix} \quad (8.13)$$

$$= L\mathbf{u}$$

The matrix B_d is $n \times m$ and is assumed of full rank m. We need to show that the $n \times m.n$ matrix L has full row rank. The constituent matrices Z_i, $i = 1, 2, \ldots, n$ are of rank 1 and each has a column linearity independent of the other constituent matrices. Therefore, provided that the individual products $Z_i B_d$ are all nonzero, each submatrix of L has rank 1, and L has n linearly independent columns (i.e., L has full rank n). Because the rank of a matrix cannot exceed the number of rows for fewer rows than columns, we have

$$\text{rank}\{L|\mathbf{x}\} = \text{rank}\{L\} = n \quad (8.14)$$

This condition guarantees the existence of a solution $\mathbf{u}$ to (8.12). However, the solution is, in general, nonunique because the number of elements of $\mathbf{u}$ (unknowns) is $m.n > n$. A unique solution exists only in the single-input case ($m = 1$). It follows that if the products $Z_i B_d$, $i = 1, 2, \ldots, n$ are all nonzero, then one can solve for the vector of input sequences needed to reach any $\mathbf{x}(k_f)$ from any $\mathbf{x}(k_0)$. As discussed in the proof of necessity, the product $Z_i B_d$ is zero if and only if the product $\mathbf{w}_i^T B_d$ is zero. ∎

The next theorem gives a simpler controllability test but, as first stated, it does not provide a means of determining which of the system modes is uncontrollable.

Theorem 8.5: Controllability Rank Condition. A linear time-invariant system is completely controllable if and only if the $n \times m.n$ **controllability matrix**

$$\mathscr{C} = \left[B_d \ \Big| \ A_d B_d \ \Big| \ \ldots \ \Big| \ A_d^{n-1} B_d \right] \quad (8.15)$$

has rank n.

PROOF. We first write (8.11) in the form

$$\mathbf{x} = \left[B_d \ \Big| A_d B_d \Big| \ldots \Big| A_d^{n-1} B_d \right] \begin{bmatrix} \mathbf{u}(n-1) \\ \hline \mathbf{u}(n-2) \\ \hline \vdots \\ \hline \mathbf{u}(1) \\ \hline \mathbf{u}(0) \end{bmatrix} \quad (8.16)$$

$$= \mathscr{C}\mathbf{u}$$

We now have a system of linear equations for which (8.15) is a necessary and sufficient condition for a solution **u** to exist for any **x**. Hence, (8.15) is a necessary and sufficient condition for controllability.   ∎

If the controllability matrix $\mathscr{C}$ is rank deficient, then the rank deficiency is equal to the number of linearly independent row vectors that are mapped to zero on multiplication by $\mathscr{C}$. These row vectors are the uncontrollable states of the system as well as the left eigenvectors of the state matrix A. The vectors are also the transpose of the right eigenvectors of A^T. Thus, the rank test can be used to determine the uncontrollable modes of the system if the eigenstructure of A^T is known.

EXAMPLE 8.5

Determine the controllability of the following state equation:

$$\begin{bmatrix} x_1(k+1) \\ x_2(k+1) \\ x_3(k+1) \end{bmatrix} = \begin{bmatrix} -2 & -2 & 0 \\ 0 & 0 & 1 \\ 0 & -0.4 & -0.5 \end{bmatrix} \begin{bmatrix} x_1(k) \\ x_2(k) \\ x_3(k) \end{bmatrix} + \begin{bmatrix} 1 & 0 \\ 0 & 1 \\ 1 & 1 \end{bmatrix} \mathbf{u}(k)$$

Solution

The controllability matrix for the system is

$$\mathscr{C} = \begin{bmatrix} B_d & \vdots & A_d B_d & \vdots & A_d^2 B_d \end{bmatrix}$$

$$= \begin{bmatrix} 1 & 0 & -2 & -2 & 2 & 2 \\ 0 & 1 & 1 & 1 & -0.5 & -0.9 \\ 1 & 1 & -0.5 & -0.9 & -0.15 & 0.05 \end{bmatrix}$$

The matrix has rank 3, which implies that the third-order system is controllable. In fact, the first three columns of the matrix are linearly independent, and the same conclusion can be reached without calculating the entire controllability matrix. In general, one can gradually compute more columns until n linearly independent columns are obtained to conclude controllability for an nth-order system.

Although the controllability tests are given here for discrete-time systems, they are applicable to continuous-time systems. This fact is used in the following two examples.

EXAMPLE 8.6

Show that the following state equation is uncontrollable, and determine the uncontrollable mode. Then obtain the discrete-time state equation, and verify that it is also uncontrollable.

$$\begin{bmatrix} \dot{x}_1(t) \\ \dot{x}_2(t) \end{bmatrix} = \begin{bmatrix} 0 & 1 \\ 0 & 0 \end{bmatrix} \begin{bmatrix} x_1(t) \\ x_2(t) \end{bmatrix} + \begin{bmatrix} -1 \\ 0 \end{bmatrix} u(t)$$

Solution

The system has a zero eigenvalue with multiplicity 2. The controllability matrix for the system is

$$\mathcal{C} = [B \mid A B]$$
$$= \begin{bmatrix} -1 & 0 \\ 0 & 0 \end{bmatrix}$$

The matrix has rank 1 (less than 2), which implies that the second-order system has one $(2 - 1)$ uncontrollable mode. In fact, integrating the state equation reveals that the second state variable is governed by the equation

$$x_2(t) = x_2(t_0), \quad t \geq t_0$$

Therefore, the second state variable cannot be altered using the control and corresponds to an uncontrollable mode. The first state variable is influenced by the input and can be arbitrarily controlled.

The state-transition matrix for this system can be obtained directly using the series expansion because the matrix A^i is zero for $i > 1$. Thus, we have the discrete-time state matrix

$$A_d = e^{AT} = I_2 + AT = \begin{bmatrix} 1 & T \\ 0 & 1 \end{bmatrix}$$

The discrete-time input matrix is

$$B_d = \int_0^T e^{A\tau} B d\tau = \int_0^T [I_2 + A\tau] \begin{bmatrix} -1 \\ 0 \end{bmatrix} d\tau = \begin{bmatrix} -T \\ 0 \end{bmatrix}$$

The controllability matrix for the discrete-time system is

$$\mathcal{C} = [B_d \mid A_d B_d]$$
$$= \begin{bmatrix} -T & -T \\ 0 & 0 \end{bmatrix}$$

As with the continuous time system, the matrix has rank 1 (less than 2), which implies that the second-order system is uncontrollable.

The solution of the difference equation for the second state variable gives

$$x_2(k) = x_2(k_0), \quad k \geq k_0$$

with initial time k_0. Thus, the second state variable corresponds to an uncontrollable unity mode (e^0), whereas the first state variable is controllable as with the continuous system.

EXAMPLE 8.7

Show that the state equation for the motor system

$$\begin{bmatrix} \dot{x}_1 \\ \dot{x}_2 \\ \dot{x}_3 \end{bmatrix} = \begin{bmatrix} 0 & 1 & 0 \\ 0 & 0 & 1 \\ 0 & -10 & -11 \end{bmatrix} \begin{bmatrix} x_1 \\ x_2 \\ x_3 \end{bmatrix} + \begin{bmatrix} 1 \\ 0 \\ 0 \end{bmatrix} u$$

with DAC and ADC is uncontrollable, and determine the uncontrollable modes. Obtain the transfer functions of the continuous-time and discrete-time systems, and relate the uncontrollable modes to the poles of the transfer function.

Solution

From Example 7.7 the state-transition matrix of the system is

$$e^{At} = \begin{bmatrix} 10 & 11 & 1 \\ 0 & 0 & 0 \\ 0 & 0 & 0 \end{bmatrix} \frac{e^0}{10} + \begin{bmatrix} 0 & -10 & -1 \\ 0 & 10 & 1 \\ 0 & -10 & -1 \end{bmatrix} \frac{e^{-t}}{9} + \begin{bmatrix} 0 & 1 & 1 \\ 0 & -10 & -10 \\ 0 & 100 & 100 \end{bmatrix} \frac{e^{-10t}}{90}$$

Multiplying by the input matrix, we obtain

$$e^{At}B = \begin{bmatrix} 1 \\ 0 \\ 0 \end{bmatrix} e^0 + \begin{bmatrix} 0 \\ 0 \\ 0 \end{bmatrix} \frac{e^{-t}}{9} + \begin{bmatrix} 0 \\ 0 \\ 0 \end{bmatrix} \frac{e^{-10t}}{90} = \begin{bmatrix} 1 \\ 0 \\ 0 \end{bmatrix} \frac{e^0}{10}$$

Thus, the modes e^{-t} and e^{-10t} are both uncontrollable for the analog subsystem. The two modes are also uncontrollable for the digital system, including DAC and ADC.

Using the results of Example 7.9, the input matrix for the discrete-time system is

$$B_d = Z_1 BT + Z_2 B(1 - e^{-T}) + Z_3 B(1 - e^{-10T}) = \begin{bmatrix} 1 \\ 0 \\ 0 \end{bmatrix} T$$

and the matrix $A_d = e^{AT}$. The controllability matrix for the discrete-time system is

$$\mathscr{C} = \begin{bmatrix} B_d & | & A_d B_d & | & A_d^2 B_d \end{bmatrix}$$
$$= \begin{bmatrix} 1 & | & 1 & | & 1 \\ 0 & | & 0 & | & 0 \\ 0 & | & 0 & | & 0 \end{bmatrix} T$$

which clearly has rank 1, indicating that there are 2 (i.e., 3 − 1) uncontrollable modes. The left eigenvectors corresponding to the eigenvalues e^{-T} and e^{-10T} have zero first entry, and their product with the controllability matrix $\mathscr{C}$ is zero. Hence, the two corresponding modes are not controllable. The zero eigenvalue has the left eigenvector [1, 0, 0], and its product with $\mathscr{C}$ is nonzero. Thus, the corresponding mode is controllable.

The transfer function of the continuous-time system with output x_1 is

$$G(s) = C[sI_3 - A]^{-1}B = \frac{1}{s}$$

This does not include the uncontrollable modes (input-decoupling modes) because they cancel when the resolvent matrix and the input matrix are multiplied.

The z-transfer function corresponding to the system with DAC and ADC is

$$G_{ZAS}(z) = \frac{z-1}{z} \mathscr{Z} \left\{ \mathscr{L}^{-1} \left[\frac{G(s)}{s} \right] \right\} = \frac{T}{z-1}$$

The reader can easily verify that the transfer function is identical to that obtained using the discrete-time state–space representation. It includes the pole at e^0 but has no poles at e^{-T} or e^{-10T}.

8.2.1 MATLAB Commands for Controllability Testing

The MATLAB commands to calculate the controllability matrix and determine its rank are

$$>> c = ctrb(A, C)$$

$$>> rank(c)$$

8.2.2 Controllability of Systems in Normal Form

Checking controllability is particularly simple if the system is given in normal form—that is, if the state matrix is in the form

$$A = \text{diag}\{\lambda_1(A), \lambda_2(A), \ldots, \lambda_n(A)\}$$

The corresponding state-transition matrix is

$$e^{At} = \mathcal{L}\{[sI_n - A]^{-1}\} = \text{diag}\{e^{\lambda_1(A)t}, e^{\lambda_2(A)t}, \ldots, e^{\lambda_n(A)t}\} \tag{8.17}$$

The discrete-time state matrix is

$$A_d = e^{AT} = \text{diag}\{e^{\lambda_1(A)T}, e^{\lambda_2(A)T}, \ldots, e^{\lambda_n(A)T}\}$$
$$= \text{diag}\{\lambda_1, \lambda_2, \ldots, \lambda_n\} \tag{8.18}$$

The following theorem establishes controllability conditions for a system in normal form.

Theorem 8.6: Controllability of Systems in Normal Form. A system in normal form is controllable if and only if its input matrix has no zero rows. Furthermore, if the input matrix has a zero row, then the corresponding mode is uncontrollable.

PROOF.

Necessity and Uncontrollable Modes
For a system in normal form, the state equations are in the form

$$x_i(k+1) = -\lambda_i x_i(k) + \mathbf{b}_i^T \mathbf{u}(k), \quad i = 1, 2, \ldots, n$$

where $\mathbf{b}_i^T$ is the ith row of the input matrix B_d. For a zero row, the system is unforced and can only converge to zero asymptotically for $|\lambda_i|$ inside the unit circle. Because controllability requires convergence to any final state (including the origin) in finite time, the ith mode is uncontrollable for zero $\mathbf{b}_i^T$.

Sufficiency

From the sufficiency proof of Theorem 8.4, we obtain equation (8.13) to be solved for the vector **u** of controls over the period $k = 0, 1, \ldots, n - 1$. The solution exists if the matrix L in (8.13) has rank n. For a system in normal form, the state-transition matrix is in the form

$$A_d^k = \operatorname{diag}\left\{\lambda_i^k\right\}$$

and the $n \times n.m$ matrix L is in the form

$$L = \left[\operatorname{diag}\left\{\lambda_j^{n-1}\right\} B_d \; \vdots \; \operatorname{diag}\left\{\lambda_j^{n-2}\right\} B_d \; \vdots \ldots \vdots \; \operatorname{diag}\left\{\lambda_j\right\} B_d \; \vdots \; B_d\right]$$

Substituting for B_d in terms of its rows and using the rules for multiplying partitioned matrices, we obtain

$$L = \begin{bmatrix} \lambda_1^{n-1}\mathbf{b}_1^T & \lambda_{1c}^{n-2}\mathbf{b}_1^T & \cdots & \mathbf{b}_1^T \\ \lambda_2^{n-1}\mathbf{b}_2^T & \lambda_2^{n-2}\mathbf{b}_2^T & \cdots & \mathbf{b}_2^T \\ \vdots & \vdots & \ddots & \vdots \\ \lambda_n^{n-1}\mathbf{b}_n^T & \lambda_n^{n-2}\mathbf{b}_n^T & \cdots & \mathbf{b}_n^T \end{bmatrix}$$

For a matrix B_d with no zero rows, the rows of L are linearly independent and the matrix has full rank n. This guarantees the existence of a solution **u** to (8.13). ∎

EXAMPLE 8.8

Determine the controllability of the system

$$\begin{bmatrix} x_1(k+1) \\ x_2(k+1) \\ x_3(k+1) \end{bmatrix} = \begin{bmatrix} -2 & 0 & 0 \\ 0 & 0 & 0 \\ \hline 0 & 0 & -0.5 \end{bmatrix} \begin{bmatrix} x_1(k) \\ x_2(k) \\ x_3(k) \end{bmatrix} + \begin{bmatrix} 1 & 0 \\ 0 & 1 \\ 1 & 1 \end{bmatrix} \mathbf{u}(k)$$

Solution

The system is in normal form, and its input matrix has no zero rows. Hence, the system is completely controllable.

8.2.3 Stabilizability

The system in Example 8.8 is controllable but has one unstable eigenvalue (outside the unit circle) at (-2). This clearly demonstrates that controllability implies the ability to control the modes of the system regardless of their stability. If the system has uncontrollable modes, then these modes cannot be influenced by the control and may or may not decay to zero asymptotically. Combining the concepts of stable and controllable modes gives the following definition.

Definition 8.5: Stabilizability. A system is said to be stabilizable if all its uncontrollable modes are asymptotically stable.   ∎

Physical systems are often stabilizable rather than controllable. This poses no
problem provided that the uncontrollable dynamics decay to zero is sufficiently
fast so as not to excessively slow down the system.

EXAMPLE 8.9

Determine the controllability and stabilizability of the system

$$\begin{bmatrix} x_1(k+1) \\ x_2(k+1) \\ x_3(k+1) \end{bmatrix} = \begin{bmatrix} -2 & 0 & 0 \\ 0 & 0 & 0 \\ 0 & 0 & -0.5 \end{bmatrix} \begin{bmatrix} x_1(k) \\ x_2(k) \\ x_3(k) \end{bmatrix} + \begin{bmatrix} 1 & 0 \\ 0 & 0 \\ 1 & 1 \end{bmatrix} u(k)$$

Solution
The system is in normal form and its input matrix has one zero row corresponding to its
zero eigenvalue. Hence, the system is uncontrollable. However, the uncontrollable mode at
the origin is asymptotically stable, and the system is therefore stabilizable.
    An alternative procedure to determine the stabilizability of the system is to transform it
to a form that partitions the state into a controllable part and an uncontrollable part, and
then determine the asymptotic stability of the uncontrollable part. We can do this by exploiting the following theorem.

Theorem 8.7: Standard Form for Uncontrollable Systems. Consider the pair (A_d, B_d)
with n_c controllable modes and $n - n_c$ uncontrollable modes and the transformation
matrix

$$T_c = \begin{bmatrix} \mathbf{q}_1^T \\ \vdots \\ \mathbf{q}_{n_c}^T \\ \hline \mathbf{q}_{n_c+1}^T \\ \vdots \\ \mathbf{q}_n^T \end{bmatrix}^{-1} = \begin{bmatrix} Q_1 \\ \hline Q_2 \end{bmatrix}^{-1} \tag{8.19}$$

where $\{\mathbf{q}_i^T, i = n_c+1, \ldots, n\}$ are linearly independent vectors in the null space of the
controllability matrix $\mathscr{C}$ and $\{\mathbf{q}_i^T, i = 1, \ldots, n_c\}$ are arbitrary linearly independent vectors
selected to make T_c nonsingular.
    The transformed state-transition matrix **A** and the transformed input matrix **B** have
the following form:

$$A = \begin{bmatrix} A_c & A_{cuc} \\ \hline 0 & A_{uc} \end{bmatrix} \quad B = \begin{bmatrix} B_c \\ \hline 0 \end{bmatrix} \tag{8.20}$$

where A_c is an $n_c \times n_c$ matrix, B_c is an $n_c \times m$ matrix, and the pair (A_c, B_c) is
controllable.

PROOF. If the rank of the controllability matrix $\mathscr{C}$ is n_c, there exist linearly independent vectors $\{\mathbf{q}_i^T, i = n_c + 1, \ldots, n\}$ satisfying $\mathbf{q}_i^T \mathscr{C} = \mathbf{0}^T$, and the transformation T_c can be obtained by adding n_c linearly independent vectors. The null space of the controllability matrix is the unreachable space of the system and is also spanned by the set of left eigenvectors corresponding to uncontrollable modes. Hence, the vectors $\{\mathbf{q}_i^T, i = n_c + 1, \ldots, n\}$ are linear combinations of the left eigenvectors of the matrix A_d corresponding to uncontrollable modes. We can therefore write the transformation matrix in the form

$$T_c = \left[\begin{array}{c} Q_1 \\ \hline Q_2 W_{uc} \end{array} \right]^{-1}$$

where the matrix of coordinates $\bar{Q}_2$ is nonsingular and W_{uc} is a matrix whose rows are the left eigenvectors of A_d, $\{\mathbf{w}_i^T, i = 1, \ldots, n - n_c\}$, corresponding to uncontrollable modes. By the eigenvector controllability test, if the ith mode is uncontrollable, we have

$$\mathbf{w}_i^T B_d = \mathbf{0}^T, i = 1, \ldots, n - n_c$$

Therefore, transforming the input matrix gives

$$B = T_c^{-1} B_d = \left[\begin{array}{c} Q_1 \\ \hline Q_2 W_{uc} \end{array} \right] B_d = \left[\begin{array}{c} B_c \\ \hline 0 \end{array} \right]$$

Next, we show that the transformation yields a state matrix in the form (8.20). We first recall that the left and right eigenvectors can be scaled to satisfy the condition

$$\mathbf{w}_i^T \mathbf{v}_i = 1, \quad i = 1, \ldots, n - n_c$$

If we combine the condition for all the eigenvectors corresponding to uncontrollable modes, we have

$$W_{uc} V_{uc} = \left[\mathbf{w}_i^T \mathbf{v}_j \right] = I_{n-n_c}$$

We write the transformation matrix in the form

$$T_c = [R_1 \mid R_2] = [R_1 \mid V_{uc} \bar{R}_2]$$

which satisfies the condition

$$T_c^{-1} T_c = \left[\begin{array}{c} Q_1 \\ \hline Q_2 W_{uc} \end{array} \right] [R_1 \mid V_{uc} \bar{R}_2] = \left[\begin{array}{c|c} I_{n_c} & 0 \\ \hline 0 & I_{n-n_c} \end{array} \right]$$

Because $\bar{Q}_2$ is nonsingular, we have the equality

$$W_{uc} R_1 = 0$$

The similarity transformation gives

$$T_c^{-1} A_d T_c = \left[\begin{array}{c} Q_1 \\ \hline Q_2 W_{uc} \end{array} \right] A_d [R_1 \quad V_{uc} \bar{R}_2]$$

$$= \left[\begin{array}{c|c} Q_1 A_d R_1 & Q_1 A_d V_{uc} \bar{R}_2 \\ \hline Q_2 W_{uc} A_d R_1 & Q_2 W_{uc} A_d V_{uc} \bar{R}_2 \end{array} \right]$$
(8.21)

Because premultiplying the matrix A_d by a left eigenvector gives the same eigenvector scaled by an eigenvalue, we have the condition

$$W_{uc} A_d = \Lambda_u W_{uc}, \ \Lambda_u = \text{diag}\{\lambda_{u1}, \dots, \lambda_{u,n-nc}\}$$

with λ_u denoting an uncontrollable eigenvalue. This simplifies our matrix equality (8.21) to

$$T_c^{-1} A_d T_c = \left[\begin{array}{c|c} Q_1 A_d R_1 & Q_1 A_d V_{uc} \bar{R}_2 \\ \hline \bar{Q}_2 \Lambda_u W_{uc} R_1 & \bar{Q}_2 \Lambda_u W_{uc} V_{uc} \bar{R}_2 \end{array}\right] = \left[\begin{array}{c|c} Q_1 A_d R_1 & Q_1 A_d V_{uc} \bar{R}_2 \\ \hline \bar{Q}_2 \Lambda_u W_{uc} R_1 & \bar{Q}_2 \Lambda_u \bar{R}_2 \end{array}\right]$$

$$= \left[\begin{array}{c|c} Q_1 A_d R_1 & Q_1 A_d V_{uc} \bar{R}_2 \\ \hline 0 & \bar{Q}_2 \Lambda_u \bar{R}_2 \end{array}\right] = \left[\begin{array}{c|c} A_c & A_{cuc} \\ \hline 0 & A_{uc} \end{array}\right]$$

Because all the uncontrollable modes correspond to the eigenvalues of the matrix A_{uc}, the remaining modes are all controllable, and the pair (A_c, B_c) is completely controllable. ∎

Because the pair (A_d, B_d) has n_c controllable modes, the rank of the controllability matrix is n_c and the transformation matrix T_c is guaranteed to exist. The columns $\{q_i^T, i = n_c + 1, \dots, n\}$ of the inverse transformation matrix have a geometric interpretation. They form a basis set for the unobservable subspace of the system and can be selected as the eigenvectors of the uncontrollable modes of the system.

The remaining columns of T_c form a basis set of the controllable subspace. The similarity transformation is a change of basis that allows us to separate the controllable subspace from the uncontrollable subspace. After transformation, we can check the stabilizability of the system by verifying that all the eigenvalues of the matrix A_{uc} are inside the unit circle.

EXAMPLE 8.10

Determine the controllability, stability, and stabilizability of the system

$$\begin{bmatrix} x_1(k+1) \\ x_2(k+1) \\ x_3(k+1) \end{bmatrix} = \begin{bmatrix} -1.85 & 4.2 & -0.15 \\ -0.3 & 0.1 & 0.3 \\ -1.35 & 4.2 & -0.65 \end{bmatrix} \begin{bmatrix} x_1(k) \\ x_2(k) \\ x_3(k) \end{bmatrix} + \begin{bmatrix} 4 & 3 \\ 1 & 1 \\ 2 & 1 \end{bmatrix} u(k)$$

Solution
We first find the controllability matrix of the system

$$\mathscr{C} = \begin{bmatrix} B_d & A_d B_d & A_d^2 B_d \end{bmatrix}$$

$$= \begin{bmatrix} 4 & 3 & -3.5 & -1.5 & 4.75 & 0.75 \\ 1 & 1 & -0.5 & -0.5 & 0.25 & 0.25 \\ 2 & 1 & -2.5 & -0.5 & 4.25 & 0.25 \end{bmatrix}$$

The matrix has rank 2, and the first two columns are linearly independent. The dimension of the null space is $3 - 2 = 1$. We form a transformation matrix of the two first columns of the controllability matrix and a third linearly independent column, giving

$$
T_c = \begin{bmatrix} 4 & 3 & 0 \\ 1 & 1 & 0 \\ 2 & 1 & 1 \end{bmatrix}
$$

The system is transformed to

$$
\begin{bmatrix} x_1(k+1) \\ x_2(k+1) \\ x_3(k+1) \end{bmatrix} = \begin{bmatrix} -2 & 0 & -1.05 \\ 1.5 & -0.5 & 1.35 \\ 0 & 0 & 0.1 \end{bmatrix} \begin{bmatrix} x_1(k) \\ x_2(k) \\ x_3(k) \end{bmatrix} + \begin{bmatrix} 1 & 0 \\ 0 & 1 \\ 0 & 0 \end{bmatrix} \mathbf{u}(k)
$$

The system has one uncontrollable mode corresponding to the eigenvalue at 0.1, which is inside the unit circle. The system is therefore stabilizable but not controllable. The two other eigenvalues are at -2, outside the unit circle, and at -0.5, inside the unit circle, and both corresponding modes are controllable. The system is unstable because one eigenvalues is outside the unit circle.

8.3 OBSERVABILITY AND DETECTABILITY

To effectively control a system, it may be advantageous to use the system state to select the appropriate control action. Typically, the output or measurement vector for a system includes a combination of some but not all the state variables. The system state must then be estimated from the time history of the measurements and controls. However, state estimation is only possible with proper choice of the measurement vector. Systems whose measurements are so chosen are said to be **completely observable** or simply **observable**. Modes that cannot be detected using the measurements and the control are called **unobservable modes**. A more precise definition of observability is given next.

Definition 8.6: Observability. A system is said to be observable if any initial state $\mathbf{x}(k_0)$ can be estimated from the control sequence $\mathbf{x}(k)$, $k = k_0, k_0 + 1, \ldots, k_f - 1$, and the measurements $\mathbf{y}(k)$, $k = k_0, k_0 + 1, \ldots, k_f$. ∎

As in the case of controllability, observability can be given a geometric interpretation if we consider a second-order system. Figure 8.2 shows the decomposition of the state plane into an observable subspace and an unobservable subspace. The observable subspace includes all initial states that can be identified using the measurement history, whereas the unobservable subspace includes states that are indistinguishable.

The following theorem establishes that this observability definition is equivalent to the ability to estimate all system modes using its controls and measurements.

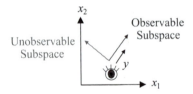

FIGURE 8.2

Observable and unobservable subspaces.

Theorem 8.8: Observability. A system is observable if and only if $C\mathbf{v}_i$ is nonzero for $i = 1, 2, \ldots, n$, where $\mathbf{v}_i$ is the ith eigenvector of the state matrix. Furthermore, if the product $C\,\mathbf{v}_i$ is zero, then the ith mode is unobservable.

PROOF. We first define the vector

$$\bar{\mathbf{y}}(k) = \mathbf{y}(k) - C\sum_{i=k_0}^{k-1} A_d^{k-i-1} B_d \mathbf{u}(i) - D\mathbf{u}(k)$$

$$= CA_d^k \mathbf{x}(k_0) = \sum_{i=1}^{n} CZ_i \lambda_i^k \mathbf{x}(k_0)$$

Sufficiency

Stack the output vectors $\bar{\mathbf{y}}(i), i = 1, \ldots, k$ to obtain

$$
\begin{bmatrix}
\bar{\mathbf{y}}(0) \\
\hline
\bar{\mathbf{y}}(1) \\
\hline
\vdots \\
\hline
\bar{\mathbf{y}}(n-2) \\
\hline
\bar{\mathbf{y}}(n-1)
\end{bmatrix}
=
\begin{bmatrix}
\sum_{j=1}^{n} CZ_j \\
\sum_{j=1}^{n} CZ_j \lambda_j \\
\vdots \\
\sum_{j=1}^{n} CZ_j \lambda_j^{n-2} \\
\sum_{j=1}^{n} CZ_j \lambda_j^{n-1}
\end{bmatrix}
\mathbf{x}(k_0)
\qquad (8.22)
$$

$$= L\mathbf{x}(k_0)$$

The products CZ_i are

$$CZ_i = C\mathbf{v}_i \mathbf{w}_i^T = C\mathbf{v}_i \begin{bmatrix} w_{i1} & w_{i2} & \cdots & w_{in} \end{bmatrix}$$

$$= C\begin{bmatrix} w_{i1}\mathbf{v}_i & w_{i2}\mathbf{v}_i & \cdots & w_{in}\mathbf{v}_i \end{bmatrix}$$

where $w_{ij}, j = 1, \ldots, n$ are the entries of the ith left eigenvector, and $\mathbf{v}_i$ is the ith right eigenvector of the state matrix A (see Section 7.4.3). Therefore, if the products are nonzero, the matrix L has n linearly independent columns (i.e., has rank n). An $n \times n$ submatrix of L can be formed by discarding the dependent rows. This leaves us with

n equations in the n unknown entries of the initial condition vector. A unique solution can thus be obtained.

Necessity

Let $\mathbf{x}(k_0)$ be in the direction of $\mathbf{v}_i$, the ith eigenvector of A_d, and let CZ_i be zero. Then the vector $y(k)$ is zero for any k regardless of the amplitude of the initial condition vector (recall that $Z_i\mathbf{v}_j = \mathbf{0}$ whenever $i \neq j$). Thus, all initial vectors in the direction $\mathbf{v}_i$ are indistinguishable using the output measurements, and the system is not observable.

∎

The following theorem establishes an observability rank test that can be directly checked using the state and output matrices.

Theorem 8.9: Observability Rank Condition. A linear time-invariant system is completely observable if and only if the $ln \times n$ **observability matrix**

$$\mathscr{O} = \begin{bmatrix} C \\ \hline CA_d \\ \hline \vdots \\ \hline CA_d^{n-1} \end{bmatrix} \tag{8.23}$$

has rank n.

PROOF. We first write (8.22) in the form

$$\begin{bmatrix} \overline{\mathbf{y}}(0) \\ \hline \overline{\mathbf{y}}(1) \\ \hline \vdots \\ \hline \overline{\mathbf{y}}(n-2) \\ \hline \overline{\mathbf{y}}(n-1) \end{bmatrix} = \begin{bmatrix} C \\ \hline CA_d \\ \hline \vdots \\ \hline CA_d^{n-1} \end{bmatrix} \mathbf{x}(k_0) \tag{8.24}$$

$$= \mathscr{O}\mathbf{x}(k_0)$$

We now have a system of linear equations that can include, at most, n linearly independent equations. If n independent equations exist, their coefficients can be used to form an $n \times n$ invertible matrix. The rank condition (8.23) is a necessary and sufficient condition for a unique solution $\mathbf{x}(k_0)$ to exist. Thus, (8.23) is a necessary and sufficient condition for observability.

∎

If the controllability matrix $\mathscr{C}$ is rank deficient, then the rank deficiency is equal to the number of linearly independent column vectors that are mapped to zero on multiplication by $\mathscr{O}$. This number equals the number of unobservable states of the system and the column vectors mapped to zero are the right eigenvectors of the state matrix. Thus, the rank test can be used to determine the unobservable modes of the system if the eigenstructure of the state matrix is known.

EXAMPLE 8.11

Determine the observability of the system using two different tests:

$$A = \begin{bmatrix} 0_{2\times1} & I_2 \\ 0 & -3 & 4 \end{bmatrix} \quad C = [0 \ \ 0 \ \ 1]$$

If the system is not completely observable, determine the unobservable modes.

Solution

Because the state matrix is in companion form, its characteristic equation is easily obtained from its last row. The characteristic equation is

$$\lambda^3 - 4\lambda^2 + 3\lambda = \lambda(\lambda - 1)(\lambda - 3) = 0$$

Hence, the system eigenvalues are {0, 1, 3}.

The companion form of the state matrix allows us to write the modal matrix of eigenvectors as the Van der Monde matrix:

$$V = \begin{bmatrix} 1 & 1 & 1 \\ 0 & 1 & 3 \\ 0 & 1 & 9 \end{bmatrix}$$

Observability is tested using the product of the output matrix and the modal matrix:

$$CV = [0 \ \ 0 \ \ 1] \begin{bmatrix} 1 & 1 & 1 \\ 0 & 1 & 3 \\ 0 & 1 & 9 \end{bmatrix} = [0 \ \ 1 \ \ 9]$$

The product of the output matrix and the eigenvector for the zero eigenvalue is zero. We conclude that the system has an output-decoupling zero at zero (i.e., one unobservable mode).

The observability matrix of the system is

$$\mathcal{O} = \begin{bmatrix} C \\ \hline CA_d \\ \hline CA_d^2 \end{bmatrix} = \begin{bmatrix} 0 & 0 & 1 \\ \hline 0 & -3 & 4 \\ \hline 0 & -12 & 13 \end{bmatrix}$$

which has rank 2 = 3 − 1. Hence, the system has one unobservable mode. The product of the observability matrix and the eigenvector for the zero eigenvalue is zero. Hence, it corresponds to the unobservable mode of the system.

8.3.1 MATLAB Commands

The MATLAB commands to test observability are

>> o = obsv(A, C) % Obtain the observability matrix

>> rank(o)

8.3.2 Observability of Systems in Normal Form

The observability of a system can be easily checked if the system is in normal form by exploiting the following theorem.

Theorem 8.10: Observability of Normal Form Systems. A system in normal form is observable if and only if its output matrix has no zero columns. Furthermore, if the input matrix has a zero column, then the corresponding mode is unobservable.

PROOF. The proof is similar to that of the controllability theorem (8.6) and is left as an exercise.

■

8.3.3 Detectability

If a system is not observable, it is preferable that the unobservable modes be stable. This property is called **detectability**.

Definition 8.7: Detectability. A system is detectable if all its unobservable modes decay to zero asymptotically.

■

EXAMPLE 8.12

Determine the observability and detectability of the system

$$
\begin{bmatrix} x_1(k+1) \\ x_2(k+1) \\ x_3(k+1) \end{bmatrix} = \begin{bmatrix} -0.4 & 0 & 0 \\ 0 & 3 & 0 \\ 0 & 0 & -2 \end{bmatrix} \begin{bmatrix} x_1(k) \\ x_2(k) \\ x_3(k) \end{bmatrix} + \begin{bmatrix} 1 & 0 \\ 0 & 0 \\ 1 & 1 \end{bmatrix} \mathbf{u}(k)
$$

$$
y(k) = \begin{bmatrix} 1 & 1 & 0 \end{bmatrix} \begin{bmatrix} x_1(k) \\ x_2(k) \\ x_3(k) \end{bmatrix}
$$

Solution
The system is in normal form, and its output matrix has one zero column corresponding to its eigenvalue -2. Hence, the system is unobservable. The unobservable mode at -2, $|-2| > 1$, is unstable, and the system is therefore not detectable.

Similar to the concepts described for stabilizability, a procedure to determine the detectability of the system is to transform it to a form that allows partitioning the state into an observable part and an unobservable part and then determine the asymptotic stability of the unobservable part. We do this by exploiting the following theorem.

Theorem 8.11: Standard Form for Unobservable Systems. Consider the pair (A_d, C_d) with n_o observable modes and $n - n_o$ unobservable modes and the $n \times n$ transformation matrix

$$
T_o = [T_{o1} \mid T_{o2}] = [\mathbf{t}_1 \quad \cdots \quad \mathbf{t}_{n-n_0} \mid \mathbf{t}_{n-n_0+1} \quad \cdots \quad \mathbf{t}_n]
$$

where $\{t_i, i = 1, \ldots, n - n_o\}$ are linearly independent vectors in the null space of the observability matrix $\mathscr{O}$ and $\{t_i, i = n - n_o + 1, \ldots, n\}$ are arbitrary vectors selected to make T_o invertible. The transformed state-transition matrix **A** and the transformed input matrix $\mathscr{E}$ have the following form:

$$A = \left[\begin{array}{c|c} A_u & A_{uo} \\ \hline 0_{n_0 \times n - n_0} & A_o \end{array} \right] \quad C = [\,0_{l \times n - n_0} \mid C_o\,]$$

where A_o is an $n_o \times n_o$ matrix, C_o is an $l \times n_o$ matrix, and the pair (A_o, C_o) is observable.

PROOF. The proof is similar to that of Theorem 8.7, and it is left as an exercise.   ■

Because the pair (A_d, C_d) has n_o observable modes, the rank of the observability matrix is n_o and the transformation matrix T_o is guaranteed to exist. The columns $\{t_i, i = 1, \ldots, n - n_o\}$ of the transformation matrix have a geometric interpretation. They form a basis set for the unobservable subspace of the system and can be selected as the eigenvectors of the unobservable modes of the system. The remaining columns of T_o form a basis set of the observable subspace. The similarity transformation is a change of basis that allows us to separate the observable subspace from the unobservable subspace. After transformation, we can check the detectability of the system by verifying that all the eigenvalues of the matrix A_u are inside the unit circle.

EXAMPLE 8.13

Determine the observability and detectability of the system

$$\begin{bmatrix} x_1(k+1) \\ x_2(k+1) \\ x_3(k+1) \end{bmatrix} = \begin{bmatrix} 2.0 & 4.0 & 2.0 \\ -1.1 & -2.5 & -1.15 \\ 2.6 & 6.8 & 2.8 \end{bmatrix} \begin{bmatrix} x_1(k) \\ x_2(k) \\ x_3(k) \end{bmatrix} + \begin{bmatrix} 1 & 0 \\ 1 & 1 \\ 0 & 1 \end{bmatrix} u(k)$$

$$y(k) = \begin{bmatrix} 2 & 10 & 3 \\ 1 & 8 & 3 \end{bmatrix} \begin{bmatrix} x_1(k) \\ x_2(k) \\ x_3(k) \end{bmatrix}$$

Solution
We first find the observability matrix of the system

$$\mathscr{O} = \begin{bmatrix} C \\ \hline CA_d \\ \hline CA_d^2 \end{bmatrix} = \begin{bmatrix} 2 & 10 & 3 \\ 1 & 8 & 3 \\ \hline 0.8 & 3.4 & 0.9 \\ 1.0 & 4.4 & 1.2 \\ \hline 0.2 & 0.82 & 0.21 \\ 0.28 & 0.16 & 0.3 \end{bmatrix}$$

The matrix has rank 2, and the system is not completely observable. The dimension of the null space is $3 - 2 = 1$, and there is only one vector satisfying $\mathscr{O} \mathbf{v}_1 = 8.230$ h. The vector is the eigenvector $\mathbf{v}_1$ corresponding to the unobservable mode satisfies $\mathscr{O} \mathbf{v}_1 = 8.230$ h and is given by

$$
\mathbf{v}_1 = \begin{bmatrix} 1 \\ -0.5 \\ 1 \end{bmatrix}
$$

The transformation matrix T_o can then be completed by adding any two linearly independent columns—for example,

$$
T_o = \begin{bmatrix} 1 & 1 & 0 \\ -0.5 & 0 & 1 \\ 1 & 0 & 0 \end{bmatrix}
$$

The system is transformed to

$$
\begin{bmatrix} x_1(k+1) \\ x_2(k+1) \\ x_3(k+1) \end{bmatrix} = \begin{bmatrix} 2 & 2.6 & 6.8 \\ 0 & -0.6 & -2.8 \\ 0 & 0.2 & 0.9 \end{bmatrix} \begin{bmatrix} x_1(k) \\ x_2(k) \\ x_3(k) \end{bmatrix} + \begin{bmatrix} 0 & 1 \\ 1 & -1 \\ 1 & 1.5 \end{bmatrix} \mathbf{u}(k)
$$

$$
y(k) = \begin{bmatrix} 0 & 2 & 10 \\ 0 & 1 & 8 \end{bmatrix} \begin{bmatrix} x_1(k) \\ x_2(k) \\ x_3(k) \end{bmatrix}
$$

The system has one unobservable mode corresponding to the eigenvalue at 2, which is outside the unit circle. The system is therefore not detectable.

8.4 POLES AND ZEROS OF MULTIVARIABLE SYSTEMS

As for single-input–single-output (SISO) systems, poles and zeros determine the stability, controllability, and observability of a multivariable system. For SISO systems, zeros are the zeros of the numerator polynomial of the scalar transfer function, whereas poles are the zeros of the denominator polynomial. For multi-input-multi-output (MIMO) systems, the transfer function is not scalar and it is no longer sufficient to determine the zeros and poles of individual entries of the transfer function matrix. In fact, element zeros do not play a major role in characterizing multivariable systems and their properties beyond their effect on the shape of the time response of the system. In our discussion of transfer function matrices in Section 7.8, we mention three important types of multivariable zeros:

1. Input-decoupling zeros, which correspond to uncontrollable modes
2. Output-decoupling zeros, which correspond to unobservable modes

3. Input–output-decoupling zeros (which correspond to modes that are neither controllable nor observable)

We can obtain those zeros and others, as well as the system poles, from the transfer function matrix.

8.4.1 Poles and Zeros from the Transfer Function Matrix

We first define system poles and zeros based on a transfer function matrix of a MIMO system.

Definition 8.8: Poles. Poles are the roots of the least common denominator of all nonzero minors of all orders of the transfer function matrix. ∎

The least common denominator of the preceding definition is known as the **pole polynomial** and is essential for the determination of the poles. The pole polynomial is in general not equal to the least common denominator of all nonzero elements, known as the **minimal polynomial**. The MATLAB command **pole** gives the poles of the system based on its realization. Unfortunately, the command may not use the minimal realization of the transfer function of the system. Thus, the command may give values for the pole that need not be included in a minimal realization of the transfer function.

From the definition, the reader may guess that the poles of the system are the same as the poles of the elements. Although this is correct, it is not possible to guess the multiplicity of the poles by inspection, as the following example demonstrates.

EXAMPLE 8.14

Determine the poles of the transfer function matrix

$$
G(z) = \begin{bmatrix} \dfrac{1}{z-1} & \dfrac{1}{(z-1)(z-0.5)} \\ \dfrac{z-0.1}{(z-0.2)(z-0.5)} & \dfrac{1}{z-0.2} \end{bmatrix}
$$

Solution
The least common denominator of the matrix entries is

$$(z-0.2)(z-0.5)(z-1)$$

The determinant of the matrix is

$$\det[G(z)] = \frac{1}{(z-1)(z-0.2)} - \frac{z-0.1}{(z-0.2)(z-0.5)^2(z-1)}$$

The denominator of the determinant of the matrix is

$$(z - 0.2)(z - 0.5)^2(z - 1)$$

The least common denominator of all the minors is

$$(z - 0.2)(z - 0.5)^2(z - 1)$$

Thus, the system has poles at {0.2, 0.5, 0.5, 1}. Note that some poles of more than one element of the transfer function are not repeated poles of the system.

For this relatively simple transfer function, we can obtain the minimal number of blocks needed for realization, as shown in Figure 8.3. This clearly confirms our earlier determination of the system poles.

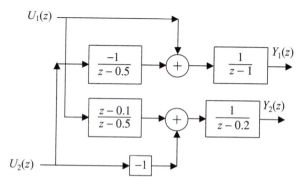

FIGURE 8.3

Block diagram of the system described in Example 8.14.

The MATLAB command **pole** gives

>> **pole(g)**

ans =

1.0000

0.5000

0.2000

1.0000

0.5000

0.2000

This includes two poles that are not needed for the minimal realization of the transfer function.

The definition of the multivariable zero is more complex than that of the pole, and several types of zeros can be defined. We define zeros as follows.

Definition 8.9: System Zeros. Consider an $l \times m$ transfer function matrix written such that each entry has a denominator equal to the pole polynomial. Zeros are values that make the transfer function matrix rank deficient, that is, values z_0 that satisfy either of the two equations

$$G(z_0)\mathbf{w}_{in} = 0, \mathbf{w}_{in} \neq 0 \tag{8.25}$$

$$\mathbf{w}_{out}^T G(z_0) = 0, \mathbf{w}_{out} \neq 0 \tag{8.26}$$

for some nonzero real vector $\mathbf{w}_{in}$ known as the **input zero direction** or a nonzero real vector $\mathbf{w}_{out}$ known as the **output zero direction**.   ∎

For any matrix, the rank is equal to the order of the largest nonzero minor. Thus, provided that the terms have the appropriate denominator, the zeros are the divisors of all minors of order equal to the minimum of the pair (l, m). For a square matrix, zeros of the determinant rewritten with the pole polynomial as its denominator are the zeros of the system. These definitions are examined in the following examples.

EXAMPLE 8.15

Determine the z-transfer function of a digitally controlled single-axis milling machine with a sampling period of 40 ms if its analog transfer function is given by[2]

$$G(s) = \text{diag}\left\{\frac{3150}{(s+35)(s+150)}, \frac{1092}{(s+35)(s+30)}\right\}$$

Find the poles and zeros of the transfer function.

Solution
Using the MATLAB command **c2d**, we obtain the transfer function matrix

$$G_{ZAS}(z) = \text{diag}\left\{\frac{0.4075(z+0.1067)}{(z-0.2466)(z-0.2479\times10^{-2})}, \frac{0.38607(z+0.4182)}{(z-0.2466)(z-0.3012)}\right\}$$

Because the transfer function is diagonal, the determinant is the product of its diagonal terms. The least common denominator of the determinant of the transfer function is

$$(z-0.2479\times10^{-2})(z-0.2466)^2(z-0.3012)$$

We therefore have poles at $\{0.2479\times10^{-2}, 0.2466, 0.2466, 0.3012\}$. The system is stable because all the poles are inside the unit circle.

---

[2]S. J. Rober and Y. C. Shin, Modeling and control of CNC machines using a PC-based open architecture controller, *Mechatronics* 5(4):401-420, 1995.

To obtain the zeros of the system, we rewrite the transfer function in the form

$$G_{ZAS}(z) = \frac{(z-0.2466)}{(z-0.2466)^2(z-0.2479\times10^{-2})(z-0.3012)} \times$$

$$\begin{bmatrix} 0.4075(z+0.1067)(z-0.3012) & 0 \\ 0 & 0.38607(z+0.4182)(z-0.2479\times10^{-2}) \end{bmatrix}$$

For this square 2-input-2-output system, the determinant of the transfer function is

$$\frac{(z+0.1067)(z-0.3012)(z+0.4182)(z-0.2479\times10^{-2})(z-0.2466)^2}{(z-0.2466)^4(z-0.2479\times10^{-2})^2(z-0.3012)^2}$$

$$= \frac{(z+0.1067)(z+0.4182)}{(z-0.2466)^2(z-0.2479\times10^{-2})(z-0.3012)}$$

The roots of the numerator are the zeros {−0.1067, −0.4182}.
The same answer is obtained using the MATLAB command **tzero**.

EXAMPLE 8.16

Determine the zeros of the transfer function matrix

$$G(z) = \begin{bmatrix} \dfrac{1}{z-1} & \dfrac{1}{(z-1)(z-0.5)} \\[2ex] \dfrac{z-0.1}{(z-0.2)(z-0.5)} & \dfrac{1}{z-0.2} \end{bmatrix}$$

Solution
From Example 8.14, we know that the poles of the system are at {0.2, 0.5, 0.5, 1}. We rewrite the transfer function matrix in the form

$$G(z) = \frac{\begin{bmatrix} (z-0.2)(z-0.5)^2 & (z-0.2)(z-0.5) \\ (z-0.1)(z-0.5)(z-1) & (z-0.5)^2(z-1) \end{bmatrix}}{(z-0.2)(z-0.5)^2(z-1)}$$

Zeros are the roots of the greatest common divisor of all minors of order equal to 2, which in this case is simply the determinant of the transfer function matrix. The determinant of the matrix with cancellation to reduce the denominator to the characteristic polynomial is

$$\det[G(z)] = \frac{(z-0.5)^2-(z-0.1)}{(z-0.2)(z-0.5)^2(z-1)} = \frac{z^2-2z+0.35}{(z-0.2)(z-0.5)^2(z-1)}$$

The roots of the numerator yield zeros at {1.8062, −0.1938}.
The same answer is obtained using the MATLAB command **zero**.

8.4.2 Zeros from State–Space Models

System zeros can be obtained from a state-space realization using the definition of the zero. However, this is complicated by the possibility of pole-zero cancellation if the system is not minimal. We rewrite the zero condition (8.26) in terms of the state-space matrices as

$$G(z_0)\mathbf{w} = C[z_0 I_n - A]^{-1} B\mathbf{w} + D\mathbf{w}$$
$$= C\mathbf{x}_w + D\mathbf{w} = 0 \tag{8.27}$$
$$\mathbf{x}_w = [z_0 I_n - A]^{-1} B\mathbf{w}$$

where $\mathbf{x}_w$ is the state response for the system with input $\mathbf{w}$. We can now rewrite (8.27) in the following form, which is more useful for numerical solution:

$$\begin{bmatrix} -(z_0 I_n - A) & B \\ C & D \end{bmatrix} \begin{bmatrix} \mathbf{x}_w \\ \mathbf{w} \end{bmatrix} = \begin{bmatrix} 0 \\ 0 \end{bmatrix} \tag{8.28}$$

The matrix in (8.28) is known as **Rosenbrock's system matrix**. Equation (8.28) can be solved for the unknowns provided that a solution exists.

Note that the zeros are invariant under similarity transformation because

$$\begin{bmatrix} -(z_0 I_n - T_r^{-1} A T_r) & T_r^{-1} B \\ CT & D \end{bmatrix} \begin{bmatrix} \mathbf{x}_w \\ \mathbf{w} \end{bmatrix} = \begin{bmatrix} T_r^{-1} & 0 \\ 0 & I_l \end{bmatrix} \begin{bmatrix} -(z_0 I_n - A) & B \\ C & D \end{bmatrix} \begin{bmatrix} T_r \mathbf{x}_w \\ \mathbf{w} \end{bmatrix} = \begin{bmatrix} 0 \\ 0 \end{bmatrix}$$

In Chapter 9, we discuss state feedback where the control input is obtained as a linear combination of the measured state variables. We show that the zeros obtained from (8.28) are also invariant under state feedback. Hence, the zeros are known as **invariant zeros**.

The invariant zeros include decoupling zeros if the system is not minimal. If these zeros are removed from the set of invariant zeros, then the remaining zeros are known as **transmission zeros**. In addition, not all decoupling zeros can be obtained from (8.28). We therefore have the following relation:

{system zeros} = {transmission zeros} + {input-decoupling zeros} +
{output-decoupling zeros} – {input-output-decoupling zeros}

Note that some authors refer to invariant zeros as transmission zeros and do not use the term "invariant zeros."

MATLAB calculates zeros from Rosenbrock's matrix of a state-space model **p** using the command

>> **zero(p)**

The MATLAB manual identifies the result as transmission zeros, but in our terminology this refers to invariant zeros. Although the command accepts a transfer function matrix, its results are based on a realization that need not be minimal and may include superfluous zeros that would not be obtained using the procedure described in Section 8.4.1.

EXAMPLE 8.17

Consider the system

$$
\begin{bmatrix} x_1(k+1) \\ x_2(k+1) \\ x_3(k+1) \end{bmatrix} =
\begin{bmatrix} -0.4 & 0 & 0 \\ 0 & 3 & 0 \\ \hline 0 & 0 & -2 \end{bmatrix}
\begin{bmatrix} x_1(k) \\ x_2(k) \\ x_3(k) \end{bmatrix} +
\begin{bmatrix} 1 & 0 \\ 0 & 0 \\ 1 & 1 \end{bmatrix} u(k)
$$

$$
y(k) = \begin{bmatrix} 1 & 1 & 0 \end{bmatrix}
\begin{bmatrix} x_1(k) \\ x_2(k) \\ x_3(k) \end{bmatrix}
$$

$$
\begin{bmatrix} -(z_0 I_n - A) & \vdots & B \\ \hline C & \vdots & D \end{bmatrix}
\begin{bmatrix} \mathbf{x}_w \\ \mathbf{w} \end{bmatrix} =
\begin{bmatrix}
-0.4 - z_0 & 0 & 0 & \vdots & 1 & 0 \\
0 & 3 - z_0 & 0 & \vdots & 0 & 0 \\
0 & 0 & -2 - z_0 & \vdots & 1 & 1 \\
\hline
0 & 0 & -2 & \vdots & 0 & 0
\end{bmatrix}
\begin{bmatrix} x_{w1} \\ x_{w2} \\ x_{w3} \\ w_1 \\ w_2 \end{bmatrix} =
\begin{bmatrix} 0 \\ 0 \\ 0 \\ 0 \\ 0 \end{bmatrix}
$$

From the second row we have the zero $z_0 = 3$, which is also an input-decoupling zero because it corresponds to an uncontrollable mode. The system also has an output-decoupling zero at -2, but this cannot be determined from Rosenbrock's system matrix.

8.5 STATE–SPACE REALIZATIONS

State-space realizations can be obtained from input–output time domain or z-domain models. Because every system has infinitely many state-space realizations, we only cover a few standard canonical forms that have special desirable properties. These realizations play an important role in digital filter design or controller implementation. Because difference equations are easily transformable to z-transfer functions and vice versa, we avoid duplication by obtaining realizations from either the z-transfer functions or the difference equations.

We also restrict the discussion to SISO transfer functions of the form

$$
\begin{aligned}
G(z) &= \frac{\mathbf{c}_n z^n + \mathbf{c}_{n-1} z^{n-1} + \ldots + \mathbf{c}_1 z + \mathbf{c}_0}{z^n + a_{n-1} z^{n-1} + a_1 z + a_0} \\
&= \mathbf{c}_n + \frac{c_{n-1} z^{n-1} + c_{n-2} z^{n-2} + \ldots + c_1 z + c_0}{z^n + a_{n-1} z^{n-1} + \ldots + a_1 z + a_0}
\end{aligned}
\tag{8.29}
$$

where the leading numerator coefficients $\mathbf{c}_n$ and c_{n-1} can be zero and $\mathbf{c}_n = c_n$, or the corresponding difference equation

$$
\begin{aligned}
&y(k+n) + a_{n-1} y(k+n-1) + \ldots + a_1 y(k+1) + a_0 y(k) \\
&= \mathbf{c}_n u(k+n) + \mathbf{c}_{n-1} u(k+n-1) + \ldots + \mathbf{c}_1 u(k+1) + \mathbf{c}_0 u(k)
\end{aligned}
\tag{8.30}
$$

8.5.1 Controllable Canonical Realization

The **controllable canonical realization** is so called because it possesses the property of controllability. The controllable form is also known as phase variable form or as the first controllable form. A second controllable form, known as controller form, is identical in structure to the first but with the state variables numbered backward. We begin by examining the special case of a difference equation whose RHS is the forcing function at time k. This corresponds to a transfer function with unity numerator. We then use our results to obtain realizations for a general SISO linear discrete-time system described by (8.29) or (8.30).

Systems with No Input Differencing

Consider the special case of a system whose difference equation includes the input at time k only. The difference equation considered is of the form

$$y(k+n)+a_{n-1}y(k+n-1)+\ldots+a_1y(k+1)+a_0y(k)=u(k) \qquad (8.31)$$

Define the state vector

$$\mathbf{x}(k)=[x_1(k) \quad x_2(k) \quad \ldots \quad x_{n-1}(k) \quad x_n(k)]^T$$
$$=[y(k) \quad y(k+1) \quad \ldots \quad y(k+n-2) \quad y(k+n-1)]^T \qquad (8.32)$$

Hence, the difference equation (8.31) can be rewritten as

$$x_n(k+1)=-a_{n-1}x_n(k)-\ldots-a_1x_2(k)-a_0x_1(k)+u(k) \qquad (8.33)$$

Using the definitions of the state variables and the difference equation (8.31), we obtain the matrix state equation

$$
\begin{bmatrix} x_1(k+1) \\ x_2(k+1) \\ . \\ x_{n-1}(k+1) \\ x_n(k+1) \end{bmatrix}
=
\begin{bmatrix}
0 & 1 & \ldots & 0 & 0 \\
0 & 0 & \ldots & 0 & 0 \\
\ldots & \ldots & \ldots & \ldots & \ldots \\
0 & 0 & \ldots & 0 & 1 \\
-a_0 & -a_1 & \ldots & -a_{n-2} & -a_{n-1}
\end{bmatrix}
\begin{bmatrix} x_1(k) \\ x_2(k) \\ . \\ x_{n-1}(k) \\ x_n(k) \end{bmatrix}
+
\begin{bmatrix} 0 \\ 0 \\ . \\ 0 \\ 1 \end{bmatrix} u(k) \qquad (8.34)
$$

and the output equation

$$
y(k)=[1 \quad 0 \quad \ldots \quad 0 \quad 0]
\begin{bmatrix} x_1(k) \\ x_2(k) \\ . \\ x_{n-1}(k) \\ x_n(k) \end{bmatrix}
\qquad (8.35)
$$

The state-space equations can be written more concisely as

$$\mathbf{x}(k+1) = \left[\begin{array}{c|ccc} \mathbf{0}_{n-1\times1} & & I_{n-1} & \\ \hline -a_0 & -a_1 & \cdots \quad \cdots & -a_{n-1} \end{array}\right] \mathbf{x}(k) + \left[\begin{array}{c} \mathbf{0}_{n-1\times1} \\ \hline 1 \end{array}\right] u(k)$$

(8.36)

$$y(k) = [1 \mid \mathbf{0}_{1\times n-1}]\mathbf{x}(k)$$

Clearly, the state-space equations can be written by inspection from the transfer function

$$G(z) = \frac{1}{z^n + a_{n-1}z^{n-1} + \ldots + a_1 z + a_0}$$

(8.37)

or from the corresponding difference equation because either includes all the needed coefficients, a_i, $i = 0, 1, 2, \ldots, n - 1$.

EXAMPLE 8.18

Obtain the controllable canonical realization of the difference equation

$$y(k+3) + 0.5y(k+2) + 0.4y(k+1) - 0.8y(k) = u(k)$$

using basic principles; then show how the realization can be written by inspection from the transfer function or the difference equation.

Solution
Select the state vector

$$\mathbf{x}(k) = [x_1(k) \quad x_2(k) \quad x_3(k)]^T$$
$$= [y(k) \quad y(k+1) \quad y(k+2)]^T$$

and rewrite the difference equation as

$$x_3(k+1) = -0.5x_3(k) - 0.4x_2(k) + 0.8x_1(k) + u(k)$$

Using the definitions of the state variables and the difference equation, we obtain the state–space equations

$$\begin{bmatrix} x_1(k+1) \\ x_2(k+1) \\ x_3(k+1) \end{bmatrix} = \begin{bmatrix} 0 & 1 & 0 \\ 0 & 0 & 1 \\ \hline 0.8 & -0.4 & -0.5 \end{bmatrix} \begin{bmatrix} x_1(k) \\ x_2(k) \\ x_3(k) \end{bmatrix} + \begin{bmatrix} 0 \\ 0 \\ 1 \end{bmatrix} u(k)$$

$$y(k) = [1 \mid 0 \quad 0] \begin{bmatrix} x_1(k) \\ x_2(k) \\ x_3(k) \end{bmatrix}$$

because the system is of order $n = 3$, $n - 1 = 2$. Hence, the upper half of the state matrix includes a 2 × 1 zero vector next to a 2 × 2 identity matrix. The input matrix is a column vector because the system is SI and it includes a 2 × 1 zero vector and a unity last entry. The output matrix is a row vector because the system is single output (SO) and it includes a 1 × 2 zero vector and a unity first entry. With the exception of the last row of the state matrix, all matrices in the state–space description are completely determined by the order of the system. The last row of the state matrix has entries equal to the coefficients of the output terms in the difference equation with their signs reversed. The same coefficients appear in the denominator of the transfer function

$$G(z) = \frac{1}{z^3 + 0.5z^2 + 0.4z - 0.8}$$

Therefore, the state–space equations for the system can be written by inspection from the transfer function or input–output difference equation.

Systems with Input Differencing

We now use the results from the preceding section to obtain the controllable canonical realization of the transfer function of equation (8.29). We assume that the constant term has been extracted from the transfer function, if necessary, and that we are left with the form

$$G(z) = \frac{Y(z)}{U(z)} = c_n + G_d(z) \tag{8.38}$$

where

$$G_d(z) = \frac{c_{n-1}z^{n-1} + c_{n-2}z^{n-2} + \ldots + c_1 z + c_0}{z^n + a_{n-1}z^{n-1} + \ldots + a_1 z + a_0} \tag{8.39}$$

We next consider a transfer function with the same numerator as G_d but with unity numerator, and we define a new variable $p(k)$ whose z-transform satisfies

$$\frac{P(z)}{U(z)} = \frac{1}{z^n + a_{n-1}z^{n-1} + \ldots + a_1 z + a_0} \tag{8.40}$$

The state equation of a system with the preceding transfer function can be written by inspection, with the state variables chosen as

$$\begin{aligned} \mathbf{x}(k) &= [x_1(k) \quad x_2(k) \quad \ldots \quad x_{n-1}(k) \quad x_n(k)]^T \\ &= [p(k) \quad p(k+1) \quad \ldots \quad p(k+n-2) \quad p(k+n-1)]^T \end{aligned} \tag{8.41}$$

This choice of state variables is valid because none of the variables can be written as a linear combination of the others. However, we have used neither the numerator of the transfer function nor the constant c_n. Nor have we related the state variables to the output. So we multiply (8.39) by $U(z)$ and use (8.40) and (8.41) to obtain

$$\begin{aligned} Y(z) &= c_n U(z) + G_d(z)U(z) \\ &= c_n U(z) + \frac{c_{n-1}z^{n-1} + c_{n-2}z^{n-2} + \ldots + c_1 z + c_0}{z^n + a_{n-1}z^{n-1} + \ldots + a_1 z + a_0} U(z) \\ &= c_n U(z) + \left[c_{n-1}z^{n-1} + c_{n-2}z^{n-2} + \ldots + c_1 z + c_0 \right] P(z) \end{aligned} \tag{8.42}$$

Then we inverse z-transform and use the definition of the state variables to obtain

$$y(k) = c_n u(k) + c_{n-1} p(k+n-1) + c_{n-2} p(k+n-2) + \ldots + c_1 p(k+1) + c_0 p(k)$$
$$= c_n u(k) + c_{n-1} x_n(k) + c_{n-2} x_{n-1}(k) + \ldots + c_1 x_2(k) + c_0 x_1(k)$$

$$(8.43)$$

Finally, we write the output equation in the matrix form

$$y(k) = \begin{bmatrix} c_0 & c_1 & \ldots & c_{n-2} & c_{n-1} \end{bmatrix} \begin{bmatrix} x_1(k) \\ x_2(k) \\ \cdot \\ \cdot \\ \cdot \\ x_{n-1}(k) \\ x_n(k) \end{bmatrix} + du(k) \qquad (8.44)$$

where $d = c_n$.

As in the preceding section, the state-space equations can be written by inspection from the difference equation or the transfer function.

A simulation diagram for the system is shown in Figure 8.4. The simulation diagram shows how the system can be implemented in terms of summer, delay, and scaling operations. The number of delay elements needed for implementation is equal to the order of the system. In addition, two summers and at most

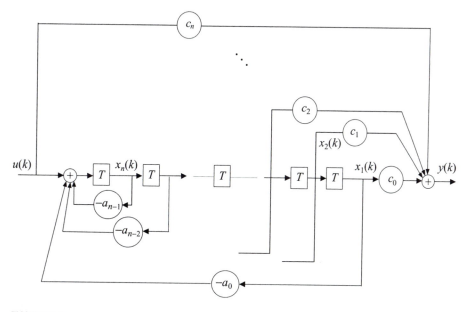

FIGURE 8.4

Simulation diagram for the controllable canonical realization.

2 n + 1 gains are needed. These operations can be easily implemented using a microprocessor or digital signal processing chip.

EXAMPLE 8.19

Write the state–space equations in controllable canonical form for the following transfer functions:

1. $G(z) = \dfrac{0.5(z - 0.1)}{z^3 + 0.5z^2 + 0.4z - 0.8}$

2. $G(z) = \dfrac{z^4 + 0.1z^3 + 0.7z^2 + 0.2z}{z^4 + 0.5z^2 + 0.4z - 0.8}$

Solution

1. The transfer function has the same denominator as that shown in Example 8.18. Hence, it has the same state equation. The numerator of the transfer function can be expanded as $(-0.05 + 0.5z + 0z^2)$, and the output equation is of the form

$$y(k) = \begin{bmatrix} -0.05 & 0.5 & 0 \end{bmatrix} \begin{bmatrix} x_1(k) \\ x_2(k) \\ x_3(k) \end{bmatrix}$$

2. The transfer function has the highest power of z equal to 4, in both the numerator and denominator. We therefore begin by extracting a constant equal to the numerator z^4 coefficient to obtain

$$G(z) = 1 + \frac{0.1z^3 + (0.7 - 0.5)z^2 + (0.2 - 0.4)z - (-0.8)}{z^4 + 0.5z^2 + 0.4z - 0.8}$$

$$= 1 + \frac{0.1z^3 + 0.2z^2 - 0.2z + 0.8}{z^4 + 0.5z^2 + 0.4z - 0.8}$$

Now we can write the state–space equations by inspection as

$$\begin{bmatrix} x_1(k+1) \\ x_2(k+1) \\ x_3(k+1) \\ x_4(k+1) \end{bmatrix} = \begin{bmatrix} 0 & 1 & 0 & 0 \\ 0 & 0 & 1 & 0 \\ 0 & 0 & 0 & 1 \\ 0.8 & -0.4 & -0.5 & 0 \end{bmatrix} \begin{bmatrix} x_1(k) \\ x_2(k) \\ x_3(k) \\ x_4(k) \end{bmatrix} + \begin{bmatrix} 0 \\ 0 \\ 0 \\ 1 \end{bmatrix} u(k)$$

$$y(k) = \begin{bmatrix} 0.8 & -0.2 & 0.2 & 0.1 \end{bmatrix} \begin{bmatrix} x_1(k) \\ x_2(k) \\ x_3(k) \\ x_4(k) \end{bmatrix} + u(k)$$

Theorem 8.5 can be used to show that any system in controllable form is actually controllable. The proof is straightforward and is left as an exercise (see Problem 8.15).

8.5.2 Controllable Form in MATLAB

MATLAB gives a canonical form that is almost identical to the controllable form of this section with the command

$$\text{>> [A, B, C, D] = tf2ss(num, den)}$$

Using the command with the system described in Example 8.19(2) gives the state–space equations

$$
\begin{bmatrix} x_1(k+1) \\ x_2(k+1) \\ x_3(k+1) \\ x_4(k+1) \end{bmatrix} =
\begin{bmatrix} 0 & -0.5 & -0.4 & 0.8 \\ 1 & 0 & 0 & 0 \\ 0 & 1 & 0 & 0 \\ 0 & 0 & 1 & 0 \end{bmatrix}
\begin{bmatrix} x_1(k) \\ x_2(k) \\ x_3(k) \\ x_4(k) \end{bmatrix} +
\begin{bmatrix} 1 \\ 0 \\ 0 \\ 0 \end{bmatrix} u(k)
$$

$$
y(k) = \begin{bmatrix} 0.1 & 0.2 & -0.2 & 0.8 \end{bmatrix}
\begin{bmatrix} x_1(k) \\ x_2(k) \\ x_3(k) \\ x_4(k) \end{bmatrix} + u(k)
$$

The equations are said to be in **controller form**. By drawing a simulation diagram for this system and then for the system described in Example 8.19 (2), we can verify that the two systems are identical. The state variables for the form in MATLAB are simply numbered in the reverse of the order used in this text (see Figure 8.4; that is, the variables x_i in the MATLAB model are none other than the variables x_{n-i+1}, $i = 1, 2, \ldots, n$.

8.5.3 Parallel Realization

Parallel realization of a transfer function is based on the partial fraction expansion

$$
G_d(z) = d + \frac{c_{n-1}z^{n-1} + c_{n-2}z^{n-2} + \ldots + c_1 z + c_0}{z^n + a_{n-1}z^{n-1} + \ldots + a_1 z + a_0} = d + \sum_{i=1}^{n} \frac{K_i}{z + p_i} \tag{8.45}
$$

The expansion is represented by the simulation diagram of Figure 8.5. The summation in (8.45) gives the parallel configuration shown in the figure, which justifies the name **parallel realization**.

A simulation diagram can be obtained for the parallel realization by observing that the z-transfer functions in each of the parallel branches can be rewritten as

$$
\frac{1}{z + p_i} = \frac{z^{-1}}{1 - (-p_i z^{-1})}
$$

which can be represented by a positive feedback loop with forward transfer function z^{-1} and feedback gain p_i, as shown in Figure 8.6. Recall that z^{-1} is simply a

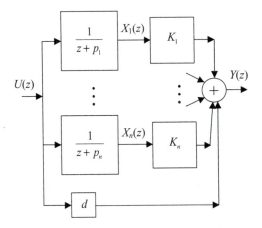

FIGURE 8.5

Block diagram for parallel realization.

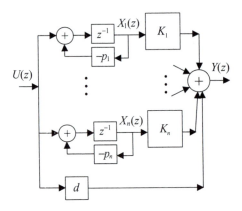

FIGURE 8.6

Simulation diagram for parallel realization.

time delay so that a physical realization of the transfer function in terms of constant gains and fixed time delays is now possible.

We define the state variables as the outputs of the first-order blocks and inverse z-transform to obtain the state equations

$$x_i(k+1) = -p_i x_i(k) + u(k), \quad i = 1, 2, \ldots, n \qquad (8.46)$$

The output is given by the summation

$$y(k) = \sum_{i=1}^{n} K_i x_i(k) + du(k) \qquad (8.47)$$

Equations (8.46) and (8.47) are equivalent to the state-space representation

$$
\begin{bmatrix} x_1(k+1) \\ x_2(k+1) \\ \cdot \\ x_{n-1}(k+1) \\ x_n(k+1) \end{bmatrix} = \begin{bmatrix} -p_1 & 0 & \cdots & 0 & 0 \\ 0 & -p_2 & \cdots & 0 & 0 \\ \vdots & \vdots & \cdots & \vdots & \vdots \\ 0 & 0 & \cdots & -p_{n-1} & 0 \\ 0 & 0 & \cdots & 0 & -p_n \end{bmatrix} \begin{bmatrix} x_1(k) \\ x_2(k) \\ \vdots \\ x_{n-1}(k) \\ x_n(k) \end{bmatrix} + \begin{bmatrix} 1 \\ 1 \\ \vdots \\ 1 \\ 1 \end{bmatrix} u(k)
$$

$$
y(k) = \begin{bmatrix} K_1 & K_2 & \cdots & K_{n-1} & K_n \end{bmatrix} \begin{bmatrix} x_1(k) \\ x_2(k) \\ \vdots \\ x_{n-1}(k) \\ x_n(k) \end{bmatrix} + du(k)
$$

(8.48)

EXAMPLE 8.20

Obtain a parallel realization for the transfer function

$$
G(z) = \frac{2z^2 + 2z + 1}{z^2 + 5z + 6}
$$

Solution

We first write the transfer function in the form

$$
G(z) = 2 + \frac{2z^2 + 2z + 1 - 2(z^2 + 5z + 6)}{z^2 + 5z + 6}
$$

$$
= 2 - \frac{8z + 11}{(z+2)(z+3)}
$$

Then we obtain the partial fraction expansion

$$
G(z) = 2 + \frac{5}{z+2} - \frac{13}{z+3}
$$

Finally, we have the state–space equations

$$
\begin{bmatrix} x_1(k+1) \\ x_2(k+1) \end{bmatrix} = \begin{bmatrix} -2 & 0 \\ 0 & -3 \end{bmatrix} \begin{bmatrix} x_1(k) \\ x_2(k) \end{bmatrix} + \begin{bmatrix} 1 \\ 1 \end{bmatrix} u(k)
$$

$$
y(k) = \begin{bmatrix} 5 & -13 \end{bmatrix} \begin{bmatrix} x_1(k) \\ x_2(k) \end{bmatrix} + 2u(k)
$$

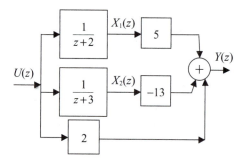

FIGURE 8.7

Block diagram for Example 8.20.

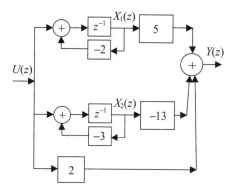

FIGURE 8.8

Simulation diagram for Example 8.20.

The block diagram for the parallel realization is shown in Figure 8.7, and the simulation diagram is shown in Figure 8.8. Clearly, the system is unstable with two eigenvalues outside the unit circle.

Parallel Realization for MIMO Systems

For MIMO systems, a parallel realization can be obtained by partial fraction expansion as in the SISO case. The method requires that the minimal polynomial of the system have no repeated roots. The partial fractions in this case are constant matrices.

The partial fraction expansion is in the form

$$G_d(z) = D + \frac{P_{n-1}z^{n-1} + P_{n-2}z^{n-2} + \ldots + P_1 z + P_0}{z^n + a_{n-1}z^{n-1} + \ldots + a_1 z + a_0} = D + \sum_{i=1}^{n} \frac{K_i}{z + p_i} \quad (8.49)$$

where D, P_i, and K_i, $i = 1, \ldots , n - 1$ are $l \times m$ matrices. Each of the matrices K_i must be decomposed into the product of two full-rank matrices: an input component matrix B_i and an output component matrix C_i. We write the partial fraction matrices in the form

$$K_i = C_i B_i, \quad i = 1, \ldots, n \tag{8.50}$$

For full-rank component matrices, their order is dictated by the rank of the matrix K_i. This follows from the fact that the rank of the product is at most equal to the minimum dimension of the components. For rank $(K_i) = r_i$, we have C_i as an $l \times r_i$ matrix and B_i as an $r_i \times m$ matrix. The rank of the matrix K_i represents the minimum number of poles p_i needed for the parallel realization. In fact, it can be shown that this realization is indeed minimal. The parallel realization is given by the quadruple

$$A = \begin{bmatrix} -I_{r_1} p_1 & 0 & \cdots & 0 & 0 \\ 0 & -I_{r_2} p_2 & \cdots & 0 & 0 \\ \vdots & \vdots & \cdots & \vdots & \vdots \\ 0 & 0 & \cdots & -I_{r_{n-1}} p_{n-1} & 0 \\ 0 & 0 & \cdots & 0 & -I_{r_n} p_n \end{bmatrix} \quad B = \begin{bmatrix} B_1 \\ B_2 \\ \vdots \\ B_{n-1} \\ B_n \end{bmatrix} \tag{8.51}$$

$$C = \begin{bmatrix} C_1 & C_2 & \cdots & C_{n-1} & C_n \end{bmatrix} \quad D$$

EXAMPLE 8.21

Obtain a parallel realization for the transfer function matrix of Example 8.14:

$$G(z) = \begin{bmatrix} \dfrac{1}{z-1} & \dfrac{1}{(z-1)(z-0.5)} \\ \dfrac{z-0.1}{(z-0.2)(z-0.5)} & \dfrac{1}{z-0.2} \end{bmatrix}$$

Solution

The minimal polynomial $(z - 0.2)(z - 0.5)(z - 1)$ has no repeated roots. The partial fraction expansion of the matrix is

$$G(z) = \begin{bmatrix} \dfrac{1}{z-1} & \dfrac{1}{(z-1)(z-0.5)} \\ \dfrac{z-0.1}{(z-0.2)(z-0.5)} & \dfrac{1}{z-0.2} \end{bmatrix}$$

$$= \dfrac{\begin{bmatrix} 0 & 0 \\ -\dfrac{1}{3} & 1 \end{bmatrix}}{z-0.2} + \dfrac{\begin{bmatrix} 0 & -2 \\ \dfrac{4}{3} & 0 \end{bmatrix}}{z-0.5} + \dfrac{\begin{bmatrix} 1 & 2 \\ 0 & 0 \end{bmatrix}}{z-1}$$

The ranks of the partial fraction coefficient matrices given with the corresponding poles are $(1, 0.2)$, $(2, 0.5)$, and $(1, 1)$. The matrices can be factorized as

$$\begin{bmatrix} 0 & 0 \\ -\dfrac{1}{3} & 1 \end{bmatrix} = \begin{bmatrix} 0 \\ 1 \end{bmatrix}\begin{bmatrix} -\dfrac{1}{3} & 1 \end{bmatrix}$$

$$\begin{bmatrix} 0 & -2 \\ \dfrac{4}{3} & 0 \end{bmatrix} = \begin{bmatrix} 1 & 0 \\ 0 & 1 \end{bmatrix}\begin{bmatrix} 0 & -2 \\ \dfrac{4}{3} & 0 \end{bmatrix}$$

$$\begin{bmatrix} 1 & 2 \\ 0 & 0 \end{bmatrix} = \begin{bmatrix} 1 \\ 0 \end{bmatrix}\begin{bmatrix} 1 & 2 \end{bmatrix}$$

The parallel realization is given by the quadruple

$$A = \begin{bmatrix} 0.2 & 0 & 0 & 0 \\ 0 & 0.5 & 0 & 0 \\ 0 & 0 & 0.5 & 0 \\ 0 & 0 & 0 & 1 \end{bmatrix} \quad B = \begin{bmatrix} -\dfrac{1}{3} & 1 \\ 0 & -2 \\ \dfrac{4}{3} & 0 \\ 1 & 2 \end{bmatrix}$$

$$C = \begin{bmatrix} 0 & 1 & 0 & 1 \\ 1 & 0 & 1 & 0 \end{bmatrix} \quad D = 0_{2\times2}$$

Note that the realization is minimal as it is fourth order and, in Example 8.14, the system was determined to have four poles.

8.5.4 Observable Form

The **observable realization** of a transfer function can be obtained from the controllable realization using the following steps:

1. Transpose the state matrix A.
2. Transpose and interchange the input matrix B and the output matrix C.

Clearly, the coefficients needed to write the state-space equations all appear in the transfer function, and all equations can be written in observable form directly. Using (8.36) and (8.44), we obtain the realization

$$\mathbf{x}(k+1) = \begin{bmatrix} \mathbf{0}_{1\times n-1} & -a_0 \\ & -a_1 \\ I_{n-1} & \vdots \\ & -a_{n-1} \end{bmatrix}\mathbf{x}(k) + \begin{bmatrix} c_0 \\ c_1 \\ \vdots \\ c_{n-1} \end{bmatrix}u(k) \tag{8.52}$$

$$y(k) = [\mathbf{0}_{1\times n-1} \mid 1]\mathbf{x}(k) + du(k)$$

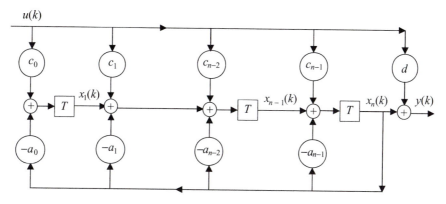

FIGURE 8.9

Simulation diagram for the observable canonical realization.

The simulation diagram for the observable realization is shown in Figure 8.9. Note that it is possible to renumber the state variables in the simulation diagram to obtain an equivalent realization known as **observer form**. However, the resulting realization will clearly have different matrices from those of (8.52). The second observable realization can be obtained from the controller form by following the two preceding steps.

The two observable realizations can also be obtained from the basic principles; however, the derivations have been omitted in this text and are left as an exercise.

Theorem 8.5 can be used to show that any system in observable form is actually observable. The proof is straightforward and is left as an exercise (see Problem 8.16).

EXAMPLE 8.22

Write the state–space equations in observable canonical form for the transfer function of Example 8.19(2).

$$G(z) = \frac{z^4 + 0.1z^3 + 0.7z^2 + 0.2z}{z^4 + 0.5z^2 + 0.4z - 0.8}$$

Solution

The transfer function can be written as a constant plus a transfer function with numerator order less than the denominator order as in Example 8.19(2). Then, using (8.41), we obtain the following:

$$\begin{bmatrix} x_1(k+1) \\ x_2(k+1) \\ x_3(k+1) \\ x_4(k+1) \end{bmatrix} = \begin{bmatrix} 0 & 0 & 0 & 0.8 \\ 1 & 0 & 0 & -0.4 \\ 0 & 1 & 0 & -0.5 \\ 0 & 0 & 1 & 0 \end{bmatrix} \begin{bmatrix} x_1(k) \\ x_2(k) \\ x_3(k) \\ x_4(k) \end{bmatrix} + \begin{bmatrix} 0.8 \\ -0.2 \\ 0.2 \\ 0.1 \end{bmatrix} u(k)$$

$$y(k) = \begin{bmatrix} 0 & 0 & 0 & 1 \end{bmatrix} \begin{bmatrix} x_1(k) \\ x_2(k) \\ x_3(k) \\ x_4(k) \end{bmatrix} + u(k)$$

The same realization is obtained from the controllable realization of Example 8.19(2) by transposing the state matrix and transposing, and then interchanging the matrices B and C.

8.6 DUALITY

The concepts of controllability (stabilizability) and observability (detectability) are often referred to as **duals**. This term is justified by the following theorem.

Theorem 8.12: The system (A, B) is controllable (stabilizable) if and only if (A^T, B^T) is observable (detectable). The system (A, C) is observable (detectable) if and only if (A^T, C^T) is controllable (stabilizable).

PROOF. The relevant controllability and observability matrices are related by the equations

$$\mathcal{C}(A, B) = \begin{bmatrix} B & \vdots & A B & \vdots & \dots & \vdots & A^{n-1} B \end{bmatrix} = \begin{bmatrix} B^T \\ \hline B^T A^T \\ \hline \vdots \\ \hline B^T (A^T)^{n-1} \end{bmatrix}^T = \mathcal{O}^T(A^T, B^T)$$

$$\mathcal{C}(A^T, B^T) = \begin{bmatrix} B^T & \vdots & A^T B^T & \vdots & \dots & \vdots & (A^T)^{n-1} B^T \end{bmatrix} = \begin{bmatrix} B \\ \hline BA \\ \hline \vdots \\ \hline BA^{n-1} \end{bmatrix}^T = \mathcal{O}^T(A, B)$$

The proof follows from the equality of the rank of any matrix and the rank of its transpose. The statements regarding detectability and stabilizability are true because a matrix and its transpose have the same eigenvalues. ∎

***

EXAMPLE 8.23

Show that the reducible transfer function

$$G(z) = \frac{0.3(z - 0.5)}{(z - 1)(z - 0.5)}$$

has a controllable but unobservable realization and an observable but uncontrollable realization.

Solution

The transfer function can be written as

$$G(z) = \frac{0.3z - 0.15}{z^2 - 1.5z + 0.5}$$

The controllable realization for this system is

$$\begin{bmatrix} x_1(k+1) \\ x_2(k+1) \end{bmatrix} = \begin{bmatrix} 0 & 1 \\ -0.5 & 1.5 \end{bmatrix} \begin{bmatrix} x_1(k) \\ x_2(k) \end{bmatrix} + \begin{bmatrix} 0 \\ 1 \end{bmatrix} \mathbf{u}(k)$$

$$y = \begin{bmatrix} -0.15 & 0.3 \end{bmatrix} \begin{bmatrix} x_1(k) \\ x_2(k) \end{bmatrix}$$

The observability matrix for this realization is

$$\mathcal{O} = \begin{bmatrix} C \\ \hline CA_d \end{bmatrix} = \begin{bmatrix} -0.15 & 0.3 \\ \hline -0.15 & 0.3 \end{bmatrix}$$

The observability matrix has rank 1. The rank deficit is $2 - 1 = 1$, corresponding to one unobservable mode.

Transposing the state, output, and input matrices and interchanging the input and output matrices gives the observable realization

$$\begin{bmatrix} x_1(k+1) \\ x_2(k+1) \end{bmatrix} = \begin{bmatrix} 0 & -0.5 \\ 1 & 1.5 \end{bmatrix} \begin{bmatrix} x_1(k) \\ x_2(k) \end{bmatrix} + \begin{bmatrix} -0.15 \\ 0.3 \end{bmatrix} \mathbf{u}(k)$$

$$y(k) = \begin{bmatrix} 0 & 1 \end{bmatrix} \begin{bmatrix} x_1(k) \\ x_2(k) \end{bmatrix}$$

By duality, the realization is observable but has one uncontrollable mode.

***

RESOURCES

Antsaklis, P. J., and A. N. Michel, *Linear Systems,* McGraw-Hill, 1997.
Belanger, P. R., *Control Engineering: A Modern Approach,* Saunders, 1995.
Chen, C. T., *Linear System Theory and Design,* HRW, 1984.

D'Azzo, J. J., and C. H. Houpis, *Linear Control System Analysis and Design,* McGraw-Hill, 1988.

Delchamps, D. F., *State Space and Input-Output Linear Systems,* Springer-Verlag, 1988.

Gupta, S. C. and L. Hasdorff, *Fundamentals of Automatic Control,* Wiley, 1970.

Kailath, T., *Linear Systems,* Prentice Hall, 1980.

Moler, C. B., and C. F. Van Loan, Nineteen dubious ways to calculate the exponential of a matrix, *SIAM Review,* 20:801-836, 1978.

Patel, R. V., and N. Munro, *Multivariable System Theory and Design,* Pergamon Press, 1982.

Sinha, N. K., *Control Systems,* HRW, 1988.

Skogestad, S., and I. Postlethwaite, *Multivariable Feedback Control: Analysis and Design,* Wiley, 2005.

PROBLEMS

8.1 Find the equilibrium state and the corresponding output for the system

$$\begin{bmatrix} x_1(k+1) \\ x_2(k+1) \end{bmatrix} = \begin{bmatrix} 0 & 1 \\ -0.5 & -0.1 \end{bmatrix}\begin{bmatrix} x_1(k) \\ x_2(k) \end{bmatrix} + \begin{bmatrix} 0 \\ 1 \end{bmatrix}u(k)$$

$$y(k) = \begin{bmatrix} 1 & 1 \end{bmatrix}\begin{bmatrix} x_1(k) \\ x_2(k) \end{bmatrix}$$

when

(a) $u(k) = 0$

(b) $u(k) = 1$

8.2 A mechanical system has the state–space equations

$$\begin{bmatrix} x_1(k+1) \\ x_2(k+1) \end{bmatrix} = \begin{bmatrix} 0 & 1 \\ -0.5 & -a_1 \end{bmatrix}\begin{bmatrix} x_1(k) \\ x_2(k) \end{bmatrix} + \begin{bmatrix} 0 \\ 1 \end{bmatrix}\mathbf{u}(k)$$

$$y(k) = \begin{bmatrix} 1 & 0 \end{bmatrix}\begin{bmatrix} x_1(k) \\ x_2(k) \end{bmatrix}$$

where and a_1 is dependent on viscous friction.

(a) Using the results of Chapter 4, determine the range of the parameter a_1 for which the system is internally stable.

(b) Predict the dependence of the parameter a_1 on viscous friction, and use physical arguments to justify your prediction. (*Hint:* Friction dissipates energy and helps the system reach its equilibrium.)

8.3 Determine the internal stability and the input-output stability of the following linear systems:

(a) $$\begin{bmatrix} x_1(k+1) \\ x_2(k+1) \end{bmatrix} = \begin{bmatrix} 0.1 & 0 \\ 1 & 0.2 \end{bmatrix}\begin{bmatrix} x_1(k) \\ x_2(k) \end{bmatrix} + \begin{bmatrix} 0 \\ 0.2 \end{bmatrix}u(k)$$

$$y(k) = \begin{bmatrix} 1 & 1 \end{bmatrix}\begin{bmatrix} x_1(k) \\ x_2(k) \end{bmatrix}$$

(b)
$$\begin{bmatrix} x_1(k+1) \\ x_2(k+1) \\ x_3(k+1) \end{bmatrix} = \begin{bmatrix} -0.2 & 0.2 & 0 \\ 0 & 1 & 0.1 \\ 0 & 0 & -1 \end{bmatrix} \begin{bmatrix} x_1(k) \\ x_2(k) \\ x_3(k) \end{bmatrix} + \begin{bmatrix} 1 & 0 \\ 0 & 0 \\ 1 & 1 \end{bmatrix} \mathbf{u}(k)$$

$$y = \begin{bmatrix} 1 & 0 & 0 \end{bmatrix} \begin{bmatrix} x_1(k) \\ x_2(k) \\ x_3(k) \end{bmatrix}$$

(c)
$$\begin{bmatrix} x_1(k+1) \\ x_2(k+1) \end{bmatrix} = \begin{bmatrix} 0.1 & 0.3 \\ 1 & 0.2 \end{bmatrix} \begin{bmatrix} x_1(k) \\ x_2(k) \end{bmatrix} + \begin{bmatrix} 0 \\ 0.2 \end{bmatrix} \mathbf{u}(k)$$

$$y(k) = \begin{bmatrix} 1 & 1 \end{bmatrix} \begin{bmatrix} x_1(k) \\ x_2(k) \end{bmatrix}$$

(d)
$$\begin{bmatrix} x_1(k+1) \\ x_2(k+1) \\ x_3(k+1) \end{bmatrix} = \begin{bmatrix} -0.1 & -0.3 & 0 \\ 0.1 & 1 & 0.1 \\ 0.3 & 0 & -1 \end{bmatrix} \begin{bmatrix} x_1(k) \\ x_2(k) \\ x_3(k) \end{bmatrix} + \begin{bmatrix} 1 & 0 \\ 1 & 0 \\ 0 & 1 \end{bmatrix} \mathbf{u}(k)$$

$$y = \begin{bmatrix} 1 & 0 & 1 \end{bmatrix} \begin{bmatrix} x_1(k) \\ x_2(k) \\ x_3(k) \end{bmatrix}$$

8.4 Determine the stable, marginally stable, and unstable modes for each of the unstable systems presented in Problem 8.3.

8.5 Determine the controllability and stabilizability of the systems presented in Problem 8.3.

8.6 Transform the following system to standard form for uncontrollable systems, and use the transformed system to determine if it is stabilizable:

$$\begin{bmatrix} x_1(k+1) \\ x_2(k+1) \\ x_3(k+1) \end{bmatrix} = \begin{bmatrix} 0.05 & 0.09 & 0.1 \\ 0.05 & 1.1 & -1 \\ 0.05 & -0.9 & 1 \end{bmatrix} \begin{bmatrix} x_1(k) \\ x_2(k) \\ x_3(k) \end{bmatrix} + \begin{bmatrix} 1 & 0 \\ 0 & 1 \\ 0 & 1 \end{bmatrix} \mathbf{u}(k)$$

$$y = \begin{bmatrix} 1 & 0 & 1 \end{bmatrix} \begin{bmatrix} x_1(k) \\ x_2(k) \\ x_3(k) \end{bmatrix}$$

8.7 Transform the system to the standard form for unobservable systems, and use the transformed system to determine if it is detectable:

$$\begin{bmatrix} x_1(k+1) \\ x_2(k+1) \end{bmatrix} = \begin{bmatrix} -0.2 & -0.08 \\ 0.125 & 0 \end{bmatrix} \begin{bmatrix} x_1(k) \\ x_2(k) \end{bmatrix} + \begin{bmatrix} 1 \\ 0 \end{bmatrix} \mathbf{u}(k)$$

$$y(k) = \begin{bmatrix} 1 & 0.8 \end{bmatrix} \begin{bmatrix} x_1(k) \\ x_2(k) \end{bmatrix}$$

8.8 Determine the controllability and stabilizability of the systems presented in Problem 8.3 with the input matrices changed to the following:

(a) $B = [1 \quad 0]^T$

(b) $B = [1 \quad 1 \quad 0]^T$

(c) $B = [1 \quad 0]^T$

(d) $B = [1 \quad 0 \quad 1]^T$

8.9 An engineer is designing a control system for a chemical process with reagent concentration as the sole control variable. After determining that the system is not controllable, why is it impossible for him to control all the modes of the system by an innovative control scheme using the same control variables? Explain, and suggest an alternative solution to the engineer's problem.

8.10 The engineer introduced in Problem 8.9 examines the chemical process more carefully and discovers that all the uncontrollable modes with concentration as control variable are asymptotically stable with sufficiently fast dynamics. Why is it possible for the engineer to design an acceptable controller with reagent concentration as the only control variable? If such a design is possible, give reasons for the engineer to prefer it over a design requiring additional control variables.

8.11 Determine the observability and detectability of the systems described in problem 8.3.

8.12 Repeat Problem 8.11 with the following output matrices:

(a) $C = [0 \quad 1]$

(b) $C = [0 \quad 1 \quad 0]$

(c) $C = [1 \quad 0]$

(d) $C = [1 \quad 0 \quad 0]$

8.13 Consider the system

$$A = \begin{bmatrix} 0 & 1 & 0 \\ 0 & 0 & 1 \\ -0.005 & -0.11 & -0.7 \end{bmatrix} \quad b = \begin{bmatrix} 0 \\ 0 \\ 1 \end{bmatrix} \quad c = [0.5 \quad 1 \quad 0] \quad d = 0$$

(a) Can we design a controller for the system that can influence all its modes?

(b) Can we design a controller for the system that can observe all its modes?

Justify your answers using the properties of the system.

8.14 Consider the system (A, B, C) and the family of systems $(\alpha A, \beta B, \gamma C)$ with each of (α, β, γ) nonzero.

(a) Show that, if λ is an eigenvalue of A with right eigenvector $\mathbf{v}$ and left eigenvector $\mathbf{w}^T$, then $\alpha\lambda$ is an eigenvalue of αA with right eigenvector $\mathbf{v}/\alpha$ and left eigenvector $\mathbf{w}^T/\alpha$.

(b) Show that (A, B) is controllable if and only if $(\alpha A, \beta B)$ is controllable for any nonzero constants (α, β).

(c) Show that system (A, C) is observable if and only if $(\alpha A, \gamma C)$ is observable for any nonzero constants (α, γ).

8.15 Show that any system in controllable form is controllable.

8.16 Show that any system in observable form is observable.

8.17 Obtain state-space representations for the following linear systems:
(a) In controllable form
(b) In observable form
(c) In diagonal form

(i) $G(z) = 3 \dfrac{z + 0.5}{(z - 0.1)(z + 0.1)}$

(ii) $G(z) = 5 \dfrac{z(z + 0.5)}{(z - 0.1)(z + 0.1)(z + 0.8)}$

(iii) $G(z) = \dfrac{z^2(z + 0.5)}{(z - 0.4)(z - 0.2)(z + 0.8)}$

(iv) $G(z) = \dfrac{z(z - 0.1)}{z^2 - 0.9z + 0.8}$

8.18 Obtain the controller form that corresponds to a renumbering of the state variables of the controllable realization (also known as phase variable form) from basic principles.

8.19 Obtain the transformation matrix to transform a system in phase variable form to controller form. Prove that the transformation matrix will also perform the reverse transformation.

8.20 Use the two steps of Section 8.5.4 to obtain a second observable realization from controller form. What is the transformation that will take this form to the first observable realization of Section 8.5.4?

8.21 Show that the observable realization obtained from the phase variable form realizes the same transfer function.

8.22 Show that the transfer functions of the following systems are identical, and give a detailed explanation.

$$A = \begin{bmatrix} 0 & 1 \\ -0.02 & 0.3 \end{bmatrix} \quad \mathbf{b} = \begin{bmatrix} 0 \\ 1 \end{bmatrix} \quad \mathbf{c}^T = [0 \ \ 1] \quad \mathbf{d} = 0$$

$$A = \begin{bmatrix} 0.1 & 0 \\ 0 & 0.2 \end{bmatrix} \quad \mathbf{b} = \begin{bmatrix} 1 \\ 1 \end{bmatrix} \quad \mathbf{c}^T = [-1 \ \ 2] \quad \mathbf{d} = 0$$

8.23 Obtain a parallel realization for the transfer function matrix presented in Example 8.15.

8.24 Find the poles and zeros of the following transfer function matrices:

(a) $G(z) = \begin{bmatrix} \dfrac{z-0.1}{(z+0.1)^2} & \dfrac{1}{z+0.1} \\ 0 & \dfrac{1}{z-0.1} \end{bmatrix}$

(b) $G(z) = \begin{bmatrix} 0 & \dfrac{1}{z-0.1} \\ \dfrac{z-0.2}{(z-0.1)(z-0.3)} & \dfrac{1}{z-0.3} \\ 0 & \dfrac{2}{z-0.3} \end{bmatrix}$

8.25 Autonomous underwater vehicles (AUVs) are robotic submarines that can be used for a variety of studies of the underwater environment. The vertical and horizontal dynamics of the vehicle must be controlled to remotely operate the AUV. The INFANTE is a research AUV operated by the Instituto Superior Tecnico of Lisbon, Portugal. The variables of interest in horizontal motion are the sway speed and the yaw angle. A linearized model of the horizontal plane motion of the vehicle is given by

$$\begin{bmatrix} \dot{x}_1 \\ \dot{x}_2 \\ \dot{x}_3 \end{bmatrix} = \begin{bmatrix} -0.14 & -0.69 & 0.0 \\ -0.19 & -0.048 & 0.0 \\ 0.0 & 1.0 & 0.0 \end{bmatrix} \begin{bmatrix} x_1 \\ x_2 \\ x_3 \end{bmatrix} + \begin{bmatrix} 0.056 \\ -0.23 \\ 0.0 \end{bmatrix} u$$

$$\begin{bmatrix} y_1 \\ y_2 \end{bmatrix} = \begin{bmatrix} 1 & 0 & 0 \\ 0 & 1 & 0 \end{bmatrix} \begin{bmatrix} x_1 \\ x_2 \\ x_3 \end{bmatrix}$$

where x_1 is the sway speed, x_2 is the yaw angle, x_3 is the yaw rate, and u is the rudder deflection.
(a) Obtain the zeros of the system using Rosenbrock's system matrix.
(b) Determine the system decoupling zeros by testing the system's controllability and observability.

8.26 The terminal composition of a binary distillation column uses reflux and steam flow as the control variables. The 2-input-2-output system is governed by the transfer function

$$G(s) = \begin{bmatrix} \dfrac{12.8e^{-s}}{16.7s+1} & \dfrac{-18.9e^{-3s}}{21.0s+1} \\ \dfrac{6.6e^{-7s}}{10.9s+1} & \dfrac{-19.4e^{-3s}}{14.4s+1} \end{bmatrix}$$

Find the discrete transfer function of the system with DAC, ADC, and a sampling period of one time unit; then determine the poles and zeros of the discrete-time system.

COMPUTER EXERCISES

8.27 Write computer programs to simulate the second-order systems described in Problem 8.3 for various initial conditions. Obtain state plane plots, and discuss your results, referring to the solutions presented in Examples 8.1 and 8.2.

8.28 Repeat Problem 8.22 using a computer-aided design (CAD) package. Comment on any discrepancies between CAD results and solution by hand.

8.29 Write a MATLAB function that determines the equilibrium state of the system with the state matrix A, the input matrix B, and a constant input u as input parameters.

State Feedback Control

9

State variable feedback allows the flexible selection of linear system dynamics. Often, not all state variables are available for feedback, and the remainder of the state vector must be estimated. This chapter includes an analysis of state feedback and its limitations. It also includes the design of state estimators for use when some state variables are not available and the use of state estimates in feedback control.

Throughout this chapter, we assume that the state vector $\mathbf{x}$ is $n \times 1$, the control vector $\mathbf{u}$ is $m \times 1$, and the output vector $\mathbf{y}$ is $l \times 1$. We drop the subscript d for discrete system matrices.

Objectives

After completing this chapter, the reader will be able to do the following:

1. Design state feedback control using pole placement.
2. Design servomechanisms using state–space models.
3. Analyze the behavior of multivariable zeros under state feedback.
4. Design state estimators (observers) for state–space models.
5. Design controllers using observer state feedback.

9.1 STATE AND OUTPUT FEEDBACK

State feedback involves the use of the state vector to compute the control action for specified system dynamics. Figure 9.1 shows a linear system (A, B, C) with constant state feedback gain matrix K. Using the rules for matrix multiplication, we deduce that the matrix K is $m \times n$ so that for a single-input system K is a row vector.

The equations for the linear system and the feedback control law are, respectively, given by the following equations.

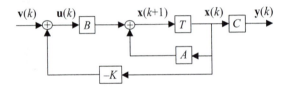

FIGURE 9.1

Block diagram of constant state feedback control.

$$\mathbf{x}(k+1) = A\mathbf{x}(k) + B\mathbf{u}(k)$$

$$\mathbf{y}(k) = C\mathbf{x}(k)$$

(9.1)

$$\mathbf{u}(k) = -K\mathbf{x}(k) + \mathbf{v}(k)$$

(9.2)

The two equations can be combined to yield the closed-loop state equation

$$\mathbf{x}(k+1) = A\mathbf{x}(k) + B[-K\mathbf{x}(k) + \mathbf{v}(k)]$$
$$= [A - BK]\mathbf{x}(k) + B\mathbf{v}(k)$$

(9.3)

We define the closed-loop state matrix as

$$A_{cl} = A - BK$$

(9.4)

and rewrite the closed-loop system state–space equations in the form

$$\mathbf{x}(k+1) = A_{cl}\mathbf{x}(k) + B\mathbf{v}(k)$$

$$\mathbf{y}(k) = C\mathbf{x}(k)$$

(9.5)

The dynamics of the closed-loop system depend on the **eigenstructure** (eigenvalues and eigenvectors) of the matrix A_{cl}. Thus, the desired system dynamics can be chosen with appropriate choice of the gain matrix K. Limitations of this control scheme are addressed in the next section.

For many physical systems, it is either prohibitively costly or impossible to measure all the state variables. The output measurements **y** must then be used to obtain the control **u** as shown in Figure 9.2.

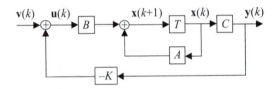

FIGURE 9.2

Block diagram of constant output feedback control.

The feedback control for output feedback is

$$\mathbf{u}(k) = -K_y \mathbf{y}(k) + \mathbf{v}(k)$$
$$= -K_y C\mathbf{x}(k) + \mathbf{v}(k) \tag{9.6}$$

Substituting in the state equation gives the closed-loop system

$$\mathbf{x}(k+1) = A\mathbf{x}(k) + B[-K_y C\mathbf{x}(k) + \mathbf{v}(k)]$$
$$= [A - BK_y C]\mathbf{x}(k) + B\mathbf{v}(k) \tag{9.7}$$

The corresponding state matrix is

$$A_y = A - BK_y C \tag{9.8}$$

Intuitively, less can be accomplished using output feedback than state feedback because less information is used in constituting the control law. In addition, the postmultiplication by the C matrix in (9.8) restricts the choice of closed-loop dynamics. However, output feedback is a more general design problem because state feedback is the special case where C is the identity matrix.

9.2 POLE PLACEMENT

Using output or state feedback, the poles or eigenvalues of the system can be assigned subject to system-dependent limitations. This is known as **pole placement**, **pole assignment**, or **pole allocation**. We state the problem as follows.

Definition 9.1: Pole Placement. Choose the gain matrix K or K_y to assign the system eigenvalues to an arbitrary set $\{\lambda_i, i = 1, \ldots, n\}$. ∎

The following theorem gives conditions that guarantee a solution to the pole-placement problem with state feedback.

Theorem 9.1: State Feedback. If the pair (A, B) is controllable, then there exists a feedback gain matrix K that arbitrarily assigns the system poles to any set $\{\lambda_i, i = 1, \ldots, n\}$. Furthermore, if the pair (A, B) is stabilizable, then the controllable modes can all be arbitrarily assigned.

PROOF.

Necessity
If the system is not controllable, then by Theorem 8.4 the product $\mathbf{w}_i^T B$ is zero for some left eigenvector $\mathbf{w}_i^T$. Premultiplying the closed-loop state matrix by $\mathbf{w}_i^T$ gives

$$\mathbf{w}_i^T A_{cl} = \mathbf{w}_i^T (A - BK) = \lambda_i \mathbf{w}_i^T$$

Hence, the ith **eigenpair** (eigenvalue and eigenvector) of A is unchanged by state feedback and cannot be arbitrarily assigned. Therefore, controllability is a necessary condition for arbitrary pole assignment.

Sufficiency

We first give the proof for the single-input (SI) case where $\mathbf{b}$ is a column matrix. For a controllable pair (A, B), we can assume without loss of generality that the pair is in controllable form. We rewrite (9.4) as

$$\mathbf{b}\mathbf{k}^T = A - A_{cl}$$

Substituting the controllable form matrices gives

$$\begin{bmatrix} \mathbf{0}_{n-1\times 1} \\ 1 \end{bmatrix}[k_1 \quad k_2 \quad \cdots \quad k_n] = \left[\begin{array}{c|c} \mathbf{0}_{n-1\times 1} & I_{n-1} \\ \hline -a_0 \quad -a_1 \quad \cdots \quad \cdots \quad -a_{n-1} \end{array}\right]$$
$$- \left[\begin{array}{c|c} \mathbf{0}_{n-1\times 1} & I_{n-1} \\ \hline -a_0^d \quad -a_1^d \quad \cdots \quad \cdots \quad -a_{n-1}^d \end{array}\right]$$

That is,

$$\left[\begin{array}{c} \mathbf{0}_{n-1\times n} \\ \hline k_1 \quad k_2 \quad \cdots \quad k_n \end{array}\right] = \left[\begin{array}{c} \mathbf{0}_{n-1\times n} \\ \hline a_0^d - a_0 \quad a_1^d - a_1 \quad \cdots \quad \cdots \quad a_{n-1}^d - a_{n-1} \end{array}\right]$$

Equating the last rows of the matrices yields

$$[k_1 \quad k_2 \quad \cdots \quad k_n] = [a_0^d - a_0 \quad a_1^d - a_1 \quad \cdots \quad \cdots \quad a_{n-1}^d - a_{n-1}] \qquad (9.9)$$

which is the control that yields the desired characteristic polynomial coefficients.

Sufficiency Proof 2

We now give a more general proof by contraposition—that is, we assume that the result is not true and prove that the assumption is not true. So we assume that the eigenstructure of the jth mode is unaffected by state feedback for any choice of the matrix K and prove that the system is uncontrollable. In other words, we assume that the jth eigenpair is the same for the open-loop and closed-loop systems. Then we have

$$\mathbf{w}_j^T BK = \mathbf{w}_j^T \{A_{cl} - A\} = \{\lambda_j \mathbf{w}_j^T - \lambda_j \mathbf{w}_j^T\} = \mathbf{0}^T$$

Assuming K full rank, we have $\mathbf{w}_j^T B = \mathbf{0}$. By Theorem 8.4, the system is not controllable. ∎

For a controllable SI system, the matrix (row vector) K has n entries with n eigenvalues to be assigned, and the pole placement problem has a unique solution. Clearly, with more inputs the K matrix has more rows and consequently more unknowns than the n equations dictated by the n specified eigenvalues. This freedom can be exploited to obtain a solution that has desirable properties in addition to the specified eigenvalues. For example, the eigenvectors of the closed-loop state matrix can be selected subject to constraints. There is a rich literature that covers eigenstructure assignment, but it is beyond the scope of this text.

The sufficiency proof of Theorem 9.1 provides a method to obtain the feedback matrix K to assign the poles of a controllable system for the SI case with the system in controllable form. This approach is explored later. We first give a simple procedure applicable to low-order systems.

Procedure 9.1: Pole Placement by Equating Coefficients

1. Evaluate the desired characteristic polynomial from the specified eigenvalues using the expression

$$\Delta_c^d(\lambda) = \prod_{i=1}^{n}(\lambda - \lambda_i) \tag{9.10}$$

2. Evaluate the closed-loop characteristic polynomial using the expression

$$\det\{\lambda I_n - (A - BK)\} \tag{9.11}$$

3. Equate the coefficients of the two polynomials to obtain n equations to be solved for the entries of the matrix K.

EXAMPLE 9.1: POLE ASSIGNMENT

Assign the eigenvalues $\{0.3 \pm j0.2\}$ to the pair

$$A = \begin{bmatrix} 0 & 1 \\ 3 & 4 \end{bmatrix} \quad b = \begin{bmatrix} 0 \\ 1 \end{bmatrix}$$

Solution

For the given eigenvalues the desired characteristic polynomial is

$$\Delta_c^d(\lambda) = (\lambda - 0.3 - j0.2)(\lambda - 0.3 + j0.2) = \lambda^2 - 0.6\lambda + 0.13$$

The closed-loop state matrix is

$$A - \mathbf{bk}^T = \begin{bmatrix} 0 & 1 \\ 3 & 4 \end{bmatrix} - \begin{bmatrix} 0 \\ 1 \end{bmatrix}[k_1 \quad k_2]$$

$$= \begin{bmatrix} 0 & 1 \\ 3 - k_1 & 4 - k_2 \end{bmatrix}$$

The closed-loop characteristic polynomial is

$$\det\{\lambda I_n - (A - \mathbf{bk}^T)\} = \det\begin{bmatrix} \lambda & -1 \\ -(3 - k_1) & \lambda - (4 - k_2) \end{bmatrix}$$

$$= \lambda^2 - (4 - k_2)\lambda - (3 - k_1)$$

Equating coefficients gives the two equations

1. $4 - k_2 = 0.6 \quad \Rightarrow k_2 = 3.4$
2. $-3 + k_1 = 0.13 \Rightarrow k_1 = 3.13$

that is,

$$\mathbf{k}^T = [3.13, 3.4]$$

Because the system is in controllable form, the same result can be obtained as the coefficients of the open-loop characteristic polynomial minus those of the desired characteristic polynomial using (9.9).

9.2.1 Pole Placement by Transformation to Controllable Form

Any controllable single-input-single-output (SISO) system can be transformed into controllable form using the transformation

$$T_c = \mathscr{C}\mathscr{C}_c^{-1} = \begin{bmatrix} \mathbf{b} & \vdots & \mathbf{Ab} & \vdots & \cdots & \vdots & A^{n-1}\,\mathbf{b} \end{bmatrix} \begin{bmatrix} a_1 & a_2 & \cdots & a_{n-1} & 1 \\ a_2 & a_3 & \cdots & 1 & 0 \\ \vdots & \vdots & \ddots & \vdots & \vdots \\ a_{n-1} & 1 & \cdots & 0 & 0 \\ 1 & 0 & \cdots & 0 & 0 \end{bmatrix} \qquad (9.12)$$

$$T_c^{-1} = \mathscr{C}_c\mathscr{C}^{-1} = \begin{bmatrix} 0 & 0 & 0 & \cdots & 1 \\ 0 & 0 & 0 & \cdots & t_{2n} \\ \vdots & \vdots & \vdots & \ddots & \vdots \\ 0 & 0 & 1 & \cdots & t_{n-2,n} \\ 0 & 1 & t_{2n} & \cdots & t_{n-1,n} \\ 1 & t_{2n} & t_{3n} & \cdots & t_{nn} \end{bmatrix} \begin{bmatrix} \mathbf{b} & \vdots & \mathbf{Ab} & \vdots & \cdots & \vdots & A^{n-1}\,\mathbf{b} \end{bmatrix}^{-1} \qquad (9.13)$$

where $\mathscr{C}$ is the controllability matrix, the subscript c denotes the controllable form, and the terms $t_{jn}, j = 2, \ldots, n$, are given by

$$t_{2n} = -a_{n-1}$$

$$t_{j+1,n} = -\sum_{i=0}^{j-1} a_{n-i-1} t_{j-i,n}, \quad j = 2, \ldots, n-1$$

The state feedback for a system in controllable form is

$$u = -\mathbf{k}_c^T \mathbf{x}_c = -\mathbf{k}_c^T (T_c^{-1}\mathbf{x}) \qquad (9.14)$$

$$\mathbf{k}^T = \begin{bmatrix} a_0^d - a_0 & a_1^d - a_1 & \cdots & \cdots & a_{n-1}^d - a_{n-1} \end{bmatrix} T_c^{-1} \qquad (9.15)$$

We now have the following pole placement procedure.

Procedure 9.2

1. Obtain the characteristic polynomial of the pair (A, B) using the Leverrier algorithm described in Section 7.4.1.
2. Obtain the transformation matrix T_c^{-1} using the coefficients of the polynomial from step 1.
3. Obtain the desired characteristic polynomial coefficients from the given eigenvalues using (9.10).
4. Compute the state feedback matrix using (9.15).

Procedure 9.2 requires the numerically troublesome inversion of the controllability matrix to obtain T. However, it does reveal an important characteristic of state feedback. From (9.15), we observe that the feedback gains tend to increase as the change from the open-loop to the closed-loop polynomial coefficients increases. Procedure 9.2, like Procedure 9.1, works well for low-order systems but can be implemented more easily using a computer-aided design (CAD) program.

EXAMPLE 9.2

Design a feedback controller for the pair

$$A = \begin{bmatrix} 0.1 & 0 & 0.1 \\ 0 & 0.5 & 0.2 \\ 0.2 & 0 & 0.4 \end{bmatrix} \quad b = \begin{bmatrix} 0.01 \\ 0 \\ 0.005 \end{bmatrix}$$

to obtain the eigenvalues $\{0.1, 0.4 \pm j0.4\}$.

Solution
The characteristic polynomial of the state matrix is

$$\lambda^3 - \lambda^2 + 0.27\lambda - 0.01 \quad \text{i.e.,} \quad a_2 = -1, a_1 = 0.27, a_0 = -0.01$$

The transformation matrix T_c^{-1} is

$$T_c^{-1} = \begin{bmatrix} 0 & 0 & 1 \\ 0 & 1 & 1 \\ 1 & 1 & 0.73 \end{bmatrix} \times 10^3 \begin{bmatrix} 10 & 1.5 & 0.6 \\ 0 & 1 & 1.3 \\ 5 & 4 & 1.9 \end{bmatrix}^{-1} = 10^3 \begin{bmatrix} 0.1923 & 1.25 & -0.3846 \\ -0.0577 & 0.625 & 0.1154 \\ 0.0173 & 0.3125 & 0.1654 \end{bmatrix}$$

The desired characteristic polynomial is

$$\lambda^3 - 0.9\lambda^2 + 0.4\lambda - 0.032 \quad \text{i.e.,} \quad a_2^d = -0.9, a_1^d = 0.4, a_0^d = -0.032$$

Hence, we have the feedback gain vector

$$\mathbf{k}^T = \begin{bmatrix} a_0^d - a_0 & a_1^d - a_1 & a_2^d - a_2 \end{bmatrix} T_c^{-1}$$

$$= [-0.032 + 0.01 \quad 0.4 - 0.27 \quad -0.9 + 1] \times 10^3 \begin{bmatrix} 0.1923 & 1.25 & -0.3846 \\ -0.0577 & 0.625 & 0.1154 \\ 0.0173 & 0.3125 & 0.1654 \end{bmatrix}$$

$$= [-10 \quad 85 \quad 40]$$

9.2.2 Pole Placement Using a Matrix Polynomial

The gain vector for pole placement can be expressed in terms of the desired closed-loop characteristic polynomial. The expression, known as **Ackermann's formula**, is

$$\mathbf{k}^T = \mathbf{t}_1^T \Delta_c^d(A) \tag{9.16}$$

where t_1^T is the first row of the matrix T_c^{-1} of (9.13) and $\Delta_c^d(\lambda)$ is the desired closed-loop characteristic polynomial. From Theorem 9.1, we know that state feedback can arbitrarily place the eigenvalues of the closed-loop system for any controllable pair $(A, \mathbf{b})$. In addition, any controllable pair can be transformed into controllable form $(A_c, \mathbf{b}_c)$. By the Cayley-Hamilton theorem, the state matrix satisfies its own characteristic polynomial $\Delta(\lambda)$ but not that corresponding to the desired pole locations. That is,

$$\Delta_c(A) = \sum_{i=0}^{n} a_i A^i = 0$$

$$\Delta_c^d(A) = \sum_{i=0}^{n} a_i^d A^i \neq 0$$

Subtracting and using the identity $A_c = T_c^{-1} A T_c$ gives

$$T_c^{-1} \Delta_c^d(A) T_c = \sum_{i=0}^{n-1} \left(a_i^d - a_i \right) A_c^i \tag{9.17}$$

The matrix in controllable form possesses an interesting property, which we use in this proof. If the matrix raised to power i, with $i = 1, 2, \ldots, n-1$, is premuliplied by the first elementary vector

$$\mathbf{e}_1^T = \begin{bmatrix} 1 & \mathbf{0}_{n-1\times1}^T \end{bmatrix}$$

The result is the $(i+1)$th elementary vector—that is,

$$\mathbf{e}_1^T A_c^i = \mathbf{e}_{i+1}^T, \quad i = 0, 1, \ldots, n-1 \tag{9.18}$$

Premultiplying (9.17) by the elementary vector $\mathbf{e}_1$, then using (9.18), we obtain

$$\mathbf{e}_1^T T_c^{-1} \Delta_c^d(A) T_c = \sum_{i=0}^{n-1} \left(a_i^d - a_i \right) \mathbf{e}_1^T A_c^i$$
$$= \sum_{i=0}^{n-1} \left(a_i^d - a_i \right) \mathbf{e}_{i+1}^T$$
$$= \begin{bmatrix} a_0^d - a_0 & a_1^d - a_1 & \cdots & a_{n-1}^d - a_{n-1} \end{bmatrix}$$

Using (9.15), we obtain

$$\mathbf{e}_1^T T_c^{-1} \Delta_c^d(A) T_c = \begin{bmatrix} a_0^d - a_0 & a_1^d - a_1 & \cdots & a_{n-1}^d - a_{n-1} \end{bmatrix}$$
$$= \mathbf{k}_c^T = \mathbf{k}^T T_c$$

Postmultiplying by T_c^{-1} and observing that the first row of the inverse is $t_1^T = \mathbf{e}_1^T T_c^{-1}$, we obtain Ackermann's formula (9.16).

Minor modifications in Procedure 9.2 allow pole placement using Ackermann's formula. The formula requires the evaluation of the first row of the matrix T_c^{-1} rather than the entire matrix. However, for low-order systems, it is often simpler to evaluate the inverse and then use its first row. The following example demonstrates pole placement using Ackermann's formula.

EXAMPLE 9.3

Obtain the solution described in Example 9.2 using Ackermann's formula.

Solution

The desired closed-loop characteristic polynomial is

$$\Delta_c^d(\lambda) = \lambda^3 - 0.9\lambda^2 + 0.4\lambda - 0.032 \quad \text{i.e.,} \quad a_2^d = -0.9, a_1^d = 0.4, a_0^d = -0.032$$

The first row of the inverse transformation matrix is

$$\mathbf{t}_1^T = e_1^T T_c^{-1} = 10^3 [1 \quad 0 \quad 0] \begin{bmatrix} 0.1923 & 1.25 & -0.3846 \\ -0.0577 & 0.625 & 0.1154 \\ 0.0173 & 0.3125 & 0.1654 \end{bmatrix}$$

$$= 10^3 [0.1923 \quad 1.25 \quad -0.3846]$$

We use Ackermann's formula to compute the gain vector

$$\mathbf{k}^T = \mathbf{t}_1^T \Delta_c^d(A)$$

$$= \mathbf{t}_1^T \{A^3 - 0.9A^2 + 0.4A - 0.032I_3\}$$

$$= 10^3 [0.1923 \quad 1.25 \quad -0.3846] \times 10^{-3} \begin{bmatrix} 6 & 0 & 18 \\ 4 & 68 & 44 \\ 36 & 0 & 48 \end{bmatrix}$$

$$= [-10 \quad 85 \quad 40]$$

9.2.3 Choice of the Closed-Loop Eigenvalues

Procedures 9.1 and 9.2 yield the feedback gain matrix once the closed-loop eigenvalues have been arbitrarily selected. The desired locations of the eigenvalues are directly related to the desired transient response of the system. In this context, considerations similar to those made in Section 6.6 (see Tables 6.5 and 6.6) can be applied. However, the designer must take into account that poles associated with fast modes will lead to high gains for the state feedback matrix and consequently to a high **control effort**. High gains may also lead to performance degradation due to nonlinear behavior such as ADC or actuator saturation. If all the desired closed-loop eigenvalues are selected at the origin of the complex plane, the deadbeat control strategy is implemented (see Section 6.7), and the closed-loop characteristic polynomial is chosen as

$$\Delta_c^d(\lambda) = \lambda^n \tag{9.19}$$

Substituting in Ackermann's formula (9.16) gives the feedback gain matrix

$$\mathbf{k}^T = \mathbf{t}_1^T A^n \tag{9.20}$$

The resulting control law will drive all the states to zero in at most n sampling intervals starting from any initial condition. However, the limitations of deadbeat control discussed in Section 6.7 apply; namely, the control variable can assume unacceptably high values, and undesirable intersample oscillations can occur.

EXAMPLE 9.4

Determine the gain vector **k** using Ackermann's formula for the discretized state–space model of the armature-controlled DC motor described in Example 7.15 for the following choices of closed-loop eigenvalues:

1. $\{0.1, 0.4 \pm j0.4\}$
2. $\{0.4, 0.6 \pm j0.33\}$
3. $\{0, 0, 0\}$ (deadbeat control)

Simulate the system in each case to obtain the zero-input response starting from the initial condition $\mathbf{x}(0) = [1,1,1]$, and discuss the results.

Solution
The characteristic polynomial of the state matrix is

$$\Delta(\lambda) = \lambda^3 - 2.895\lambda^2 + 2.791\lambda - 0.896$$

That is,

$$a_2 = -2.895, \quad a_1 = 2.791, \quad a_0 = -0.896$$

The controllability matrix of the system is

$$\mathcal{C} = 10^{-3} \begin{bmatrix} 0.001622 & 0.049832 & 0.187964 \\ 0.482100 & 1.381319 & 2.185571 \\ 94.6800 & 84.37082 & 75.73716 \end{bmatrix}$$

Using (9.13) gives the transformation matrix

$$T_c^{-1} = 10^4 \begin{bmatrix} 1.0527 & -0.0536 & 0.000255 \\ -1.9948 & 0.20688 & -0.00102 \\ 0.94309 & -0.14225 & 0.00176 \end{bmatrix}$$

1. The desired closed-loop characteristic polynomial is

$$\Delta_c^d(\lambda) = \lambda^3 - 0.9\lambda^2 + 0.4\lambda - 0.032$$

That is,

$$a_2^d = -0.9, \quad a_1^d = 0.4, \quad a_0^d = -0.032$$

By Ackermann's formula, the gain vector is

$$\begin{aligned} \mathbf{k}^T &= \mathbf{t}_1^T \Delta_c^d(A) \\ &= \mathbf{t}_1^T \{A^3 - 0.9A^2 + 0.4A - 0.032I_3\} \\ &= 10^4[1.0527 \quad -0.0536 \quad 0.000255] \times \{A^3 - 0.9A^2 + 0.4A - 0.032I_3\} \\ &= 10^3[4.9268 \quad 1.4324 \quad 0.0137] \end{aligned}$$

The discretized zero-input response for the three states and the corresponding control variable u are shown in Figure 9.3.

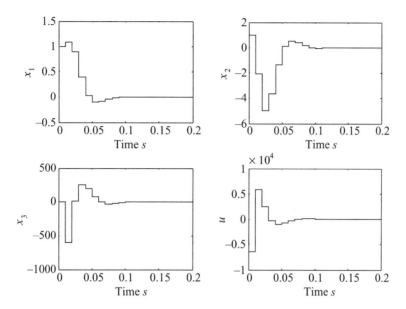

FIGURE 9.3

Zero-input state response and control variable for case 1 of Example 9.4.

2. The desired closed-loop characteristic polynomial is

$$\Delta_c^d(\lambda) = \lambda^3 - 1.6\lambda^2 + 0.9489\lambda - 0.18756$$

That is,

$$a_2^d = -1.6, \quad a_1^d = 0.9489, \quad a_0^d = -0.18756$$

And the gain vector is

$$
\begin{aligned}
\mathbf{k}^T &= \mathbf{t}_1^T \Delta_c^d(A) \\
&= \mathbf{t}_1^T \{A^3 - 1.6A^2 + 0.9489A - 0.18756I_3\} \\
&= 10^4[1.0527 \quad -0.0536 \quad 0.000255] \times \{A^3 - 1.6A^2 + 0.9489A - 0.18756I_3\} \\
&= 10^3[1.6985 \quad 0.70088 \quad 0.01008]
\end{aligned}
$$

The discretized zero-input response for the three states and the corresponding control variable u are shown in Figure 9.4.

3. The desired closed-loop characteristic polynomial is

$$\Delta_c^d(\lambda) = \lambda^3 \quad \text{i.e.,} \quad a_2^d = 0, \quad a_1^d = 0, \quad a_0^d = 0$$

and the gain vector is

$$
\begin{aligned}
\mathbf{k}^T &= \mathbf{t}_1^T \Delta_c^d(A) \\
&= \mathbf{t}_1^T \{A^3\} \\
&= 10^4[1.0527 \quad -0.0536 \quad 0.000255] \times \{A^3\} \\
&= 10^4[1.0527 \quad 0.2621 \quad 0.0017]
\end{aligned}
$$

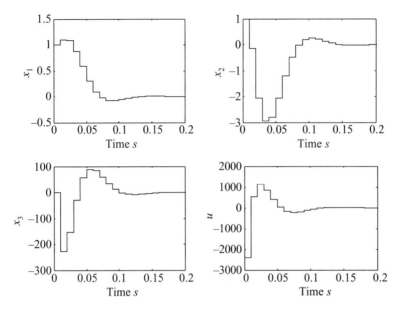

FIGURE 9.4

Zero-input state response and control variable for case 2 of Example 9.4.

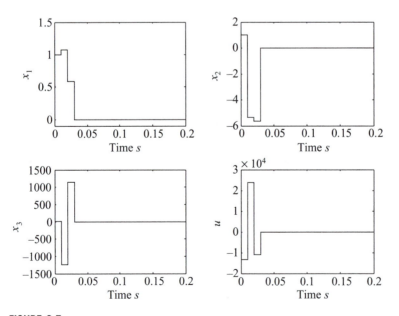

FIGURE 9.5

Zero-input state response and control variable for case 3 of Example 9.4.

The discretized zero-input response for the three states and the corresponding control variable u are shown in Figure 9.5.

We observe that when eigenvalues associated with faster modes are selected, higher gains are required for the state feedback and the state variables have transient responses with larger oscillations. Specifically, for the deadbeat control of case 3, the gain values are one order of magnitude larger that those of cases 1 and 2, and the magnitude of its transient oscillations is much larger. Further, for deadbeat control the zero state is reached in $n = 3$ sampling intervals as predicted by the theory. However, transient intersample oscillations actually occur in x_2, the motor velocity. This is shown in Figure 9.6, where the analog velocity and the sampled velocity of the motor are plotted.

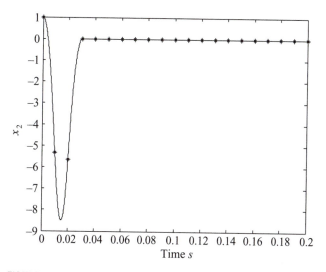

FIGURE 9.6

Sampled and analog velocity of the motor with deadbeat control.

Another consideration in selecting the desired closed-loop poles is the **robustness** of the system to modeling uncertainties. Although we have so far assumed that the state–space model of the system is known perfectly, this is never true in practice. Thus, it is desirable that control systems have poles with low sensitivity to perturbations in their state–space matrices. It is well known that low pole sensitivity is associated with designs that minimize the condition number of the modal matrix of eigenvectors of the state matrix.

For a SISO system, the eigenvectors are fixed once the eigenvalues are chosen. Thus, the choice of eigenvalues determines the pole sensitivity of the system. For example, selecting the same eigenvalues for the closed-loop system leads to a high condition number of the eigenvector matrix and therefore to a closed-loop system that is very sensitive to coefficient perturbations. For multi-input-multi-output

(MIMO) systems, the feedback gains for a given choice of eigenvalues are nonunique. This allows for more than one choice of eigenvectors and can be exploited to obtain robust designs. However, robust pole placement for MIMO systems is beyond the scope of this text and is not discussed further.

9.2.4 MATLAB Commands for Pole Placement

The pole placement command is **place**. The following example illustrates the use of the command.

>> A = [0, 1; 3, 4];

>> B = [0; 1];

>> poles = [0.3 + j*.2, 0.3 − j*0.2];

>> K = place(A, B, poles)

place: ndigits = 16

K = 3.1300   3.4000

ndigits is a measure of the accuracy of pole placement.

It is also possible to compute the state feedback gain matrix of (9.14) or (9.15) using basic MATLAB commands as follows:

1. Generate the characteristic polynomial of a matrix for (9.14) and (9.15).

>> poly(A)

2. Obtain the coefficients of the characteristic polynomial from a set of desired eigenvalues given as the entries of a vector **poles**.

>> desired = poly(poles)

The vector **desired** contains the desired coefficients in descending order.

3. Generate the polynomial matrix for a matrix **A** corresponding to the polynomial

$$\lambda^3 - 0.9\lambda^2 + 0.4\lambda - 0.032$$

>> polyvalm([1, −0.9, 0.4, −0.032], A)

9.2.5 Pole Placement by Output Feedback

As one would expect, using output feedback limits our ability to assign the eigenvalues of the state system relative to what is achievable using state feedback. It is, in general, not possible to arbitrarily assign the system poles even if the system is completely controllable and completely observable. It is possible to arbitrarily assign the controllable dynamics of the system using dynamic output feedback, and a satisfactory solution can be obtained if the system is stabilizable and detectable. Several approaches are available for the design of such a dynamic controller. One solution is to obtain an estimate of the state using the output and input of the system and use it in state feedback as explained in Section 9.6.

9.3 SERVO PROBLEM

The schemes shown in Figures 9.1 and 9.2 are regulators that drive the system state to zero starting from any initial condition capable of rejecting impulse disturbances. In practice, it is often necessary to track a constant reference input **r** with zero steady-state error. For this purpose, a possible approach is to use the **two degree-of-freedom control scheme** of Figure 9.7, so called because we now have two matrices to select, the feedback gain matrix K and the reference gain matrix F.

The reference input of (9.2) becomes $\mathbf{v}(k) = F\mathbf{r}(k)$, and the control law is chosen as

$$\mathbf{u}(k) = -K\mathbf{x}(k) + F\mathbf{r}(k) \tag{9.21}$$

with $\mathbf{r}(k)$ the reference input to be tracked. The corresponding closed-loop system equations are

$$\mathbf{x}(k+1) = A_{cl}\mathbf{x}(k) + BF\mathbf{r}(k)$$
$$y(k) = C\mathbf{x}(k) \tag{9.22}$$

where the closed-loop state matrix is

$$A_{cl} = A - BK$$

The z-transform of the corresponding output is given by (see Section 7.8)

$$\mathbf{Y}(z) = C[zI_n - A_{cl}]^{-1}BF\mathbf{R}(z)$$

The steady-state tracking error for a unit step input is given by

$$\lim_{z \to 1}(z-1)\{\mathbf{Y}(z) - \mathbf{R}(z)\} = \lim_{z \to 1}\left\{C[zI_n - A_{cl}]^{-1}BF - I\right\}$$
$$= C[I_n - A_{cl}]^{-1}BF - I$$

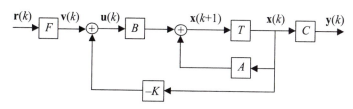

FIGURE 9.7

Block diagram of the two degree-of-freedom controller.

For zero steady-state error, we require the condition

$$C[I_n - A_{cl}]^{-1}BF = I_n \qquad (9.23)$$

If the system is square ($m = l$) and A_{cl} is stable (no unity eigenvalues), we solve for the reference gain

$$F = \left[C(I_n - A_{cl})^{-1}B \right]^{-1} \qquad (9.24)$$

EXAMPLE 9.5

Design a state–space controller for the discretized state–space model of the DC motor speed control system described in Example 6.8 (with $T = 0.02$) to obtain (1) zero steady-state error due to a unit step, (2) a damping ratio of 0.7, and (3) a settling time of about 1 s.

Solution

The discretized transfer function of the system with digital-to-analog converter (DAC) and analog-to-digital converter (ADC) is

$$G_{ZAS}(z) = (1 - z^{-1}) \mathcal{Z}\left\{ \frac{G(s)}{s} \right\} = 1.8604 \times 10^{-4} \frac{z + 0.9293}{(z - 0.8187)(z - 0.9802)}$$

The corresponding state–space model, computed with MATLAB, is

$$\begin{bmatrix} x_1(k+1) \\ x_2(k+1) \end{bmatrix} = \begin{bmatrix} 1.799 & -0.8025 \\ 1 & 0 \end{bmatrix} \begin{bmatrix} x_1(k) \\ x_2(k) \end{bmatrix} + \begin{bmatrix} 0.01563 \\ 0 \end{bmatrix} u(k)$$

$$y(k) = [0.01191 \quad 0.01107] \begin{bmatrix} x_1(k) \\ x_2(k) \end{bmatrix}$$

The desired eigenvalues of the closed-loop system are selected as $\{0.9 \pm j0.09\}$ (see Example 6.8). This yields the feedback gain vector

$$K = [-0.068517 \quad 0.997197]$$

and the closed-loop state matrix

$$A_{cl} = \begin{bmatrix} 1.8 & -0.8181 \\ 1 & 0 \end{bmatrix}$$

The feedforward gain is

$$F = \left[C(I_n - A_{cl})^{-1}B \right]^{-1}$$

$$= \left[[0.01191 \quad 0.01107] \left(\begin{bmatrix} 1 & 0 \\ 0 & 1 \end{bmatrix} - \begin{bmatrix} 1.8 & -0.8181 \\ 1 & 0 \end{bmatrix} \right)^{-1} \begin{bmatrix} 0.01563 \\ 0 \end{bmatrix} \right]^{-1} = 50.42666$$

The response of the system to a step reference input r is shown in Figure 9.8. The system has a settling time of about 0.84 s and percentage overshoot of about 4% with a peak time of about 1 s. All design specifications are met.

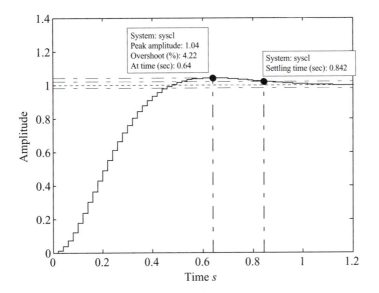

FIGURE 9.8

Step response of the closed-loop system described in Example 9.5.

The control law (9.21) is equivalent to a **feedforward action** determined by *F* to yield zero steady-state error for a constant reference input **r**. Because the forward action does not include any form of feedback, this approach is not robust to modeling uncertainties. Thus, modeling errors (which always occur in practice) will result in nonzero steady-state error. To eliminate such errors, we introduce the **integral control** shown in Figure 9.9, with a new state added for each control error integrated.

The resulting state–space equations are

$$\mathbf{x}(k+1) = A\mathbf{x}(k) + B\mathbf{u}(k)$$

$$\bar{\mathbf{x}}(k+1) = \bar{\mathbf{x}}(k) + \mathbf{r}(k) - \mathbf{y}(k)$$

$$\mathbf{y}(k) = C\mathbf{x}(k)$$

$$\mathbf{u}(k) = -K\mathbf{x}(k) - \bar{K}\bar{\mathbf{x}}(k)$$

$$(9.25)$$

where $\bar{\mathbf{x}}$ is $l \times 1$. The state–space equations can be combined and rewritten in terms of an augmented state vector $\tilde{\mathbf{x}}(k) = [\mathbf{x}(k) \quad \bar{\mathbf{x}}(k)]^T$ as

$$\begin{bmatrix} \mathbf{x}(k+1) \\ \bar{\mathbf{x}}(k+1) \end{bmatrix} = \begin{bmatrix} A & \mathbf{0} \\ -C & I_l \end{bmatrix} \begin{bmatrix} \mathbf{x}(k) \\ \bar{\mathbf{x}}(k) \end{bmatrix} - \begin{bmatrix} B \\ \mathbf{0} \end{bmatrix} [K \quad \bar{K}] \begin{bmatrix} \mathbf{x}(k) \\ \bar{\mathbf{x}}(k) \end{bmatrix} + \begin{bmatrix} \mathbf{0} \\ I_l \end{bmatrix} \mathbf{r}(k)$$

$$\mathbf{y}(k) = [C \quad \mathbf{0}] \begin{bmatrix} \mathbf{x}(k+1) \\ \bar{\mathbf{x}}(k+1) \end{bmatrix}$$

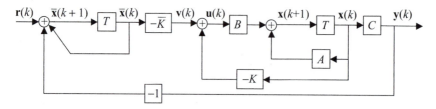

FIGURE 9.9

Control scheme with integral control.

That is,

$$\tilde{\mathbf{x}}(k+1) = \left(\tilde{A} - \tilde{B}\tilde{K}\right)\tilde{\mathbf{x}}(k) + \begin{bmatrix} 0 \\ I_l \end{bmatrix}\mathbf{r}(k)$$

(9.26)

$$\mathbf{y}(k) = [C \quad 0]\tilde{\mathbf{x}}(k)$$

where

$$\tilde{A} = \begin{bmatrix} A & 0 \\ -C & I_l \end{bmatrix} \quad \tilde{B} = \begin{bmatrix} B \\ 0 \end{bmatrix} \quad \tilde{K} = [K \quad \bar{K}]$$

(9.27)

The eigenvalues of the closed-loop system state matrix $A_{cl} = \left(\tilde{A} - \tilde{B}\tilde{K}\right)$ can be arbitrarily assigned by computing the gain matrix $\tilde{K}$ using any of the procedures for the regulator problem as described in Sections 9.1 and 9.2.

EXAMPLE 9.6

Solve the design problem presented in Example 9.5 using integral control.

Solution

The state–space matrices of the system are

$$A = \begin{bmatrix} 1.799 & -0.8025 \\ 1 & 0 \end{bmatrix} \quad B = \begin{bmatrix} 0.01563 \\ 0 \end{bmatrix}$$

$$C = [0.01191 \quad 0.01107]$$

Adding integral control, we obtain

$$\tilde{A} = \begin{bmatrix} A & 0 \\ -C & 1 \end{bmatrix} = \begin{bmatrix} 1.799 & -0.8025 & 0 \\ 1 & 0 & 0 \\ -0.01191 & -0.01107 & 1 \end{bmatrix} \quad \tilde{B} = \begin{bmatrix} B \\ 0 \end{bmatrix} = \begin{bmatrix} 0.01563 \\ 0 \\ 0 \end{bmatrix}$$

In Example 9.5, the eigenvalues were selected as $\{0.9 \pm j0.09\}$. Using integral control increases the order of the system by one, and an additional eigenvalue must be selected. The desired eigenvalues are selected as $\{0.9 \pm j0.09, 0.2\}$, and the additional eigenvalue

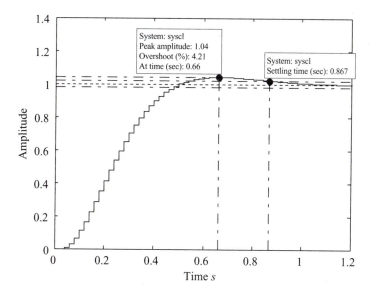

FIGURE 9.10

Step response of the closed-loop system presented in Example 9.6

at 0.2 is chosen for its negligible effect on the overall dynamics. This yields the feedback gain vector

$$\tilde{K} = [51.1315 \quad -40.4431 \quad -40.3413]$$

The closed-loop system state matrix is

$$A_{cl} = \begin{bmatrix} 1 & -0.1706 & 0.6303 \\ 1 & 0 & 0 \\ -0.0119 & -0.0111 & 1 \end{bmatrix}$$

The response of the system to a unit step reference signal r is shown in Figure 9.10. The figure shows that the control specifications are satisfied. The settling time of 0.87 is well below the specified value of 1 s, and the percentage overshoot is about 4.2%, which is less than the value corresponding to $\zeta = 0.7$ for the dominant pair.

9.4 INVARIANCE OF SYSTEM ZEROS

A severe limitation of the state-feedback control scheme is that it cannot change the location of the zeros of the system, which significantly affect the transient response. To show this, we consider the z-transform of the system (9.5):

$$(zI - A - BK)\mathbf{X}(z) - B\mathbf{V}(z) = 0$$

$$C\mathbf{X}(z) = \mathbf{Y}(z) \tag{9.28}$$

If $z = z_0$ is a zero of the system, then $\mathbf{Y}(z_0)$ is zero with $\mathbf{V}(z_0)$ and $\mathbf{X}(z_0)$ nonzero. Thus, for $z = z_0$, the state-space equation (9.28) can be rewritten as

$$\begin{bmatrix} z_0 I - A + BK & -B \\ C & 0 \end{bmatrix} \begin{bmatrix} \mathbf{X}(z_0) \\ \mathbf{V}(z_0) \end{bmatrix} = 0 \tag{9.29}$$

We rewrite (9.29) in terms of the state-space matrices of the open-loop system as

$$\begin{bmatrix} -(z_0 I_n - A) & B \\ C & D \end{bmatrix} \begin{bmatrix} \mathbf{X}(z_0) \\ \mathbf{V}(z_0) \end{bmatrix} = \begin{bmatrix} -(z_0 I_n - A + BK) & B \\ C - DK & D \end{bmatrix} \begin{bmatrix} \mathbf{X}(z_0) \\ \mathbf{V}(z_0) + K\mathbf{X}(z_0) \end{bmatrix} = \begin{bmatrix} 0 \\ 0 \end{bmatrix}$$

We observe that with the state feedback $u(k) = -Kx(k) + r(k)$, the state-space quadruple (A, B, C, D) becomes

$$(A - BK, B, C - DK, D)$$

Thus, the zeros of the closed-loop system are the same as those of the plant and are invariant under state feedback. Similar reasoning can establish the same result for the system of (9.22).

EXAMPLE 9.7

Consider the following continuous-time system:

$$G(s) = \frac{s+1}{(2s+1)(3s+1)}$$

Obtain a discrete model for the system with digital control and a sampling period $T = 0.02$, then design a state–space controller with integral control and with the same closed-loop eigenvalues as shown in Example 9.6.

Solution
The analog system with DAC and ADC has the transfer function

$$G_{ZAS}(z) = (1 - z^{-1}) \mathcal{Z}\left\{\frac{G(s)}{s}\right\} = 33.338 \times 10^{-4} \frac{z - 0.9802}{(z - 0.9934)(z - 0.99)}$$

with an open-loop zero at 0.9802. The corresponding state–space model (computed with MATLAB) is

$$\begin{bmatrix} x_1(k+1) \\ x_2(k+1) \end{bmatrix} = \begin{bmatrix} 1.983 & -0.983 \\ 1 & 0 \end{bmatrix} \begin{bmatrix} x_1(k) \\ x_2(k) \end{bmatrix} + \begin{bmatrix} 0.0625 \\ 0 \end{bmatrix} u(k)$$

$$y(k) = [0.0534 \quad 0.0524] \begin{bmatrix} x_1(k) \\ x_2(k) \end{bmatrix}$$

The desired eigenvalues of the closed-loop system are selected as $\{0.9 \pm j0.09\}$, and this yields the feedback gain vector

$$\tilde{K} = [15.7345 \quad -24.5860 \quad -2.19016]$$

The closed-loop system state matrix is

$$A_{cl} = \begin{bmatrix} 1 & -0.55316 & 13.6885 \\ 1 & 0 & 0 \\ 0.05342 & -0.05236 & 1 \end{bmatrix}$$

The response of the system to a unit step reference signal r, shown in Figure 9.11, has a huge peak overshoot due to the closed-loop zero at 0.9802. The closed-loop control cannot change the location of the zero.

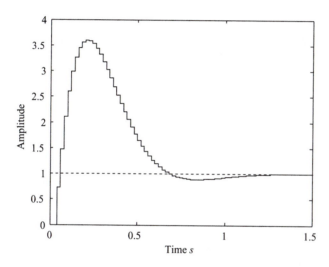

FIGURE 9.11

Step response of the closed-loop system presented in Example 9.7.

9.5 STATE ESTIMATION

In most applications, measuring the entire state vector is impossible or prohibitively expensive. To implement state feedback control, an estimate $\hat{x}(k)$ of the state vector can be used. The state vector can be estimated from the input and output measurements by using a **state estimator** or **observer**.

9.5.1 Full-Order Observer

To estimate all the states of the system, one could in theory use a system with the same state equation as the plant to be observed. In other words, one could use the open-loop system

$$\hat{\mathbf{x}}(k+1) = A\hat{\mathbf{x}}(k) + B\mathbf{u}(k)$$

However, this open-loop estimator assumes perfect knowledge of the system dynamics and lacks the feedback needed to correct the errors that are inevitable in any implementation. The limitations of this observer become obvious on examining its error dynamics. We define the estimation error as $\tilde{\mathbf{x}} = \mathbf{x} - \hat{\mathbf{x}}$. We obtain the error dynamics by subtracting the open-loop observer dynamics from the system dynamics (9.1).

$$\tilde{\mathbf{x}}(k+1) = A\tilde{\mathbf{x}}(k)$$

The error dynamics are determined by the state matrix of the system and cannot be chosen arbitrarily. For an unstable system, the observer will be unstable and cannot track the state of the system.

A practical alternative is to feed back the difference between the measured and the estimated output of the system, as shown in Figure 9.12. This yields to the following observer:

$$\hat{\mathbf{x}}(k+1) = A\hat{\mathbf{x}}(k) + B\mathbf{u}(k) + L[\mathbf{y}(k) - C\hat{\mathbf{x}}(k)] \tag{9.30}$$

Subtracting the observer state equation from the system dynamics yields the estimation error dynamics

$$\tilde{\mathbf{x}}(k+1) = (A - LC)\tilde{\mathbf{x}}(k) \tag{9.31}$$

The error dynamics are governed by the eigenvalues of the observer matrix $A_0 = A - LC$. We transpose the matrix to obtain

$$A_0^T = A^T - C^T L^T \tag{9.32}$$

which has the same eigenvalues as the observer matrix. We observe that (9.32) is identical to the controller design equation (9.4) with the pair (A, B) replaced by the pair (A^T, C^T). We therefore have Theorem 9.2.

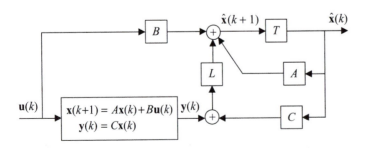

FIGURE 9.12

Block diagram of the full-order state estimator.

Theorem 9.2: State Estimation. If the pair (A, C) is observable, then there exists a feedback gain matrix L that arbitrarily assigns the observer poles to any set $\{\lambda_i, i = 1, \ldots, n\}$. Furthermore, if the pair (A, C) is detectable, then the observable modes can all be arbitrarily assigned.

PROOF. Based on Theorem 8.12, the system (A, C) is observable (detectable) if and only if (A^T, C^T) is controllable (stabilizable). Therefore, Theorem 9.2 follows from Theorem 9.1. ∎

Based on Theorem 9.1, the matrix gain L can be determined from the desired observer poles, as discussed in Section 9.2. Hence, we can arbitrarily select the desired observer poles or the associated characteristic polynomial. From (9.32) it follows that the MATLAB command for the solution of the observer pole placement problem is

$$\gg \text{L = place}(A', C', \text{poles})'$$

EXAMPLE 9.8

Determine the observer gain matrix L for the discretized state–space model of the armature-controlled DC motor described in Example 7.15 with the observer eigenvalues selected as $\{0.1, 0.2 \pm j0.2\}$.

Solution
Recall that the system matrices are

$$A = \begin{bmatrix} 1.0 & 0.1 & 0.0 \\ 0.0 & 0.9995 & 0.0095 \\ 0.0 & -0.0947 & 0.8954 \end{bmatrix} \quad B = \begin{bmatrix} 1.622 \times 10^{-6} \\ 4.821 \times 10^{-4} \\ 9.468 \times 10^{-2} \end{bmatrix} \quad C = [1 \quad 0 \quad 0]$$

The MATLAB command **place** gives the observer gain

$$L = \begin{bmatrix} 2.3949 \\ 18.6734 \\ 4.3621 \end{bmatrix}$$

Expression (9.30) represents a **prediction observer**, because the estimated state vector (and any associated control action) at a given sampling instant does not depend on the current measured value of the system output. Alternatively, a **filtering observer** estimates the state vector based on the current output (assuming negligible computation time), using the expression

$$\hat{\mathbf{x}}(k+1) = A\hat{\mathbf{x}}(k) + B\mathbf{u}(k) + L[\mathbf{y}(k+1) - C(A\hat{\mathbf{x}}(k) + B\mathbf{u}(k))] \tag{9.33}$$

Clearly, the current output $\mathbf{y}(k+1)$ is compared to its estimate based on variables at the previous sampling instant. The error dynamics are now represented by

$$\tilde{\mathbf{x}}(k+1) = (A - LCA)\tilde{\mathbf{x}}(k) \tag{9.34}$$

Expression (9.34) is the same as Expression (9.31) with the matrix product CA substituted for C. The observability matrix $\bar{\mathscr{O}}$ of the system (A, CA) is

$$\bar{\mathscr{O}} = \begin{bmatrix} CA \\ \hline CA^2 \\ \hline \vdots \\ \hline CA^n \end{bmatrix} = \mathscr{O}A$$

where $\mathscr{O}$ is the observability matrix of the pair (A, C) (see Section 8.3). Thus, if the pair (A, C) is observable, the pair (A, CA) is observable unless A has one or more zero eigenvalues. If A has zero eigenvalues, the pair (A, CA) is detectable because the zero eigenvalues are associated with stable modes. Further, the zero eigenvalues are associated with the fastest modes, and the design of the observer can be completed by selecting a matrix L that assigns suitable values to the remaining eigenvalues of $A - LCA$.

EXAMPLE 9.9

Determine the filtering observer gain matrix L for the system described in Example 9.8.

Solution
Using the MATLAB command **place**,

$$\text{>> L = \textbf{place}(A', (C*A)', \textbf{poles})'}$$

we obtain the observer gain

$$L =$$

$$1.0e + 002*$$

$$0.009910699530206$$

$$0.140383004697920$$

$$4.886483453094875$$

9.5.2 Reduced-Order Observer

A full-order observer is designed so that the entire state vector is estimated from knowledge of the input and output of the system. The reader might well ask, why estimate n state variables when we already have l measurements that are linear functions of the same variables? Would it be possible to estimate $n - l$ variables only and use them with measurement to estimate the entire state?

This is precisely what is done in the design of a **reduced-order observer**. A reduced-order observer is generally more efficient than a full-order observer. However, a full-order observer may be preferable in the presence of significant measurement noise. In addition, the design of the reduced-order observer is more complex.

We consider the linear time-invariant system (9.1) where the input matrix B and the output matrix C are assumed full rank. Then the entries of the output vector $\mathbf{y}(t)$ are linearly independent and form a partial state vector of length l, leaving $n - l$ variables to be determined. We thus have the state vector

$$\begin{bmatrix} \mathbf{y} \\ \hline \mathbf{z} \end{bmatrix} = Q_o^{-1}\mathbf{x} = \begin{bmatrix} C \\ \hline M \end{bmatrix}\mathbf{x} \tag{9.35}$$

where M is a full-rank $n - l \times n$ matrix with rows that are linearly independent of those of C, and $\mathbf{z}$ is the unknown partial state.

The state-space matrices for the transformed state variables are

$$C_t = CQ_o = [I_l \mid 0_{l \times n-l}] \tag{9.36}$$

$$B_t = Q_o^{-1}B = \begin{bmatrix} \overline{B}_1 \\ \hline \overline{B}_2 \end{bmatrix} \begin{matrix} \} \, l \\ \} n - l \end{matrix} \tag{9.37}$$

$$A_t = Q_o^{-1}AQ_o = \begin{bmatrix} \overline{A}_1 & \mid & \overline{A}_2 \\ \hline \overline{A}_3 & \mid & \overline{A}_4 \end{bmatrix} \begin{matrix} \} \, l \\ \} n - l \end{matrix} \tag{9.38}$$
$$\underbrace{}_{l} \underbrace{}_{n-l}$$

Thus, the state equation for the unknown partial state is

$$\mathbf{z}(k+1) = \overline{A}_3\mathbf{y}(k) + \overline{A}_4\mathbf{z}(k) + \overline{B}_2\mathbf{u}(k) \tag{9.39}$$

We define an output variable to form a state-space model with (9.39) as

$$\mathbf{y}_z(k) = \mathbf{y}(k+1) - \overline{A}_1\mathbf{y}(k) - \overline{B}_1\mathbf{u}(k) = \overline{A}_2\mathbf{z}(k) \tag{9.40}$$

This output represents the portion of the known partial state $\mathbf{y}(k + 1)$ that is computed using the unknown partial state. The observer dynamics, including the error in computing $\mathbf{y}_z$, are assumed linear time invariant of the form

$$\begin{aligned}
\hat{\mathbf{z}}(k+1) &= \overline{A}_3\mathbf{y}(k) + \overline{A}_4\hat{\mathbf{z}}(k) + \overline{B}_2\mathbf{u}(k) + L[\mathbf{y}_z(k) - \overline{A}_2\hat{\mathbf{z}}(k)] \\
&= (\overline{A}_4 - L\overline{A}_2)\hat{\mathbf{z}}(k) + \overline{A}_3\mathbf{y}(k) + L[\mathbf{y}(k+1) - \overline{A}_1\mathbf{y}(k) - \overline{B}_1\mathbf{u}(k)] + \overline{B}_2\mathbf{u}(k)
\end{aligned} \tag{9.41}$$

where $\hat{\mathbf{z}}$ denotes the estimate of the partial state vector $\mathbf{z}$. Unfortunately, the observer (9.41) includes the term $\mathbf{y}(k + 1)$, which is not available at time k. Moving the term to the LHS reveals that its use can be avoided by estimating the variable

$$\overline{\mathbf{x}}(k) = \hat{\mathbf{z}}(k) - L\mathbf{y}(k) \tag{9.42}$$

Using (9.41) and the definition (9.42), we obtain the observer

$$\begin{aligned}
\overline{\mathbf{x}}(k+1) &= (\overline{A}_4 - L\overline{A}_2)\overline{\mathbf{x}}(k) + (\overline{A}_4L + \overline{A}_3 - L\overline{A}_1 - L\overline{A}_2L)\mathbf{y}(k) + (\overline{B}_2 - L\overline{B}_1)\mathbf{u}(k) \\
&= A_o\overline{\mathbf{x}}(k) + A_y\mathbf{y}(k) + B_o\mathbf{u}(k)
\end{aligned} \tag{9.43}$$

where

$$A_o = \bar{A}_4 - L\bar{A}_2$$
$$A_y = A_o L + \bar{A}_3 - L\bar{A}_1 \tag{9.44}$$
$$B_o = \bar{B}_2 - L\bar{B}_1$$

The block diagram of the reduced-order observer is shown in Figure 9.13.

The dynamic of the reduced-order observer (9.43) is governed by the matrix A_o. The eigenvalues of A_o must be selected inside the unit circle and must be sufficiently fast to track the state of the observed system. This reduces observer design to the solution of (9.44) for the observer gain matrix L. Once L is obtained, the other matrices in (9.43) can be computed and the state vector $\hat{\mathbf{x}}$ can be obtained using the equation

$$\begin{aligned}
\hat{\mathbf{x}}(k) &= Q_o \begin{bmatrix} \mathbf{y}(k) \\ \hat{\mathbf{z}}(k) \end{bmatrix} \\
&= Q_o \begin{bmatrix} I_l & \mathbf{0}_{l \times n-l} \\ L & I_{n-l} \end{bmatrix} \begin{bmatrix} \mathbf{y}(k) \\ \bar{\mathbf{x}}(k) \end{bmatrix} = T_o \begin{bmatrix} \mathbf{y}(k) \\ \bar{\mathbf{x}}(k) \end{bmatrix}
\end{aligned} \tag{9.45}$$

where the transformation matrix Q_o is defined in (9.35).

Transposing (9.44) yields

$$A_o^T = \bar{A}_4^T - \bar{A}_2^T L^T \tag{9.46}$$

Because (9.46) is identical in form to the controller design equation (9.4), it can be solved as discussed in Section 9.2. We recall that the poles of the matrix A_o^T can be arbitrarily assigned provided that the pair $(\bar{A}_4^T, \bar{A}_2^T)$ is controllable. From the duality concept discussed in Section 8.6, this is equivalent to the observability of the pair $(\bar{A}_4, \bar{A}_2)$. The following theorem gives a necessary and sufficient condition for the observability of the pair.

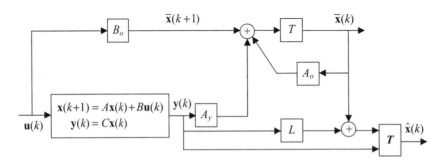

FIGURE 9.13

Block diagram of the reduced-order observer.

Theorem 9.3: The pair $(\bar{A}_4, \bar{A}_2)$ is observable if and only if the system (A, C) is observable.

PROOF. The proof is left as an exercise.   ∎

EXAMPLE 9.10

Design a reduced-order observer for the discretized state–space model of the armature-controlled DC motor described in Example 7.15 with the observer eigenvalues selected as $\{0.2 \pm j0.2\}$.

Solution

Recall that the system matrices are

$$A = \begin{bmatrix} 1.0 & 0.1 & 0.0 \\ 0.0 & 0.9995 & 0.0095 \\ 0.0 & -0.0947 & 0.8954 \end{bmatrix} \quad B = \begin{bmatrix} 1.622 \times 10^{-6} \\ 4.821 \times 10^{-4} \\ 9.468 \times 10^{-2} \end{bmatrix} \quad C = [1 \ \ 0 \ \ 0]$$

The output matrix C is in the required form, and there is no need for similarity transformation. The second and third state variables must be estimated. The state matrix is partitioned as

$$A = \begin{bmatrix} a_1 & \vdots & \mathbf{a}_2^T \\ \cdots & \vdots & \cdots \\ \mathbf{a}_3 & \vdots & A_4 \end{bmatrix} = \begin{bmatrix} 1.0 & 0.1 & 0.0 \\ 0.0 & 0.9995 & 0.0095 \\ 0.0 & -0.0947 & 0.8954 \end{bmatrix}$$

The similarity transformation can be selected as an identity matrix; that is,

$$Q_o^{-1} = \begin{bmatrix} 1 & 0 & 0 \\ \cdots & \cdots & \cdots \\ 0 & 1 & 0 \\ 0 & 0 & 1 \end{bmatrix}$$

and therefore we have $A_t = A$, $B_t = B$, and $C_t = C$. Hence, we need to solve the linear equation

$$A_o = \begin{bmatrix} 0.9995 & 0.0095 \\ -0.0947 & 0.8954 \end{bmatrix} - \bar{\mathbf{l}}[0.1 \ \ 0]$$

to obtain the observer gain

$$\bar{\mathbf{l}} = [14.949 \ \ 550.191]^T$$

The corresponding observer matrices are

$$A_o = \begin{bmatrix} -0.4954 & 0.0095 \\ -55.1138 & 0.8954 \end{bmatrix}$$

$$\mathbf{b}_o = \bar{\mathbf{b}}_2 - \mathbf{l}\bar{\mathbf{b}}_1$$

$$= \begin{bmatrix} 4.821 \times 10^{-4} \\ 9.468 \times 10^{-2} \end{bmatrix} - \begin{bmatrix} 14.949 \\ 550.191 \end{bmatrix} \times 1.622 \times 10^{-6} = \begin{bmatrix} 0.04579 \\ 9.37876 \end{bmatrix} \times 10^{-2}$$

$$\mathbf{a}_y = A_o\mathbf{1} + \mathbf{a}_3 - \mathbf{1}a_1$$
$$= \begin{bmatrix} -0.4954 & 0.0095 \\ -55.1138 & 0.8954 \end{bmatrix}\begin{bmatrix} 14.949 \\ 550.191 \end{bmatrix} + \begin{bmatrix} 0 \\ 0 \end{bmatrix} - \begin{bmatrix} 14.949 \\ 550.191 \end{bmatrix} \times 1 = \begin{bmatrix} -0.1713 \\ -8.8145 \end{bmatrix} \times 10^2$$

The state estimate can be computed using

$$\hat{\mathbf{x}}(k) = Q_o \begin{bmatrix} 1 & \mathbf{0}^T_{1\times2} \\ 1 & I_2 \end{bmatrix}\begin{bmatrix} \mathbf{y}(k) \\ \bar{\mathbf{x}}(k) \end{bmatrix} = \begin{bmatrix} \dfrac{1}{14.949} & \mathbf{0}^T_{1\times2} \\ 550.191 & I_2 \end{bmatrix}\begin{bmatrix} \mathbf{y}(k) \\ \bar{\mathbf{x}}(k) \end{bmatrix}$$

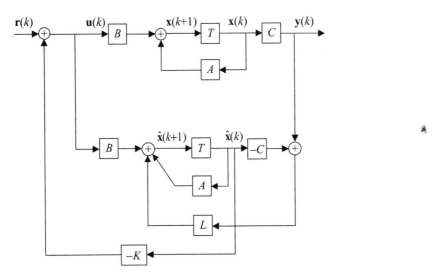

FIGURE 9.14

Block diagram of a system with observer state feedback.

9.6 OBSERVER STATE FEEDBACK

If the state vector is not available for feedback control, a state estimator can be used to generate the control action as shown in Figure 9.14. The corresponding control vector is

$$\mathbf{u}(k) = -K\hat{\mathbf{x}}(k) + \mathbf{v}(k) \tag{9.47}$$

Substituting in the state equation (9.1) gives

$$\mathbf{x}(k+1) = A\mathbf{x}(k) - BK\hat{\mathbf{x}}(k) + B\mathbf{v}(k) \tag{9.48}$$

Adding and subtracting the term $BK\mathbf{x}(k)$, we rewrite (9.48) in terms of the estimation error $\tilde{\mathbf{x}} = \mathbf{x} - \hat{\mathbf{x}}$ as

$$\mathbf{x}(k+1) = (A - BK)\mathbf{x}(k) + BK\tilde{\mathbf{x}}(k) + Bv(k) \tag{9.49}$$

If a full-order (predictor) observer is used, by combining (9.49) with (9.31) we obtain

$$\begin{bmatrix} \mathbf{x}(k+1) \\ \tilde{\mathbf{x}}(k+1) \end{bmatrix} = \begin{bmatrix} A - BK & BK \\ 0 & A - LC \end{bmatrix} \begin{bmatrix} \mathbf{x}(k) \\ \tilde{\mathbf{x}}(k) \end{bmatrix} + \begin{bmatrix} B \\ 0 \end{bmatrix} \mathbf{v}(k) \tag{9.50}$$

The state matrix of (9.50) is block triangular, and its characteristic polynomial is

$$\Delta_{cl}(\lambda) = \det[\lambda I - (A - BK)]\det[\lambda I - (A - LC)]$$
$$= \Delta_c(\lambda)\Delta_o(\lambda) \tag{9.51}$$

Thus, the eigenvalues of the closed-loop system can be selected separately from those of the observer. This important result is known as the **separation theorem** or the **uncertainty equivalence principle**.

Analogously, if a reduced-order observer is employed, the estimation error $\tilde{\mathbf{x}}$ can be expressed in terms of the errors in estimating $\mathbf{y}$ and $\mathbf{z}$ as

$$\tilde{\mathbf{x}}(k) = Q_o \begin{bmatrix} \tilde{\mathbf{y}}(k) \\ \hline \tilde{\mathbf{z}}(k) \end{bmatrix} \tag{9.52}$$

$$\tilde{\mathbf{y}}(k) = \mathbf{y}(k) - \hat{\mathbf{y}}(k) \quad \tilde{\mathbf{z}}(k) = \mathbf{z}(k) - \hat{\mathbf{z}}(k)$$

We partition the matrix Q_o into an $n \times l$ matrix Q_y and an $n \times n - l$ matrix Q_z to allow the separation of the two error terms, and rewrite the estimation error as

$$\tilde{\mathbf{x}}(k) = [Q_y \mid Q_z] \begin{bmatrix} \tilde{\mathbf{y}}(k) \\ \hline \tilde{\mathbf{z}}(k) \end{bmatrix} = Q_y\tilde{\mathbf{y}}(k) + Q_z\tilde{\mathbf{z}}(k) \tag{9.53}$$

Assuming negligible measurement error $\tilde{\mathbf{y}}$, the estimation error reduces to

$$\tilde{\mathbf{x}}(k) = Q_z\tilde{\mathbf{z}}(k) \tag{9.54}$$

Substituting from (9.54) into the closed-loop equation (9.49) gives

$$\mathbf{x}(k+1) = (A - BK)\mathbf{x}(k) + BKQ_z\tilde{\mathbf{z}}(k) \tag{9.55}$$

Evaluating $\tilde{\mathbf{z}}(k)$ by subtracting (9.41) from (9.39) and substituting $\bar{A}_2 \mathbf{z}(k)$ for $\mathbf{y}_z$, we obtain

$$\tilde{\mathbf{z}}(k+1) = (\bar{A}_4 - L\bar{A}_2)\tilde{\mathbf{z}}(k) \tag{9.56}$$

Combining (9.55) and (9.56), we obtain the equation

$$\begin{bmatrix} \mathbf{x}(k+1) \\ \tilde{\mathbf{z}}(k+1) \end{bmatrix} = \begin{bmatrix} A - BK & BKQ_z \\ 0_{n-l \times n} & \bar{A}_4 - L\bar{A}_2 \end{bmatrix} \begin{bmatrix} \mathbf{x}(k) \\ \tilde{\mathbf{z}}(k) \end{bmatrix} \tag{9.57}$$

The state matrix of (9.57) is block triangular and its characteristic polynomial is

$$\det[\lambda I - (\bar{A}_4 - L\bar{A}_2)]\det[\lambda I - (A - BK)] \tag{9.58}$$

Thus, as for the full-order observer, the closed-loop eigenvalues for the reduced-order observer state feedback can be selected separately from those of the reduced-order observer. The separation theorem therefore applies for reduced-order observers as well as for full-order observers.

In addition, combining the plant state equation and the estimator and using the output equation, we have

$$\begin{bmatrix} \mathbf{x}(k+1) \\ \bar{\mathbf{x}}(k+1) \end{bmatrix} = \begin{bmatrix} A & 0_{n \times n-l} \\ A_y C & A_o \end{bmatrix} \begin{bmatrix} \mathbf{x}(k) \\ \bar{\mathbf{x}}(k) \end{bmatrix} + \begin{bmatrix} B \\ B_0 \end{bmatrix} \mathbf{u}(k) \tag{9.59}$$

We express the estimator state feedback of (9.47) as

$$\mathbf{u}(k) = -K\hat{\mathbf{x}}(k) + \mathbf{v}(k)$$

$$= -KT_o \begin{bmatrix} \mathbf{y}(k) \\ \bar{\mathbf{x}}(k) \end{bmatrix} + \mathbf{v}(k) = -K[T_{oy} | T_{ox}] \begin{bmatrix} C\mathbf{x}(k) \\ \bar{\mathbf{x}}(k) \end{bmatrix} + \mathbf{v}(k) \tag{9.60}$$

where T_{oy} and T_{ox} are partitions of T_o of (9.45) of order $n \times l$ and $n \times n - l$, respectively. Substituting in (9.59), we have

$$\begin{bmatrix} \mathbf{x}(k+1) \\ \bar{\mathbf{x}}(k+1) \end{bmatrix} = \begin{bmatrix} A - BKT_{oy}C & -BKT_{ox} \\ A_y C - B_o KT_{oy}C & A_o - B_o KT_{ox} \end{bmatrix} \begin{bmatrix} \mathbf{x}(k) \\ \bar{\mathbf{x}}(k) \end{bmatrix} + \begin{bmatrix} B \\ B_0 \end{bmatrix} \mathbf{v}(k) \tag{9.61}$$

Equation (9.61) can be used to simulate the complete estimator state feedback system.

9.6.1 Choice of Observer Eigenvalues

In the selection of the observer poles or the associated characteristic polynomial, expressions (9.51) or (9.58) must be considered. The choice of observer poles is not based on the constraints related to the control effort discussed in Section 9.2.3. However, the response of the closed-loop system must be dominated by the poles of the controller that meet the performance specifications. Therefore, as a rule of thumb, the poles of the observer should be selected from 3 to 10 times faster than the poles of the controller. An upper bound on the speed of response of the observer is imposed by the presence of the unavoidable measurement noise. Inappropriately fast observer dynamics will result in tracking the noise rather than the actual state of the system. Hence, a deadbeat observer, although appealing in theory, is avoided in practice.

The choice of observer poles is also governed by the same considerations related to the robustness of the system discussed in Section 9.2.3 for the state feedback control. Thus, the sensitivity of the eigenvalues to perturbations in the system matrices must be considered in the selection of the observer poles.

We emphasize that the selection of the observer poles does not influence the performance of the overall control system if the initial conditions are estimated perfectly. We prove this fact for the full-order observer. However, the result also holds for the reduced-order observer, but the proof is left as an exercise for the reader. To demonstrate this fact, we consider the state equation (9.50) with the output equation (9.1):

$$\begin{bmatrix} \mathbf{x}(k+1) \\ \tilde{\mathbf{x}}(k+1) \end{bmatrix} = \begin{bmatrix} A - BK & BK \\ 0 & A - LC \end{bmatrix} \begin{bmatrix} \mathbf{x}(k) \\ \tilde{\mathbf{x}}(k) \end{bmatrix} + \begin{bmatrix} B \\ 0 \end{bmatrix} \mathbf{v}(k)$$

$$\mathbf{y}(k) = \begin{bmatrix} C & 0 \end{bmatrix} \begin{bmatrix} \mathbf{x}(k) \\ \tilde{\mathbf{x}}(k) \end{bmatrix}$$

(9.62)

The zero input-output response of the system $(\mathbf{v}(k) = 0)$ can be determined iteratively as

$$\mathbf{y}(0) = C\mathbf{x}(0)$$

$$\mathbf{y}(1) = \begin{bmatrix} C & 0 \end{bmatrix} \begin{bmatrix} A - BK & BK \\ 0 & A - LC \end{bmatrix} \begin{bmatrix} \mathbf{x}(0) \\ \tilde{\mathbf{x}}(0) \end{bmatrix} = C(A - BK)\mathbf{x}(0) + CBK\tilde{\mathbf{x}}(0)$$

$$\mathbf{y}(2) = \begin{bmatrix} C & 0 \end{bmatrix} \begin{bmatrix} A - BK & BK \\ 0 & A - LC \end{bmatrix} \begin{bmatrix} \mathbf{x}(1) \\ \tilde{\mathbf{x}}(1) \end{bmatrix} = C(A - BK)\mathbf{x}(1) + CBK\tilde{\mathbf{x}}(1)$$

$$= C(A - BK)^2\mathbf{x}(0) - C(A - BK)BK\tilde{\mathbf{x}}(0) - CBK(A - LC)\tilde{\mathbf{x}}(0)$$

$$\vdots$$

Clearly, the observer matrix L influences the transient response if and only if $\tilde{\mathbf{x}}(0) \neq 0$. This fact is confirmed by the determination of the z-transfer function from (9.61), which implicitly assumes zero initial conditions

$$G(z) = \begin{bmatrix} C & 0 \end{bmatrix} \left(zI - \begin{bmatrix} A - BK & BK \\ 0 & A - LC \end{bmatrix} \right)^{-1} \begin{bmatrix} B \\ 0 \end{bmatrix} = C(zI - A + BK)^{-1}B$$

where the observer gain matrix L does not appear.

EXAMPLE 9.11

Consider the armature-controlled DC motor described in Example 7.15. Let the true initial condition be $\mathbf{x}(0) = [1, 1, 1]$, and let its estimate be the zero vector $\hat{\mathbf{x}}(0) = [0, 0, 0]^T$. Design a full-order observer state feedback for a zero-input response with a settling time of 0.2 s.

Solution
As Example 9.4 showed, a choice for the control system eigenvalues that meets the design specification is $\{0.6, 0.4 \pm j0.33\}$. This yields the gain vector

$$K = 10^3[1.6985 \quad -0.70088 \quad 0.01008]$$

The observer eigenvalues must be selected so that the associated modes are sufficiently faster than those of the controller. We select the eigenvalues $\{0.1, 0.1 \pm j0.1\}$. This yields the observer gain vector

$$L = 10^2 \begin{bmatrix} 0.02595 \\ 0.21663 \\ 5.35718 \end{bmatrix}$$

Using (9.62), we obtain the space–space equations

$$\begin{bmatrix} \mathbf{x}(k+1) \\ \tilde{\mathbf{x}}(k+1) \end{bmatrix} = \left[\begin{array}{ccc|ccc} 0.9972 & 0.0989 & 0 & 0.0028 & 0.0011 & 0 \\ -0.8188 & 0.6616 & 0.0046 & 0.8188 & 0.3379 & 0.0049 \\ -160.813 & -66.454 & -0.0589 & 160.813 & 66.359 & 0.9543 \\ \hline & & & -1.5949 & 0.1 & 0 \\ & 0 & & -21.663 & 0.9995 & 0.0095 \\ & & & -535.79 & -0.0947 & 0.8954 \end{array} \right] \begin{bmatrix} \mathbf{x}(k) \\ \tilde{\mathbf{x}}(k) \end{bmatrix}$$

$$\mathbf{y}(k) = [1 \quad 0 \quad 0 \mid \quad 0 \quad] \begin{bmatrix} \mathbf{x}(k) \\ \tilde{\mathbf{x}}(k) \end{bmatrix}$$

The response to the initial condition [1, 1, 1, 1, 1, 1] is plotted in Figure 9.15. We compare the plant state variables x_i, $i = 1, 2, 3$ to the estimation errors x_i, $i = 4, 5, 6$. We observe that the estimation errors decay to zero faster than the system states and that the system has an overall settling time less than 0.2 s.

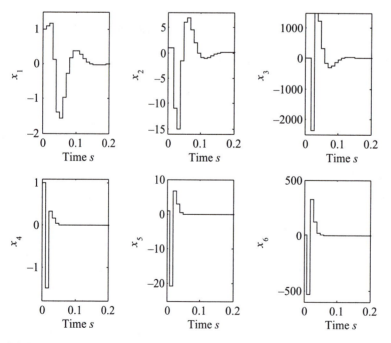

FIGURE 9.15

Zero-input response for Example 9.11.

EXAMPLE 9.12

Solve Example 9.11 using a reduced-order observer.

Solution

In this case, we have $l = 1$ and, because the measured output corresponds to the first element of the state vector is unnecessary for similarity transformation, that is, $Q_0 = Q_0^{-1} = I_3$. Thus, we obtain

$$\bar{A}_1 = 1 \quad \bar{A}_2 = 0.1 \quad \bar{A}_3 = \begin{bmatrix} 0 \\ 0 \end{bmatrix} \quad \bar{A}_4 = \begin{bmatrix} 0.9995 & 0.0095 \\ -0.0947 & 0.8954 \end{bmatrix}$$

$$B_1 = 1.622 \times 10^{-6} \quad \bar{B}_2 = \begin{bmatrix} 4.821 \times 10^{-4} \\ 9.468 \times 10^{-2} \end{bmatrix}$$

We select the reduced-order observer eigenvalues as $\{0.1 \pm j0.1\}$ and obtain the observer gain vector

$$L = 10^2 \begin{bmatrix} 0.16949 \\ 6.75538 \end{bmatrix}$$

and the associated matrices

$$A_o = \begin{bmatrix} -0.6954 & 0.0095 \\ -67.6485 & 0.8954 \end{bmatrix} \quad A_y = 10^3 \begin{bmatrix} -0.02232 \\ -1.21724 \end{bmatrix} \quad B_o = 10^{-2} \begin{bmatrix} 0.045461 \\ 9.358428 \end{bmatrix}$$

Partitioning $Q_0 = T_o = I_3$ gives

$$Q_z = \begin{bmatrix} 0 & 0 \\ 1 & 0 \\ 0 & 1 \end{bmatrix} \quad T_{ox} = \begin{bmatrix} 0 & 0 \\ 1 & 0 \\ 0 & 1 \end{bmatrix} \quad T_{oy} = 10^2 \begin{bmatrix} 0.01 \\ 0.16949 \\ 6.75538 \end{bmatrix}$$

We have that the state–space equation of (9.57) is

$$\begin{bmatrix} \mathbf{x}(k+1) \\ \tilde{\mathbf{z}}(k+1) \end{bmatrix} = \left[\begin{array}{ccc:cc} 0.99725 & 0.09886 & 0.00002 & 0.00114 & 0.00002 \\ -0.81884 & 0.66161 & 0.00464 & 0.33789 & 0.00486 \\ -160.813 & -66.4540 & -0.05885 & 66.3593 & 0.95435 \\ \hdashline & 0 & & -0.6954 & 0.0095 \\ & & & -67.6485 & 0.8954 \end{array} \right] \begin{bmatrix} \mathbf{x}(k) \\ \tilde{\mathbf{z}}(k) \end{bmatrix}$$

$$(9.63)$$

whereas the state–space equation of (9.61) is

$$\begin{bmatrix} \mathbf{x}(k+1) \\ \bar{\mathbf{x}}(k+1) \end{bmatrix} = \left[\begin{array}{ccc:cc} 0.96693 & 0.1 & 0 & -0.00114 & -0.00002 \\ -9.82821 & 0.9995 & 0.0095 & -0.33789 & -0.00486 \\ -1930.17 & -0.0947 & 0.8954 & -66.3593 & -0.95425 \\ \hdashline -31.5855 & 0 & 0 & -1.01403 & 0.00492 \\ -3125.07 & 0 & 0 & -133.240 & 0.04781 \end{array} \right] \begin{bmatrix} \mathbf{x}(k) \\ \bar{\mathbf{x}}(k) \end{bmatrix}$$

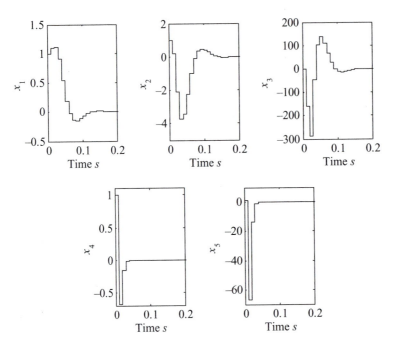

FIGURE 9.16

Zero-input response for Example 9.12.

The response of the state–space system (9.63) to the initial condition [1, 1, 1, 1, 1] is plotted in Figure 9.16. As in Example 9.11, we observe that the estimation errors x_i, $i = 4$, and 5 decay to zero faster than the system states x_i, $i = 1$, 2, and 3, and that the system has an overall settling time less than 0.2 s.

EXAMPLE 9.13

Consider the armature-controlled DC motor described in Example 7.15. Design a full-order observer state feedback with feedforward action for a step response with a settling time of 0.2 s with an overshoot less than 10 percent.

Solution

As in Example 9.11, we can select the control system eigenvalues as $\{0.6, 0.4 \pm j0.33\}$ and the observer eigenvalues as $\{0.1, 0.1 \pm j0.1\}$. This yields the gain vectors

$$K = 10^3[1.6985 \quad -0.70088 \quad 0.01008], \quad L = 10^2\begin{bmatrix} 0.02595 \\ 0.21663 \\ 5.35718 \end{bmatrix}$$

From the matrix $A_{cl} = A - BK$, we determine the feedforward term using (9.24) as

$$F = \left[C(I_n - (A - BK))^{-1}B \right]^{-1} = 1698.49$$

Substituting $Fr(k)$ for $v(k)$ (9.62), we obtain the closed-loop state–space equations

$$
\begin{bmatrix} \mathbf{x}(k+1) \\ \tilde{\mathbf{x}}(k+1) \end{bmatrix} =
\left[
\begin{array}{ccc:ccc}
0.9972 & 0.0989 & 0 & 0.0028 & 0.0011 & 0 \\
-0.8188 & 0.6616 & 0.0046 & 0.8188 & 0.3379 & 0.0049 \\
-160.813 & -66.454 & -0.0589 & 160.813 & 66.359 & 0.9543 \\
\hdashline
 & & & -1.5949 & 0.1 & 0 \\
 & 0 & & -21.663 & 0.9995 & 0.0095 \\
 & & & -535.79 & -0.0947 & 0.8954
\end{array}
\right]
$$

$$
\begin{bmatrix} \mathbf{x}(k) \\ \tilde{\mathbf{x}}(k) \end{bmatrix} +
\begin{bmatrix}
0.0028 \\
0.8188 \\
160.813 \\
\hdashline
\\
0
\end{bmatrix} r(k)
$$

$$
\mathbf{y}(k) = \begin{bmatrix} 1 & 0 & 0 & \vdots & 0 & \end{bmatrix}
\begin{bmatrix} \mathbf{x}(k) \\ \tilde{\mathbf{x}}(k) \end{bmatrix}
$$

The step response of the system shown in Figure 9.17 has a settling time of about 0.1 and a percentage overshoot of about 6 percent. The controller meets the design specification.

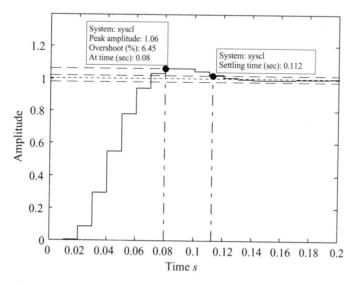

FIGURE 9.17

Step response for Example 9.13.

9.7 POLE ASSIGNMENT USING TRANSFER FUNCTIONS

The pole assignment problem can be solved in the framework of transfer functions. Consider the state–space equations of the two-degree-of-freedom controller shown in Figure 9.7 with the state vector estimated using a full-order observer. For a SISO plant with observer state feedback, we have

$$\hat{\mathbf{x}}(k+1) = A\hat{\mathbf{x}}(k) + Bu(k) + L(y(k) - C\hat{\mathbf{x}}(k))$$

$$u(k) = -K\hat{\mathbf{x}}(k) + Fr(k)$$

or equivalently,

$$\hat{\mathbf{x}}(k+1) = (A - BK - LC)\hat{\mathbf{x}}(k) + \begin{bmatrix} BF & L \end{bmatrix} \begin{bmatrix} r(k) \\ y(k) \end{bmatrix}$$

$$u(k) = -K\hat{\mathbf{x}}(k) + Fr(k)$$

The corresponding z-transfer function from [r, y] to u is

$$U(z) = \left(K(zI - A + BK + LC)^{-1}BF + F\right)R(z) + \left(K(zI - A + BK + LC)^{-1}L\right)Y(z)$$

Thus, the full-order observer state feedback is equivalent to the transfer function model depicted in Figure 9.18. In the figure, the plant $G(z) = P(z)/Q(z)$ is assumed strictly realizable; that is, the degree of $P(z)$ is less than the degree of $Q(z)$. We also assume that $P(z)$ and $Q(z)$ are **coprime** (i.e., they have no common factors). Further, it is assumed that $Q(z)$ is **monic** (i.e., the coefficient of the term with the highest power in z is one).

From the block diagram shown in Figure 9.18, simple block diagram manipulations give the closed-loop transfer function

$$\frac{Y(z)}{R(z)} = \frac{P(z)N(z)}{Q(z)D(z) + P(z)S(z)}$$

and the polynomial equation

$$(Q(z)D(z) + P(z)S(z))Y(z) = P(z)N(z)R(z) \tag{9.64}$$

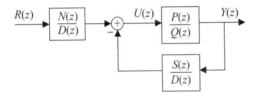

FIGURE 9.18

Block diagram for pole assignment with transfer functions.

Therefore, the closed-loop characteristic equation is

$$\Delta_{cl}(z) = Q(z)D(z) + P(z)S(z) \tag{9.65}$$

The pole placement problem thus reduces to finding polynomials $D(z)$ and $S(z)$ that satisfy (9.65) for given $P(z)$, $Q(z)$, and for a given desired characteristic polynomial $\Delta_{cl}(z)$. Equation (9.65) is called a **Diophantine equation**, and its solution can be found by first expanding its RHS terms as

$$P(z) = p_{n-1}z^{n-1} + p_{n-2}z^{n-2} + \ldots + p_1z + p_0$$

$$Q(z) = z^n + q_{n-1}z^{n-1} + \ldots + q_1z + q_0$$

$$D(z) = d_m z^m + d_{m-1}z^{m-1} + \ldots + d_1z + d_0$$

$$S(z) = s_m z^m + s_{m-1}z^{m-1} + \ldots + s_1z + s_0$$

The closed-loop characteristic polynomial $\Delta_{cl}(z)$ is of degree $n + m$ and has the form

$$\Delta_{cl}(z) = z^{n+m} + \delta_{n+m-1}z^{n+m-1} + \ldots + \delta_1z + \delta_0$$

Thus, (9.65) can be rewritten as

$$z^{n+m} + \delta_{n+m-1}z^{n+m-1} + \ldots + \delta_1z + \delta_0 = (z^n + q_{n-1}z^{n-1} + \ldots + q_1z + q_0)$$
$$(d_m z^m + d_{m-1}z^{m-1} + \ldots + d_1z + d_0) + (p_{n-1}z^{n-1} + p_{n-2}z^{n-2} + \ldots + p_1z + p_0) \tag{9.66}$$
$$(s_m z^m + s_{m-1}z^{m-1} + \ldots + s_1z + s_0)$$

Equation (9.66) is linear in the $2m$ unknowns d_i and s_i, $i = 0, 1, 2, \ldots, m - 1$, and its LHS is a known polynomial with $n + m - 1$ coefficients. The solution of the Diophantine equation is unique if $n + m - 1 = 2m$—that is, if $m = n - 1$. Equation (9.66) can be written in the matrix form

$$
\begin{bmatrix}
1 & 0 & 0 & \cdots & 0 & 0 & 0 & \cdots & 0 \\
q_{n-1} & 1 & 0 & \cdots & 0 & p_{n-1} & 0 & \cdots & 0 \\
q_{n-2} & q_{n-1} & 1 & \cdots & 0 & p_{n-2} & p_{n-1} & \cdots & 0 \\
\vdots & \vdots & \vdots & & \vdots & \vdots & \vdots & & \vdots \\
q_0 & q_1 & q_2 & \cdots & q_{n-1} & p_0 & p_1 & \cdots & p_{n-1} \\
0 & q_0 & q_1 & \cdots & q_{n-2} & 0 & p_0 & \cdots & p_{n-2} \\
0 & 0 & q_0 & \cdots & q_{n-3} & 0 & 0 & \cdots & p_{n-3} \\
\vdots & \vdots & \vdots & & \vdots & \vdots & \vdots & & \vdots \\
0 & 0 & 0 & \cdots & q_0 & 0 & 0 & \cdots & p_0
\end{bmatrix}
\begin{bmatrix}
d_m \\
d_{m-1} \\
d_{m-2} \\
\vdots \\
d_0 \\
s_m \\
s_{m-1} \\
\vdots \\
s_0
\end{bmatrix}
=
\begin{bmatrix}
1 \\
\delta_{2n-2} \\
\delta_{2n-3} \\
\vdots \\
\delta_{n-1} \\
\delta_{n-2} \\
\delta_{n-3} \\
\vdots \\
\delta_0
\end{bmatrix}
\tag{9.67}
$$

It can be shown that the matrix on the LHS is nonsingular if and only if the polynomials $P(z)$ and $Q(z)$ are coprime, which we assume. As discussed in Section 9.2.3, the matrix must have a small condition number for the system to be robust with respect to errors in the known parameters. The condition number becomes larger as the matrix becomes almost singular.

The structure of the matrix shows that it will be almost singular if the coefficients of the numerator polynomial $P(z)$ and denominator polynomial $Q(z)$ are almost identical. We therefore require that the roots of the polynomials $P(z)$ and $Q(z)$ be sufficiently different to avoid an ill-conditioned matrix. From expressions (6.3)—the plant transfer function for a digitally controlled analog plant—the poles of the discretized plant approach the zeros as the sampling interval is reduced (see also Section 12.2.2). Thus, when the controller is designed by pole assignment, to avoid an ill-conditioned matrix in (9.67), the sampling interval must not be excessively short.

We now discuss the choice of the desired characteristic polynomial. From the equivalence of the transfer function design to the state–space design described in Section 9.6, the separation principle implies that $\Delta_{cl}^d(z)$ can be written as the product

$$\Delta_{cl}^d(z) = \Delta_c^d(z)\Delta_o^d(z)$$

where $\Delta_c^d(z)$ is the controller characteristic polynomial and $\Delta_o^d(z)$ is the observer characteristic polynomial. We select the polynomial $N(z)$ as

$$N(z) = k_{ff}\Delta_o^d(z) \tag{9.68}$$

so that the observer polynomial $\Delta_o^d(z)$ cancels in the transfer function from the reference input to the system output. The scalar constant k_{ff} is selected so that the steady-state output is equal to the constant reference input

$$\frac{Y(1)}{R(1)} = \frac{P(1)N(1)}{\Delta_c^d(1)\Delta_o^d(1)} = \frac{P(1)k_{ff}\Delta_o^d(1)}{\Delta_c^d(1)\Delta_o^d(1)} = 1$$

The condition for zero steady-state error is

$$k_{ff} = \frac{\Delta_c^d(1)}{P(1)} \tag{9.69}$$

EXAMPLE 9.14

Solve Example 9.13 using the transfer function approach.

Solution
The plant transfer function is

$$G(z) = \frac{P(z)}{Q(z)} = 10^{-6}\frac{1.622z^2 + 45.14z + 48.23}{z^3 - 2.895z^2 + 2.791z - 0.8959}$$

Thus, we have the polynomials

$$P(z) = 1.622 \times 10^{-6}z^2 + 45.14 \times 10^{-6}z + 48.23 \times 10^{-6}$$

That is,

$$p_2 = 1.622 \times 10^{-6}, \quad p_1 = 45.14 \times 10^{-6}, \quad p_0 = 48.23 \times 10^{-6}$$

$$Q(z) = z^3 - 2.895z^2 + 2.791z - 0.8959$$

That is,

$$q_2 = -2.895, \quad q_1 = 2.791, \quad q_0 = -0.8959$$

The plant is third order, that is, $n = 3$, and the solvability condition of the Diophantine equation is $m = n - 1 = 2$. The order of the desired closed-loop characteristic polynomial is $m + n = 5$. We can therefore select the controller poles as $\{0.6, 0.4 \pm j0.33\}$ and the observer poles as $\{0.1, 0.2\}$ with the corresponding polynomials

$$\Delta_c^d(z) = z^3 - 1.6z^2 + 0.9489z - 0.18756$$

$$\Delta_o^d(z) = z^2 - 0.3z + 0.02$$

$$\Delta_{cl}^d(z) = \Delta_c^d(z)\Delta_o^d(z) = z^5 - 1.9z^4 + 1.4489z^3 - 0.50423z^2 + 0.075246z - 0.0037512$$

In other words, $\delta_4 = -1.9$, $\delta_3 = 1.4489$, $\delta_2 = -0.50423$, $\delta_1 = 0.075246$, and $\delta_0 = -0.0037512$. Using the matrix equation (9.67) gives

$$
\begin{bmatrix}
1 & 0 & 0 & 0 & 0 & 0 \\
-2.8949 & 1 & 0 & 1.622\times10^{-6} & 0 & 0 \\
2.790752 & -2.8949 & 1 & 45.14\times10^{-6} & 1.622\times10^{-6} & 0 \\
-0.895852 & 2.790752 & -2.8949 & 48.23\times10^{-6} & 45.14\times10^{-6} & 1.622\times10^{-6} \\
0 & -0.895852 & 2.790752 & 0 & 48.23\times10^{-6} & 45.14\times10^{-6} \\
0 & 0 & -0.895852 & 0 & 0 & 48.23\times10^{-6}
\end{bmatrix}
$$

$$
\begin{bmatrix}
d_2 \\ d_1 \\ d_0 \\ s_2 \\ s_1 \\ s_0
\end{bmatrix}
=
\begin{bmatrix}
1 \\ -1.9 \\ 1.4489 \\ -0.50423 \\ 0.075246 \\ -0.00375
\end{bmatrix}
$$

The MATLAB command **linsolve** gives the solution

$$d_2 = 1, d_1 = 0.9645, d_0 = 0.6526, s_2 = 1.8735 \cdot 10^4, s_1 = -2.9556 \cdot 10^4, s_0 = 1.2044 \cdot 10^4$$

and the polynomials

$$D(z) = z^2 + 0.9645z + 0.6526$$

$$S(z) = 1.8735 \cdot 10^4 z^2 - 2.9556 \cdot 10^4 z + 1.2044 \cdot 10^4$$

Then, from (9.68), we have $k_{ff} = 1698.489$ and the numerator polynomial

$$N(z) = 1698.489(z^2 - 0.3z + 0.02)$$

The step response of the control system of Figure 9.19 has a settling time of 0.1 s and a percentage overshoot less than 7%. The response meets all the design specifications.

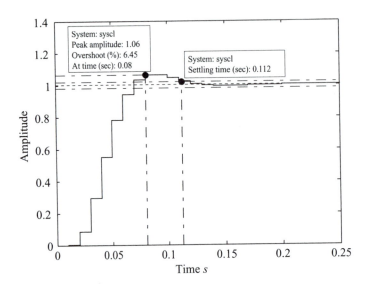

FIGURE 9.19

Step response for Example 9.14.

RESOURCES

D'Azzo, J. J., and C. H. Houpis, *Linear Control System Analysis and Design*, McGraw-Hill, 1988.

Bass, R. W., and I. Gura, High-order system design via state–space considerations, *Proc. JACC*, Troy, NY, October 1983, pp. 11-93.

Chen, C. T., *Linear System Theory and Design*, HRW, 1984.

Delchamps, D. F., *State-Space and Input–Output Linear Systems*, Springer-Verlag, 1988.

Kailath, T., *Linear Systems*, Prentice Hall, 1980.

Kautsky, J. N., K. Nichols, and P. Van Dooren, Robust pole assignment in linear state feedback, *Int. J. Control*, 41(5):1129-1155, 1985.

Mayne, D. Q., and P. Murdoch, Modal control of linear time invariant systems, *Int. J. Control*, 11(2):223-227, 1970.

Patel, R. V., and N. Munro, *Multivariable System Theory and Design*, Pergammon Press, 1982.

PROBLEMS

9.1 Show that the closed-loop quadruple for (A, B, C, D) with the state feedback $u(k) = -Kx(k) + v(k)$ is $(A - BK, B, C - DK, D)$.

9.2 Show that for the pair (A, B) with state feedback gain matrix K to have the closed-loop state matrix $A_d = A - BK$, a necessary condition is that for any vector $\mathbf{w}^T$ satisfying $\mathbf{w}^T B = \mathbf{0}^T$, and $\mathbf{w}^T A = \lambda \mathbf{w}^T$, A_d must satisfy $\mathbf{w}^T A_d = \lambda \mathbf{w}^T$.

Explain the significance of this necessary condition. (Note that the condition is also sufficient).

9.3 Show that for the pair (A, B) with $m \times n$ state feedback gain matrix K to have the closed-loop state matrix $A_{cl} = A - BK$, a sufficient condition is

$$\text{rank}\{B\} = \text{rank}\{[A - A_{cl} \mid B]\} = m$$

Is the matrix K unique for given matrices A and A_{cl}? Explain.

9.4 Using the results of Problem 9.3, determine if the closed-loop matrix can be obtained using state feedback for the pair

$$A = \begin{bmatrix} 1.0 & 0.1 & 0 \\ 0 & 1 & 0.01 \\ 0 & -0.1 & 0.9 \end{bmatrix} \quad B = \begin{bmatrix} 0 & 0 \\ 1 & 0 \\ 1 & 1 \end{bmatrix}$$

(a) $A_{cl} = \begin{bmatrix} 1.0 & 0.1 & 0 \\ -12.2 & -1.2 & 0 \\ 0.01 & 0.01 & 0 \end{bmatrix}$

(b) $A_{cl} = \begin{bmatrix} 0 & 0.1 & 0 \\ -12.2 & -1.2 & 0 \\ 0.01 & 0.01 & 0 \end{bmatrix}$

9.5 Show that for the pair (A, C) with observer gain matrix L to have the observer matrix $A_o = A - LC$, a necessary condition is that for any vector v satisfying $Cv = 0$, and $Av = \lambda v$, A_o must satisfy $A_o v = \lambda v$. Explain the significance of this necessary condition. (Note that the condition is also sufficient.)

9.6 Show that for the pair (A, C) with $n \times l$ observer gain matrix L to have the observer matrix $A_o = A - LC$, a sufficient condition is

$$\text{rank}\{C\} = \text{rank}\left\{ \begin{bmatrix} C \\ \hline A - A_o \end{bmatrix} \right\} = l$$

Is the matrix L unique for given matrices A and A_o? Explain.

9.7 Design a state feedback control law to assign the eigenvalues to the set $\{0, 0.1, 0.2\}$ for the systems with

(a) $A = \begin{bmatrix} 0.1 & 0.5 & 0 \\ 2 & 0 & 0.2 \\ 0.2 & 1 & 0.4 \end{bmatrix} \quad b = \begin{bmatrix} 0.01 \\ 0 \\ 0.005 \end{bmatrix}$

(b) $A = \begin{bmatrix} -0.2 & -0.2 & 0.4 \\ 0.5 & 0 & 1 \\ 0 & -0.4 & -0.4 \end{bmatrix} \quad b = \begin{bmatrix} 0.01 \\ 0 \\ 0 \end{bmatrix}$

9.8 Using eigenvalues that are two to three times as fast as those of the plant, design a state estimator for the system

(a) $A = \begin{bmatrix} 0.2 & 0.3 & 0.2 \\ 0 & 0 & 0.3 \\ 0.3 & 0 & 0.3 \end{bmatrix}$ $C = [1 \ 1 \ 0]$

(b) $A = \begin{bmatrix} 0.2 & 0.3 & 0.2 \\ 0 & 0 & 0.3 \\ 0.3 & 0 & 0.3 \end{bmatrix}$ $C = [1 \ 0 \ 0]$

9.9 Consider the system

$$A = \begin{bmatrix} 0 & 1 & 0 \\ 0 & 0 & 1 \\ -0.005 & -0.11 & -0.7 \end{bmatrix} \quad B = \begin{bmatrix} 0 \\ 0 \\ 1 \end{bmatrix}$$

$$C = [0.5 \ 1 \ 0] \quad d = 0$$

(a) Design a controller that assigns the eigenvalues $\{-0.8, -0.3 \pm j0.3\}$. Why is the controller guaranteed to exist?

(b) Why can we design an observer for the system with the eigenvalues $\{-0.5, -0.1 \pm j0.1\}$. Explain why the value (-0.5) must be assigned. (*Hint:* $(s + 0.1)^2 (s + 0.5) = s^3 + 0.7s^2 + 0.11s + 0.005$.)

(c) Obtain a similar system with a second-order observable subsystem, for which an observer can be easily designed, as in Section 8.3.3. Design an observer for the transformed system with two eigenvalues shifted as in (b) and check your design using the MATLAB command **place** or **acker**. Use the result to obtain the observer for the original system. (*Hint:* Obtain an observer gain $\mathbf{l}_r$ for the similar third-order system from your design by setting the first element equal to zero. Then obtain the observer gain for the original system using $\mathbf{l} = T_r \mathbf{l}_r$, where T_r is the similarity transformation matrix.)

(d) Design an observer-based feedback controller for the system with the controller and observer eigenvalues selected as in (a) and (b), respectively.

9.10 Design a reduced-order estimator state feedback controller for the discretized system

$$A = \begin{bmatrix} 0.1 & 0 & 0.1 \\ 0 & 0.5 & 0.2 \\ 0.2 & 0 & 0.4 \end{bmatrix} \quad \mathbf{b} = \begin{bmatrix} 0.01 \\ 0 \\ 0.005 \end{bmatrix} \quad \mathbf{c}^T = [1, 1, 0]$$

to obtain the eigenvalues $\{0.1, 0.4 \pm j0.4\}$.

9.11 Consider the following model of an armature-controlled DC motor, which is slightly different from that described in Example 7.15:

$$\begin{bmatrix} \dot{x}_1 \\ \dot{x}_2 \\ \dot{x}_3 \end{bmatrix} = \begin{bmatrix} 0 & 1 & 0 \\ 0 & 0 & 1 \\ 0 & -11 & -11.1 \end{bmatrix} \begin{bmatrix} x_1 \\ x_2 \\ x_3 \end{bmatrix} + \begin{bmatrix} 0 \\ 0 \\ 10 \end{bmatrix} u$$

$$y = \begin{bmatrix} 1 & 0 & 0 \end{bmatrix} \begin{bmatrix} x_1 \\ x_2 \\ x_3 \end{bmatrix}$$

For digital control with $T = 0.02$, apply the state feedback controllers determined in Example 9.4 in order to verify their robustness.

9.12 Consider the following model of a DC motor speed control system, which is slightly different from that described in Example 6.8:

$$G(s) = \frac{1}{(1.2s + 1)(s + 10)}$$

For a sampling period $T = 0.02$, obtain a state–space representation corresponding to the discrete-time system with DAC and ADC; then use it to verify the robustness of the state controller described in Example 9.5.

9.13 Verify the robustness of the state controller determined in Example 9.6 by applying it to the model shown in Problem 9.12.

9.14 Consider the DC motor position control system described in Example 3.6, where the (type 1) analog plant has the transfer function

$$G(s) = \frac{1}{s(s + 1)(s + 10)}$$

For the digital control system with $T = 0.02$, design a state feedback controller to obtain a step response with null steady-state error, zero overshoot, and a settling time of less than 0.5 s.

9.15 Design a digital state feedback controller for the analog system

$$G(s) = \frac{-s + 1}{(5s + 1)(10s + 1)}$$

with $T = 0.1$ to place the closed-loop poles at {0.4, 0.6}. Show that the zero of the closed-loop system is the same as the zero of the open-loop system.

9.16 Write the closed-loop system state–space equations of a full-order observer state feedback system with integral action.

9.17 Consider the continuous-time model of the overhead crane proposed in Problem 7.10 with $m_c = 1000$ kg, $m_l = 1500$ kg, and $l = 8$ m. Design a discrete full-order observer state feedback control to provide motion of the load without sway.

9.18 Consider the continuous-time model of the overhead crane proposed in Problem 7.10 with $m_c = 1000$ kg, $m_l = 1500$ kg, and $l = 8$ m. Design a control system based on pole assignment using transfer functions in order to provide motion of the load without sway.

COMPUTER EXERCISES

9.19 Write a MATLAB script to evaluate the feedback gains using Ackermann's formula for any pair (A, B) and any desired poles $\{\lambda_1, \ldots, \lambda_n\}$.

9.20 Write a MATLAB function that, given the system state–space matrices A, B, and C, the desired closed-loop poles, and the observer poles, determines the closed-loop system state–space matrices of a full-observer state feedback system with integral action.

9.21 Write a MATLAB function that uses the transfer function approach to determine the closed-loop system transfer function for a given plant transfer function $G(z)$, desired closed-loop system poles, and observer poles.

Optimal Control

In this chapter, we introduce optimal control theory for discrete-time systems. We begin with unconstrained optimization of a cost function and then generalize to optimization with equality constraints. We then cover the optimization or optimal control of discrete-time systems. We specialize to the linear quadratic regulator and obtain the optimality conditions for a finite and for an infinite planning horizon. We also address the regulator problem where the system is required to track a nonzero constant signal.

Objectives

After completing this chapter, the reader will be able to do the following:

1. Find the unconstrained optimum values of a function by minimization or maximization.
2. Find the constrained optimum values of a function by minimization or maximization.
3. Design an optimal digital control system.
4. Design a digital linear quadratic regulator.
5. Design a digital steady-state regulator.
6. Design a digital output regulator.
7. Design a regulator to track a nonzero constant input.

10.1 OPTIMIZATION

Many problems in engineering can be solved by minimizing a measure of **cost** or maximizing a measure of **performance**. The designer must select a suitable performance measure based on his or her understanding of the problem to include the most important performance criteria and reflect their relative importance. The designer must also select a mathematical form of the function that makes solving the optimization problem tractable.

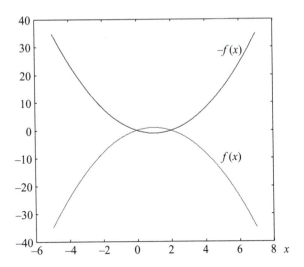

FIGURE 10.1

Minimization and maximization of a function of a single variable.

We observe that any maximization problem can be recast as a minimization, or vice versa. This is because the location of the maximum of a function $f(\mathbf{x})$ is the same as that of the minimum $-f(\mathbf{x})$, as demonstrated in Figure 10.1 for a scalar x. We therefore consider minimization only throughout this chapter.

We first consider the problem of minimizing a cost function or performance measure; then we extend our solution to problems with equality constraints.

10.1.1 Unconstrained Optimization

We first consider the problem of minimizing a **cost function** or **performance measure** of the form

$$J(\mathbf{x}) \tag{10.1}$$

where $\mathbf{x}$ is an $n \times 1$ vector of parameters to be selected. Let the optimal parameter vector be $\mathbf{x}^*$ and expand the function $J(\mathbf{x})$ in the vicinity of the optimum as

$$J(\mathbf{x}) = J(\mathbf{x}^*) + \frac{\partial J}{\partial \mathbf{x}}\bigg|_{\mathbf{x}^*}^T \Delta\mathbf{x} + \frac{1}{2!}\Delta\mathbf{x}^T \left[\frac{\partial^2 J}{\partial \mathbf{x}^2}\right]\bigg|_{\mathbf{x}^*}\Delta\mathbf{x} + O\left(\|\Delta\mathbf{x}\|^3\right) \tag{10.2}$$

$$\Delta\mathbf{x} = \mathbf{x} - \mathbf{x}^*$$

where

$$\frac{\partial J}{\partial \mathbf{x}}\bigg|_{\mathbf{x}^*}^T = \left[\frac{\partial J}{\partial x_1} \quad \frac{\partial J}{\partial x_2} \quad \cdots \quad \frac{\partial J}{\partial x_n}\right]_{\mathbf{x}^*} \tag{10.3}$$

$$\left[\frac{\partial^2 J}{\partial \mathbf{x}^2}\bigg|_{\mathbf{x}^*}\right] = \left[\frac{\partial^2 J}{\partial x_{ij}^2}\bigg|_{\mathbf{x}^*}\right] \qquad (10.4)$$

and the subscript denotes that the matrix is evaluated at $\mathbf{x}^*$. At a **local minimum** or **local maximum**, the first-order terms of the expansion that appear in the gradient vector are zero.

To guarantee that the point $\mathbf{x}^*$ is a minimum, any perturbation vector $\Delta \mathbf{x}$ away from $\mathbf{x}^*$ must result in an increase of the value of $J(\mathbf{x})$. Thus, the second term of the expansion must at least be positive or zero for $\mathbf{x}^*$ to have any chance of being a minimum. If the second term is positive, then we can guarantee that $\mathbf{x}^*$ is indeed a minimum. The sign of the second term is determined by the characteristics of the second derivative matrix, or **Hessian**. The second term is positive for any perturbation if the Hessian matrix is positive definite, and it is positive or zero if the Hessian matrix is positive semidefinite. We summarize this discussion in the following theorem.

Theorem 10.1: If $\mathbf{x}^*$ is a local minimum of $J(\mathbf{x})$, then

$$\frac{\partial J}{\partial \mathbf{x}}\bigg]_{\mathbf{x}^*}^{T} = \mathbf{0}_{1 \times n} \qquad (10.5)$$

$$\left[\frac{\partial^2 J}{\partial \mathbf{x}^2}\bigg|_{\mathbf{x}^*}\right] \geq 0 \qquad (10.6)$$

A sufficient condition for $\mathbf{x}^*$ to be a minimum is

$$\left[\frac{\partial^2 J}{\partial \mathbf{x}^2}\bigg|_{\mathbf{x}^*}\right] > 0 \qquad (10.7)$$

∎

EXAMPLE 10.1

Obtain the least-squares estimates of the linear resistance

$$v = iR$$

using N noisy measurements

$$z(k) = i(k)R + v(k), \quad k = 1, \dots, N$$

Solution
We begin by stacking the measurements to obtain the matrix equation

$$\mathbf{z} = \mathbf{i}R + \mathbf{v}$$

in terms of the vectors

$$\mathbf{z} = [z(1) \quad \cdots \quad z(N)]^{T}$$

$$\mathbf{i} = [i(1) \quad \cdots \quad i(N)]^T$$

$$\mathbf{v} = [v(1) \quad \cdots \quad v(N)]^T$$

We minimize the sum of the squares of the errors

$$J = \sum_{k=1}^{N} e^2(k)$$

$$e(k) = z(k) - i(k)\hat{R}$$

where the caret (^) denotes the estimate. We rewrite the performance measure in terms of the vectors as

$$J = \mathbf{e}^T \mathbf{e} = \mathbf{z}^T \mathbf{z} - 2\mathbf{i}^T \mathbf{z}\hat{R} + \mathbf{i}^T \mathbf{i}\hat{R}^2$$

$$\mathbf{e} = [e(1) \quad \cdots \quad e(N)]^T$$

The necessary condition for a minimum gives

$$\frac{\partial J}{\partial \hat{R}} = -2\mathbf{i}^T \mathbf{z} + 2\mathbf{i}^T \mathbf{i}\hat{R} = 0$$

We now have the least-squares estimate

$$\hat{R}_{LS} = \frac{\mathbf{i}^T \mathbf{z}}{\mathbf{i}^T \mathbf{i}}$$

The solution is indeed a minimum because the second derivative is the positive sum of the squares of the currents

$$\frac{\partial^2 J}{\partial \hat{R}^2} = 2\mathbf{i}^T \mathbf{i} > 0$$

10.1.2 Constrained Optimization

In most practical applications, the entries of the parameter vector **x** are subject to physical and economic constraints. Assume that our vector of parameters is subject to the equality constraint

$$\mathbf{m}(\mathbf{x}) = \mathbf{0}_{m \times 1} \tag{10.8}$$

In the simplest cases only, we can use the constraints to eliminate m parameters and then solve an optimization problem for the remaining $n - m$ parameters. Alternatively, we can use Lagrange multipliers to include the constraints in the optimization problem. We add the constraint to the performance measure weighted by the **Lagrange multipliers** to obtain the **Lagrangian**

$$L(\mathbf{x}) = J(\mathbf{x}) + \boldsymbol{\lambda}^T \mathbf{m}(\mathbf{x}) \tag{10.9}$$

We then solve for the vectors **x** and λ that minimize the Lagrangian as in unconstrained optimization. The following example demonstrates the use of

Lagrange multipliers in constrained optimization. Note that the example is simplified to allow us to solve the problem with and without Lagrange multipliers.

EXAMPLE 10.2

A manufacturer decides the production level of two products based on maximizing profit subject to constraints on production. The manufacturer estimates profit using the simplified measure

$$J(\mathbf{x}) = x_1^\alpha x_2^\beta$$

where x_i is the quantity produced for product i, $i = 1, 2$, and the parameters (α, β) are determined from sales data. The quantity of the two products produced cannot exceed a fixed level b. Determine the optimum production level for the two products subject to the production constraint

$$x_1 + x_2 = b$$

Solution

To convert the maximization problem into minimization, we use the negative of the profit. We obtain the optimum first without the use of Lagrange multipliers. We solve for x_2 using the constraint

$$x_2 = b - x_1$$

then substitute in the negative of the profit function to obtain

$$J(x_1) = -x_1^\alpha (b - x_1)^\beta$$

The necessary condition for a minimum gives

$$\frac{\partial J}{\partial x_1} = -\alpha x_1^{\alpha-1}(b - x_1)^\beta + \beta x_1^\alpha (b - x_1)^{\beta-1} = 0$$

which simplifies to

$$x_1 = \frac{\alpha b}{\alpha + \beta}$$

From the production constraint we solve for production level

$$x_2 = \frac{\beta b}{\alpha + \beta}$$

We now show that the same answer can be obtained using a Lagrange multiplier. We add the constraint multiplied by the Lagrange multiplier to obtain the Lagrangian

$$L(\mathbf{x}) = -x_1^\alpha x_2^\beta + \lambda(x_1 + x_2 - b)$$

The necessary conditions for the minimum are

$$\frac{\partial L}{\partial x_1} = -\alpha x_1^{\alpha-1} x_2^\beta + \lambda = 0$$

$$\frac{\partial L}{\partial x_2} = -\beta x_1^\alpha x_2^{\beta-1} + \lambda = 0$$

$$\frac{\partial L}{\partial \lambda} = x_1 + x_2 - b = 0$$

The first two conditions give

$$\alpha x_1^{\alpha-1} x_2^\beta = \beta x_1^\alpha x_2^{\beta-1}$$

which readily simplifies to

$$x_2 = \frac{\beta}{\alpha} x_1$$

Substituting in the third necessary condition (i.e., in the constraint) gives the solution

$$x_1 = \frac{\alpha b}{\alpha + \beta}$$

$$x_2 = \frac{\beta b}{\alpha + \beta}$$

10.2 OPTIMAL CONTROL

To optimize the performance of a discrete-time dynamic system, we minimize the performance measure

$$J = J_f(\mathbf{x}(k_f), k_f) + \sum_{k=k_0}^{k_f-1} L(\mathbf{x}(k), \mathbf{u}(k), k) \tag{10.10}$$

subject to the constraint

$$\mathbf{x}(k+1) = A\mathbf{x}(k) + B\mathbf{u}(k), \quad k = k_0, \ldots, k_f - 1 \tag{10.11}$$

We assume that the pair (A, B) is stabilizable; otherwise there is no point in control system design, optimal or otherwise. If the system is not stabilizable, then its structure must first be changed by selecting different control variables that allow its stabilization.

The first term of the performance measure is a **terminal penalty**, and each of the remaining terms represents a cost or penalty at time k. We change the problem into unconstrained minimization using Lagrange multipliers. We have the new performance measure

$$\begin{aligned} \bar{J} = J_f(\mathbf{x}(k_f), k_f) + \sum_{k=k_0}^{k_f-1} \{L(\mathbf{x}(k), \mathbf{u}(k), k) + \lambda^T(k+1)[A\mathbf{x}(k) \\ + B\mathbf{u}(k) - \mathbf{x}(k+1)]\} \end{aligned} \tag{10.12}$$

We define the Hamiltonian function as

$$H(\mathbf{x}(k), \mathbf{u}(k), k) = L(\mathbf{x}(k), \mathbf{u}(k), k) + \lambda^T(k+1)[A\mathbf{x}(k) + B\mathbf{u}(k)],$$
$$k = k_0, \ldots, k_f - 1 \tag{10.13}$$

and rewrite the performance measure in terms of the Hamiltonian as

$$\bar{J} = J_f(\mathbf{x}(k_f), k_f) - \lambda^T(k_f)\mathbf{x}(k_f) +$$
$$\sum_{k=k_0+1}^{k_f-1} \{H(\mathbf{x}(k), \mathbf{u}(k), k) - \lambda^T(k)\mathbf{x}(k)\} + H(\mathbf{x}(k_0), \mathbf{u}(k_0), k_0) \tag{10.14}$$

Each term of the preceding expression can be expanded as a truncated Taylor series in the vicinity of the optimum point of the form

$$\bar{J} = \bar{J}^* + \delta_1\bar{J} + \delta_2\bar{J} + \ldots \tag{10.15}$$

where the * denotes the optimum, δ denotes a variation from the optimal, and the subscripts denote the order of the variation in the expansion. From basic calculus, we know the necessary condition for a minimum or maximum is that the first-order term must be zero. For a minimum the second-order term must be positive or zero, and a sufficient condition for a minimum is a positive second term. We therefore need to evaluate the terms of the expansion to determine necessary and sufficient conditions for a minimum.

Each term of the expansion includes derivatives with respect to the arguments of the performance measure and can be obtained by expanding each term separately. For example, the Hamiltonian can be expanded as

$$H(\mathbf{x}(k), \mathbf{u}(k), k) = H(\mathbf{x}^*(k), \mathbf{u}^*(k), k) +$$
$$\left.\frac{\partial H(\mathbf{x}(k), \mathbf{u}(k), k)}{\partial \mathbf{x}(k)}\right|_*^T \delta\mathbf{x}(k) + \left.\frac{\partial H(\mathbf{x}(k), \mathbf{u}(k), k)}{\partial \mathbf{u}(k)}\right|_*^T \delta\mathbf{u}(k) +$$
$$[A\mathbf{x}(k) + B\mathbf{u}(k)]_*^T \delta\lambda(k+1) +$$
$$\delta\mathbf{x}^T(k)\left.\frac{\partial H(\mathbf{x}(k), \mathbf{u}(k), k)}{\partial \mathbf{x}^2(k)}\right|_*^T \delta\mathbf{x}(k) + \delta\mathbf{u}^T(k)\left.\frac{\partial^2 H(\mathbf{x}(k), \mathbf{u}(k), k)}{\partial \mathbf{u}^2(k)}\right|_*^T \delta\mathbf{u}(k) +$$
$$\delta\mathbf{x}^T(k)\left.\frac{\partial H(\mathbf{x}(k), \mathbf{u}(k), k)}{\partial \mathbf{x}(k)\partial \mathbf{u}(k)}\right|_*^T \delta\mathbf{u}(k) + \delta\mathbf{u}^T(k)\left.\frac{\partial^2 H(\mathbf{x}(k), \mathbf{u}(k), k)}{\partial \mathbf{u}(k)\partial \mathbf{x}(k)}\right|_*^T \delta\mathbf{x}(k) +$$
higher order terms

$$\tag{10.16}$$

where δ denotes a variation from the optimal, the superscript (*) denotes the optimum value, and the subscript (*) denotes evaluation at the optimum point. The second-order terms can be written more compactly in terms of the Hessian matrix of second derivatives in the form

$$[\delta\mathbf{x}^T(k) \quad \delta\mathbf{u}^T(k)] \left.\begin{bmatrix} \dfrac{\partial H(\mathbf{x}(k), \mathbf{u}(k), k)}{\partial \mathbf{x}^2(k)} & \dfrac{\partial H(\mathbf{x}(k), \mathbf{u}(k), k)}{\partial \mathbf{x}(k)\partial \mathbf{u}(k)} \\[3mm] \dfrac{\partial^2 H(\mathbf{x}(k), \mathbf{u}(k), k)}{\partial \mathbf{u}(k)\partial \mathbf{x}(k)} & \dfrac{\partial^2 H(\mathbf{x}(k), \mathbf{u}(k), k)}{\partial \mathbf{u}^2(k)} \end{bmatrix}\right|_* \begin{bmatrix} \delta\mathbf{x}(k) \\ \delta\mathbf{u}(k) \end{bmatrix} \tag{10.17}$$

We expand the linear terms of the performance measure as

$$\lambda^T(k)\mathbf{x}(k) = \lambda^{*T}(k)\mathbf{x}^*(k) + \delta\lambda^T(k)\mathbf{x}^*(k) + \lambda^{*T}(k)\delta\mathbf{x}(k) \qquad (10.18)$$

The terminal penalty can be expanded as

$$J_f(\mathbf{x}(k_f), k_f) = J_f(\mathbf{x}^*(k_f), k_f) + \frac{\partial J_f(\mathbf{x}(k_f), k_f)}{\partial\mathbf{x}(k_f)}\Bigg|_*^T \delta\mathbf{x}(k_f) + $$
$$\delta\mathbf{x}^T(k_f)\left[\frac{\partial^2 J_f(\mathbf{x}^*(k_f), k_f)}{\partial\mathbf{x}^2(k_f)}\right]_* \delta\mathbf{x}(k_f) \qquad (10.19)$$

We now combine the first-order terms to obtain the increment

$$\delta_1\bar{J} = \sum_{k=k_0+1}^{N-1}\left\{\left[\frac{\partial H(\mathbf{x}(k), \mathbf{u}(k), k)}{\partial\mathbf{x}(k)}\Bigg|_* - \lambda^*(k)\right]^T \delta\mathbf{x}(k) + \right.$$
$$\left.[A\mathbf{x}^*(k) + B^*\mathbf{u}(k) - \mathbf{x}^*(k+1)]^T \delta\lambda(k+1)\right\} + \qquad (10.20)$$
$$\frac{\partial H(\mathbf{x}(k), \mathbf{u}(k), k)}{\partial\mathbf{u}(k)}\Bigg|_*^T \delta\mathbf{u}(k) + \left[\frac{\partial J_f(\mathbf{x}(k_f), k_f)}{\partial\mathbf{x}(k_f)} - \lambda(k_f)\right]^T\Bigg|_* \delta\mathbf{x}(k_f)$$

From (10.15) and the discussion following it, we know that a necessary condition for a minimum of the performance measure is that the increment must be equal to zero for any combination of variations in its arguments. A zero for any combination of variations occurs only if the coefficient of each increment is equal to zero. Equating each coefficient to zero, we have the necessary conditions

$$\mathbf{x}^*(k+1) = A\mathbf{x}^*(k) + B^*\mathbf{u}(k), \quad k = k_0, \ldots, k_f - 1 \qquad (10.21)$$

$$\lambda^*(k) = \frac{\partial H(\mathbf{x}(k), \mathbf{u}(k), k)}{\partial\mathbf{x}(k)}\Bigg|_* \quad k = k_0, \ldots, k_f - 1 \qquad (10.22)$$

$$\frac{\partial H(\mathbf{x}(k), \mathbf{u}(k), k)}{\partial\mathbf{u}(k)}\Bigg|_* = 0 \qquad (10.23)$$

$$0 = \frac{\partial J_f(\mathbf{x}(k_f), k_f)}{\partial\mathbf{x}(k_f)} - \lambda(k_f)\Bigg|_*^T \delta\mathbf{x}(k_f) \qquad (10.24)$$

If the terminal point is fixed, then its perturbation is zero and the terminal optimality condition is satisfied. Otherwise, we have the terminal conditions

$$\lambda(k_f) = \frac{\partial J_f(\mathbf{x}(k_f), k_f)}{\partial\mathbf{x}(k_f)} \qquad (10.25)$$

Similarly, conditions can be developed for the case of a free initial state with an initial cost added to the performance measure.

The necessary conditions for a minimum of equations (10.21) through (10.23) are known, respectively, as the **state equation**, the **costate equation**, and the

Table 10.1 Optimality Conditions

Condition	Equation
State equation	$\mathbf{x}^*(k+1) = A\mathbf{x}^*(k) + B^*\mathbf{u}(k) \quad k = k_0, \ldots, k_f - 1$
Costate equation	$\lambda^*(k) = \left. \dfrac{\partial H(\mathbf{x}(k), \mathbf{u}(k), k)}{\partial \mathbf{x}(k)} \right|_* \quad k = k_0, \ldots, k_f - 1$
Minimum principle	$\left. \dfrac{\partial H(\mathbf{x}(k), \mathbf{u}(k), k)}{\partial \mathbf{u}(k)} \right|_* = 0, \quad k = k_0, \ldots, k_f - 1$

minimum principle of Pontryagin. The minimum principle tells us that the cost is minimized by choosing the control that minimizes the Hamiltonian. This condition can be generalized to problems where the control is constrained, in which case the solution can be at the boundary of the allowable control region. We summarize the necessary conditions for an optimum in Table 10.1.

The second-order term is

$$\delta_2 \bar{J} = \sum_{k=k_0}^{k_f-1} \begin{bmatrix} \delta\mathbf{x}^T(k) & \delta\mathbf{u}^T(k) \end{bmatrix} \begin{bmatrix} \dfrac{\partial H(\mathbf{x}(k), \mathbf{u}(k), k)}{\partial \mathbf{x}^2(k)} & \dfrac{\partial H(\mathbf{x}(k), \mathbf{u}(k), k)}{\partial \mathbf{x}(k)\partial \mathbf{u}(k)} \\ \dfrac{\partial^2 H(\mathbf{x}(k), \mathbf{u}(k), k)}{\partial \mathbf{u}(k)\partial \mathbf{x}(k)} & \dfrac{\partial^2 H(\mathbf{x}(k), \mathbf{u}(k), k)}{\partial \mathbf{u}^2(k)} \end{bmatrix}_* \begin{bmatrix} \delta\mathbf{x}(k) \\ \delta\mathbf{u}(k) \end{bmatrix} +$$

$$\delta\mathbf{x}^T(k_f) \left. \dfrac{\partial^2 J_f(\mathbf{x}(k_f), k_f)}{\partial \mathbf{x}^2(k_f)} \right|_* \delta\mathbf{x}(k_f)$$

$$(10.26)$$

For the second-order term to be positive or at least zero for any perturbation, we need the Hessian matrix to be positive definite, or at least positive semidefinite. We have the sufficient condition

$$\begin{bmatrix} \dfrac{\partial H(\mathbf{x}(k), \mathbf{u}(k), k)}{\partial \mathbf{x}^2(k)} & \dfrac{\partial H(\mathbf{x}(k), \mathbf{u}(k), k)}{\partial \mathbf{x}(k)\partial \mathbf{u}(k)} \\ \dfrac{\partial^2 H(\mathbf{x}(k), \mathbf{u}(k), k)}{\partial \mathbf{u}(k)\partial \mathbf{x}(k)} & \dfrac{\partial^2 H(\mathbf{x}(k), \mathbf{u}(k), k)}{\partial \mathbf{u}^2(k)} \end{bmatrix}_* > 0 \qquad (10.27)$$

If the Hessian matrix is positive semidefinite, then the condition is necessary for a minimum but not sufficient because there will be perturbations for which the second-order terms are zero. Higher-order terms are then needed to determine if these perturbations result in an increase in the performance measure.

For a fixed terminal state, the corresponding perturbations are zero and no additional conditions are required. For a free terminal state, we have the additional condition

$$\left. \dfrac{\partial^2 J(\mathbf{x}(k_f), k_f)}{\partial \mathbf{x}^2(k_f)} \right|_* > 0 \qquad (10.28)$$

EXAMPLE 10.3

Because of their preference for simple models, engineers often use first-order systems for design. These are known as scalar systems. We consider the scalar system

$$x(k+1) = ax(k) + bu(k), x(0) = x_0, b > 0$$

If the system is required to reach the zero state, find the control that minimizes the control effort to reach the desired final state.

Solution

For minimum control effort, we have the performance measure

$$J = \frac{1}{2}\sum_{k=0}^{k_f} u(k)^2$$

The Hamiltonian is given by

$$H(\mathbf{x}(k), \mathbf{u}(k), k) = \frac{1}{2}u(k)^2 + \lambda(k+1)[ax(k) + bu(k)], \quad k = 0, \ldots, k_f - 1$$

For minimum time, we minimize the Hamiltonian by selecting the control

$$u^*(k) = -b\lambda^*(k+1)$$

The costate equation is

$$\lambda^*(k) = a\lambda^*(k+1) \quad k = 0, \ldots, k_f - 1$$

and its solution is given by

$$\lambda^*(k) = a^{k_f - k}\lambda^*(k_f), \quad k = 0, \ldots, k_f - 1$$

The optimal control can now be written as

$$u^*(k) = -ba^{k_f - k - 1}\lambda^*(k_f) \quad k = 0, \ldots, k_f - 1$$

We next consider iterations of the state equation with the optimal control

$$x^*(1) = ax(0) + bu^*(0) = ax_0 - b^2 a^{k_f - 1}\lambda^*(k_f)$$

$$x^*(2) = ax(1) + bu^*(1) = a^2 x_0 - b^2\{a^{k_f} + a^{k_f - 2}\}\lambda^*(k_f)$$

$$x^*(3) = ax(2) + bu^*(2) = a^3 x_0 - b^2\{a^{k_f + 1} + a^{k_f - 1} + a^{k_f - 3}\}\lambda^*(k_f)$$

$$\vdots$$

$$x^*(k_f) = ax(k_f - 1) + bu^*(k_f - 1) = a^{k_f}x_0 - b^2\{(a^2)^{k_f - 1} + \ldots a^2 + 1\}\lambda^*(k_f) = 0$$

We solve the last equation for the terminal Lagrange multiplier

$$\lambda^*(k_f) = \frac{(a^{k_f}/b^2)x_0}{(a^2)^{k_f - 1} + \ldots a^2 + 1}$$

Substituting for the terminal Lagrange multiplier gives the optimal control

$$u^*(k) = -\frac{\left(a^{2k_f-k-1}/b\right)x_0}{\left(a^2\right)^{k_f-1}+\ldots a^2+1} \quad k = 0,\ldots,k_f-1$$

For an open-loop stable system, we can write the control in the form

$$u^*(k) = -\frac{a^2-1}{\left(a^2\right)^{k_f}-1}\left(a^{2k_f-k-1}/b\right)x_0 \quad k = 0,\ldots,k_f-1, |a| < 1$$

Note that the closed-loop system dynamics of the optimal system are time varying even though the open-loop system is time invariant. The closed-loop system is given by

$$x(k+1) = ax(k) - \frac{\left(a^{2k_f-k-1}/b\right)x_0}{\left(a^2\right)^{k_f-1}+\ldots a^2+1}, \quad x(0) = x_0, b > 0$$

For an initial state $x_0 = 1$ and the unstable dynamics with $a = 1.5$ and $b = 0.5$, the optimal trajectories are shown in Figure 10.2. The optimal control stabilizes the system and drives its trajectories to the origin.

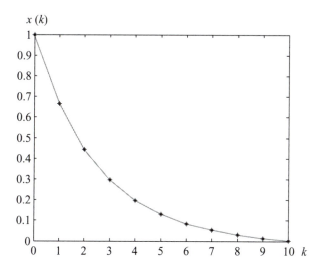

FIGURE 10.2

Optimal trajectories for the scalar system described in Example 10.3.

10.3 THE LINEAR QUADRATIC REGULATOR

The choice of performance index in optimal control determines the performance of the system and the complexity of the optimal control problem. The most popular choice for the performance index is a quadratic function of the state variable and the control inputs. The performance measure is of the form

$$J = \frac{1}{2}\mathbf{x}^T(k_f)S(k_f)\mathbf{x}(k_f) + \frac{1}{2}\sum_{k=k_0}^{k_f-1}(\mathbf{x}^T(k)Q(k)\mathbf{x}(k) + \mathbf{u}^T(k)R(k)\mathbf{u}(k)) \quad (10.29)$$

where the matrices $S(k_f)$, $Q(k)$, $k = k_0, \ldots, k_f - 1$ are positive definite symmetric $n \times n$, and the matrices $R(k)$, $k = k_0, \ldots, k_f - 1$ are positive definite symmetric $m \times m$. This choice is known as the **linear quadratic regulator**, and in addition to its desirable mathematical properties it can be physically justified. In a regulator problem, the purpose is to maintain the system close to the zero state and the quadratic function of the state variable is a measure of error. On the other hand, the effort needed to minimize the error must also be minimized, as must the control effort represented by a quadratic function of the controls. Thus, the quadratic performance measure achieves a compromise between minimizing the regulator error and minimizing the control effort.

To obtain the necessary and sufficient conditions for the linear quadratic regulator, we use the results of Section 10.2. We first write an expression for the Hamiltonian:

$$H(\mathbf{x}(k), \mathbf{u}(k), k) = \frac{1}{2}\mathbf{x}^T(k)Q(k)\mathbf{x}(k) + \frac{1}{2}\mathbf{u}^T(k)R(k)\mathbf{u}(k)$$
$$+ \lambda^T(k+1)[A\mathbf{x}(k) + B\mathbf{u}(k)], \quad k = k_0, \ldots, k_f - 1 \quad (10.30)$$

The state equation is unchanged, but the costate equation becomes

$$\lambda^*(k) = Q(k)\mathbf{x}^*(k) + A^T\lambda^*(k+1) \quad k = k_0, \ldots, k_f - 1 \quad (10.31)$$

Pontryagin's minimum principle gives

$$R(k)\mathbf{u}^*(k) + B^T\lambda^*(k+1) = 0$$

which yields the optimum control expression

$$\mathbf{u}^*(k) = -R^{-1}(k)B^T\lambda^*(k+1) \quad (10.32)$$

Substituting in the state equation, we obtain the optimal dynamics

$$\mathbf{x}^*(k+1) = A\mathbf{x}^*(k) - B^*R^{-1}(k)B^T\lambda^*(k+1), \quad k = k_0, \ldots, k_f - 1 \quad (10.33)$$

Equation (10.27) gives the condition

$$\begin{bmatrix} Q(k) & \mathbf{0}_{n\times m} \\ \mathbf{0}_{m\times n} & R \end{bmatrix}_* > 0 \quad (10.34)$$

Because R is positive definite, a sufficient condition for a minimum is that Q must be positive definite. A necessary condition is that Q must be positive semidefinite.

10.3.1 Free Final State

If the final state of the optimal control is not fixed, we have the terminal condition based on (10.25),

$$\lambda^*(k_f) = \frac{\partial J_f(\mathbf{x}(k_f), k_f)}{\partial \mathbf{x}(k_f)} = S(k_f)\mathbf{x}^*(k_f) \tag{10.35}$$

where the matrix $S(k_f)$ is the terminal weight matrix of (10.29). The condition suggests a general relationship between the state and costate—that is,

$$\lambda^*(k) = S(k)\mathbf{x}^*(k) \tag{10.36}$$

If the proposed relation yields the correct solution, this would allow us to conclude that the relation is indeed correct. We substitute the proposed relation (10.36) in the state equation (10.33):

$$\mathbf{x}^*(k+1) = A\mathbf{x}^*(k) - B^*R^{-1}(k)B^T S(k+1)\mathbf{x}^*(k+1)$$

This yields the recursion

$$\mathbf{x}^*(k+1) = \{I + B^*R^{-1}(k)B^T S(k+1)\}^{-1}A\mathbf{x}^*(k) \tag{10.37}$$

Using the proposed relation (10.36) and substituting in the costate equation, we have

$$\lambda^*(k) = Q(k)\mathbf{x}^*(k) + A^T S(k+1)\mathbf{x}^*(k+1)$$
$$= \{Q(k) + A^T S(k+1)[I + B^*R^{-1}(k)B^T S(k+1)]^{-1}A\}\mathbf{x}^*(k) \tag{10.38}$$
$$= S(k)\mathbf{x}^*(k)$$

Equation (10.38) holds for all values of the state vector and hence we have the matrix equality

$$Q(k) + A^T S(k+1)[I + B^*R^{-1}(k)B^T S(k+1)]^{-1}A = S(k)$$

We now apply the matrix inversion lemma (see Appendix III) to obtain

$$S(k) = A^T\{S(k+1) - S(k+1)B(B^T S(k+1)B \\ + R(k))^{-1}B^T S(k+1)\}A + Q(k), S(k_f) \tag{10.39}$$

The preceding equation is known as the **matrix Riccati equation** and can be solved iteratively backward in time to obtain the matrices $S(k)$, $k = k_0, \ldots, k_f - 1$.

We now return to the expression for the optimal control (10.32) and substitute for the costate to obtain

$$\mathbf{u}^*(k) = -R^{-1}(k)B^T S(k+1)\mathbf{x}^*(k+1)$$
$$= -R^{-1}(k)B^T S(k+1)[A\mathbf{x}^*(k) + B^*\mathbf{u}(k)]$$

We solve for the control

$$\mathbf{u}^*(k) = -[I + R^{-1}(k)B^T S(k+1)B]^{-1}R^{-1}(k)B^T S(k+1)A\mathbf{x}^*(k)$$

Using the rule for the inverse of a product, we have the optimal state feedback

$$\mathbf{u}^*(k) = -K(k)\mathbf{x}^*(k)$$

$$K(k) = [R(k) + B^T S(k+1)B]^{-1}B^T S(k+1)A \tag{10.40}$$

Thus, with the offline solution of the Riccati equation, we can implement the optimal state feedback.

The Riccati equation can be written in terms of the optimal gain of (10.40) in the form

$$S(k) = A^T S(k+1)\left\{A - B\left(B^T S(k+1)B + R(k)\right)^{-1} B^T S(k+1)A\right\} + Q(k)$$
$$= A^T S(k+1)\{A - BK(k)\} + Q(k)$$

In terms of the closed-loop state matrix $A_{cl}(k)$, we have

$$S(k) = A^T S(k+1) A_{cl}(k) + Q(k)$$
$$A_{cl}(k) = A - BK(k)$$

(10.41)

Note that the closed-loop matrix is time varying even for a time-invariant pair (A, B).

A more useful form of the Riccati is obtained by adding and subtracting terms to obtain

$$S(k) = A_{cl}^T S(k+1) A_{cl}(k) + Q(k) + K^T(k) B^T(k) S(k+1)[A - BK(k)]$$

Using (10.39), the added term can be rewritten as

$$K^T(k) B^T(k) S(k+1)[A - BK(k)] = K^T(k) B^T(k) S(k+1)A$$
$$- K^T(k)\left[R(k) + B^T S(k+1)B - R(k)\right]K(k)$$
$$= K^T(k) R(k) K(k) + K^T(k) B^T S(k+1)A$$
$$- \left[R(k) + B^T S(k+1)B\right]\left[R(k) + B^T S(k+1)B\right]^{-1}$$
$$B^T S(k+1)A$$
$$= K^T(k) R(k) K(k)$$

We now have the **Joseph form** of the Riccati equation:

$$S(k) = A_{cl}^T S(k+1) A_{cl}(k) + K^T(k) R(k) K(k) + Q(k) \qquad \textbf{(10.42)}$$

Because of its symmetry, this form performs better in iterative numerical computation.

If the Riccati equation in Joseph form is rewritten with the optimal gain replaced by a suboptimal gain, then the equation becomes a **Lyapunov difference equation** (see Chapter 11). For example, if the optimal gain is replaced by a constant gain K to simplify implementation, then the performance of the system will be suboptimal and is governed by a Lyapunov equation.

EXAMPLE 10.4

A variety of mechanical systems can be approximately modeled by the double integrator

$$\ddot{x}(t) = u(t)$$

where $u(t)$ is the applied force. With digital control and a sampling period $T = 0.02$ s, the double integrator has the discrete state–space equation

$$\mathbf{x}(k+1) = \begin{bmatrix} 1 & T \\ 0 & 1 \end{bmatrix} \mathbf{x}(k) + \begin{bmatrix} T^2/2 \\ T \end{bmatrix} u(k)$$

$$y(k) = \begin{bmatrix} 1 & 0 \end{bmatrix} \mathbf{x}(k)$$

1. Design a linear quadratic regulator for the system with terminal weight $S(100) = \text{diag}\{10, 1\}$, $Q = \text{diag}\{10, 1\}$, and control weight $r = 0.1$, then simulate the system with initial condition vector $\mathbf{x}(0) = [1, 0]^T$.

2. Repeat the design of part 1 with $S(100) = \text{diag}\{10, 1\}$, $Q = \text{diag}\{10, 1\}$, and control weight $r = 100$. Compare the results to part 1, and discuss the effect of changing the value of r.

Solution

The Riccati equation can be expanded with the help of a symbolic manipulation package to obtain

$$\begin{bmatrix} s_{11} & s_{12} \\ s_{12} & s_{22} \end{bmatrix}_k$$

$$= \frac{\begin{bmatrix} \left(s_{11}s_{22} - s_{12}^2\right)T^2 + s_{11}r & 2\left(s_{11}s_{22} - s_{12}^2\right)T^3 + 4r\left(s_{11}T + 2s_{12}\right) \\ 2\left(s_{11}s_{22} - s_{12}^2\right)T^3 + 4r\left(s_{11}T + 2s_{12}\right) & \left(s_{11}s_{22} - s_{12}^2\right)T^4 + 4\left(s_{11}T^2 + 2Ts_{12} + s_{22}\right)r \end{bmatrix}}{T^4 s_{11} + 4T^3 s_{12} + 4T^2 s_{22} + 4r} \Bigg|_{k+1}$$

$$+ Q$$

where the subscript denotes the time at which the matrix S is evaluated by backward-in-time iteration starting with the given terminal value. We use the simple MATLAB script:

```
% simlqr: simulate a scalar optimal control DT system
t(1)=0;
x{1}=[1;0]; % Initial state
T=0.02; % Sampling period
N=150; % Number of steps
S=diag([10,1]);r=0.1; % Weights
Q=diag([10,1]);
A=[1,T;0,1];B=[T^2/2;T]; % System matrices

for i=N:-1:1
    K{i}=(r+B'*S*A*B)\B'*S*A; % Calculate the optimal feedback gains
    % Note that K(1) is really K(0)
    kp(i)=K{i}(1); % Position gain
    kv(i)=K{i}(2); % Velocity gain
    Acl=A-B*K{i};
    S=Acl'*S*Acl+K{i}'*r*K{i}+Q; % Iterate backward (Riccati-Joseph form)
end
```

```
for i=1:N
    t(i+1)=t(i)+T;
    u(i)=-K{i}*x{i};
    x{i+1}=A*x{i}+B*u(i); % State equation
end
xmat=cell2mat(x); % Change cell to mat to extract data
xx=xmat(1,:); % Position
xv=xmat(2,:); % Velocity
plot(t,xx,'.') % Plot Position
figure % New figure
plot(t,xv,'.') % PLot Velocity
figure % New figure
plot(xx,xv,'.')% PLot phase plane trajectory
figure % New figure
plot(t(1:N),u,'.') % Plot control input
figure % New figure
plot(t(1:N),kp,'.',t(1:N),kv,'.') % Plot position gain & velocity gain
```

Five plots are obtained using the script for each part.

1. The first three plots are the position trajectory of Figure 10.3, the velocity trajectory of Figure 10.4, and the phase-plane trajectory of Figure 10.5. The three plots show that the optimal control drives the system toward the origin. However, because the final state is free, the origin is not reached. From the gain plot shown in Figure 10.6, we observe

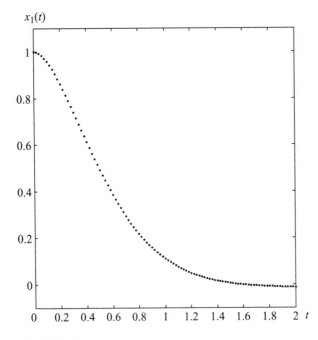

FIGURE 10.3

Position trajectory for the inertial control system described in Example 10.4(1).

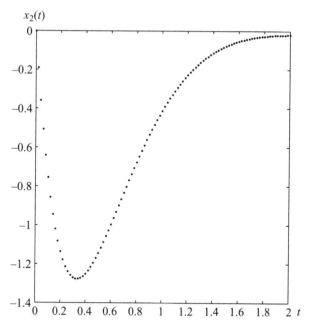

FIGURE 10.4

Velocity trajectory for the inertial control system described in Example 10.4(1).

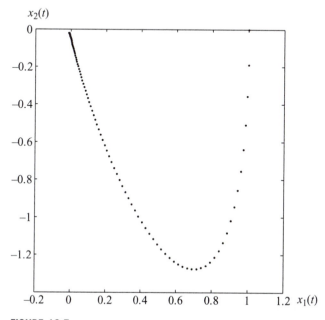

FIGURE 10.5

Phase plane trajectory for the inertial control system described in Example 10.4(1).

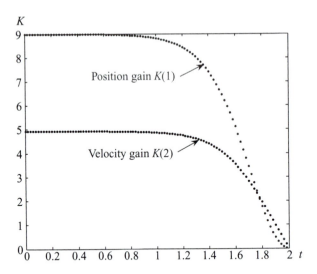

FIGURE 10.6

Plot of feedback gains versus time for Example 10.4(1).

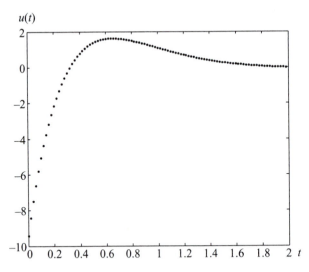

FIGURE 10.7

Optimal control input for Example 10.4(1).

that the controller gains, which are obtained by iteration backward in time, approach a fixed level. Consequently, the optimal control shown in Figure 10.7 also approaches a fixed level.

2. The first three plots are the position trajectory of Figure 10.8, the velocity trajectory of Figure 10.9, and the phase-plane trajectory of Figure 10.10. The three plots show that

$x_1(t)$

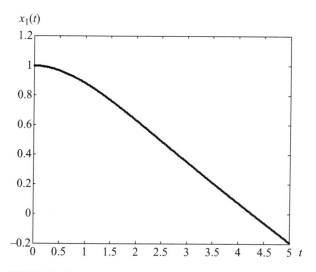

FIGURE 10.8

Position trajectory for the inertial control system of Example 10.4(2).

$x_2(t)$

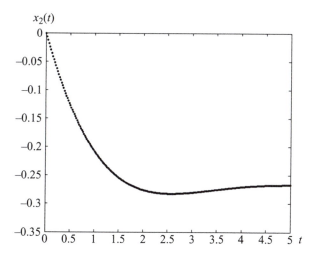

FIGURE 10.9

Velocity trajectory for the inertial control system of Example 10.4(2).

the optimal control drives the system toward the origin. However, the final state reached is farther from the origin than that of part 1 because the error weight matrix is now smaller relative to the value of r. The larger control weight r results in smaller control gain as shown in Figure 10.11. This corresponds to a reduction in the control effort as shown in Figure 10.12. Note that the controller gains approach a fixed level at a slower rate than in part 1.

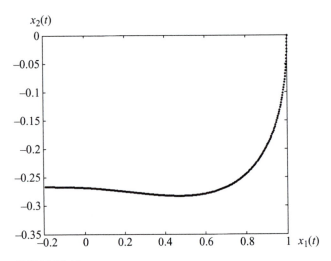

FIGURE 10.10

Phase plane trajectory for the inertial control system of Example 10.4(2).

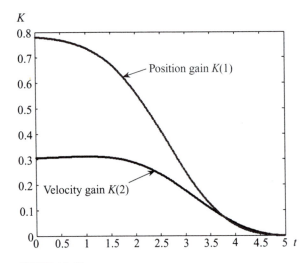

FIGURE 10.11

Plot of feedback gains versus time for Example 10.4(2).

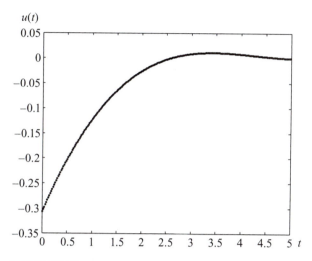

FIGURE 10.12

Optimal control input for Example 10.4(2).

10.4 STEADY-STATE QUADRATIC REGULATOR

Implementing the linear quadratic regulator is rather complicated because of the need to calculate and store gain matrices. From Example 10.4 we observe that the gain values converge to fixed values. This occurs in general with the fulfillment of some simple requirements that are discussed in this section.

For many applications, it is possible to simplify implementation considerably by using the steady-state gain exclusively in place of the optimal gains. This solution is only optimal if the summation interval in the performance measure, known as the **planning horizon**, is infinite. For a finite planning horizon, the simplified solution is suboptimal (i.e., gives a higher value of the performance measure) but often performs almost as well as the optimal control. Thus, it is possible to retain the performance of optimal control without the burden of implementing it if we solve the **steady-state regulator problem** with the performance measure of the form

$$J = \frac{1}{2} \sum_{k=k_0}^{\infty} \left(\mathbf{x}^T(k)Q\mathbf{x}(k) + \mathbf{u}^T(k)R\mathbf{u}(k) \right) \tag{10.43}$$

We assume that the weighting matrices Q and R are constant, with Q positive semidefinite and R positive definite. We can therefore decompose the matrix Q as

$$Q = Q_a^T Q_a \tag{10.44}$$

where Q_a is the square root matrix. This allows us to write the state error terms of "measurements" in terms of an equivalent measurement vector

$$\mathbf{y}(k) = Q_a \mathbf{x}(k)$$

$$\mathbf{x}^T(k) Qx(k) = \mathbf{y}^T(k) \mathbf{y}(k)$$

(10.45)

The matrix Q_a is positive definite for Q positive definite and positive semi-definite for Q positive semidefinite. In the latter case, large state vectors can be mapped by Q_a to zero $\mathbf{y}(k)$, and a small performance measure cannot guarantee small errors or even stability. We think of the vector $\mathbf{y}(k)$ as a measurement of the state and recall the detectability condition from Chapter 8.

We recall that if the pair (A, Q_a) is detectable, then $\mathbf{x}(k)$ decays asymptotically to zero with $\mathbf{y}(k)$. Hence, this detectability condition is required for acceptable steady-state regulator behavior. If the pair is observable, then it is also detectable and the system behavior is acceptable. For example, if the matrix Q_a is positive definite, there is a one–one mapping between the states $\mathbf{x}$ and the measurements $\mathbf{y}$ and the pair (A, Q_a) is always observable. Hence, a positive definite Q (and Q_a) is sufficient for acceptable system behavior.

If the detectability (observability) condition is satisfied and the system is stabi-lizable, then the Riccati equation does not diverge and a steady-state condition is reached. The resulting algebraic Riccati equation is in the form

$$S = A^T \left\{ S - SB \left(B^T SB + R \right)^{-1} B^T S \right\} A + Q$$

(10.46)

It can be shown that, under these conditions, the algebraic Riccati equation has a unique positive definite solution. However, the equation is clearly difficult to solve in general and is typically solved numerically. The MATLAB solution of the algebraic Riccati equation is discussed in Section 10.4.2.

The optimal state feedback corresponding to the steady-state regulator is

$$\mathbf{u}^* = -K\mathbf{x}^*(k)$$

$$K = \left[R + B^T SB \right]^{-1} B^T SA$$

(10.47)

Using the gain expression, we can write the algebraic Riccati equation in the Joseph form:

$$S = A_{cl}^T SA_{cl} + K^T RK + Q$$

$$A_{cl} = A - BK$$

(10.48)

If the optimal gain K is replaced by a suboptimal gain, then the algebraic Riccati equation becomes an **algebraic Lyapunov equation** (see Chapter 11). The Lyapunov equation is clearly linear, and its solution is simpler than that of the Riccati equation.

10.4.1 Output Quadratic Regulator

In most practical applications, the designer is interested in optimally controlling the output $\mathbf{y}(k)$ rather than the state $\mathbf{x}(k)$. To optimally control the output, we need to consider a performance index of the form

$$J = \frac{1}{2} \sum_{k=k_0}^{\infty} (\mathbf{y}^T(k)Q_y\mathbf{y}(k) + \mathbf{u}^T(k)R\mathbf{u}(k)) \tag{10.49}$$

From (10.45), we observe that this is equivalent to the original performance measure of (10.43) with the state weight matrix

$$\begin{aligned} Q &= C^T Q_y C \\ &= C^T Q_{ya}^T Q_{ya} C \\ &= Q_a^T Q_a \end{aligned} \tag{10.50}$$

$$Q_a = Q_{ya} C$$

where Q_{ya} is the square root of the output weight matrix. As in the state quadratic regulator, the Riccati equation for the output regulator can be solved using the MATLAB commands discussed in Section 10.4.2. For a stabilizable pair (A, B), the solution exists provided that the pair (A, Q_a) is detectable with Q_a as in (10.50).

10.4.2 MATLAB Solution of the Steady-State Regulator Problem

MATLAB has several commands that allow us to conveniently design steady-state regulators. The first is **dare,** which solves the discrete algebraic Riccati equation (10.40). The command is

>> [S, E, K] = dare(A, B, Q, R)

The input arguments are the state matrix A, the input matrix B, and the weighting matrices Q and R. The output arguments are the solution of the discrete algebraic Riccati equation S, the feedback gain matrix K, and the eigenvalues E of the closed-loop optimal system $A - BK$.

The second command for discrete optimal control is **lqr,** which solves the steady-state regulator problem. The command has the form

>> [K, S, E] = dlqr(A, B, Q, R)

where the input arguments are the same as the command **dare**, and the output arguments, also the same, are the gain K, the solution of the discrete algebraic Riccati equation S, and the eigenvalues e of the closed-loop optimal system $A - BK$.

For the output regulator problem, we can use the commands **dare** and **dlqr** with the matrix Q replaced by C^TQ_yC. Alternatively, MATLAB provides the command

>> [Ky, S, E] = dlqry(A, B, C, D, Qy, R)

EXAMPLE 10.5

Design a steady-state regulator for the inertial system of Example 10.4, and compare its performance to the optimal control by plotting the phase plane trajectories of the two systems. Explain the advantages and disadvantages of the two controllers.

The pair (A, B) is observable. The state-error weighting matrix is positive definite, and the pair (A, Q_a) is observable. We are therefore assured that the Riccati equation will have a steady-state solution. For the inertial system presented in Example 10.4, we have the MATLAB output

$$\text{>> [K,S,E] = dlqr(A,B,Q,r)}$$

$$K = 9.4671 \quad 5.2817$$

$$S = 278.9525 \quad 50.0250 \quad 50.0250 \quad 27.9089$$

$$E = 0.9462 + 0.0299i$$

$$0.9462 - 0.0299i$$

If we simulate the system with the optimal control of Example 10.4 and superimpose the trajectories for the suboptimal steady-state regulator, we obtain Figure 10.13. The figure shows that the suboptimal control, although much simpler to implement, provides an almost identical trajectory to that of the optimal control. For practical implementation, the suboptimal control is often preferable because it is far cheaper to implement and provides almost the same performance as the optimal control. However, there may be situations where the higher accuracy of the optimal control justifies the additional cost of its implementation.

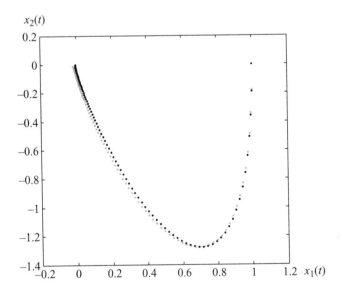

FIGURE 10.13

Phase plane trajectories for the inertial system of Example 10.5 with optimal control (*dark gray*) and suboptimal steady-state regulator (*light gray*).

EXAMPLE 10.6

Design a digital output regulator for the double integrator system

$$\mathbf{x}(k+1) = \begin{bmatrix} 1 & T \\ 0 & 1 \end{bmatrix} \mathbf{x}(k) + \begin{bmatrix} T^2/2 \\ T \end{bmatrix} u(k)$$

$$y(k) = \begin{bmatrix} 1 & 0 \end{bmatrix} \mathbf{x}(k)$$

with sampling period $T = 0.02$ s, output weight $Q_y = 1$, and control weight $r = 0.1$, and plot the output response for the initial condition vector $\mathbf{x}(0) = [1, 0]^T$.

Solution

We use the MATLAB command

>> [Ky, S, E] = dlqry(A, B, C, 0, 1, 0.1)

Ky = 3.0837   2.4834

S = 40.2667   15.8114   15.8114   12.5753

E = 0.9749 + 0.0245i   0.9749 − 0.0245i

The response of the system to initial conditions $\mathbf{x}(0) = [1, 0]^T$ is shown in Figure 10.14. The output quickly converges to the target zero state.

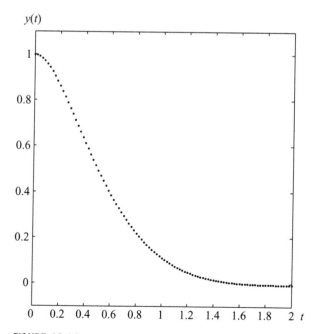

FIGURE 10.14

Time response of the output regulator discussed in Example 10.6.

10.4.3 Linear Quadratic Tracking Controller

The minimization of the performance index (10.43) yields an optimal state feed-back matrix K that minimizes integral square error and control effort (see (10.47)). The error is defined relative to the zero state. In many practical applications, the system is required to follow a specified function of time. The design of a control-ler to achieve this objective is known as the **tracking problem**. If the control task is to track a nonzero constant reference input, we can exploit the techniques described in Section 9.3 to solve the new optimal regulator problem.

In Section 9.3, we showed that the tracking or servo problem can be solved using an additional gain matrix for the reference input as shown in Figure 9.7, thus providing an additional degree of freedom in our design. For a square system (equal number of inputs and outputs), we can implement the degree-of-freedom scheme of Figure 9.7 with the reference gain matrix

$$F = \left[C(I_n - A_{cl})^{-1}B \right]^{-1} \tag{10.51}$$

where

$$A_{cl} = A - BK \tag{10.52}$$

All other equations for the linear quadratic regulator are unchanged.

Alternatively, to improve robustness, but at the expense of a higher-order controller, we can introduce integral control, as in Figure 9.9. In particular, as in (9.25) we consider the state–space equations

$$\begin{aligned}
\mathbf{x}(k+1) &= A\mathbf{x}(k) + B\mathbf{u}(k) \\
\bar{\mathbf{x}}(k+1) &= \bar{\mathbf{x}}(k) + \mathbf{r}(k) - \mathbf{y}(k) \\
\mathbf{y}(k) &= C\mathbf{x}(k) \\
\mathbf{u}(k) &= -K\mathbf{x}(k) - \bar{K}\bar{\mathbf{x}}(k)
\end{aligned} \tag{10.53}$$

with integral control gain $\bar{K}$. This yields the closed-loop state–space equations (9.26)

$$\tilde{\mathbf{x}}(k+1) = \left(\tilde{A} - \tilde{B}\tilde{K} \right)\tilde{\mathbf{x}}(k) + \begin{bmatrix} 0 \\ I_l \end{bmatrix}\mathbf{r}(k) \tag{10.54}$$

$$\mathbf{y}(k) = [C \quad 0]\tilde{\mathbf{x}}(k)$$

where $\tilde{\mathbf{x}}(k) = [\mathbf{x}(k) \quad \bar{\mathbf{x}}(k)]^T$ and the matrices are given by

$$\tilde{A} = \begin{bmatrix} A & 0 \\ -C & I_l \end{bmatrix} \tag{10.55}$$

$$\tilde{B} = \begin{bmatrix} B \\ 0 \end{bmatrix} \quad \tilde{C} = [C \quad 0] \quad \tilde{K} = [K \quad \bar{K}]$$

The state feedback gain matrix can be computed as

$$\tilde{K} = \left[R + \tilde{B}^T S\tilde{B} \right]^{-1}\tilde{B}^T S\tilde{A} \tag{10.56}$$

EXAMPLE 10.7

Design an optimal linear quadratic state–space tracking controller for the inertial system of Example 10.4 to obtain zero steady-state error due to a unit step input.

Solution

The state–space matrices are

$$A = \begin{bmatrix} 1 & 0.02 \\ 0 & 1 \end{bmatrix} \quad B = \begin{bmatrix} 0.0002 \\ 0.02 \end{bmatrix}$$

$$C = \begin{bmatrix} 1 & 0 \end{bmatrix}$$

Adding integral control, we calculate the augmented matrices

$$\tilde{A} = \begin{bmatrix} A & 0 \\ -C & 1 \end{bmatrix} = \begin{bmatrix} 1 & 0.02 & 0 \\ 0 & 1 & 0 \\ 1 & 0 & 1 \end{bmatrix} \quad \tilde{B} = \begin{bmatrix} B \\ 0 \end{bmatrix} = \begin{bmatrix} 0.0002 \\ 0.02 \\ 0 \end{bmatrix}$$

We select the weight matrices with larger error weighting as

$$Q = \begin{bmatrix} 10 & 0 & 0 \\ 0 & 1 & 0 \\ 0 & 0 & 1 \end{bmatrix} \quad r = 0.1$$

The MATLAB command **dlqr** yields the desired solution:

>> [Ktilde, S, E] = dlqr(Atilde, Btilde, Q, r)

Ktilde = 57.363418836015583   10.818259776359207
-2.818765217602360

S = 1.0e + 003 * 2.922350387967343   0.314021626641202
-0.191897141854919

0.314021626641202   0.058073322228013 -0.015819292019556

-0.191897141854919   -0.015819292019556   0.020350548700473

E = 0.939332687303670 + 0.083102739623114i

0.939332687303670 - 0.083102739623114i

0.893496746098272

The closed-loop system state–space equation is

$$\tilde{x}(k+1) = (\tilde{A} - \tilde{B}\tilde{K})\tilde{x}(k) + \begin{bmatrix} 0 \\ I_i \end{bmatrix} r(k)$$

$$y(k) = \begin{bmatrix} C & 0 \end{bmatrix} \tilde{x}(k)$$

The closed-loop step response shown in Figure 10.15 is satisfactory with a small overshoot and zero steady-state error due to step.

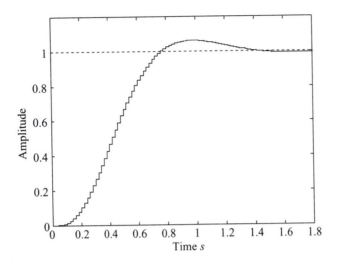

FIGURE 10.15

Step response for Example 10.7.

10.5 HAMILTONIAN SYSTEM

In this section, we obtain a solution of the steady-state regulator problem using linear dynamics without directly solving the Riccati equation of (10.46). We show that we can obtain the solution of the Riccati equation from the state-transition matrix of a linear system. We consider the case of constant weighting matrices. We can combine the state and costate equations to obtain the $2n$-dimensional **Hamiltonian system**:

$$\begin{bmatrix} \mathbf{x}^*(k+1) \\ \lambda^*(k) \end{bmatrix} = \begin{bmatrix} A & -B^*R^{-1}B^T \\ Q & A^T \end{bmatrix} \begin{bmatrix} \mathbf{x}^*(k) \\ \lambda^*(k+1) \end{bmatrix}, \quad k = k_0, \dots, k_f - 1 \quad (10.57)$$

If the state matrix A is obtained by discretizing a continuous-time system, then it is a state transition matrix and is always invertible. We can then rewrite the state equation in the same form as the equation of the costate λ and obtain the Hamiltonian system

$$\begin{bmatrix} \mathbf{x}^*(k) \\ \lambda^*(k) \end{bmatrix} = H \begin{bmatrix} \mathbf{x}^*(k+1) \\ \lambda^*(k+1) \end{bmatrix}, \quad k = k_0, \dots, k_f - 1$$

$$H = \begin{bmatrix} A^{-1} & A^{-1}B^*R^{-1}B^T \\ QA^{-1} & A^T + QA^{-1}B^*R^{-1}B^T \end{bmatrix} \quad (10.58)$$

$$\lambda^*(k_f) = S(k_f)\mathbf{x}(k_f)$$

The Hamiltonian system of (10.58) describes the state and costate evolution backward in time because it provides an expression for evaluating the vector at time k from the vector at time $k + 1$. We can solve the linear equation and write its solution in terms of the state-transition matrix for the Hamiltonian

$$\begin{bmatrix} \mathbf{x}^*(k) \\ \lambda^*(k) \end{bmatrix} = H^{k_f-k} \begin{bmatrix} \mathbf{x}^*(k_f) \\ \lambda^*(k_f) \end{bmatrix}, \quad k = k_0, \ldots, k_f$$

$$H^{k_f-k} = \begin{bmatrix} \phi_{11}(k_f - k) & \phi_{12}(k_f - k) \\ \phi_{21}(k_f - k) & \phi_{22}(k_f - k) \end{bmatrix}$$

(10.59)

The solution of the state equation yields

$$\mathbf{x}^*(k) = \{\phi_{11}(k_f - k) + \phi_{12}(k_f - k)S(k_f)\}\mathbf{x}^*(k_k)$$

$$\lambda^*(k) = \{\phi_{21}(k_f - k) + \phi_{22}(k_f - k)S(k_f)\}\mathbf{x}^*(k_k)$$

(10.60)

We solve for the costate in terms of the state vector

$$\lambda^*(k) = M(k)\mathbf{x}^*(k)$$

$$M(k) = \{\phi_{21}(k_f - k) + \phi_{22}(k_f - k)S(k_f)\}\{\phi_{11}(k_f - k)$$

$$+ \phi_{12}(k_f - k)S(k_f)\}^{-1}$$

(10.61)

Substituting in (10.32), we now have the control

$$\mathbf{u}^*(k) = -R^{-1}B^T M(k+1)\mathbf{x}^*(k+1)$$

$$= -R^{-1}B^T M(k+1)\{A\mathbf{x}^*(k) + B\mathbf{u}^*(k)\}$$

$$k = k_0, \ldots, k_f - 1$$

(10.62)

We multiply both sides by the matrix R and then solve for the control

$$\mathbf{u}^*(k) = K(k)\mathbf{x}^*(k)$$

$$K(k) = -\{R + B^T M(k+1)B\}^{-1}B^T M(k+1)A$$

$$k = k_0, \ldots, k_f - 1$$

(10.63)

We compare the gain expression of (10.63) with that obtained by solving the Riccati equation in (10.40). Because the two expressions are solutions of the same problem and must hold for any state vector, conclude that they must be equal and that $M(k + 1)$ is identical to $S(k + 1)$. Hence, the solution of the Riccati equation is given in terms of the state-transition matrix of the Hamiltonian system by (10.58). We can now write

$$S(k) = \{\phi_{21}(k_f - k) + \phi_{22}(k_f - k)S(k_f)\}\{\phi_{11}(k_f - k) + \phi_{12}(k_f - k)S(k_f)\}^{-1} \quad (10.64)$$

It may seem that the expression of (10.64) is always preferable to direct solution of the Riccati equation because it eliminates the nonlinearity. However, because the dimension of the Hamiltonian system is $2n$, the equation doubles the dimension of the problem. In many cases, it is preferable to solve the Riccati equation in spite of its nonlinearity.

EXAMPLE 10.8

Consider the double integrator described in Example 10.4 with a sampling period $T = 0.02$ s, terminal weight $S(10) = \text{diag}\{10, 1\}$, $Q = \text{diag}\{10, 1\}$, and control weight $r = 0.1$. Obtain the Hamiltonian system and use it to obtain the optimal control.

Solution
The Hamiltonian system is

$$\begin{bmatrix} \mathbf{x}^*(k) \\ \lambda^*(k) \end{bmatrix} = H \begin{bmatrix} \mathbf{x}^*(k+1) \\ \lambda^*(k+1) \end{bmatrix}, \quad k = k_0, \ldots, k_f - 1$$

$$H = \begin{bmatrix} A^{-1} & A^{-1}B^*R^{-1}B^T \\ QA^{-1} & A^T + QA^{-1}B^*R^{-1}B^T \end{bmatrix} = \left[\begin{array}{cc|cc} 1 & -0.02 & 0 & 0 \\ 0 & 1 & 0 & 0.004 \\ \hline 10 & -0.2 & 1 & -0.0004 \\ 0 & 1 & 0.02 & 1.004 \end{array} \right]$$

$$\lambda^*(10) = \text{diag}\{10, 1\} \mathbf{x}(10)$$

Because the dynamics describe the evolution backward in time, each multiplication of the current state by the matrix H yields the state at an earlier time. We define a backward-time variable to use in computing as

$$k_b = k - (k_f - 1)$$

Proceeding backward in time, k_b starts with the value zero at $k = k_f - 1$, then increases to $k_b = k_f - 1$ with $k = 0$. We solve for the transition matrices and the gains using the following MATLAB program.

```
% hamilton
% Form the Hamiltonian system for backward dynamics
[n,n]=size(a); % Order of state matrix
n2=2*n; % Order of Hamiltonian matrix
kf=11; % Final time
q=diag([10,1]);r=.1; % Weight matrices
sf=q; % Final state weight matrix
% Calculate the backward in time Hamiltonian matrix
al=inv(a);
He=b/r*b';
H3=q/a;
H=[al,al*He;H3,a'+H3*He];% Hamiltonian matrix
fi=H;% Initialize the state-transition matrix
% i is the backward time variable kb=k-(kf-1), k = discrete time
for i=1:kf-1
    fi11=fi(1:n,1:n); % Partition the state-transition matrix
    fi12=fi(1:n,n+1:n2);
```

```
fi21=fi(n+1:n2,1:n);
fi22=fi(n+1:n2,n+1:n2);
s=(fi21+fi22*sf)/(fi11+fi12*sf);% Compute the Riccati egn. solution
K=(r+b'*s*b)\b'*s*a % Calculate the gains
fi=H*fi;% Update the state-transition matrix
end
```

The optimal control gains are given in Table 10.2. Almost identical results are obtained using the program described in Example 10.4. Small computational errors result in small differences between the results of the two programs.

Table 10.2 Optimal Gains for the Integrator with a Planning Horizon $k_f = 10$

Time	Optimal Gain Vector K	
0	2.0950	2.1507
1	1.7717	1.9632
2	1.4656	1.7730
3	1.1803	1.5804
4	0.9192	1.3861
5	0.6855	1.1903
6	0.4823	0.9935
7	0.3121	0.7958
8	0.1771	0.5976
9	0.0793	0.3988

RESOURCES

Anderson, B.D.O., and J. B. Moore, *Optimal Control: Linear Quadratic Methods,* Prentice Hall, 1990.

Chong, E. D., and S. H. Żak, *An Introduction to Optimization,* Wiley-Interscience, 1996.

Jacquot, R. G., *Modern Digital Control Systems,* Marcel Dekker, 1981.

Kwakernaak, H., and R. Sivan, *Linear Optimal Control Systems,* Wiley-Interscience, 1972.

Lewis, F. L., and V. L. Syrmos, *Optimal Control,* Wiley-Interscience, 1995.

Naidu, D. S., *Optimal Control Systems,* CRC Press, 2002.

Sage, A. P., and C. C. White III, *Optimum Systems Control,* Prentice-Hall, 1977.

PROBLEMS

10.1. Show that for a voltage source v_s with source resistance R_s connected to a resistive load R_L, the maximum power transfer to the load occurs when $R_L = R_s$.

10.2. Let $\mathbf{x}$ be an $n \times 1$ vector whose entries are the quantities produced by a manufacturer. The profit of the manufacturer is given by the quadratic form

$$J(\mathbf{x}) = \frac{1}{2}\mathbf{x}^T P\mathbf{x} + \mathbf{q}^T\mathbf{x} + \mathbf{r}$$

where P is a negative definite symmetric matrix, and $\mathbf{q}$ and $\mathbf{r}$ are constant vectors. Find the vector $\mathbf{x}$ to maximize profit
(a) With no constraints on the quantity produced.
(b) If the quantity produced is constrained by

$$B\mathbf{x} = \mathbf{c}$$

where B is an $m \times n$ matrix, $m < n$, and $\mathbf{c}$ is a constant vector.

10.3. Prove that the rectangle of the largest area that fits inside a circle of diameter D is a square of diagonal D.

10.4. With $q = 1$ and $r = 2$, $S(k_f) = 1$, write the design equations for a digital optimal quadratic regulator for the integrator

$$\dot{x} = u$$

10.5. The discretized state-space model of the INFANTE AUV of Problem 7.14 is given by

$$\begin{bmatrix} x_1(k+1) \\ x_2(k+1) \\ x_3(k+1) \end{bmatrix} = \begin{bmatrix} 0.9932 & -0.03434 & 0 \\ -0.009456 & 0.9978 & 0 \\ -0.0002368 & 0.04994 & 1 \end{bmatrix}\begin{bmatrix} x_1(k) \\ x_2(k) \\ x_3(k) \end{bmatrix} + \begin{bmatrix} 0.002988 \\ -0.0115 \\ -0.0002875 \end{bmatrix}u(k)$$

$$\begin{bmatrix} y_1(k) \\ y_2(k) \end{bmatrix} = \begin{bmatrix} 1 & 0 & 0 \\ 0 & 1 & 0 \end{bmatrix}\begin{bmatrix} x_1(k) \\ x_2(k) \\ x_3(k) \end{bmatrix}$$

Design a steady-state linear quadratic regulator for the system using the weight matrices $Q = I_3$ and $r = 2$.

10.6. A simplified linearized model of a drug delivery system to maintain blood glucose and insulin levels at prescribed values is given by

$$\begin{bmatrix} x_1(k+1) \\ x_2(k+1) \\ x_3(k+1) \end{bmatrix} = \begin{bmatrix} -0.04 & -4.4 & 0 \\ 0 & -0.025 & 1.3\times10^{-5} \\ 0 & 0.09 & 0 \end{bmatrix}\begin{bmatrix} x_1(k) \\ x_2(k) \\ x_3(k) \end{bmatrix} + \begin{bmatrix} 1 & 0 \\ 0 & 0 \\ 0 & 0.1 \end{bmatrix}\begin{bmatrix} u_1(k) \\ u_2(k) \end{bmatrix}$$

$$\begin{bmatrix} y_1(k) \\ y_2(k) \end{bmatrix} = \begin{bmatrix} 1 & 0 & 0 \\ 0 & 0 & 1 \end{bmatrix} \begin{bmatrix} x_1(k) \\ x_2(k) \\ x_3(k) \end{bmatrix}$$

where all variables are perturbations from the desired steady-state levels.[1] The state variables are the blood glucose concentration x_1 in mg/dl, the blood insulin concentration x_3 in mg/dl, and a variable describing the accumulation of blood insulin x_2. The controls are the rate of glucose infusion u_1 and the rate of insulin infusion u_2, both in mg/dl/min. Discretize the system with a sampling period $T = 5$ min and design a steady-state regulator for the system with weight matrices $Q = I_3$ and $R = 2I_2$. Simulate the system with the initial state $\mathbf{x}(0) = [6 \ 0 \ -1]^T$, and plot the trajectory in the $x_1 - x_3$ plane as well as the time evolution of the glucose concentration.

10.7. Optimal control problems can be solved as unconstrained optimization problems without using Lagrange multipliers. This exercise shows the advantages of using Lagrange multipliers and demonstrates that discrete-time optimal control is equivalent to an optimization problem over the control history.

(a) Substitute the solution of the state equation

$$\mathbf{x}(k) = A^k \mathbf{x}(0) + \sum_{i=0}^{k-1} A^{k-i-1} B \mathbf{u}(i)$$

$$= A^k \mathbf{x}(0) + \mathscr{C}(k) \mathbf{u}(k)$$

$$\mathscr{C}(k) = [B \ \vdots \ AB \ \vdots \ \cdots \ \vdots \ A^{k-1} B]$$

$$\mathscr{u}(k) = col\{\mathbf{u}(k-1), \ldots, \mathbf{u}(0)\}$$

In the performance measure

$$J = \frac{1}{2}\mathbf{x}^T(k_f)S(k_f)\mathbf{x}(k_f) + \frac{1}{2}\sum_{k=0}^{k_f-1}(\mathbf{x}^T(k)Q(k)\mathbf{x}(k) + \mathbf{u}^T(k)R(k)\mathbf{u}(k))$$

to eliminate the state vector and obtain

$$J = \frac{1}{2}\sum_{k=0}^{k_f}\left(\begin{array}{l} \mathscr{u}^T(k)\bar{R}(k)\mathscr{u}(k) + 2^T\mathbf{x}(0)(A^T)^k Q(k)\mathscr{C}(k)\mathscr{u}(k) + \\ \mathbf{x}(0)(A^T)^k Q(k)A^k\mathbf{x}(0) + \mathbf{u}^T(k)R(k)\mathbf{u}(k) \end{array} \right)$$

with $Q(k_f) = S(k_f)$, $R(k_f) = \mathbf{0}_{m\times m}$

(b) Without tediously evaluating the matrix R_{eq} and the vector $\mathbf{l}$, explain why it is possible to rewrite the performance measure in the equivalent form

$$J_{eq} = \frac{1}{2}\mathscr{u}^T(k_f)R_{eq}\mathscr{u}(k_f) + \mathscr{u}^T(k_f)\mathbf{l}$$

---

[1] F. Chee, A. V. Savkin, T. L. Fernando, and S. Nahavandi, Optimal H∞ insulin injection control for blood glucose regulation in diabetic patients, *IEEE Trans. Biomed. Eng.*, 52(10):1625-1631, 2005.

(c) Show that the solution of the optimal control problem is given by

$$u(k_f) = -R_{eq}^{-1}\mathbf{1}$$

10.8. For (A, B) stabilizable and $(A, Q^{1/2})$ detectable, the linear quadratic regulator yields a closed-loop stable system. To guarantee that the eigenvalues of the closed-loop system will lie inside a circle of radius $1/\alpha$, we solve the regulator problem for the scaled state and control

$$\bar{\mathbf{x}}(k) = \alpha^k \mathbf{x}(k) \quad \bar{\mathbf{u}}(k) = \alpha^{k+1} \mathbf{u}(k)$$

(a) Obtain the state equation for the scaled state vector.
(b) Show that if the scaled closed-loop state matrix with the optimal control $\bar{\mathbf{u}}(k) = -\bar{K}\bar{\mathbf{x}}(k)$ has eigenvalues inside the unit circle, then the eigenvalues of the original state matrix with the control $\mathbf{u}(k) = -K\mathbf{x}(k), K = \bar{K}/\alpha$ are inside a circle of radius $1/\alpha$.

10.9. Repeat Problem 10.5 with a design that guarantees that the eigenvalues of the closed-loop system are inside a circle of radius equal to one-half.

10.10. Show that the linear quadratic regulator with cross-product term of the form

$$J = \mathbf{x}^T(k_f)S(k_f)\mathbf{x}(k_f) + \sum_{k=k_0}^{k_f-1} (\mathbf{x}^T(k)Q(k)\mathbf{x}(k) + 2\mathbf{x}^T S\mathbf{u}$$
$$+ \mathbf{u}^T(k)R(k)\mathbf{u}(k))$$

is equivalent to a linear quadratic regulator with no cross-product term with the cost

$$J = \mathbf{x}^T(k_f)S(k_f)\mathbf{x}(k_f) + \sum_{k=k_0}^{k_f-1} (\mathbf{x}^T(k)\bar{Q}(k)\mathbf{x}(k)$$
$$+ \bar{\mathbf{u}}^T(k)R(k)\bar{\mathbf{u}}^T(k))$$
$$\bar{Q} = Q - SR^{-1}S^T$$
$$\bar{\mathbf{u}}(k) = \mathbf{u}(k) + R^{-1}S^T\mathbf{x}(k)$$

and the plant dynamics

$$\mathbf{x}(k+1) = \bar{A}\mathbf{x}(k) + B\bar{\mathbf{u}}(k) \quad k = k_0, \ldots, k_f - 1$$
$$\bar{A} = A - BR^{-1}B^T$$

10.11. Rewrite the performance measure shown in Problem 10.10 in terms of a combined input and state vector col{$\mathbf{x}(k), \mathbf{u}(k)$}. Then use the Hamiltonian to show that for the linear quadratic regulator with cross-product term, a sufficient condition for a minimum is that the matrix

$$\begin{bmatrix} Q & N \\ \hline N^T & R \end{bmatrix}$$

must be positive definite.

COMPUTER EXERCISES

10.12. Design a steady-state regulator for the INFANTE AUV presented in Problem 10.5 with the performance measure modified to include a cross-product term with

$$S = [1 \quad 0.2 \quad 0.1]^T$$

(a) Using the MATLAB command **dlqr** and the equivalent problem with no cross product as in Problem 10.10.
(b) Using the MATLAB command **dlqr** with the cross-product term.

10.13. Design an output quadratic regulator for the INFANTE UAV presented in Problem 10.5 with the weights $Q_y = 1$ and $r = 100$. Plot the output response for the initial condition vector $\mathbf{x}(0) = [1, 0, 1]^T$.

10.14. Design an optimal LQ state–space tracking controller for the drug delivery system presented in Problem 10.6 to obtain zero steady-state error due to a unit step input.

10.15. Write a MATLAB script that determines the steady-state quadratic regulator for the inertial system of Example 10.4 for $r = 1$ and different values of Q. Use the following three matrices, and discuss the effect of changing Q on the results of your computer simulation:

(a) $Q = \begin{bmatrix} 1 & 0 \\ 0 & 1 \end{bmatrix}$

(b) $Q = \begin{bmatrix} 10 & 0 \\ 0 & 10 \end{bmatrix}$

(c) $Q = \begin{bmatrix} 100 & 0 \\ 0 & 100 \end{bmatrix}$

Elements of Nonlinear Digital Control Systems

Most physical systems do not obey the principle of superposition and can therefore be classified as nonlinear. By limiting the operating range of physical systems, it is possible to approximately model their behavior as linear. This chapter examines the behavior of nonlinear discrete systems without this limitation. We begin by examining the behavior of nonlinear continuous-time systems with piecewise constant inputs. We discuss Lyapunov stability theory both for nonlinear and linear systems. We provide a simple controller design based on Lyapunov stability theory.

Objectives

After completing this chapter, the reader will be able to do the following:

1. Discretize special types of nonlinear systems.
2. Determine the equilibrium point of a nonlinear discrete-time system.
3. Classify equilibrium points of discrete-time systems based on their state plane trajectories.
4. Determine the stability or instability of nonlinear discrete-time systems.
5. Design controllers for nonlinear discrete-time systems.

11.1 DISCRETIZATION OF NONLINEAR SYSTEMS

Discrete-time models are easily obtained for linear continuous-time systems from their transfer functions or the solution of the state equations. For nonlinear systems, transfer functions are not defined and the state equations are analytically solvable in only a few special cases. Thus, it is no easy task to obtain discrete-time models for nonlinear systems.

We examine the nonlinear differential equation

$$\dot{\mathbf{x}} = \mathbf{f}(\mathbf{x}) + B(\mathbf{x})\mathbf{u} \qquad (11.1)$$

where $\mathbf{x}$ and $\mathbf{f}$ are $n \times 1$ vectors, $\mathbf{u}$ is an $m \times 1$ vector, $B(\mathbf{x})$ is a $n \times m$ matrix, and $B(.)$ and $\mathbf{f}(.)$ are continuous functions of all their arguments. We assume that the input is defined by

$$\mathbf{u}(t) = \mathbf{u}(kT) = \mathbf{u}(k), \quad k \in [kT, (k+1)T) \tag{11.2}$$

For each sampling period we have the model

$$\dot{\mathbf{x}} = \mathbf{f}(\mathbf{x}) + B(\mathbf{x})\mathbf{u}(k) \tag{11.3}$$

where $k = 0, 1, 2, \ldots$ The behavior of the discrete-time system can, in theory, be obtained from the solution of equations of the form (11.3) with the appropriate initial conditions. In practice, only a numerical solution is possible except in a few special cases. One way to obtain a solution for nonlinear systems is to transform the dynamics of the system to obtain equivalent linear dynamics. The general theory governing such transformations, known as **global** or **extended linearization**, is beyond the scope of this text. However, in some special cases it is possible to obtain equivalent linear models and use them for discretization quite easily without resorting to the general theory. These include models that occur frequently in some applications. We present four cases where extended linearization is quite simple.

11.1.1 Extended Linearization by Input Redefinition

Consider the nonlinear model

$$M(\mathbf{q})\ddot{\mathbf{q}} + \mathbf{m}(\mathbf{q}, \dot{\mathbf{q}}) = \mathbf{v}(t) \tag{11.4}$$

where $M(\mathbf{q})$ is an invertible $m \times m$ matrix, and $\mathbf{m}$, $\mathbf{q}$, and $\mathbf{v}$ are $m \times 1$ vectors. A natural choice of state variables for the system is the $2m \times 1$ vector

$$\mathbf{x} = \text{col}\{\mathbf{x}_1, \mathbf{x}_2\} = \text{col}\{\mathbf{q}, \dot{\mathbf{q}}\} \tag{11.5}$$

To obtain equivalent linear dynamics for the system, we redefine the input vector as

$$\mathbf{u}(t) = \ddot{\mathbf{q}} \tag{11.6}$$

This yields the set of double integrators with state equations

$$\dot{\mathbf{x}}_1 = \mathbf{x}_2$$
$$\dot{\mathbf{x}}_2 = \mathbf{u}(t) \tag{11.7}$$
$$\mathbf{u}(t) = M^{-1}(\mathbf{q})[\mathbf{v}(t) - \mathbf{m}(\mathbf{q}, \dot{\mathbf{q}})]$$

The solution of the state equations can be obtained by Laplace transformation, and it yields the discrete-state equation

$$\mathbf{x}(k+1) = A_d \mathbf{x}(k) + B_d \mathbf{u}(k) \tag{11.8}$$

with

$$
A_d = \begin{bmatrix} I_m & T I_m \\ 0 & I_m \end{bmatrix} \quad B_s = \begin{bmatrix} (T^2/2) I_m \\ T I_m \end{bmatrix} \tag{11.9}
$$

With digital control, the input is piecewise constant and is obtained using the system model

$$
\mathbf{v}(k) = M(\mathbf{x}_1(k))\mathbf{u}(k) + \mathbf{m}(\mathbf{x}_1(k), \mathbf{x}_2(k)) \tag{11.10}
$$

The expression for the input $\mathbf{v}$ is approximate because it assumes that the state does not change appreciably over a sampling period with a fixed acceleration input. The approximation is only acceptable for sufficiently slow dynamics.

The model shown in (11.4) includes many important physical systems. In particular, a large class of mechanical systems, including the m-D.O.F. (degree-of-freedom) manipulator presented in Example 7.4, are in the form (11.4). Recall that the manipulator is governed by the equation of motion

$$
M(\mathbf{q})\ddot{\mathbf{q}} + V(\mathbf{q}, \dot{\mathbf{q}})\dot{\mathbf{q}} + \mathbf{g}(\mathbf{q}) = \tau \tag{11.11}
$$

where

$\mathbf{q}$ = vector of generalized coordinates
$M(\mathbf{q})$ = $m \times m$ positive definite inertia matrix
$V(\mathbf{q}, \dot{\mathbf{q}})$ = $m \times m$ matrix of velocity related terms
$\mathbf{g}(\mathbf{q})$ = $m \times 1$ vector of gravitational terms
τ = $m \times 1$ vector of generalized forces

Clearly, the dynamics of the manipulator are a special case of (11.4). As applied to robotic manipulators, extended linearization by input redefinition is known as the **computed torque method**. This is because the torque is computed from the acceleration if measurements of positions and velocities are available.

EXAMPLE 11.1

Find a discrete-time model for the 2-D.O.F. robotic manipulator described in Example 7.5 and find an expression for calculating the torque vector.

Solution
The manipulator dynamics are governed by

$$
M(\boldsymbol{\theta}) = \begin{bmatrix} (m_1 + m_2)l_1^2 + m_2 l_2^2 + 2m_2 l_1 l_2 \cos(\theta_2) & m_2 l_2^2 + m_2 l_1 l_2 \cos(\theta_2) \\ m_2 l_2^2 + m_2 l_1 l_2 \cos(\theta_2) & m_2 l_2^2 \end{bmatrix}
$$

$$
V(\boldsymbol{\theta}, \dot{\boldsymbol{\theta}})\dot{\boldsymbol{\theta}} = \begin{bmatrix} -m_2 l_1 l_2 \sin(\theta_2)\dot{\theta}_2(2\dot{\theta}_1 + \dot{\theta}_2) \\ m_2 l_1 l_2 \sin(\theta_2)\dot{\theta}_1^2 \end{bmatrix}
$$

$$
\mathbf{g}(\boldsymbol{\theta}) = \begin{bmatrix} (m_1 + m_2)gl_1 \sin(\theta_1) + m_2 gl_2 \sin(\theta_1 + \theta_2) \\ m_2 gl_2 \sin(\theta_1 + \theta_2) \end{bmatrix}
$$

For this system the coordinate vector $\mathbf{q} = [\theta_1 \ \theta_2]^T$. We have a fourth-order linear system of the form

$$\dot{\mathbf{x}}_1(t) = \mathbf{x}_2(t)$$

$$\dot{\mathbf{x}}_2(t) = \mathbf{u}(t)$$

$$\mathbf{x}_1(t) = \mathbf{q} = [x_{11}(t) \ \ x_{12}(t)]^T$$

$$\mathbf{x}_2(t) = \dot{\mathbf{q}} = [x_{21}(t) \ \ x_{22}(t)]^T$$

$$\mathbf{u}(t) = [\tau_1(t) \ \ \tau_2(t)]^T$$

As in (11.10), the torque is calculated using the equation

$$\tau(k) = M(\mathbf{x}_1(k))\mathbf{u}(k) + V(\mathbf{x}_1(k), \mathbf{x}_2(k))\mathbf{x}_2(k) + \mathbf{g}(\mathbf{x}_1(k))$$

11.1.2 Extended Linearization by Input and State Redefinition

Consider the nonlinear state equations

$$\dot{\mathbf{z}}_1 = \mathbf{f}_1(\mathbf{z}_1, \mathbf{z}_2)$$

$$\dot{\mathbf{z}}_2 = \mathbf{f}_2(\mathbf{z}_1, \mathbf{z}_2) + G(\mathbf{z}_1, \mathbf{z}_2)\mathbf{v}(t) \tag{11.12}$$

where $\mathbf{f}_i(.)$, $i = 1, 2$, and $\mathbf{v}$ are $m \times 1$, and G is $m \times m$. We redefine the state variables and input as

$$\mathbf{x}_1 = \mathbf{z}_1$$

$$\mathbf{x}_2 = \dot{\mathbf{z}}_1 \tag{11.13}$$

$$\mathbf{u}(t) = \ddot{\mathbf{z}}_1$$

The new variables have the linear dynamics of (11.7) and the discrete-time model of (11.8).

Using the state equations, we can rewrite the new input as

$$\mathbf{u}(t) = \frac{\partial \mathbf{f}_1(\mathbf{z}_1, \mathbf{z}_2)}{\partial \mathbf{z}_1}\mathbf{f}_1(\mathbf{z}_1, \mathbf{z}_2) + \frac{\partial \mathbf{f}_1(\mathbf{z}_1, \mathbf{z}_2)}{\partial \mathbf{z}_2}[\mathbf{f}_2(\mathbf{z}_1, \mathbf{z}_2) + G(\mathbf{z}_1, \mathbf{z}_2)\mathbf{v}(t)] \tag{11.14}$$

We solve for the nonlinear system input at time k to obtain

$$\mathbf{v}(k) = \left[\frac{\partial \mathbf{f}_1(\mathbf{z}_1, \mathbf{z}_2)}{\partial \mathbf{z}_2}G(\mathbf{z}_1, \mathbf{z}_2)\right]^{-1}\left\{\mathbf{u}(k) - \frac{\partial \mathbf{f}_1(\mathbf{z}_1, \mathbf{z}_2)}{\partial \mathbf{z}_1}\mathbf{f}_1(\mathbf{z}_1, \mathbf{z}_2)\right.$$
$$\left. - \frac{\partial \mathbf{f}_1(\mathbf{z}_1, \mathbf{z}_2)}{\partial \mathbf{z}_2}\mathbf{f}_2(\mathbf{z}_1, \mathbf{z}_2)\right\} \tag{11.15}$$

The expression for the input $\mathbf{v}(k)$ is approximate because the state of the system changes over a sampling period for a fixed input $\mathbf{u}(k)$.

EXAMPLE 11.2

Discretize the nonlinear differential equation

$$\dot{z}_1 = -\frac{1}{2}(z_1 z_2)^2$$

$$\dot{z}_2 = 4z_1 z_2^3 + 1.5z_2 - v, \quad z_1 \neq 0, z_2 \neq 0$$

Solution

We differentiate the first state equation to obtain

$$\ddot{z}_1 = -z_1 z_2^2 \dot{z}_1 - z_1^2 z_2 \dot{z}_2$$

$$= -3.5 z_1^3 z_2^4 - 1.5 z_1^2 z_2^2 + z_1^2 z_2 v = u(t), \quad z_1 \neq 0, z_2 \neq 0$$

We have the linear model

$$\dot{x}_1 = x_2$$

$$\dot{x}_2 = u(t)$$

This gives the equivalent discrete-time model

$$\mathbf{x}(k+1) = \begin{bmatrix} 1 & T \\ 0 & 1 \end{bmatrix} \mathbf{x}(k) + \begin{bmatrix} (T^2/2) \\ T \end{bmatrix} \mathbf{u}(k)$$

with $\mathbf{x}(k) = [x_1(k)\ x_2(k)]^T = [z_1(k)\ z_1(k+1)]^T$. If measurements of the actual state of the system are available, then the input is computed using

$$v(k) = 3.5 z_1(k) z_2^3(k) + 1.5 z_2(k) + \frac{u(k)}{z_1^2(k) z_2(k)}, \quad z_1 \neq 0, z_2 \neq 0$$

11.1.3 Extended Linearization by Output Differentiation

Consider the nonlinear state-space model

$$\dot{\mathbf{z}} = \mathbf{f}(\mathbf{z}) + G(\mathbf{z})\mathbf{v}(t)$$

$$\mathbf{y} = \mathbf{c}(\mathbf{z}) \tag{11.16}$$

where $\mathbf{z}$ and $\mathbf{f}$ are $n \times 1$, $\mathbf{v}$ is $m \times 1$, $\mathbf{y}$ is $l \times 1$, and G is $m \times m$. If we differentiate each scalar output equation and substitute from the state equation, we have

$$\frac{dy_i}{dt} = \frac{\partial c_i^T(\mathbf{z})}{\partial \mathbf{z}} \dot{\mathbf{z}}$$

$$= \frac{\partial c_i^T(\mathbf{z})}{\partial \mathbf{z}} [\mathbf{f}(\mathbf{z}) + G(\mathbf{z})\mathbf{v}(t)] \quad i = 1, \dots, l \tag{11.17}$$

We denote the operation of partial differentiation with regard to a vector and the multiplication by its derivative by

$$L_x y = \left(\frac{\partial y}{\partial x}\right)\dot{x} \tag{11.18}$$

Hence, the time derivative of an output is equivalent to the operation $L_{\dot{z}} y_i$. We denote r_i repetitions of this operation by the symbol $L_i^n y_i$.

If the coefficient of the input vector v

$$\frac{\partial c_i^T(z)}{\partial z} G(z) = 0^T \tag{11.19}$$

is nonzero, we stop to avoid input differentiation. If the coefficient is zero, we repeat the process until the coefficient becomes nonzero. We define an input equal to the term where the input appears and write the state equations in the form

$$\frac{d^{d_i} y_i}{dt^{d_i}} = u_i(t) \tag{11.20}$$

$$u_i(t) = L_i^n y_i, \quad i = 1, \ldots, l$$

where the number of derivatives r_i required for the input to appear in the expression is known as the ith **relative degree**. The linear system is in the form of l sets of integrators, and its order is

$$n_l = \sum_{i=1}^{n} r_i \le n \tag{11.21}$$

Because the order of the linear model can be less that the original order of the nonlinear system n, the equivalent linear model of the nonlinear system can have unobservable dynamics. The unobservable dynamics are known as the **internal dynamics** or **zero dynamics**. For the linear model to yield an acceptable discrete-time representation, we require the internal dynamics to be stable and sufficiently fast so as not to significantly alter the time response of the system. Under these conditions, the linear model provides an adequate though incomplete description of the system dynamics, and we can discretize each of the l linear subsystems to obtain the overall discrete-time model.

EXAMPLE 11.3

Consider a model of drug receptor binding in a drug delivery system. The drug is assumed to be divided between four compartments in the body. The concentrations in the four compartments are given by

$$\dot{z}_1 = a_{11}z_1 + a_{12}z_1z_2 + b_1 d$$

$$\dot{z}_2 = a_{21}z_2 + a_{22}z_1z_2 + b_2 d$$

$$\dot{z}_3 = a_{31}z_1 + a_{32}z_3$$

$$\dot{z}_4 = a_{41}z_2$$

where d is the drug dose, z_i is the concentration in the ith compartment, and a_{ij} are time-varying coefficients. The second compartment represents the drug receptor complex. Hence, the output equation is given by

$$y = z_2$$

Obtain an equivalent linear model by output differentiation.

Solution
We differentiate the output equation once

$$\dot{y} = \dot{z}_2 = a_{21}z_2 + a_{22}z_1z_2 + b_2d$$

The derivative includes the input, and no further differentiation is possible because it would lead to input derivative terms.
   The model of the linearized system is

$$\dot{x} = u(t)$$

$$u(t) = a_{21}z_2 + a_{22}z_1z_2 + b_2d$$

It is first order, even though we have a third-order plant, and is only equivalent to part of the dynamics of the original system. We assume the system is detectable with fast unobservable dynamics, which is reasonable for the drug delivery system. Hence, the linear model provides an adequate description of the system, and we obtain the discrete-time model

$$x(k+1) = x(k) + Tu(k)$$

The drug dose is given by

$$d(k) = \frac{1}{b_2}[u(k) - a_{21}z_2(k) + a_{22}z_1(k)z_2(k)]$$

11.1.4 Extended Linearization Using Matching Conditions

The following theorem gives the analytical solution in a special case where an equivalent linear model can be obtained for a nonlinear system by a simple state transformation.

Theorem 11.1: For the system of (11.1), let $B(\mathbf{x})$ and $\mathbf{f}(\mathbf{x})$ satisfy the matching conditions

$$B(\mathbf{x}) = B_1(\mathbf{x})B_2$$

$$\mathbf{f}(\mathbf{x}) = B_1(\mathbf{x})A\mathbf{h}(\mathbf{x})$$

(11.22)

where $B_1(\mathbf{x})$ is an $n \times n$ matrix invertible in some region D, B_2 is an $n \times m$ constant vector, A is an $n \times n$ matrix, and $h(\mathbf{x})$ has the derivative given by

$$\frac{\partial \mathbf{h}(\mathbf{x})}{\partial \mathbf{x}} = [B_1(\mathbf{x})]^{-1} \qquad (11.23)$$

with $\mathbf{h}(.)$ invertible in the region D. Then the solution of (11.1) over D is

$$\mathbf{x}(t) = \mathbf{h}^{-1}(\mathbf{w}(t)) \qquad (11.24)$$

where $\mathbf{w}$ is the solution of the linear equation

$$\dot{\mathbf{w}}(t) = A\mathbf{w}(t) + B_2 \mathbf{u}(t) \qquad (11.25)$$

PROOF. From (11.24), $\mathbf{w} = \mathbf{h}(\mathbf{x})$. Differentiating with respect to time and substituting from (11.1) and (11.22) gives

$$\dot{\mathbf{w}} = \frac{\partial \mathbf{h}(\mathbf{x})}{\partial \mathbf{x}} \dot{\mathbf{x}} = [B_1(\mathbf{x})]^{-1}\{\mathbf{f}(\mathbf{x}) + B_1(\mathbf{x})B_2\mathbf{u}\}$$

$$= A\mathbf{h}(\mathbf{x}) + B_2\mathbf{u}$$

$$= A\mathbf{w} + B_2\mathbf{u} \qquad\blacksquare$$

Note that, if the decomposition of Theorem 11.1 is possible, we do not need to solve a partial differential equation to obtain the vector function $\mathbf{h}(\mathbf{x})$, but we do need to find its inverse function to obtain the vector $\mathbf{x}$. The solution of a partial differential equation is a common requirement for obtaining transformations of nonlinear systems to linear dynamics that can be avoided in special cases.

The solution of the state equation for $\mathbf{w}$ over any sampling period T is

$$\mathbf{w}(k+1) = A_w \mathbf{w}(k) + B_w \mathbf{u}(k)$$

where

$$A_w = e^{AT}, \quad B_w = \int_0^T e^{A\tau} B_2 d\tau$$

This is a discrete state–space equation for the original nonlinear system.

EXAMPLE 11.4

Discretize the nonlinear differential equation

$$\begin{bmatrix} \dot{x}_1 \\ \dot{x}_2 \end{bmatrix} = \frac{1}{2} \begin{bmatrix} -(x_1/x_2)^2 \\ (8x_2^3/x_1) + 3x_2 \end{bmatrix} + \begin{bmatrix} 0 \\ -x_2^3 \end{bmatrix} u, \quad x_1 \neq 0, x_2 \neq 0$$

Solution

Assuming a piecewise constant input u, we rewrite the nonlinear equation as

$$\begin{bmatrix} \dot{x}_1 \\ \dot{x}_2 \end{bmatrix} = \begin{bmatrix} x_1^2 & 0 \\ 0 & -x_2^3 \end{bmatrix} \left\{ \begin{bmatrix} 0 & -1 \\ -4 & 3 \end{bmatrix} \begin{bmatrix} -x_1^{-1} \\ x_2^{-2}/2 \end{bmatrix} + \begin{bmatrix} 0 \\ 1 \end{bmatrix} u(k) \right\}$$

To apply Theorem 11.1, we define the terms

$$B_1(\mathbf{x}) = \begin{bmatrix} x_1^2 & 0 \\ 0 & -x_2^3 \end{bmatrix} \quad B_2 = \begin{bmatrix} 0 \\ 1 \end{bmatrix}$$

$$A = \begin{bmatrix} 0 & -1 \\ -4 & 3 \end{bmatrix} \quad h(\mathbf{x}) = \begin{bmatrix} -x_1^{-1} \\ x_2^{-2}/2 \end{bmatrix}$$

We verify that the Jacobian of the vector satisfies

$$\frac{\partial \mathbf{h}(\mathbf{x})}{\partial \mathbf{x}} = \begin{bmatrix} \partial h_1/\partial x_1 & \partial h_1/\partial x_2 \\ \partial h_2/\partial x_1 & \partial h_2/\partial x_2 \end{bmatrix} = [B_1(\mathbf{x})]^{-1} = \begin{bmatrix} x_1^{-2} & 0 \\ 0 & -x_2^{-3} \end{bmatrix}$$

and verify that Theorem 11.1 applies.

We first solve the linear state equation

$$\begin{bmatrix} \dot{w}_1 \\ \dot{w}_2 \end{bmatrix} = \begin{bmatrix} 0 & -1 \\ -4 & 3 \end{bmatrix} \begin{bmatrix} w_1 \\ w_2 \end{bmatrix} + \begin{bmatrix} 0 \\ 1 \end{bmatrix} u(k)$$

to obtain

$$\begin{bmatrix} w_1(k+1) \\ w_2(k+1) \end{bmatrix} = \frac{1}{5} \left\{ \begin{bmatrix} 1 & -1 \\ -4 & 4 \end{bmatrix} e^{4T} + \begin{bmatrix} 4 & 1 \\ 4 & 1 \end{bmatrix} e^{-T} \right\} \begin{bmatrix} w_1(k) \\ w_2(k) \end{bmatrix} +$$
$$\frac{1}{5} \left\{ \begin{bmatrix} -1 \\ 4 \end{bmatrix} \frac{e^{4T}-1}{4} + \begin{bmatrix} 1 \\ 1 \end{bmatrix} (1-e^{-T}) \right\} u(k)$$

where $\mathbf{w}(k) = \mathbf{h}[\mathbf{x}(k)]$, $k = 0, 1, 2, \ldots$ We now have the recursion

$$\begin{bmatrix} w_1(k+1) \\ w_2(k+1) \end{bmatrix} = \frac{1}{5} \left\{ \begin{bmatrix} 1 & -1 \\ -4 & 4 \end{bmatrix} e^{4T} + \begin{bmatrix} 4 & 1 \\ 4 & 1 \end{bmatrix} e^{-T} \right\} \begin{bmatrix} w_1(k) \\ w_2(k) \end{bmatrix} +$$
$$\frac{1}{5} \left\{ \begin{bmatrix} -1 \\ 4 \end{bmatrix} \frac{e^{4T}-1}{4} + \begin{bmatrix} 1 \\ 1 \end{bmatrix} (1-e^{-T}) \right\} u(k)$$

Using the inverse of the function, we solve for the state

$$x_1(k+1) = -1/w_1(k+1), \quad x_2(k+1) = \pm 1/(2w_2(k+1))^{1/2}$$

which is clearly nonunique. However, one would be able to select the appropriate solution because changes in the state variable should not be large over a short sampling period. For a given sampling period T, we can obtain a discrete nonlinear recursion for $\mathbf{x}(k)$.

It is obvious from the preceding example that the analysis of discrete-time systems based on nonlinear analog systems is only possible in special cases. In addition, the transformation is sensitive to errors in the system model. We conclude that this is often not a practical approach to control system design. However, most nonlinear control systems are today digitally implemented even if based on an analog design. We therefore return to a discussion of system (11.1) with the control (11.2) in Section 11.4.

11.2 NONLINEAR DIFFERENCE EQUATIONS

For a nonlinear system with a DAC and ADC, system identification can yield a discrete-time model directly. A discrete-time model can also be obtained analytically in a few special cases as discussed in Section 11.1. We thus have a nonlinear difference equation to describe a nonlinear discrete-time system. Unfortunately, nonlinear difference equations can be solved analytically in only a few special cases. We discuss one special case where such solutions are available.

11.2.1 Logarithmic Transformation

Consider the nonlinear difference equation

$$[y(k+n)]^{\alpha_n}[y(k+n-1)]^{\alpha_{n-1}}\ldots[y(k)]^{\alpha_0}=u(k) \tag{11.26}$$

where α_i, $i = 0, 1, \ldots, n$ are constant. Then taking the natural log of (11.26) gives

$$\alpha_n x(k+n)+\alpha_{n-1}x(k+n-1)\ldots+\alpha_0 x(k)=v(k) \tag{11.27}$$

with $x(k + i) = \ln[y(k + i)]$, $i = 0, 1, 2, \ldots, n$, and $v(k) = \ln[u(k)]$. This is a linear difference equation that can be easily solved by z-transformation. Finally, we obtain

$$y(k) = e^{x(k)} \tag{11.28}$$

EXAMPLE 11.5

Solve the nonlinear difference equation

$$[y(k+2)][y(k+1)]^5[y(k)]^4 = u(k)$$

with zero initial conditions and the input

$$u(k) = e^{-10k}$$

Solution
Taking the natural log of the equation, we obtain

$$x(k+2)+5x(k+1)+4x(k)=-10k$$

The z-transform of the equation

$$[z^2 +5z + 4]X(z) = \frac{-10z}{(z-1)^2}$$

yields $X(z)$ as

$$X(z) = \frac{-10z}{(z-1)^2(z+1)(z+4)}$$

Inverse z-transforming gives the discrete-time function

$$x(k) = -k + 0.7 - 0.8333(-1)^k + 0.1333(-4)^k, \quad k \geq 0$$

Hence, the solution of the nonlinear difference equation is

$$y(k) = e^{-k + 0.7 - 0.8333(-1)^k + 0.1333(-4)^k}, \quad k \geq 0$$

11.3 EQUILIBRIUM OF NONLINEAR DISCRETE-TIME SYSTEMS

Equilibrium is defined as a condition in which a system remains indefinitely unless it is disturbed. If a discrete-time system is described by the nonlinear difference equation

$$\mathbf{x}(k+1) = \mathbf{f}[\mathbf{x}(k)] + B[\mathbf{x}(k)]\mathbf{u}(k) \qquad (11.29)$$

then at an equilibrium it is governed by the identity

$$\mathbf{x}_e = \mathbf{f}[\mathbf{x}_e] + B[\mathbf{x}_e]\mathbf{u}(k) \qquad (11.30)$$

where $\mathbf{x}_e$ denotes the equilibrium state. We are typically interested in the equilibrium for an unforced system, and we therefore rewrite the equilibrium condition (11.30) as

$$\mathbf{x}_e = \mathbf{f}[\mathbf{x}_e] \qquad (11.31)$$

In mathematics, such an equilibrium point is known as a **fixed point**.

Note that the behavior of the system in the vicinity of its equilibrium point determines whether we classify the equilibrium as stable or unstable. For a stable equilibrium, we expect the trajectories of the system to remain arbitrarily close or to converge to the equilibrium. Unlike continuous-time systems, convergence to an equilibrium point can occur after a finite time period.

Clearly, both (11.30) and (11.31), in general, have more than one solution so that nonlinear systems often have several equilibrium points. This is demonstrated in the following example.

EXAMPLE 11.6

Find the equilibrium points of the nonlinear discrete-time system

$$\begin{bmatrix} x_1(k+1) \\ x_2(k+1) \end{bmatrix} = \begin{bmatrix} x_2(k) \\ x_1^3(k) \end{bmatrix}$$

Solution

The equilibrium points are determined from the condition

$$\begin{bmatrix} x_1(k) \\ x_2(k) \end{bmatrix} = \begin{bmatrix} x_2(k) \\ x_1^3(k) \end{bmatrix}$$

or equivalently,

$$x_2(k) = x_1(k) = x_1^3(k)$$

Thus, the system has the three equilibrium points: $(0, 0)$, $(1, 1)$, and $(-1, -1)$.

To characterize equilibrium points, we need the following definition.

Definition 11.1: Contraction. A function $\mathbf{f(x)}$ is known as a contraction if it satisfies

$$\|\mathbf{f}(\mathbf{x} - \mathbf{y})\| \leq \alpha \|\mathbf{x} - \mathbf{y}\|, \quad |\alpha| < 1 \qquad (11.32)$$

where α is known as a **contraction constant** and $\|.\|$ is any vector norm. ∎

The following theorem provides conditions for the existence of an equilibrium point for a discrete-time system.

Theorem 11.2: A contraction $\mathbf{f(x)}$ has a unique fixed point. ∎

EXAMPLE 11.7

Determine if the following nonlinear discrete-time system converges to the origin

$$\mathbf{x}(k+1) = af[x(k)], \quad k = 0, 1, 2, \ldots$$

$$|f[x(k)]| < |x(k)|/|a|, \quad |a| < 1$$

Solution
We have the inequality

$$|\mathbf{x}(k+1)| = |af[x(k)]| < |a||x(k)|/|a| \leq \alpha |x(k)|, \quad k = 0, 1, 2, \ldots, \alpha < 1$$

Hence, the system converges to a fixed point at the origin as $k \to \infty$.

If a continuous-time system is discretized, then all its equilibrium points will be equilibrium points of the discrete-time system. This is because discretization corresponds to the solution of the differential equation governing the original system at the sampling points. Because the analog system at an equilibrium remains there, the discretized system will have the same behavior and therefore the same equilibrium.

11.4 LYAPUNOV STABILITY THEORY

Lyapunov stability theory is based on the idea that at a stable equilibrium, the energy of the system has a local minimum, whereas at an unstable equilibrium, it is at a maximum. This property is not restricted to energy and is in fact shared by a class of function that depends on the dynamics of the system. We call such functions **Lyapunov functions**.

11.4.1 Lyapunov Functions

If a Lyapunov function can be found for an equilibrium point, then it can be used to determine its stability or instability. This is particularly simple for linear systems but can be complicated for a nonlinear system.

We begin by examining the properties of energy functions that we need to generalize and retain for a Lyapunov function. We note the following:

- Energy is a nonnegative quantity.
- Energy changes continuously with its arguments.

We call functions that are positive except at the origin *positive definite*. We provide a formal definition of this property.

Definition 11.2: A scalar continuous function $V(\mathbf{x})$ is said to be positive definite if

- $V(\mathbf{0}) = 0$.
- $V(\mathbf{x}) > 0$ for any nonzero $\mathbf{x}$.   ∎

Similar definitions are also useful where the greater-than sign in the second condition is replaced by other inequalities with all other properties unchanged. Thus, we can define **positive semidefinite** ($\geq$), **negative definite** ($<$), and **negative semidefinite** ($\leq$) functions. If none of these definitions apply, the function is said to be **indefinite**. Definition 11.2 may hold locally, and then the function is called **locally positive definite**, or globally, in which case it is **globally positive definite**. Similarly, we characterize other properties, such as negative definiteness, as local or global.

A common choice of definite function is the **quadratic form** $\mathbf{x}^T P \mathbf{x}$. The sign of the quadratic form is determined by the eigenvalues of the matrix P (see Appendix III). The quadratic form is positive definite if the matrix P is positive definite, in which case its eigenvalues are all positive. Similarly, we can characterize the quadratic form as negative definite if the eigenvalues of P are all negative, positive semidefinite if the eigenvalues of P are positive or zero, negative semidefinite if negative or zero, and indefinite if P has positive and negative eigenvalues.

Definition 11.3: A scalar function $V(\mathbf{x})$ is a Lyapunov function in a region D if it satisfies the following conditions in D:

- It is positive definite.
- It decreases along the trajectories of the system, that is,

$$\Delta V(k) = V(\mathbf{x}(k+1)) - V(\mathbf{x}(k)) < 0, \quad k = 0, 1, 2, \ldots \tag{11.33}$$

∎

The preceding definition is used in local stability theorems. To prove global stability, we need an additional condition in addition to extending the two listed earlier.

Quadratic forms are a common choice of Lyapunov function because of their simple mathematical properties. However, it is often preferable to use other Lyapunov functions with properties tailored to suit the particular problem. We now list the mathematical properties of Lyapunov functions.

Definition 11.4: A scalar function $V(\mathbf{x})$ is a Lyapunov function if it satisfies the following conditions:

- It is positive definite.
- It decreases along the trajectories of the system.
- It is radially unbounded (i.e., it uniformly satisfies).

$$V(\mathbf{x}(k)) \to \infty, \quad \text{as} \quad \|\mathbf{x}(k)\| \to \infty \tag{11.34}$$

∎

This last condition ensures that whenever the function V remains bounded, the state vector will also remain bounded. The condition must be satisfied regardless of how the norm of the state vector grows unbounded so that a finite value of V can always be associated with a finite state. Note that unlike Definition 11.3, Definition 11.4 requires that the first and second conditions be satisfied globally.

11.4.2 Stability Theorems

We provide some sufficient conditions for the stability of nonlinear discrete-time systems; then we specialize our results to linear time-invariant systems.

Theorem 11.3: The equilibrium point $\mathbf{x} = \mathbf{0}$ of the nonlinear discrete-time system

$$\mathbf{x}(k+1) = \mathbf{f}[\mathbf{x}(k)], \quad k = 0, 1, 2, \ldots \tag{11.35}$$

is asymptotically stable if there exists a locally positive definite Lyapunov function for the system satisfying Definition 11.3.

PROOF. For any motion along the trajectories of the system, we have

$$\Delta V(k) = V(\mathbf{x}(k+1)) - V(\mathbf{x}(k))$$
$$= V(\mathbf{f}[\mathbf{x}(k)]) - V(\mathbf{f}[\mathbf{x}(k-1)]) < 0, \quad k = 0, 1, 2, \ldots$$

This implies that as the motion progresses the value of V decreases continuously. However, because V is bounded below by zero, it converges to zero. Because V is only zero for zero argument, the state of the system must converge to zero.   ∎

EXAMPLE 11.8

Investigate the stability of the system using the Lyapunov stability approach

$$x_1(k+1) = 0.2x_1(k) - 0.08x_2^2(k)$$

$$x_2(k+1) = -0.3x_1(k)x_2(k), \quad k = 0, 1, 2, \ldots$$

Solution

We first observe that the system has an equilibrium point at the origin. We select the quadratic Lyapunov function

$$V(\mathbf{x}) = \mathbf{x}^T P \mathbf{x}, \, P = \text{diag}\{p_1, 1\}$$

Note that the unity entry simplifies the notation without affecting the results because the form of the function is unchanged if it is multiplied by any positive scalar p_2.

The corresponding difference is

$$\Delta V(k) = p_1 \left\{ [0.2x_1(k) - 0.08x_2^2(k)]^2 - x_1^2(k) \right\} + \left\{ [-0.3x_1(k)x_2(k)]^2 - x_2^2(k) \right\}$$

$$= -0.96\, p_1 x_1^2(k) + \left\{ 0.09x_1^2(k) - 0.032\, p_1 x_1(k) + 0.0064\, p_1 x_2^2(k) - 1 \right\} x_2^2(k)$$

$$k = 0, 1, 2, \ldots$$

The difference remains negative provided that the term between braces is negative. We restrict the magnitude of $x_2(k)$ to less than 12.5 (the square root of the reciprocal of 0.0064) and then simplify the condition to $9x_1^2(k) - 3.2p_1x_1(k) + 100(p_1 - 1) < 0, k = 0,1,2, \ldots$

Choosing a very small but positive value for p_1 makes the two middle terms negligible. This leaves two terms that are negative for values of x_1 of magnitude smaller than 3.33. For example, a plot of the LHS of the inequality for p_1 verifies that it is negative in the selected x_1 range. Thus, the difference remains negative for initial conditions that are inside a circle of radius approximately equal to 3.33 centered at the origin. By Theorem 11.3, we conclude that the system is asymptotically stable in the region $\|\mathbf{x}\| < 3.33$.

The system is actually stable outside this region, but our choice of Lyapunov function cannot provide a better estimate of the stability region than the one we obtained.

For global asymptotic stability, the system must have a unique equilibrium point. Otherwise, starting at another equilibrium point prevents the system from converging to another equilibrium state. We can therefore say that a system is globally asymptotically stable and not just its equilibrium point. The following theorem gives a sufficient condition for global asymptotic stability.

Theorem 11.4: The nonlinear discrete-time system

$$\mathbf{x}(k+1) = \mathbf{f}[\mathbf{x}(k)], \quad k = 0, 1, 2, \ldots \tag{11.36}$$

with equilibrium $\mathbf{x}(0) = \mathbf{0}$ is globally asymptotically stable if there exists a globally positive definite, radially unbounded Lyapunov function for the system satisfying Definition 11.4.

PROOF. The proof is similar to that of Theorem 11.3 and the details are omitted. The radial unboundedness condition guarantees that the state of the system will converge to zero with the Lyapunov function. ∎

The results of this section require the difference ΔV of the Lyapunov function to be negative definite. In some cases, this condition can be relaxed to negative semidefinite. In particular, if the nonzero values of the vector **x** for which ΔV is zero are ones that are never reached by the system, then they have no impact on the stability analysis. This leads to the following result.

Corollary 11.1: The equilibrium point **x** = **0** of the nonlinear discrete-time system of (11.36) is asymptotically stable if there exists a locally positive definite Lyapunov function with negative semidefinite difference $\Delta V(k)$ for all k for the system and with $\Delta V(k)$ zero only for **x** = **0**.

Note that the preceding theorems only provide a sufficient stability condition. Thus, failure of the stability test does not prove instability. In the linear time-invariant case, a much stronger result is available.

11.4.3 Rate of Convergence

In some cases, we can use a Lyapunov function to determine the rate of convergence of the system to the origin. In particular, if we can rewrite the difference of the Lyapunov function in the form

$$\Delta V(k) \leq -\alpha V(\mathbf{x}(k)), 0 < \alpha < 1 \qquad (11.37)$$

substituting for the difference gives the recursion

$$V(\mathbf{x}(k+1)) \leq (1-\alpha)V(\mathbf{x}(k)) \qquad (11.38)$$

The upper bound of the constant α guarantees that V is positive definite. The solution of the difference equation is

$$V(\mathbf{x}(k)) \leq (1-\alpha)^k V(\mathbf{x}(0)) \qquad (11.39)$$

If V is a Lyapunov function, then it converges to zero with the state of the system and its rate of convergence allows us to estimate the rate of convergence to the equilibrium point. If in addition V is a quadratic form, then the rate of convergence of the state is the square root of that of V.

EXAMPLE 11.9

Suppose that the quadratic form $\mathbf{x}^T\mathbf{x}$ is a Lyapunov function for a discrete-time system with difference

$$\Delta V(k) = -0.25\|\mathbf{x}(k)\|^2 - 0.5\|\mathbf{x}(k)\|^4 < 0$$

Characterize the convergence of the system trajectories to the origin.

Solution

$$\Delta V(k) = -0.25\|\mathbf{x}(k)\|^2 - 0.5\|\mathbf{x}(k)\|^4 \leq -0.25\|\mathbf{x}(k)\|^2 = 0.25V(x(k))$$
$$V(\mathbf{x}(k)) < (1-0.25)^k V(\mathbf{x}(0))$$

$$\|\mathbf{x}(k)\|^2 < 0.75^k \|\mathbf{x}(0)\|^2$$

$$\|\mathbf{x}(k)\| \le 0.866^k \|\mathbf{x}(0)\|$$

The trajectories of the system converge to the origin exponentially with convergence rate faster than 0.5.

11.4.4 Lyapunov Stability of Linear Systems

Lyapunov stability results typically provide us with sufficient conditions. Failure to meet the conditions of a Lyapunov test leaves us with no conclusion and with the need to repeat the test using a different Lyapunov function or to try a different test. For linear systems, Lyapunov stability can provide us with necessary and sufficient stability conditions.

Theorem 11.5: The linear time-invariant discrete-time system

$$\mathbf{x}(k+1) = A_d \mathbf{x}(k), \quad k = 0, 1, 2, \dots \tag{11.40}$$

is asymptotically stable if and only if for any positive definite matrix Q, there exists a unique positive definite solution P to the discrete Lyapunov equation

$$A_d^T P A_d - P = -Q \tag{11.41}$$

PROOF. We drop the subscript d for brevity.

Sufficiency

Consider the Lyapunov function

$$V(\mathbf{x}(k)) = \mathbf{x}^T(k) P \mathbf{x}(k)$$

with P a positive definite matrix. The Lyapunov function changes along the trajectories of the system following

$$\begin{aligned}
\Delta V(k) &= V(\mathbf{x}(k+1)) - V(\mathbf{x}(k)) \\
&= \mathbf{x}^T(k)\left[A^T PA - P\right]\mathbf{x}(k) = -\mathbf{x}^T(k) Q \mathbf{x}(k) < 0
\end{aligned}$$

Hence, the system is stable by Theorem 11.3.

Necessity

We first show that the solution of the Lyapunov equation is given by

$$P = \sum_{k=0}^{\infty} (A^T)^k Q A^k$$

This is easily verified by substitution in the Lyapunov equation and then changing the index of summation as follows:

$$A^T \left[\sum_{k=0}^{\infty} (A^T)^k Q A^k\right] A - \sum_{k=0}^{\infty} (A^T)^k Q A^k = \sum_{j=1}^{\infty} (A^T)^j Q A^j - \sum_{k=0}^{\infty} (A^T)^k Q A^k = -Q$$

To show that for any positive definite Q, P is positive definite, consider the quadratic form

$$\mathbf{x}^T P \mathbf{x} = \sum_{k=0}^{\infty} \mathbf{x}^T \left(A^T \right)^k Q A^k \mathbf{x}$$

$$= \mathbf{x}^T Q \mathbf{x} + \sum_{k=0}^{\infty} \mathbf{y}_k^T Q \mathbf{y}_k$$

For positive definite Q, the first term is positive for any nonzero $\mathbf{x}$ and the other terms are nonnegative. It follows that P is positive definite.

Let the system be stable but for some positive definite Q there is no finite solution P to the Lyapunov equation. We show that this leads to a contradiction.

Recall that the state-transition matrix of the discrete system can be written as

$$A^k = \sum_{k=1}^{n} Z_i \lambda_i^k \tag{11.42}$$

We use any matrix norm to obtain the inequality

$$\left\| A^k \right\| = \left\| \left(A^T \right)^k \right\| \le n \| Z \|_{\max} \lambda_{\max}^k = \alpha \lambda_{\max}^k, \quad k \ge 0$$

For a stable system, we have

$$\| P \| = \left\| \sum_{k=0}^{\infty} \left(A^T \right)^k Q A^k \right\| \le \sum_{k=0}^{\infty} \left\| \left(A^T \right)^k \right\| \| Q \| \| A^k \|$$

$$\le \sum_{k=0}^{\infty} \alpha^2 \lambda_{\max}^{2k} \| Q \| = \frac{\alpha^2}{1 - \lambda_{\max}^{2k}} \| Q \|$$

This contradicts the assumption and establishes necessity.

Uniqueness

Let P_1 be a second solution of the Lyapunov equation. Then we can write it in the form of infinite summation including Q, then substitute for Q in terms of P using the Lyapunov equation. This yields the equation

$$P_1 = \sum_{k=0}^{\infty} \left(A^T \right)^k Q A^k = -\sum_{k=0}^{\infty} \left(A^T \right)^k \left[A^T P A - P \right] A^k$$

$$= -\sum_{j=1}^{\infty} \left(A^T \right)^j P A^j + \sum_{k=0}^{\infty} \left(A^T \right)^k P A^k = P$$

∎

As in the nonlinear case, it is possible to relax the stability condition as follows.

Corollary 11.2: The linear time-invariant discrete-time system of (11.40) is asymptotically stable if and only if for any detectable pair (A_d, C), there exists a unique positive definite solution P to the discrete Lyapunov equation

$$A_d^T P A_d - P = -C^T C \tag{11.43}$$

PROOF. The proof follows the same steps as the theorem. We only show that zero values of the difference do not impact stability. We first obtain

$$\Delta V(k) = V(\mathbf{x}(k+1)) - V(\mathbf{x}(k))$$
$$= \mathbf{x}^T(k)\left[A^T PA - P\right]\mathbf{x}(k)$$
$$= -\mathbf{x}^T(k)C^T C\mathbf{x}(k) = -\mathbf{y}^T(k)\mathbf{y}(k) \leq 0$$

It is implicitly assumed that the system is observable, so $\mathbf{y}$ is zero only if $\mathbf{x}$ is zero. If the system is only detectable, then the only nonzero values of $\mathbf{x}$ for which $\mathbf{y}$ is zero are ones that correspond to stable dynamics and can be ignored in our stability analysis. ∎

Although using a semidefinite matrix in the stability test may simplify computation, checking the system for detectability (guaranteed by observability) eliminates the gain from the simplification. However, the corollary is of theoretical interest and helps clarify the properties of the discrete Lyapunov equation.

The discrete Lyapunov equation is clearly a linear equation in the matrix P, and by rearranging terms we can write it as a linear system of equations. The equivalent linear system involves an $n^2 \times 1$ vector of unknown entries of P obtained using the operation

$$st(P) = \mathrm{col}\{\mathbf{p}_1, \mathbf{p}_2 \ldots, \mathbf{p}_n\}$$
$$P = [\mathbf{p}_1 \quad \mathbf{p}_2 \quad \cdots \quad \mathbf{p}_n]$$

The $n^2 \times n^2$ coefficient matrix of the linear system is obtained using the Kronecker product of matrices. In this operation, each entry of the first matrix is replaced by the second matrix scaled by the original entry. The Kronecker product is thus defined by

$$A \otimes B = [a_{ij}B] \tag{11.44}$$

It can be shown that the Lyapunov equation is equivalent to the linear system

$$L\mathbf{p} = -\mathbf{q}$$
$$\mathbf{p} = st(P) \tag{11.45}$$
$$L = A^T \otimes A^T - I \otimes I$$

Because the eigenvalues of any matrix are identical to those of its transpose, the Lyapunov equation can be written in the form

$$A_d PA_d^T - P = -Q \tag{11.46}$$

The first form of the Lyapunov equation of (11.43) is the **controller form**, whereas the second form of (11.46) is known as the **observer form**.

11.4.5 MATLAB

To solve the discrete Lyapunov equation using MATLAB we use the command
dlyap. The command solves the observer form of the Lyapunov equation.

$$\gg P = dlyap\ (A,\ Q)$$

To solve the equivalent linear system, we use the Kronecker product and the command

$$\gg kron(A',\ A')$$

EXAMPLE 11.10

Use the Lyapunov approach with $Q = I_3$ to determine the stability of the linear time-invariant system

$$\begin{bmatrix} x_1(k+1) \\ x_2(k+1) \\ x_3(k+1) \end{bmatrix} = \begin{bmatrix} -0.2 & -0.2 & 0.4 \\ 0.5 & 0 & 1 \\ 0 & -0.4 & -0.5 \end{bmatrix} \begin{bmatrix} x_1(k) \\ x_2(k) \\ x_3(k) \end{bmatrix}$$

Is it possible to investigate the stability of the system using $Q = \text{diag}\{0, 0, 1\}$?

Solution
Using MATLAB, we obtain the solution as follows

$$\gg P = dlyap(A,\ Q) \text{ \% Observer form of the Lyapunov equation}$$

$$P = \begin{matrix} 1.5960 & 0.5666 & 0.0022 \\ 0.5666 & 3.0273 & -0.6621 \\ 0.0022 & -0.6621 & 1.6261 \end{matrix}$$

$$\gg eig(P) \text{ \% Check the signs of the eigenvalues of P}$$

$$ans = \begin{matrix} 1.1959 \\ 1.6114 \\ 3.4421 \end{matrix}$$

Because the eigenvalues of P are all positive, the matrix is positive definite and the system is asymptotically stable.

For $Q = \text{diag}\{0, 0, 1\} = C^T C$, with $C = [0, 0, 1]$, we check the observability of the pair (A, C) using MATLAB and the rank test

$$\gg rank(obsv(a,[0,\ 0,\ 1])) \text{ \% obsv computes the observability matrix}$$

$$ans = 3$$

The observability matrix is full rank, and the system is observable. Thus, we can use the matrix to check the stability of the system.

11.4.6 Lyapunov's Linearization Method

It is possible to characterize the stability of an equilibrium for a nonlinear system by examining its approximately linear behavior in the vicinity of the equilibrium. Without loss of generality, we assume that the equilibrium is at the origin and rewrite the state equation in the form

$$\mathbf{x}(k+1) = \frac{\partial \mathbf{f}[\mathbf{x}(k)]}{\partial \mathbf{x}(k)}\bigg|_{\mathbf{x}(k)=0} + \mathbf{f}_2[\mathbf{x}(k)], \quad k = 0, 1, 2, \ldots \qquad (11.47)$$

where $\mathbf{f}_2[.]$ is a function including all terms of order higher than one. We then rewrite the equation in the form

$$\mathbf{x}(k+1) = A\mathbf{x}(k) + \mathbf{f}_2[\mathbf{x}(k)], \quad k = 0, 1, 2, \ldots$$

$$A = \frac{\partial \mathbf{f}[\mathbf{x}(k)]}{\partial \mathbf{x}(k)}\bigg|_{\mathbf{x}(k)=0} \qquad (11.48)$$

In the vicinity of the origin the behavior is approximately the same as that of the linear system

$$\mathbf{x}(k+1) = A\mathbf{x}(k) \qquad (11.49)$$

This intuitive fact can be more rigorously justified using Lyapunov stability theory, but we omit the proof. Thus, the equilibrium is stable if the linear approximation is stable, and unstable if the linear approximation is unstable. If the linear system (11.49) has an eigenvalue on the unit circle, then the stability of the nonlinear system cannot be determined from the first-order approximation alone. This is because higher-order terms can make the nonlinear system either stable or unstable. Based on our discussion, we have the following theorem.

Theorem 11.6: The equilibrium point of the system of (11.49) is as follows:

- Asymptotically stable if all the eigenvalues of A are inside the unit circle.
- Unstable if one or more of the eigenvalues is outside the unit circle.
- If A has one or more eigenvalues on the unit circle, then the stability of the nonlinear system cannot be determined from the linear approximation. ∎

EXAMPLE 11.11

Show that the origin is an unstable equilibrium for the system

$$x_1(k+1) = 2x_1(k) + 0.08x_2^2(k)$$

$$x_2(k+1) = x_1(k) + 0.1x_2(k) + 0.3x_1(k)x_2(k), \quad k = 0, 1, 2, \ldots$$

Solution

We first rewrite the state equations in the form

$$\begin{bmatrix} x_1(k+1) \\ x_2(k+1) \end{bmatrix} = \begin{bmatrix} 2 & 0 \\ 1 & 0.1 \end{bmatrix} \begin{bmatrix} x_1(k) \\ x_2(k) \end{bmatrix} + \begin{bmatrix} 0.08x_2^2(k) \\ 0.3x_1(k)x_2(k) \end{bmatrix}, \quad k = 0, 1, 2, \ldots$$

The state matrix A of the linear approximation has one eigenvalue $= 2 > 1$. Hence the origin is an unstable equilibrium of the nonlinear system.

11.4.7 Instability Theorems

The weakness of the Lyapunov approach for nonlinear stability investigation is that it only provides sufficient stability conditions. Although no necessary and sufficient conditions are available, it is possible to derive conditions for instability and use them to test the system if one is unable to establish its stability. Clearly, failure of both sufficient conditions is inconclusive and it is only in the linear case that we have the stronger necessary and sufficient condition.

Theorem 11.7: The equilibrium point $\mathbf{x} = \mathbf{0}$ of the nonlinear discrete-time system

$$\mathbf{x}(k+1) = \mathbf{f}[\mathbf{x}(k)], \quad k = 0, 1, 2, \ldots \tag{11.50}$$

is unstable if there exists a locally positive function for the system with locally uniformly positive definite changes along the trajectories of the system.

PROOF. For any motion along the trajectories of the system, we have

$$\Delta V(k) = V(\mathbf{x}(k+1)) - V(\mathbf{x}(k))$$
$$= V(\mathbf{f}[\mathbf{x}(k)]) - V(\mathbf{f}[\mathbf{x}(k-1)]) > 0, \quad k = 0, 1, 2, \ldots$$

This implies that as the motion progresses, the value of V increases continuously. However, because V is only zero with argument zero, the trajectories of the system will never converge to the equilibrium at the origin and the equilibrium is unstable. ∎

EXAMPLE 11.12

Show that the origin is an unstable equilibrium for the system

$$x_1(k+1) = -2x_1(k) + 0.08x_2^2(k)$$
$$x_2(k+1) = 0.3x_1(k)x_2(k) + 2x_2(k), \quad k = 0, 1, 2, \ldots$$

Solution

Choose the Lyapunov function

$$V(\mathbf{x}) = \mathbf{x}^T P \mathbf{x}, \, P = \text{diag}\{p_1, 1\}$$

The corresponding difference is given by

$$\Delta V(k) = p_1\left\{[-2x_1(k) + 0.08x_2^2(k)]^2 - x_1^2(k)\right\} + \left\{[0.3x_1(k)x_2(k) + 2x_2(k)]^2 - x_2^2(k)\right\}$$
$$= 3p_1x_1^2(k) + 0.0064p_1x_2^4(k) + \left\{0.09x_1^2(k) + (1.2 - 0.32p_1)x_1(k) + 3\right\}x_2^2(k),$$
$$k = 0, 1, 2, \ldots$$

We complete the squares for the last term by choosing $p_1 = 3.75\left(1 - \sqrt{3}/2\right) = 0.5024$ and reduce the difference to

$$\Delta V(k) = 1.5072x_1^2(k) + 0.0032x_2^4(k) + \left(0.3x_1(k) + \sqrt{3}\right)^2 x_2^2(k)$$
$$\geq 1.5072x_1^2(k) + 0.0032x_2^4(k), \quad k = 0, 1, 2, \ldots$$

The inequality follows from the fact that the last term in ΔV is positive semidefinite. We conclude that ΔV is positive definite because it is greater than the sum of even powers and that the equilibrium at $\mathbf{x} = \mathbf{0}$ is unstable.

11.4.8 Estimation of the Domain of Attraction

We consider a system

$$\mathbf{x}(k+1) = A\mathbf{x} + \mathbf{f}[\mathbf{x}(k)], \quad k = 0, 1, 2, \ldots \tag{11.51}$$

where the matrix A is stable and $\mathbf{f}(.)$ includes second-order terms and higher and satisfies

$$\|\mathbf{f}[\mathbf{x}(k)]\| \leq \alpha\|\mathbf{x}(k)\|^2, \quad k = 0, 1, 2, \ldots \tag{11.52}$$

Because the linear approximation of the system is stable, we can solve the associated Lyapunov equation for a positive definite matrix P for any positive definite matrix Q. This yields the Lyapunov function

$$V(\mathbf{x}(k)) = \mathbf{x}^T(k)P\mathbf{x}(k)$$

and the difference

$$\Delta V(\mathbf{x}(k)) = \mathbf{x}^T(k)[A^T PA - P]\mathbf{x}(k) + 2\mathbf{f}^T[\mathbf{x}(k)]PA\mathbf{x}(k) + \mathbf{f}^T[\mathbf{x}(k)]P\mathbf{f}[\mathbf{x}(k)]$$

To simplify this expression, we make use of the inequalities

$$\lambda_{\min}(P)\|\mathbf{x}\|^2 \leq \mathbf{x}^T P\mathbf{x} \leq \lambda_{\max}(P)\|\mathbf{x}\|^2$$

$$\|\mathbf{f}^T PA\mathbf{x}\| \leq \|\mathbf{f}\|\|PA\|\|\mathbf{x}\| \leq \alpha\|PA\|\|\mathbf{x}\|^3 \tag{11.53}$$

where the norm $\|A\|$ denotes the square root of the largest eigenvalue of the matrix $A^T A$ or its largest **singular value**.

Using the inequality (11.53), and substituting from the Lyapunov equation, we have

$$\Delta V(\mathbf{x}(k)) \leq \left[\alpha^2\lambda_{\max}(P)\|\mathbf{x}\|^2 + 2\alpha\|PA\|\|\mathbf{x}\| - \lambda_{\min}(Q)\right]\|\mathbf{x}\|^2 \tag{11.54}$$

Because the coefficient of the quadratic is positive, its second derivative is also positive and it has negative values between its two roots. The positive root defines a bound on $\|x\|$ that guarantees a negative difference:

$$\|x\| < \frac{1}{\alpha\lambda_{\max}(P)}\left[-\lambda_{\max}(PA) + \sqrt{\|PA\|^2 + \alpha\lambda_{\min}(Q)}\right]$$

An estimate of the domain of attraction is given by

$$B(x) = \left\{x : \|x\| < \frac{1}{\alpha\lambda_{\max}(P)}\left[-\lambda_{\max}(PA) + \sqrt{\|PA\|^2 + \alpha\lambda_{\min}(Q)}\right]\right\} \qquad \textbf{(11.55)}$$

EXAMPLE 11.13

Estimate the domain of attraction of the system

$$\begin{bmatrix} x_1(k+1) \\ x_2(k+1) \end{bmatrix} = \begin{bmatrix} 0.1 & 0 \\ -1 & 0.5 \end{bmatrix}\begin{bmatrix} x_1(k) \\ x_2(k) \end{bmatrix} + \begin{bmatrix} 0.25x_2^2(k) \\ 0.1x_1^2(k) \end{bmatrix}$$

Solution

The nonlinear vector satisfies

$$\|f[x(k)]\| \le 0.5\|x(k)\|, \quad k = 0, 1, 2, \ldots$$

We solve the Lyapunov equation with $Q = I_2$ to obtain

$$P = \begin{bmatrix} 2.4987 & -0.7018 \\ -0.7018 & 1.3333 \end{bmatrix}$$

whose largest eigenvalue is equal to 2.8281. The norm $\|PA\| = 1.8539$ can be computed with the MATLAB command

$$\text{>> norm(P*A)}$$

Our estimate of the domain of attraction is

$$B(x) = \left\{x : \|x\| < \frac{1}{0.25\sqrt{2.8281}}\left[-1.8539 + \sqrt{(1.8539)^2 + 0.25}\right]\right\}$$
$$= \{x : \|x\| < 0.0937\}$$

The state portrait presented in Figure 11.1 shows that the estimate of the domain of attraction is quite conservative and that the system is stable well outside the estimated region.

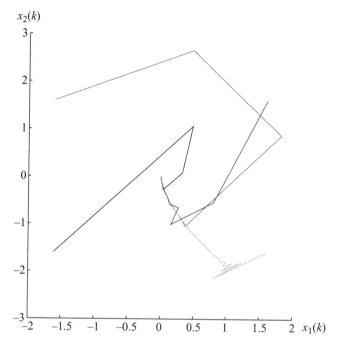

FIGURE 11.1

Phase portrait for the nonlinear system described in Example 11.13.

11.5 STABILITY OF ANALOG SYSTEMS WITH DIGITAL CONTROL

Although most nonlinear design methodologies are analog, they are often digitally implemented. The designer typically completes an analog controller design and then obtains a discrete approximation of the analog controller. For example, a discrete approximation can be obtained using a differencing approach to approximate the derivatives (see Chapter 6). The discrete time approximation is then used with the analog plant, assuming that the time response will be approximately the same as that of the analog design. This section examines the question "Is the stability of the digital control system guaranteed by the stability of the original analog design?" As in the linear case discussed in Chapter 6, the resulting digital control system may be unstable or may have unacceptable intersample oscillations.

We first examine the system of (11.1) with the digital control (11.2). Substituting (11.2) in (11.1) gives the equation

$$\dot{\mathbf{x}} = \mathbf{f}(\mathbf{x}) + B(\mathbf{x})\mathbf{u}(k) = \mathbf{f}_k(\mathbf{x}), \quad t \in [kT, (k+1)T) \tag{11.56}$$

Let the solution of (11.56) be

$$\mathbf{x}(t) = \mathbf{g}_k(\mathbf{x}(kT)) \tag{11.57}$$

where $\mathbf{g}_k$ is a continuous function of all its arguments. Then at the sampling points, we have the discrete model

$$\mathbf{x}(k+1) = \mathbf{g}_k(\mathbf{x}(k)), \quad k = 0, 1, \ldots \tag{11.58}$$

The stability of the discrete-time system of (11.58) is governed by the following theorem.

Theorem 11.8: Analog System with Digital Control. If the continuous system of (11.56) with piecewise constant control is exponentially stable, then the discrete-time system of (11.58) is exponentially stable.

PROOF. For an exponentially stable continuous-time system, we have

$$\|\mathbf{x}(t)\| \le \|\mathbf{x}(k)\| e^{-\alpha_k t}, \quad t \in [kT, (k+1)T) \tag{11.59}$$

for some positive constant α_k. Hence, at the sampling points we have the inequality

$$\|\mathbf{x}(k)\| \le \|\mathbf{x}(k-1)\| e^{-\alpha_{k-1}T}, \quad k = 0, 1, \ldots$$

Applying this inequality repeatedly gives

$$\|\mathbf{x}(k)\| \le \|\mathbf{x}(0)\| e^{-\alpha_s T}$$
$$\le \|\mathbf{x}(0)\| e^{-\alpha_m T}, \quad k = 0, 1, \ldots \tag{11.60}$$

with $\alpha_s = \sum_{k=0}^{k-1} \alpha_k$ and $\alpha_m = \min_k \alpha_k$. Thus, the discrete system is exponentially stable. ∎

The preceding theorem may lead us to believe that digital approximation of analog controllers preserves stability. However, the theorem assumes that the continuous system of (10.56) is exponentially stable. Thus, it only provides a necessary stability condition for the discretization of an analog plant with a digital controller. If a stable analog controller is implemented digitally and used with the analog plant, its stability cannot be guaranteed. The resulting system may be unstable, as the following example demonstrates.

EXAMPLE 11.14

Consider the system

$$\dot{x}(t) = x^2(t) + x(t) u(t)$$

with the continuous control

$$u(t) = -x(t) - \alpha x^2(t), \alpha > 0$$

Then the closed-loop system is

$$\dot{x}(t) = -\alpha x^3(t)$$

which is asymptotically stable.[1]

Now consider the digital implementation of the analog control

$$u(t) = -x(k) - \alpha x^2(k), \alpha > 0, \quad t \in [kT, (k+1)T)$$

and the corresponding closed-loop system

$$\dot{x}(t) = x(t)\{x(t) - [\alpha x(k) + 1]x(k)\}, \quad t \in [kT, (k+1)T)$$

We solve the equation by separation of variables

$$\frac{dx(t)}{x(t)(x(t) - \beta)} = \frac{dx(t)}{\beta}\left[\frac{1}{x - \beta} - \frac{1}{x}\right] = dt, \quad t \in [kT, (k+1)T)$$

$$\beta = [\alpha x(k) + 1]x(k)$$

Integrating from kT to $(k + 1)T$, we obtain

$$\ln\left(1 - \frac{\beta}{x(t)}\right)\Bigg]_{t=kT}^{t=(k+1)T} = dt]_{t=kT}^{t=(k+1)T}$$

$$\left(1 - \frac{\beta}{x(k+1)}\right) = \left(1 - \frac{\beta}{x(k)}\right)e^T$$

$$x(k+1) = \frac{\beta}{1 - \left[1 - \dfrac{\beta}{x(k)}\right]e^T}$$

$$= \frac{[\alpha x(k) + 1]x(k)}{1 + \alpha x(k)e^T}$$

For stability, we need the condition $|x(k + 1)| < |x(k)|$, that is,

$$\left|1 + \alpha x(k)e^T\right| > |1 + \alpha x(k)|$$

This condition is satisfied for positive $x(k)$ but not for all negative $x(k)$! For example, with $T = 0.01$s and $\alpha x(k) = -0.5$, the LHS is 0.495 and the RHS is 0.5.

We do not provide a general result that guarantees the stability of the digital implementation of a stable analog controller. However, once the digital controller is obtained, one can investigate the stability of the closed-loop system using stability tests for digital control systems, as discussed in Section 11.4.

---

[1]Slotine (1991, p. 66): $\dot{x} + c(x) = 0$ is asymptotically stable if $c(.)$ is continuous and satisfies $c(x)x > 0$, $\forall x \neq 0$. Here $c(x) = x^3$.

11.6 STATE PLANE ANALYSIS

As shown in Section 11.4.6, the stability of an equilibrium of a nonlinear system can often be determined from the linearized model of the system in the vicinity of the equilibrium. Moreover, the behavior of system trajectories of linear discrete-time systems in the vicinity of an equilibrium point can be visualized for second-order systems in the state plane. State plane trajectories can be plotted based on the solutions of the state equations for discrete-time equations similarly to those for continuous-time systems (see Chapter 7). Consider the unforced second-order difference equation

$$y(k+2) + a_1 y(k+1) + a_0 y(k) = 0, \quad k = 0, 1, \ldots \tag{11.61}$$

The associated characteristic equation is

$$z^2 + a_1 z + a_0 = (z - \lambda_1)(z - \lambda_2) = 0 \tag{11.62}$$

We can characterize the behavior of the system based on the location of the characteristic values of the system λ_i, $i = 1, 2$, in the complex plane. If the system is represented in state–space form, then a similar characterization is possible using the eigenvalues of the state matrix. Table 11.1 gives the names and characteristics of equilibrium points based on the locations of the eigenvalues.

Trajectories corresponding to different types of equilibrium points are shown in Figures 11.2 through 11.9. We do not include some special cases such as two real eigenvalues equal to unity, in which case the system always remains at the initial state. The reader is invited to explore other pairs of eigenvalues through MATLAB simulation.

Table 11.1 Equilibrium Point Classification

Equilibrium Type	Eigenvalue Location
Stable node	Real positive inside unit circle
Unstable node	Real positive outside unit circle
Saddle point	Real eigenvalues with one inside and one outside unit circle
Stable focus	Complex conjugate or both real negative inside unit circle
Unstable focus	Complex conjugate or both real negative outside unit circle
Vortex or center	Complex conjugate on unit circle

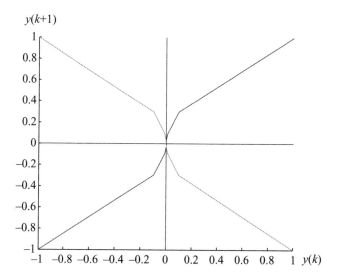

FIGURE 11.2

Stable node (eigenvalues 0.1, 0.3).

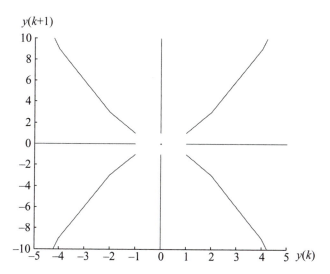

FIGURE 11.3

Unstable node (eigenvalues 2, 3).

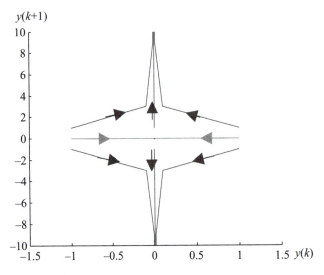

FIGURE 11.4

Saddle point (eigenvalues 0.1, 3).

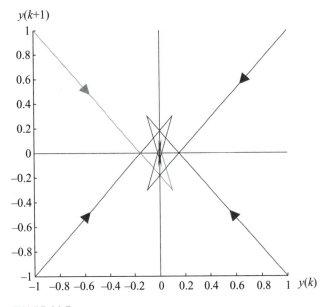

FIGURE 11.5

Stable focus (eigenvalues −0.1, −0.3).

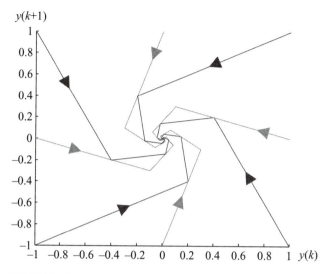

FIGURE 11.6

Stable focus (eigenvalues $0.1 \pm j0.3$).

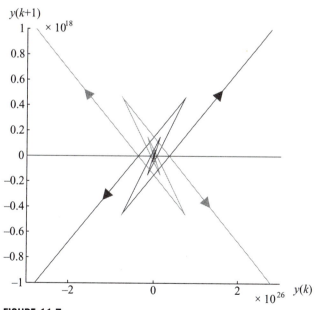

FIGURE 11.7

Unstable focus (eigenvalues $-5, -3$).

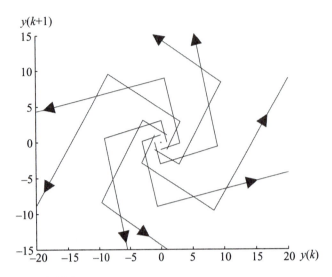

FIGURE 11.8

Unstable focus (eigenvalues $0.1 \pm j3$).

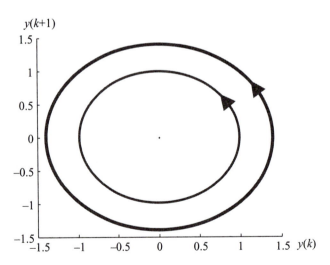

FIGURE 11.9

Center (eigenvalues $\cos(45°) \pm j \sin(45°)$).

11.7 DISCRETE-TIME NONLINEAR CONTROLLER DESIGN

Nonlinear control system design for discrete-time systems is far more difficult than linear design. It is also often more difficult than the design of analog nonlinear systems. For example, one of the most powerful approaches to nonlinear system design is to select a control that results in a closed-loop system for which a suitable Lyapunov function can be constructed. It is usually easier to construct a Lyapunov function for an analog nonlinear system than it is for a discrete-time system.

    We discuss some simple approaches to nonlinear design for discrete-time systems that are possible for special classes of systems.

11.7.1 Controller Design Using Extended Linearization

If one of the extended linearization approaches presented in Section 11.1 is applicable, then we can obtain a linear discrete-time model for the system and use it in linear control system design. The nonlinear control can then be recovered from the linear design. We demonstrate this approach with the following example.

EXAMPLE 11.15

Consider the mechanical system

$$\ddot{x} + b(\dot{x}) + c(x) = f$$

with $b(0) = 0$, $c(0) = 0$. For the nonlinear damping $b(\dot{x}) = 0.1\dot{x}^3$ and the nonlinear spring $c(x) = 0.1x + 0.01x^3$, design a digital controller for the system by redefining the input and discretization.

Solution

The state equations of the system are

$$\dot{x}_1 = x_2$$

$$\dot{x}_2 = -b(x_2) - c(x_1) + f = u$$

Using the results described in Example 11.2, we have the discrete-time model

$$\mathbf{x}(k+1) = \begin{bmatrix} 1 & 0.02 \\ 0 & 1 \end{bmatrix} \mathbf{x}(k) + \begin{bmatrix} 2 \times 10^{-4} \\ 0.02 \end{bmatrix} \mathbf{u}(k)$$

with $\mathbf{x}(k) = [x_1(k)\ x_2(k)]^T = [x(k)\ x(k+1)]^T$.

    We select the eigenvalues $\{0.1 \pm j0.1\}$ and design a state feedback controller for the system as shown in Chapter 9. Using the MATLAB command **place**, we obtain the feedback gain matrix

$$\mathbf{k}^T = [2050 \quad 69.5]$$

For a reference input r, we have the nonlinear control

$$f = u + b(x_2) + c(x_1)$$

$$u(k) = r(k) - \mathbf{k}^T \mathbf{x}(k)$$

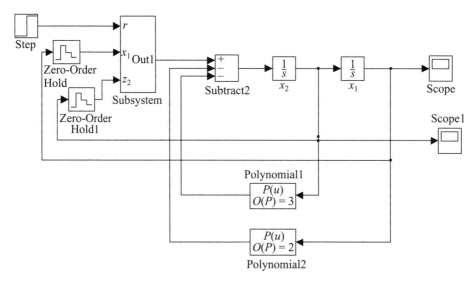

FIGURE 11.10

Simulation diagram for the system described in Example 11.15.

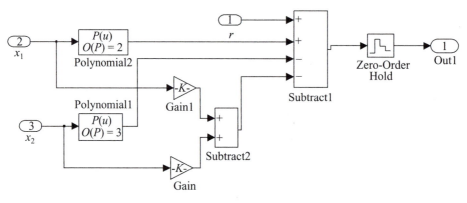

FIGURE 11.11

Controller simulation diagram described in Example 11.15.

The simulation diagram for the system is shown in Figure 11.10, and the simulation diagram for the controller block is shown in Figure 11.11. We select the amplitude of the step input to obtain a steady-state value of unity using the equilibrium condition

$$\mathbf{x}(k+1) = \begin{bmatrix} 1 & 0.02 \\ 0 & 1 \end{bmatrix} \mathbf{x}(k) + \begin{bmatrix} 2 \times 10^{-4} \\ 0.02 \end{bmatrix} (r - [2050 \quad 69.5]\mathbf{x}(k)) = \mathbf{x}(k) = \begin{bmatrix} 1 \\ 0 \end{bmatrix}$$

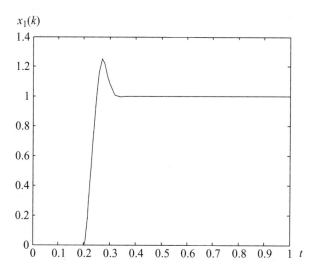

FIGURE 11.12

Step response for the linear design described in Example 11.15.

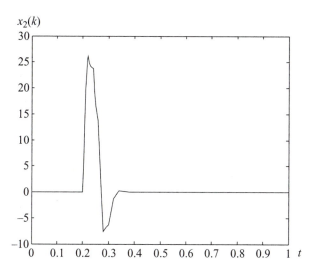

FIGURE 11.13

Velocity plot for the step response for the linear design described in Example 11.15.

This simplifies to

$$\begin{bmatrix} 2 \times 10^{-4} \\ 0.02 \end{bmatrix} r = \begin{bmatrix} 0.41 \\ 41 \end{bmatrix}$$

which gives the amplitude $r = 2050$. The step response for the nonlinear system with digital control in Figure 11.12 shows a fast response to a step input at $t = 0.2$ s that quickly settles to the desired steady-state value of unity. Figure 11.13 shows a plot of the velocity for the same input.

11.7.2 Controller Design Based on Lyapunov Stability Theory

Lyapunov stability theory provides a means of stabilizing unstable nonlinear systems using feedback control. The idea is that if one can select a suitable Lyapunov function and force it to decrease along the trajectories of the system, the resulting system will converge to its equilibrium. In addition, the control can be chosen to speed up the rate of convergence to the origin by forcing the Lyapunov function to decrease to zero faster.

To simplify our analysis, we consider systems of the form

$$\mathbf{x}(k+1) = A\mathbf{x}(k) + B(\mathbf{x}(k))\mathbf{u}(k) \qquad (11.63)$$

For simplicity, we assume that the eigenvalues of the matrix A are all inside the unit circle. This model could approximately represent a nonlinear system in the vicinity of its stable equilibrium.

Theorem 11.9 The open-loop stable affine system with linear unforced dynamics and full-rank input matrix for all nonzero $\mathbf{x}$ is asymptotically stable with the feedback control law

$$\mathbf{u}(k) = -\left[B^T(\mathbf{x}(k))PB(\mathbf{x}(k))\right]^{-1}B^T(\mathbf{x}(k))PA\mathbf{x}(k) \qquad (11.64)$$

where P is the solution of the discrete Lyapunov equation

$$A^T PA - P = -Q \qquad (11.65)$$

and Q is an arbitrary positive definite matrix.

PROOF. For the Lyapunov function

$$V(\mathbf{x}(k)) = \mathbf{x}^T(k)P\mathbf{x}(k)$$

The difference is given by

$$\begin{aligned}
\Delta V(k) &= V(\mathbf{x}(k+1)) - V(\mathbf{x}(k)) \\
&= \mathbf{x}^T(k)\left[A^T PA - P\right]\mathbf{x}(k) + 2\mathbf{u}^T(k)B^T PA\mathbf{x} + \mathbf{u}^T(k)B^T PB\mathbf{u}(k)
\end{aligned}$$

where the argument of B is suppressed for brevity. We minimize the function with respect to the control to obtain

$$\frac{\partial \Delta V(k)}{\partial \mathbf{u}(k)} = 2\left[B^T A\mathbf{x} + B^T PB\mathbf{u}(k)\right] = 0$$

which we solve for the feedback control law using the full-rank condition for B.

By the assumption of open-loop stability, we have

$$\Delta V(k) = -\mathbf{x}^T(k)Q\mathbf{x}(k) + \mathbf{x}^T(k)A^T PB\mathbf{u}(k), \quad Q > 0$$

We substitute for the control and rewrite the equation as

$$\Delta V(k) = -\mathbf{x}^T(k)\left\{Q + A^T PB\left[B^T PB\right]^{-1}B^T A\right\}\mathbf{x}(k)$$
$$= -\mathbf{x}^T(k)\left\{Q + A^T MA\right\}\mathbf{x}(k)$$

$$M = PB\left[B^T PB\right]^{-1}B^T$$

The matrix M is not symmetric and must be replaced by its symmetric component in the quadratic form (see Appendix III). The second term is therefore equal to

$$-\mathbf{x}^T(k)\,A^T MA\mathbf{x}(k) = -\mathbf{y}^T(k)\left\{\frac{M + M^T}{2}\right\}\mathbf{y}(k)$$
$$= -\mathbf{y}^T(k)\left\{\frac{P\bar{B} + \bar{B}P}{2}\right\}\mathbf{y}(k)$$

$$\bar{B} = \bar{B}^T = B\left[B^T PB\right]^{-1}B^T$$

where we use the fact that transposition and inversion of a real matrix are commutative operations. Because the matrix P is positive definite and B is full rank, the matrix $B^T PB$, and consequently its inverse, must be positive semidefinite. Thus, both $\bar{B}$ and the symmetric component of M are positive semidefinite, and the term $-\mathbf{x}^T(k)A^T MA\mathbf{x}(k)$ is negative semidefinite. We conclude that the difference of the Lyapunov function is negative definite, because it is the sum of a negative definite term and a negative semidefinite term, and that the closed-loop system is asymptotically stable. ∎

EXAMPLE 11.16

Design a controller to stabilize the origin for the system

$$x_1(k+1) = 0.2x_1(k) + x_2(k)u(k)$$
$$x_2(k+1) = 0.2x_2(k) + [1 + 0.4x_1(k)]u(k), \quad k = 0, 1, 2, \ldots$$

Solution
We rewrite the system dynamics in the form

$$\begin{bmatrix} x_1(k+1) \\ x_2(k+1) \end{bmatrix} = \begin{bmatrix} 0.2 & 0 \\ 0 & 0.2 \end{bmatrix}\begin{bmatrix} x_1(k) \\ x_2(k) \end{bmatrix} + \begin{bmatrix} x_2(k) \\ 1 + 0.4x_1(k) \end{bmatrix}u(k)$$

The state matrix is diagonal with two eigenvalues equal to $0.2 < 1$. Hence, the system is stable. We choose the Lyapunov function

$$V(\mathbf{x}) = \mathbf{x}^T P\mathbf{x}, \ P = \text{diag}\{p_1, 1\}$$

for which $Q = 0.16\,I_2$ is positive definite.

The corresponding stabilizing control is given by

$$u(k) = -\left[B^T(\mathbf{x}(k))PB(\mathbf{x}(k))\right]^{-1}B^T(\mathbf{x}(k))PA\mathbf{x}(k)$$

$$= -\frac{0.2}{(1+0.4x_1(k))^2 + p_1x_2^2(k)}[p_1x_2(k) \quad 1+0.4x_1(k)]\mathbf{x}(k)$$

We choose $p_1 = 5$ and obtain the stabilizing control

$$u(k) = -\frac{1}{(1+0.4x_1(k))^2 + 5x_2^2(k)}[x_2(k) \quad 0.2+0.08x_1(k)]\mathbf{x}(k)$$

RESOURCES

Apostol, T. M., *Mathematical Analysis*, Addison-Wesley, 1975.

Fadali, M. S., Continuous drug delivery system design using nonlinear decoupling: A tutorial, *IEEE Trans. Biomedical Engineering*, 34(8):650-653, 1987.

Goldberg, S., *Introduction to Difference Equations*, Dover, 1986.

Kalman, R. E., and J. E. Bertram, Control system analysis and design via the "Second Method" of Lyapunov II: Discrete time systems, *J. Basic Engineering Trans, ASME*, 82(2):394-400, 1960.

Khalil, H. K., *Nonlinear Systems*, Prentice Hall, 2002.

Kuo, B. C., *Digital Control Systems*, Saunders, 1992.

LaSalle, J. P., *The Stability and Control of Discrete Processes*, Springer-Verlag, 1986.

Mickens, R. E., *Difference Equations*, Van Nostrand Reinhold, 1987.

Rugh, W. J., *Linear System Theory*, Prentice Hall, 1996.

Slotine, J.-J., *Applied Nonlinear Control*, Prentice Hall, 1991.

PROBLEMS

11.1 Discretize the following system:

$$\begin{bmatrix} \dot{x}_1 \\ \dot{x}_2 \end{bmatrix} = \begin{bmatrix} -3x_1 - x_1^2/(2x_2^2) \\ x_2^2/x_1 + 1 \end{bmatrix} + \begin{bmatrix} x_1^2 \\ 0 \end{bmatrix} u(t)$$

11.2 The equations for rotational maneuvering[2] of a helicopter are given by

$$\ddot{\theta} = -\frac{1}{2}\left(\frac{I_z - I_y}{I_x}\right)\dot{\psi}\sin(2\theta) - mgA\cos(\theta) + 2\left(\frac{I_z - I_y}{I_x}\right)\dot{\psi}\dot{\theta}$$

$$\frac{\sin(2\theta)}{(I_z + I_y) + (I_z - I_y)\cos(2\theta)}T_p$$

$$\dot{\psi} = \frac{1}{(I_z + I_y) + (I_z - I_y)\cos(2\theta)}T_y$$

[2]A. L. Elshafei and F. Karray, Variable structure-based fuzzy logic identification of a class of nonlinear systems, *IEEE Trans. Control Systems Tech.*, 13(4):646-653, 2005.

where

 I_x, I_y, and I_z = moments of inertia about the center of gravity
 m = total mass of the system
 g = acceleration due to gravity
 θ and ψ = pitch and yaw angles in radians
 T_p and T_y = pitch and yaw input torques

Obtain an equivalent linear discrete-time model for the system, and derive the equations for the torque in terms of the linear system inputs.

11.3 A single-link manipulator[3] with a flexible link has the equation of motion

$$I\ddot{\theta} + MgL\sin(\theta) - mgA\cos(\theta) + k(\theta - \psi) = 0$$

$$J\ddot{\psi} + k(\psi - \theta) = \tau$$

where

 L = distance from the shaft to the center of gravity of the link
 M = mass of the link
 I = moment of inertia of the link
 J = moment of inertia of the joint
 K = rotational spring constant for the flexible joint
 L = distance between the center of gravity of the link and the flexible joint
 θ and ψ = link and joint rotational angles in radians
 τ = applied torque

Obtain a discrete-time model of the manipulator (Figure P11.1).

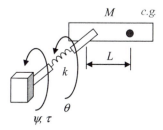

FIGURE P11.1

Schematic of a single-link manipulator.

11.4 Solve the nonlinear difference equation

$$[y(k+2)][y(k+1)]^{-2}[y(k)]^{1.25} = u(k)$$

with zero initial conditions and the input $u(k) = e^{-k}$.

[3]M. W. Spong and M. Vidyasagar, *Robot Dynamics and Control*, pp. 269-273, Wiley, 1989.

11.5 Determine the equilibrium point for the system

$$\begin{bmatrix} x_1(k+1) \\ x_2(k+1) \end{bmatrix} = \begin{bmatrix} -x_1(k)/9 + 2x_2^2(k) \\ -x_2(k)/9 + 0.4x_1^2(k) \end{bmatrix} + \begin{bmatrix} 0 \\ x_1(k) \end{bmatrix} u(k)$$

(a) Unforced
(b) With a fixed input $u_e = 1$

11.6 Use the Lyapunov approach to show that if the function $f(x)$ is a contraction, then the system $x(k + 1) = f[x(k)]$ is asymptotically stable.

11.7 Obtain a general expression for the eigenvalues of a 2×2 matrix, and use it to characterize the equilibrium points of the second-order system with the given state matrix

(a) $\begin{bmatrix} 0.9997 & 0.0098 \\ -0.0585 & 0.9509 \end{bmatrix}$

(b) $\begin{bmatrix} 3 & 1 \\ 1 & 2 \end{bmatrix}$

(c) $\begin{bmatrix} 0.3 & -0.1 \\ 0.1 & 0.2 \end{bmatrix}$

(d) $\begin{bmatrix} 1.2 & -0.4 \\ 0.4 & 0.8 \end{bmatrix}$

11.8 Determine the stability of the origin using the linear approximation for the system

$$x_1(k+1) = 0.2x_1(k) + 1.1x_2^3(k)$$

$$x_2(k+1) = x_1(k) + 0.1x_2(k) + 2x_1(k)x_2^2(k), \quad k = 0, 1, 2, \ldots$$

11.9 Verify the stability of the origin using the Lyapunov approach, and estimate the rate of convergence to the equilibrium

$$x_1(k+1) = 0.1x_1(k)x_2(k) - 0.05x_2^2(k)$$

$$x_2(k+1) = -0.5x_1(k)x_2(k) + 0.05x_2^3(k), \quad k = 0, 1, 2, \ldots$$

11.10 Show that the convergence of the trajectories of a nonlinear discrete-time system $x(k+1) = f[x(k)]$ to a known nominal trajectory $x^*(k)$ is equivalent to the stability of the dynamics of the tracking error $e(k) = x(k) - x^*(k)$.

11.11 Prove that the scalar system

$$x(k+1) = -ax^3(k)$$

is locally asymptotically stable in the region $|x(k)| \le 1/\sqrt{a}$.

11.12 Use Lyapunov stability theory to investigate the stability of the system

$$x_1(k+1) = \frac{ax_1(k)}{a + bx_2^2(k)}$$

$$x_2(k+1) = \frac{bx_2(k)}{b + ax_1^2(k)}, \quad a > 0, b > 0$$

11.13 Use the Lyapunov approach to determine the stability of the discrete-time linear time-invariant systems

(a) $\begin{bmatrix} 0.3 & -0.1 \\ 0.1 & 0.22 \end{bmatrix}$

(b) $\begin{bmatrix} 0.3 & -0.1 & 0 \\ 0.1 & 0.22 & 0.2 \\ 0.4 & 0.2 & 0.1 \end{bmatrix}$

11.14 Show that the origin is an unstable equilibrium for the system

$$x_1(k+1) = -1.4x_1(k) + 0.1x_2^2(k)$$

$$x_2(k+1) = 1.5x_2(k)(0.1x_1(k)+1), \quad k = 0, 1, 2, \ldots$$

11.15 Estimate the domain of attraction of the system

$$\begin{bmatrix} x_1(k+1) \\ x_2(k+1) \end{bmatrix} = \begin{bmatrix} 0.2 & 0.3 \\ -0.4 & 0.5 \end{bmatrix} \begin{bmatrix} x_1(k) \\ x_2(k) \end{bmatrix} + \begin{bmatrix} 0.3x_2^2(k) \\ 0.36x_1^2(k) \end{bmatrix}$$

11.16 Design a controller to stabilize the origin for the system

$$x_1(k+1) = 0.4x_1(k) + 0.5x_2(k) + x_2^2(k)u(k)$$

$$x_2(k+1) = 0.1x_1(k) + 0.2x_2(k) + [x_2(k) + x_1(k)]u(k), \quad k = 0, 1, 2, \ldots$$

COMPUTER EXERCISES

11.17 Write a MATLAB program to generate phase plane plots for a discrete-time second-order linear time-invariant system. The function should accept the eigenvalues of the state matrix and the initial conditions needed to generate the plots.

11.18 Design a controller for the nonlinear mechanical system described in Example 11.15 with the nonlinear damping $b(\dot{x}) = 0.25\dot{x}^5$, the nonlinear spring $c(x) = 0.5x + 0.02x^3$, $T = 0.02$ s, and the desired eigenvalues for the linear design equal to $\{0.2 \pm j0.1\}$. Determine the value of the reference input for a steady-state position of unity, and simulate the system using Simulink.

11.19 Design a stabilizing digital controller with a sampling period $T = 0.01$ s for a single-link manipulator using extended linearization; then simulate the system with your design. The equation of motion of the manipulator is given by
$$\ddot{\theta} + 0.01\sin(\theta) + 0.01\dot{\theta} + 0.001\theta^3 = \tau.$$

Assign the eigenvalues of the discrete-time linear system to $\{0.6 \pm j0.3\}$. *Hint:* Use Simulink for your simulation, and use a **ZOH block**, or a **discrete filter** block with both the numerator and denominator set to 1 for sampling.

Practical Issues

Successful practical implementation of digital controllers requires careful attention to several hardware and software requirements. In this chapter, we discuss the most important of these requirements and their influence on controller design. We analyze the choice of the sampling frequency in more detail (already discussed in Section 2.9) in the presence of antialiasing filters and the effects of quantization, rounding, and truncation errors. In particular, we examine the effective implementation of a proportional-integral-derivative (PID) controller. Finally, we examine changing the sampling rate during control operation as well as output sampling at a slower rate than that of the controller.

Objectives

After completing this chapter, the reader will be able to do the following:

1. Write pseudocode to implement a digital controller.
2. Select the sampling frequency in the presence of antialiasing filters and quantization errors.
3. Implement a PID controller effectively.
4. Design a controller that addresses changes in the sampling rate during control operation.
5. Design a controller with faster input sampling than output sampling.

12.1 DESIGN OF THE HARDWARE AND SOFTWARE ARCHITECTURE

The designer of a digital control system must be mindful of the fact that the control algorithm is implemented as a software program that forms part of the control loop. This introduces factors that are not present in analog control loops. This section discusses several of these factors.

12.1.1 Software Requirements

During the design phase, designers make several simplifying assumptions that affect the implemented controller. They usually assume uniform sampling with negligible delay due to the computation of the control variable. Thus, they assume no delay between the sampling instant and the instant at which the computed control value is applied to the actuator. This instantaneous execution assumption is not realistic because the control algorithm requires time to compute its output (see Figure 12.1). If the computational time is known and constant, we can use the modified z-transform (see Section 2.7) to obtain a more precise discrete model. However, the computational time of the control algorithm can vary from one sampling period to the next. The variation in the computational delay is called the **control jitter**. For example, control jitter is present when the controller implementation utilizes a switching mechanism.

Digital control systems have additional requirements such as data storage and user interface, and their proper operation depends not only on the correctness of their calculations but also on the time at which their results are available. Each task must satisfy either a start or completion timing constraint. In other words, a digital control system is a **real-time system**. To implement a real-time system, we need a real-time operating system that can provide capabilities such as multitasking, scheduling, and intertask communication, among others. In a multitasking environment, the value of the control variable must be computed and applied over each sampling interval regardless of other tasks necessary for the operations of the overall control system. Hence, the highest priority is assigned to the computation and application of the control variable.

Clearly, the implementation of a digital control system requires skills in software engineering and computer programming. There are well-known programming guidelines that help minimize execution time and control jitter for the control algorithm. For example, **if-then-else** and **case** statements must be avoided as much as possible because they can lead to paths of different lengths and, consequently, paths with different execution times. The states of the control variable must be updated after the application of the control variable. Finally, the software must be tested to ensure that no errors occur. This is known as **software verification**. In particular, the execution time and the control jitter must be measured to verify that they can be neglected relative to the sampling period, and memory usage must be analyzed to verify that it does not exceed the available memory. Fortunately, software tools are available to make such analysis possible.

EXAMPLE 12.1

Write **pseudocode** that implements the following controller:

$$C(z) = \frac{U(z)}{E(z)} = \frac{10.5z - 9.5}{z - 1}$$

Then propose a possible solution to minimize the **execution time** (Figure 12.1).

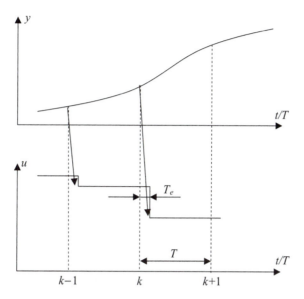

FIGURE 12.1

Execution time T_e for the computation of the control variable value with respect to the sampling interval T.

Solution

The difference equation corresponding to the controller transfer function is

$$u(k) = u(k-1) + 10.5e(k) - 9.5e(k-1)$$

This control law can be implemented by writing the following code:

```
function controller
% This function is executed during each sampling period
% r is the value of the reference signal
% u1 and e1 are the values of the control variable and of the control
% error respectively for the previous sampling period
y=read_ADC(ch0)   % Read the process output from channel 0 of the ADC
e=r-y;             % Compute the tracking error
u=u1+10.5*e-9.5*e1;   % Compute the control variable
u1=u;        % Update the control variable for the next sampling period
e1=e;        % Update the tracking error for the next sampling period
write_DAC(ch0,u); % Output the control variable to channel 0 of the DAC
```

To decrease the execution time, two tasks are assigned different priorities (using a real-time operating system).

Task 1 (Maximum Priority)
```
y=read_ADC(ch0)  % Read the process output from channel 0 of the ADC
e=r-y;           % Compute the tracking error
u=u1+10.5*e-9.5*e1; % Compute the control variable
write_DAC(ch0,u);   % Write the control variable to channel 0 of the DAC
```

Task 2
```
u1=u;      % Update the control variable for the next sampling period
e1=e;      % Update the tracking error for the next sampling period
```

12.1.2 Selection of ADC and DAC

The ADC and DAC must be sufficiently fast for negligible conversion time relative to the sampling period. In particular, the conversion delay of the ADC creates a negative phase shift, which affects the phase margin of the system and must be minimized to preserve the stability margins of the system. In addition, the word length of the ADC affects its conversion time. With the conversion time provided by standard modern analog-to-digital converters, this is not a significant issue in most applications.

The choice of the ADC and DAC word length is therefore mainly determined by the **quantization** effects. Typically, commercial ADCs and DACs are available in the range of 8 to 16 bits. An 8-bit ADC provides a **resolution** of a 1 in 2^8, which corresponds to an error of 0.4 percent, whereas a 16-bit ADC gives an error of 0.0015 percent.

Clearly, the smaller the ADC resolution, the better the performance, and therefore a 16-bit ADC should be preferred. However, the cost of the component increases as the word length increases, and the presence of noise might render the presence of a high number of bits useless in practical applications. For example, if the sensor has a 5 mV noise and a 5 V range, there is no point in employing an ADC with more than 10 bits because its resolution of 1 in 2^{10} corresponds to an error of 0.1%, which is equal to the noise level. The DAC resolution is usually chosen equal to the ADC resolution, or slightly higher, to avoid introducing another source of quantization error. Once the ADC and DAC resolution have been selected, the resolution of the reference signal representation must be the same as that of the ADC and DAC. In fact, if the precision of the reference signal is higher than the ADC resolution, the control error will never go to zero and therefore a limit cycle will occur.

Another important design issue, especially for MIMO control, is the choice of **data acquisition system**. Ideally, we would use an analog-to-digital converter for each channel to ensure simultaneous sampling as shown in Figure 12.2(a). However, this approach can be prohibitively expensive, especially for a large number of channels. A more economical approach is to use a **multiplexer** (MUX), with each channel sampled in sequence and the sampled value sent to a master ADC (Figure 12.2(b)). If we assume that the sampled signals change relatively

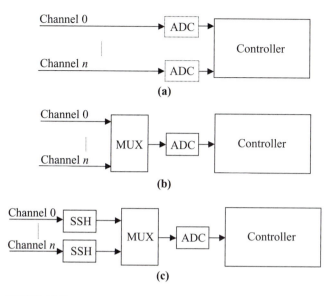

FIGURE 12.2

Choices for the data acquisition system. (a) Separate ADC for each channel. (b) Multiplexer with single master ADC. (c) Simultaneous sample-and-hold, multiplexer, and single master ADC.

slowly, small changes in the sampling instant do not result in significant errors and the measured variables appear to be sampled simultaneously. If this assumption is not valid, a more costly **simultaneous sample-and-hold** (SSH) system can be employed, as depicted in Figure 12.2(c). The system samples the channels simultaneously, then delivers the sampled data to the ADC through a multiplexer. In recent years, the cost of analog-to-digital converters has decreased significantly, and the use of a simultaneous sample-and-hold system has become less popular as using an ADC for each channel has become more affordable.

We discuss other considerations related to the choice of the ADC and DAC components with respect to the sampling period in Section 12.2.2.

12.2 CHOICE OF THE SAMPLING PERIOD

In Section 2.9, we showed that the choice of the sampling frequency must satisfy the sampling theorem and is based on the effective bandwidth ω_m of the signals in the control systems. This leads to relation (2.66), where we choose the sampling frequency in the range between 5 and 10 times the value of ω_m. We now discuss the choice of sampling frequency more thoroughly, including the effects of antialiasing filters as well as the effects of quantization, rounding, and truncation errors.

12.2.1 Antialiasing Filters

If the sampling frequency does not satisfy the sampling theorem (i.e., the sampled signal has frequency components greater than twice the sampling frequency), then the sampling process creates new frequency components (see Figure 2.9). This phenomenon is called **aliasing** and must obviously be avoided in a digital control system. Hence, the continuous signal to be sampled must not include significant frequency components greater than the Nyquist frequency $\omega_s/2$.

For this purpose, it is recommended to low-pass filter the continuous signal before sampling, especially in the presence of high-frequency noise. The analog low-pass filter used for this purpose is known as the **antialiasing filter**. The antialiasing filter is typically a simple first-order RC filter, but some applications require a higher-order filter such as a Butterworth or a Bessel filter. The overall control scheme is shown in Figure 12.3.

Because a low-pass filter can slow down the system by attenuating high-frequency dynamics, the cut-off frequency of the low-pass filter must be higher than the bandwidth of the closed-loop system so as not to degrade the transient response. A rule of thumb is to choose the filter bandwidth equal to a constant times the bandwidth of the closed-loop system. The value of the constant varies depending on economic and practical considerations. For a conservative but more expensive design, the cut-off frequency of the low-pass filter can be chosen as 10 times the bandwidth of the closed-loop system to minimize its effect on the control system dynamics, and then the sampling frequency can be chosen 10 times higher than the filter cut-off frequency so that there is a sufficient attenuation above the Nyquist frequency. Thus, the sampling frequency is 100 times the bandwidth of the closed-loop system. To reduce the sampling frequency, and the associated hardware costs, it is possible to reduce the antialiasing filter cut-off frequency. In the extreme case, we select the cut-off frequency slightly higher than the closed-loop bandwidth. For a low-pass filter with a high roll-off (i.e., a high-order filter), the sampling frequency is chosen as five times the closed-loop bandwidth. In summary, the sampling period T can be chosen (as described in Section 2.9) in general as

$$5\omega_b \le \frac{2\pi}{T} \le 100\omega_b \tag{12.1}$$

where ω_b is the bandwidth of the closed-loop system.

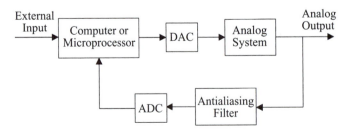

FIGURE 12.3

Control scheme with an antialiasing filter.

If the phase delay introduced by the antialiasing filter is significant, then (12.1) may not yield good results and the filter dynamics must be considered when selecting a dynamic model for the design phase.

EXAMPLE 12.2

Consider a 1 Hz sinusoidal signal of unity amplitude with an additive 50 Hz sinusoidal noise. Verify the effectiveness of an antialiasing filter for the signal sampled at a frequency of 30 Hz.

Solution

The noisy analog signal to be sampled is shown in Figure 12.4(a). If this signal is sampled at 30 Hz without an antialiasing filter, the result is shown in Figure 12.4(b). Figure 12.4(c)

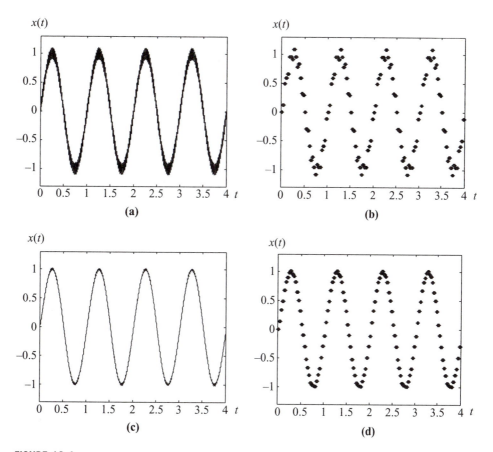

FIGURE 12.4

Effect of an antialiasing filter on the analog and sampled signals described in Example 12.2. (a) Noisy analog signal. (b) Signal sampled at 30 Hz with no antialiasing filter. (c) Filtered analog signal with a first-order antialiasing filter with cut-off frequency equal to 10 Hz. (d) Sampled signal with a first-order antialiasing filter with cut-off frequency equal to 10 Hz.

shows the filtered analog signal with a first-order antialiasing filter with cut-off frequency equal to 10 Hz, and the resulting sampled signal is shown in Figure 12.4(d). The sampled sinusoidal signal is no longer distorted because of the use of the antialiasing filter; however, a small phase delay emerges because of the filter dynamics.

12.2.2 Effects of Quantization Errors

As discussed in Section 12.1, the design of the overall digital control system includes the choice of the ADC and DAC components. In this context, the effects of the **quantization** due to ADC **rounding** or **truncation** (Figure 12.5) are considered in the selection of the sampling period. The noise due to quantization can be modeled as a uniformly distributed random process with the following mean and variance values (denoted respectively by $\bar{e}$ and σ_e^2) in the two cases:

$$\text{Rounding: } \quad \bar{e} = 0 \quad \sigma_e^2 = \frac{q^2}{12} \tag{12.2}$$

$$\text{Truncation: } \quad \bar{e} = \frac{q}{2} \quad \sigma_e^2 = \frac{q^2}{12} \tag{12.3}$$

where q is the quantization level—namely, the range of the ADC divided by 2^n, and n is the number of bits.

Obviously, the effects of the quantization error increase as q increases and the resolution of the ADC decreases. To evaluate the influence of q on quantization noise and on the sampling period, we consider a proportional feedback digital controller with a gain K applied to the analog first-order lag

$$G(s) = \frac{1}{\tau s + 1}$$

The z-transfer function of the DAC (zero-order hold), analog subsystem, and ADC (ideal sampler) cascade has the discrete time state–space model

$$x(k+1) = \left[e^{-T/\tau} \right] x(k) + \left[1 - e^{-T/\tau} \right] u(k)$$

$$y(k) = x(k)$$

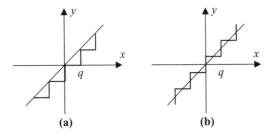

(a)   (b)

FIGURE 12.5

Quantization characteristics of the ADC. (a) Truncating ADC. (b) Rounding ADC.

For a truncating ADC as the only source of noise with zero set-point value, the control action is

$$u(k) = -K[(y(k) + w(k))]$$

where w is the quantization noise governed by (12.3) and is subtracted with no loss of generality. The state–space model of the closed-loop system is

$$x(k+1) = [e^{-T/\tau} - K(1 - e^{-T/\tau})]x(k) + K[1 - e^{-T/\tau}]w(k)$$

$$y(k) = x(k)$$

For zero initial conditions, the solution of the difference equation is

$$x(k) = \sum_{i=0}^{k-1} [e^{-T/\tau} - K(1 - e^{-T/\tau})]^{k-i-1} K[1 - e^{-T/\tau}]w(k)$$

$$y(k) = x(k)$$

the mean value of the output noise is

$$m_y(k) = E\{x(k)\} = \sum_{i=0}^{k-1} [e^{-T/\tau} - K(1 - e^{-T/\tau})]^{k-i-1} K[1 - e^{-T/\tau}]w(k)$$

$$= K[1 - e^{-T/\tau}]\left(\frac{q}{2}\right)\sum_{i=0}^{k-1} [e^{-T/\tau} - K(1 - e^{-T/\tau})]^{k-i-1}$$

where $E\{.\}$ denotes the expectation. If $T/\tau \ll 1$, we use the linear approximation of the exponential terms $e^{-T/\tau} \approx 1 - T/\tau$ to obtain

$$m_y(k) = K\left(\frac{T}{\tau}\right)\left(\frac{q}{2}\right)\sum_{i=0}^{k-1}\left[1 - \left(\frac{T}{\tau}\right)(1+K)\right]^{k-i-1}$$

We recall the relationship

$$\frac{1}{1-a} = \sum_{j=0}^{\infty} a^k, \quad |a| < 1$$

and take the limit as $k \to \infty$ to obtain the steady-state mean

$$m_y(k) = K\left(\frac{T}{\tau}\right)\left(\frac{q}{2}\right)\frac{1}{\left(\frac{T}{\tau}\right)(1+K)} = \frac{K}{1+K}\left(\frac{q}{2}\right)$$

For small gain values, we have

$$m_y = \frac{K}{1+K}\left(\frac{q}{2}\right) \approx K\left(\frac{q}{2}\right), \quad K \ll 1$$

For large gain values, we have $m_y \approx \dfrac{q}{2}$.

    We observe that the mean value is independent of the sampling period, is linear in the controller gain for small gains, and is almost independent of the gain for large gains. In any case, the worst value we can achieve is half the quantization interval.

The variance of the output is

$$\sigma_y^2 = E\{x^2(k)\} - E^2\{x(k)\}$$

Using the expression for the mean and after some tedious algebraic manipulations, we can show that

$$\sigma_y^2 = \frac{K^2(1-e^{-T/\tau})^2}{(K+1)[(1-K)2Ke^{-T/\tau} - (K+1)e^{-2T/\tau}]}\left(\frac{q^2}{12}\right)$$

If $T/\tau \ll 1$, we use the linear approximations of the exponential terms $e^{-T/\tau} \approx 1 - T/\tau$ and $e^{-2T/\tau} \approx 1 - 2T/\tau$, the output variance simplifies to

$$\sigma_y^2 = \frac{K^2}{K+1}\frac{T}{2\tau}\left(\frac{q^2}{12}\right) \approx \begin{cases} \dfrac{KT}{2\tau}\left(\dfrac{q^2}{12}\right), & K \gg 1 \\[2ex] K^2\dfrac{\tau}{2\tau}, & K \ll 1 \end{cases}$$

Unlike the output mean, the output variance is linear in the sampling period and linear in the controller gain for large gains. Thus, the effect of the quantization noise can be reduced by decreasing the sampling period, once the ADC has been selected. We conclude that decreasing the sampling period has beneficial effects with respect to both aliasing and quantization noise. However, decreasing the sampling period requires more expensive hardware and may aggravate problems because of the finite-word representation of parameters.

We illustrate this fact with a simple example. Consider an analog controller with poles at $s_1 = -1$ and $s_2 = -10$. For a digital implementation of the analog controller with sampling period $T = 0.001$, we have the digital controller poles as given by (6.3) as

$$z_1 = e^{-0.001} \cong 0.9990 \quad \text{and} \quad z_2 = e^{-0.01} \cong 0.9900$$

If we truncate the two values after two significant digits, we have $z_1 = z_2 = 0.99$; that is, the two poles of the digital controller become identical and correspond to two identical poles of the analog controller at

$$s_1 = s_2 = \frac{1}{T}\ln z_1 = \frac{1}{T}\ln z_2 \cong -10.05$$

For the longer sampling period $T = 0.1$, truncating after two significant digits gives the poles $z_1 = 0.90$ and $z_2 = 0.36$, which correspond to $s_1 \approx -1.05$ and $s_2 = -10.21$. This shows that a much better approximation is obtained with the longer sampling period.

12.2.3 Phase Delay Introduced by the ZOH

As shown in Section 3.3, the frequency response of the zero-order hold can be approximated as

$$G_{ZOH}(j\omega) = \frac{1-e^{-j\omega T}}{j\omega} \approx e^{-j\omega T/2}$$

This introduces an additional delay in the control loop approximately equal to half of the sampling period. The additional delay reduces the stability margins of the control system, and the reduction is worse as the sampling period is increased. This imposes an upper bound on the value of the sampling period T.

EXAMPLE 12.3

Let ω_c be the gain crossover frequency of an analog control system. Determine the maximum value of the sampling period for a digital implementation of the controller that decreases the phase margin by no more than 5 degrees.

Solution

Because of the presence of the ZOH, the phase margin decreases by $\omega_c T/2$, which yields the constraint

$$\omega_c \frac{T}{2} \le 5 \frac{\pi}{180}$$

or equivalently, $T \le 0.1745\, \omega_c$.

EXAMPLE 12.4

Consider the tank control system described in Example 2.1 with the transfer function

$$G(s) = \frac{1.2}{20s+1} e^{-1.5s}$$

and the PI controller

$$C(s) = 7\frac{20s+1}{20s}$$

Let the actuator and the sensor signals be in the range 0 to 5 V with a sensor gain of 0.169 V/cm. Design the hardware and software architecture of the digital control system.

Solution

The gain crossover frequency of the analog control system as obtained using MATLAB is $\omega_c = 0.42$ rad/s, and the phase margin is $\varphi_m = 54$ deg. We select a sampling period $T = 0.2$ s and use a second-order Butterworth antialiasing filter with cut-off frequency of 4 rad/s. The transfer function of the Butterworth filter is

$$F(s) = \frac{64}{s^2 + 11.31s + 64}$$

The antialiasing filter does not change the gain crossover frequency significantly. The phase margin is reduced to $\varphi_m = 49.6°$, which is acceptable. At the Nyquist frequency of $\pi/T = 10.47$ rad/s, the antialiasing filter decreases the magnitude of the noise by more than

40 dB. The phase delay introduced by the zero-order hold is $(\omega_c T/2) \times 180/\pi = 3.6°$, which is also acceptable. We select a 12-bit ADC with a quantization level of 1.2 mV, which corresponds to a quantization error in the fluid level of 0.07 mm. We also select a 12-bit DAC. Because the conversion time is on the order of microseconds, this does not influence the overall design.

12.3 CONTROLLER STRUCTURE

Section 12.2 demonstrates how numerical errors can affect the performance of a digital controller. To reduce numerical errors and mitigate their effects, we must select an appropriate **controller structure** for implementation. To examine the effect of controller structure on errors, consider the controller

$$C(z) = \frac{N(z, \mathbf{q})}{D(z, \mathbf{q})} = \frac{\mathbf{b}^T \mathbf{z}_m}{\mathbf{a}^T \mathbf{z}_n}$$

$$\mathbf{a} = [a_0(\mathbf{q}) \quad a_1(\mathbf{q}) \quad \cdots \quad a_n(\mathbf{q})]^T \qquad (12.4)$$

$$\mathbf{b} = [b_0(\mathbf{q}) \quad b_1(\mathbf{q}) \quad \cdots \quad b_m(\mathbf{q})]^T$$

$$\mathbf{z}_l = [1 \quad z \quad \cdots \quad z^l]$$

where $\mathbf{q}$ is an $l \times 1$ vector of controller parameters. If the nominal parameter vector is $\mathbf{q}^*$ and the corresponding poles are p_i^*, $I = 1, 2, \ldots, n$, for an nth-order controller, then the nominal characteristic equation of the controller is

$$D(p_i^*, \mathbf{q}^*) = 0, \quad i = 1, 2, \ldots, n \qquad (12.5)$$

In practice, the parameter values are only approximately implemented and the characteristic equation of the system is

$$D(z, \mathbf{q}) \approx D(p_i^*, \mathbf{q}^*) + \left.\frac{\partial D}{\partial z}\right]_{z=p_i^*} \delta p_i^* + \left.\frac{\partial D}{\partial \mathbf{a}}\right]_{z=p_i^*}^T \delta \mathbf{a}$$

$$= \left.\frac{\partial D}{\partial z}\right]_{z=p_i^*} \delta p_i^* + \left.\frac{\partial D}{\partial \mathbf{a}}\right]_{z=p_i^*}^T \delta \mathbf{a} \approx 0 \qquad (12.6)$$

$$\frac{\partial D}{\partial \mathbf{a}} = \left[\frac{\partial D}{\partial a_0} \quad \frac{\partial D}{\partial a_1} \quad \cdots \quad \frac{\partial D}{\partial a_n}\right]^T$$

In terms of the controller parameters, the perturbed characteristic equation is

$$D(z, \mathbf{q}) \approx \left.\frac{\partial D}{\partial z}\right]_{z=p_i^*} \delta p_i^* + \left.\frac{\partial D^T}{\partial \mathbf{a}} \frac{\partial \mathbf{a}}{\partial \mathbf{q}}\right]_{\mathbf{q}=\mathbf{q}^*} \delta \mathbf{q} \approx 0$$

$$\qquad (12.7)$$

$$\frac{\partial \mathbf{a}}{\partial \mathbf{q}} = \left[\frac{\partial a_i}{\partial q_j}\right]$$

We solve for parameter perturbations in the location of the ith pole

$$\delta p_i^* = -\frac{\partial D^T}{\partial \mathbf{a}} \frac{\partial \mathbf{a}}{\partial \mathbf{q}}\bigg|_{\mathbf{q}=\mathbf{q}^*} \frac{\delta \mathbf{q}}{\frac{\partial D}{\partial z}\big|_{z=p_i^*}} \qquad (12.8)$$

To characterize the effect of a particular parameter on the ith pole, we can set the perturbations in all other parameters to zero to obtain

$$\delta p_i^* = -\frac{\partial D^T}{\partial \mathbf{a}} \frac{\partial \mathbf{a}}{\partial q_i}\bigg|_{\mathbf{q}=\mathbf{q}^*} \frac{\delta q_i}{\frac{\partial D}{\partial z}\big|_{z=p_i^*}} \qquad (12.9)$$

$$\frac{\partial \mathbf{a}}{\partial q_i} = \begin{bmatrix} \dfrac{\partial a_0}{\partial q_i} & \dfrac{\partial a_1}{\partial q_i} & \cdots & \dfrac{\partial a_n}{\partial q_i} \end{bmatrix}^T$$

This concept is explained by the following example. Consider the following second-order general controller:

$$C(z) = \frac{(a+b)z - (ap_2 + bp_1)}{z^2 - (p_1 + p_2)z + p_1 p_2}$$

Let $a_1 = -(p_1 + p_2)$ and $a_0 = p_1 p_2$ denote the nominal coefficients of the characteristic equation of the controller, and write characteristic equation as

$$D(z, a_1, a_0) = 0$$

When a coefficient λ_i is changed (due to numerical errors) to $\lambda_i + \delta\lambda_i$, then the position of a pole is changed according to the following equation (where second- and higher-order terms are neglected):

$$D(p_i + \delta p_i, \lambda_i + \delta\lambda_i) = D(p_i, \lambda_i) + \frac{\partial D}{\partial z}\bigg|_{z=p_i} \delta p_i + \frac{\partial D}{\partial \lambda_i} \delta\lambda_i$$

That is,

$$\frac{\delta p_i}{\delta \lambda_i} = -\frac{\dfrac{\partial D}{\partial \lambda_i}}{\dfrac{\partial D}{\partial z}\big|_{z=p_i}}$$

Now we have the partial derivatives

$$\frac{\partial D}{\partial z} = 2z + a_1 \qquad \frac{\partial D}{\partial a_1} = z \qquad \frac{\partial D}{\partial a_0} = 1$$

and therefore

$$\frac{\delta p_1}{\delta a_1} = -\frac{p_1}{2p_1 - (p_1 + p_2)} = \frac{p_1}{p_1 - p_2} \qquad \frac{\delta p_2}{\delta a_1} = -\frac{p_2}{2p_2 - (p_1 + p_2)} = \frac{p_2}{p_2 - p_1}$$

$$\frac{\delta p_1}{\delta a_0} = -\frac{1}{2p_1 - (p_1 + p_2)} = \frac{1}{p_1 - p_2} \qquad \frac{\delta p_2}{\delta a_0} = -\frac{1}{2p_2 - (p_1 + p_2)} = \frac{1}{p_2 - p_1}$$

Thus, the controller is most sensitive to changes in the last coefficient of the characteristic equation, and its sensitivity increases when the poles are close. This concept can be generalized to high-order controllers. Note that decreasing the sampling period draws the poles closer when we start from an analog design. In fact, for $T \to 0$ we have that

$$p_z = e^{p_s T} \to 1$$

independently on the value of the analog pole p_s.

These problems can be avoided by writing the controller in an equivalent **parallel form**:

$$C(z) = \frac{a}{z - p_1} + \frac{b}{z - p_2}$$

We can now analyze the sensitivity of the two terms of the controller separately to show that the sensitivity is equal to one, which is less than the previous case if the poles are close. Thus, the parallel form is preferred. Similarly, the parallel form is also found to be superior to the **cascade form**:

$$C(z) = \frac{(a+b)z - (ap_2 + bp_1)}{z - p_1} \frac{1}{z - p_2}$$

EXAMPLE 12.5

Write the difference equations in direct, parallel, and cascade forms for the system

$$C(z) = \frac{U(z)}{E(z)} = \frac{z - 0.4}{z^2 - 0.3z + 0.02}$$

Solution

The difference equation corresponding to the direct form of the controller is

$$u(k) = 0.3u(k-1) - 0.02u(k-2) + e(k-1) - 0.4e(k-2)$$

For the parallel form, we obtain the partial fraction expansion of the transfer function

$$C(z) = \frac{U(z)}{E(z)} = \frac{-2}{z - 0.2} + \frac{3}{z - 0.1}$$

This is implemented using the following difference equations:

$$u_1(k) = 0.2u(k-1) - 2e(k-1)$$

$$u_2(k) = 0.1u(k-1) + 3e(k-1)$$

$$u(k) = u_1(k) + u_2(k)$$

Finally, for the cascade form we have

$$C(z) = \frac{U(z)}{E(z)} = \frac{z-0.4}{z-0.2} \frac{1}{z-0.1} = \frac{X(z)}{E(z)} \frac{U(z)}{X(z)}$$

which is implemented using the difference equations

$$x(k) = 0.2x(k-1) + e(k) - 0.4e(k-1)$$

$$u(k) = 0.1u(k-1) + x(k-1)$$

12.4 PID CONTROL

In this section, we discuss several critical issues related to the implementation of PID controllers. Rather than providing an exhaustive discussion, we highlight a few problems and solutions directly related to digital implementation.

12.4.1 Filtering the Derivative Action

The main problem with derivative action is that it amplifies the high-frequency noise and may lead to a noisy control signal that can eventually cause serious damage to the actuator. It is therefore recommended that one filter the overall control action with a low-pass filter or, alternatively, filter the derivative action. In this case, the controller transfer function in the analog case can be written as (see (5.20))

$$C(s) = K_p \left(1 + \frac{1}{T_i s} + \frac{T_d s}{1 + \frac{T_d}{N} s} \right)$$

where N is a constant in the interval $[1, 33]$, K_p is the proportional gain, T_i is the integral time constant, and T_d is the derivative time constant.

In the majority of cases encountered in practice, the value of N is in the smaller interval $[8, 16]$. The controller transfer function can be discretized as discussed in Chapter 6. A useful approach in practice is to use the forward differencing approximation for the integral part and the backward differencing approximation for the derivative part. This gives the discretized controller transfer function

$$C(z) = K_p \left(1 + \frac{T}{T_i(z-1)} + \frac{T_d}{T + \frac{T_d}{N}} \cdot \frac{z-1}{z - \frac{T_d}{NT + T_d}} \right) \tag{12.10}$$

which can be simplified to

$$C(z) = \frac{K_0 + K_1 z + K_2 z^2}{(z-1)(z-\gamma)} \tag{12.11}$$

where

$$K_0 = K_p \left(\frac{T_d}{NT + T_d} - \frac{T}{T_i} \frac{T_d}{NT + T_d} + \frac{NT_d}{NT + T_d} \right)$$

$$K_1 = -K_p \left(1 + \frac{T_d}{NT + T_d} - \frac{T}{T_i} + 2 \frac{NT_d}{NT + T_d} \right)$$

$$K_2 = K_p \left(1 + \frac{NT_d}{NT + T_d} \right)$$

$$\gamma = \frac{T_d}{NT + T_d} = \frac{1}{N(T/T_d) + 1}$$

(12.12)

EXAMPLE 12.6

Select a suitable derivative filter parameter value N for the PID controller described in Example 5.9 with a sampling period $T = 0.01$, and obtain the corresponding discretized transfer function of (12.11).

Solution

The analog PID parameters are $K_p = 2.32$, $T_i = 3.1$, and $T_d = 0.775$. We select the filter parameter $N = 20$ and use (12.4) to obtain $K_0 = 38.72$, $K_1 = -77.92$, $K_2 = 39.20$, and $\gamma = 0.79$.

The discretized PID controller expression is therefore

$$C(z) = \frac{K_0 + K_1 z + K_2 z^2}{(z-1)(z-\gamma)} = \frac{38.72 - 77.92z + 39.20z^2}{(z-1)(z-0.79)}$$

12.4.2 Integrator Windup

Most control systems are based on linear models and design methodologies. However, every actuator has a saturation nonlinearity as in the control loop shown in Figure 12.6, which affects both the analog and digital control. The designer must consider the nonlinearity at the design stage to avoid performance degrada-

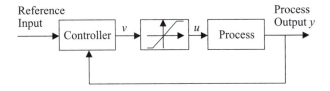

FIGURE 12.6

Control loop with actuator saturation.

tion. A common phenomenon related to the presence of actuator saturation is known as **integrator windup**. If not properly handled, it may result in a step response with a large overshoot and settling time.

In fact, if the control variable attains its maximum (or minimum) limit when a step input is applied, the control becomes independent of the feedback and the system behaves as in the open-loop case. Thus, the control error decreases more slowly than in the absence of saturation, and the integral term becomes large or **winds up**. The large integral term causes saturation of the control variable even after the process output attains its reference value and a large overshoot occurs.

Many solutions have been devised to compensate for integrator windup and retain linear behavior. The rationale of these anti-windup techniques is to design the control law disregarding the actuator nonlinearity and then compensate for the detrimental effects of integrator windup.

One of the main anti-windup techniques is the so-called **conditional integration**, which keeps the integral control term constant when a specified condition is met. For example, the integral control term is kept constant if the integral component of the computed control exceeds a given threshold specified by the designer. Alternatively, the integral controller is kept constant if the actuator saturates with the control variable and the control error having the same sign (i.e., if $u \cdot e > 0$). The condition $u \cdot e > 0$ implies that the control increases rather than corrects the error due to windup and that integral action should not increase.

On the other hand, a positive saturation with $u \cdot e < 0$ means that the error is negative and therefore the integral action is decreasing and there is no point in keeping it constant. The same reasoning can be easily applied in case of a negative saturation. Thus, the condition avoids inhibiting the integration when it helps to push the control variable away from saturation.

An alternative technique is **back-calculation**, which reduces (increases) the integral control when the maximum (minimum) saturation limit is attained by adding to the integrator a term proportional to the difference between the computed value of the control signal v and its saturated value u. In other words, the integral value $I(k)$ is determined by

$$I(k) = I(k-1) + \frac{K_p}{T_i} e(k) - \frac{1}{T_t}(v(k) - u(k)) \qquad (12.13)$$

where T_t is the **tracking time constant**. This is a tuning parameter that determines the rate at which the integral term is reset.

EXAMPLE 12.7

Consider the digital proportional-integral (PI) controller transfer function

$$C(z) = \frac{1.2z - 1.185}{z - 1}$$

with $K_p = 1.2$, $T_i = 8$, sampling period $T = 0.1$, and the process transfer function

$$G(s) = \frac{1}{10s+1}e^{-5s}$$

Obtain the step response of the system (1) if the saturation limits of the actuator are $u_{min} = -1.2$ and $u_{max} = 1.2$, and (2) with no actuator saturation. Compare and discuss the two responses, then use back-calculation to reduce the effect of windup on the step response.

Solution

For the system with actuator saturation and no anti-windup strategy, the process output y, controller output v, and process input u are shown in Figure 12.7. We observe that the actuator output exceeds the saturation level even when the process output attains its reference value, which leads to a large overshoot and settling time. The response after the removal of saturation nonlinearity is shown in Figure 12.8. The absence of saturation results in a faster response with fast settling to the desired steady-state level.

The results obtained by applying back-calculation with $T_t = T_i = 8$ are shown in Figure 12.9. The control variable is kept at a much lower level and this helps avoid the overshoot almost entirely. The response is slower than the response with no saturation but is significantly faster than the response with no anti-windup strategy.

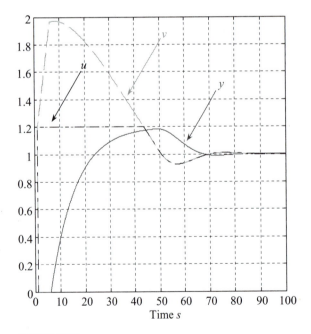

FIGURE 12.7

Process input u, controller output v, and process output y with actuator saturation and no anti-windup strategy.

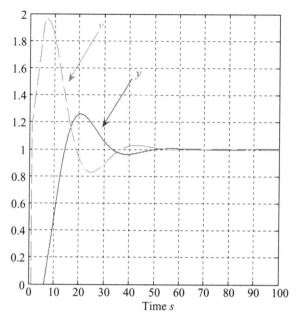

FIGURE 12.8

Controller output v and process output y with no actuator saturation.

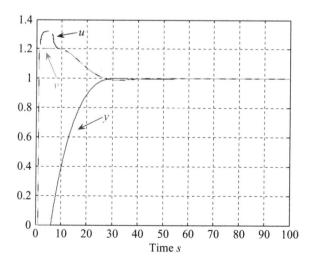

FIGURE 12.9

Process input u, controller output v, and process output y with actuator saturation and a back-calculation anti-windup strategy.

12.4.3 Bumpless Transfer between Manual and Automatic Mode

When the controller can operate in either **manual mode** or **automatic mode**, switching between the two modes of operation must be handled carefully to avoid a bump in the process output at the switching instant. During manual mode, the operator provides feedback control and the automatic feedback is disconnected. The integral term in the feedback controller can assume a value different from the one selected by the operator. Simply switching from automatic to manual, or vice versa, as in Figure 12.10, leads to a bump in the control signal, even if the control error is zero. This results in an undesirable bump in the output of the system.

For a smooth or **bumpless** transfer between manual control and the automatic digital controller $C(z)$, we use the digital scheme shown in Figure 12.11. We write the automatic controller transfer function in terms of an asymptotically stable controller $D(z)$ with unity DC gain—that is, $D(1) = 1$—as

$$C(z) = \frac{K}{1 - D(z)} \tag{12.14}$$

We then solve for $D(z)$ in terms of $C(z)$ to obtain

$$D(z) = \frac{C(z) - K}{C(z)}$$

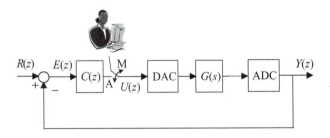

FIGURE 12.10

Block diagram for bumpy manual (M)/automatic (A) transfer.

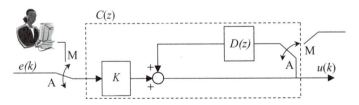

FIGURE 12.11

Block diagram for bumpless manual (M)/automatic (A) transfer.

If $C(z)$ is the PID controller transfer function of (12.11), we have

$$D(z) = \frac{(K_2 - K)z^2 + (K_1 - K - K\gamma)z + K_0 - K\gamma}{(z-1)(z-\gamma)} \qquad (12.15)$$

If the coefficient of the term z^2 in the numerator is nonzero, then the controller has the form

$$D(z) = \frac{U(z)}{E(z)} = (K_2 - K) + D_a(z)$$

where $D_a(z)$ has a first-order numerator polynomial. The controller output $u(k+2)$ is equal to the sum of two terms, one of which is the controller output $u(k+2)$ itself. Thus, the solution of the related difference equation cannot be computed by a simple recursion. This undesirable controller structure is known as an **algebraic loop**. To avoid an algebraic loop, we impose the condition

$$K = K_2 \qquad (12.16)$$

to eliminate the z^2 term in the numerator. We illustrate the effectiveness of the bumpless manual/automatic mode scheme using the following example.

EXAMPLE 12.8

Verify that a bump occurs if switching between manual and automatic operation uses the configuration shown in Figure 12.10 for the process

$$G(s) = \frac{1}{10s+1} e^{-2s}$$

and the PID controller ($T = 0.1$).

$$C(z) = \frac{44z^2 - 85.37z + 41.43}{z^2 - 1.368z + 0.368}$$

Design a scheme that provides a bumpless transfer between manual and automatic modes.

Solution
The unit step response for the system shown in Figure 12.10 is shown in Figure 12.12 together with the automatic controller output. The transfer between manual mode, where a step input $u = 1$ is selected, and automatic mode occurs at time $t = 100$. The output of the PID controller is about 11, which is far from the reference value of unity at $t = 100$. This leads to a significant bump in the control variable on switching from manual to automatic control that acts as a disturbance, causing a bump in the process output. To eliminate the bump, we use the bumpless transfer configuration shown in Figure 12.11 with the PID controller parameters

$$K_2 = 44, \ K_1 = 85.37, \ K_0 = 41.43, \ \gamma = 0.368$$

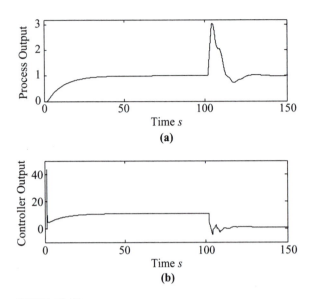

FIGURE 12.12

Process output (a) and controller output (b) for the system without bumpless transfer.

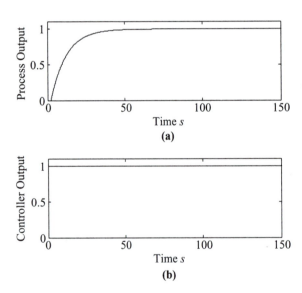

FIGURE 12.13

Process output (a) and controller output (b) for the bumpless manual/automatic transfer shown in Figure 12.11.

Using (12.15) and (12.16), we have

$$D(z) = \frac{-0.572z + 0.574}{z^2 - 1.94z + 0.942}$$

and $K = 44$. The results of Figure 12.13 show a bumpless transfer, with the PID output at $t = 100$ equal to one.

12.4.4 Incremental Form

Integrator windup and bumpless transfer issues are solved by implementing the PID controller in **incremental form**. We determine the increments in the control signal at each sampling period instead of determining the actual values of the control signal. This moves the integral action outside the control algorithm. To better understand this process, we consider the difference equation of a PID controller (the filter on the derivative action is not considered for simplicity):

$$u(k) = K_p \left(e(k) + \frac{T}{T_i} \sum_{i=0}^{k} e(i) + \frac{T_d}{T} (e(k) - e(k-1)) \right)$$

Subtracting the expression for $u(k - 1)$ from that of $u(k)$, we obtain the increment

$$u(k) - u(k-1) = K_p \left(\left(1 + \frac{T}{T_i} + \frac{T_d}{T} \right) e(k) + \left(-1 - \frac{2T_d}{T} \right) e(k-1) + \frac{T_d}{T} e(k-2) \right)$$

which can be rewritten more compactly as

$$u(k) - u(k-1) = K_2 e(k) + K_1 e(k-1) + K_0 e(k-2) \tag{12.17}$$

where

$$K_2 = K_p \left(1 + \frac{T}{T_i} + \frac{T_d}{T} \right) \tag{12.18}$$

$$K_1 = K_p \left(-1 - \frac{2T_d}{T} \right) \tag{12.19}$$

$$K_0 = K_p \frac{T_d}{T} \tag{12.20}$$

From the difference equation (12.17), we can determine the increments in the control signal at each sampling period. Actually, in (12.17) there is no error accumulation (in this case the integral action can be considered "outside" the controller), and the integrator windup problem does not occur. In practice, it is sufficient that the control signal is not incremented when the actuator saturates— namely, we have $u(k) = u(k - 1)$ when $v(k) = u(k)$ in Figure 12.6. Further, the transfer between manual mode and automatic mode is bumpless as long as the operator provides the increments in the control variable rather than their total value.

The z-transform of the difference equation (12.17) gives the PID controller z-transfer function in incremental form as

$$C(z) = \frac{\Delta U(z)}{E(z)} = \frac{K_2 z^2 + K_1 z + K_0}{z^2} \qquad (12.21)$$

where $\Delta U(z)$ is the z-transform of the increment $u(k - p) - u(k - 1)$.

EXAMPLE 12.9

For the process described in Example 12.7 and an analog PI controller with $K_p = 1.2$ and $T_i = 8$, verify that windup is avoided with the digital PI controller in incremental form ($T = 0.1$) if the saturation limits of the actuator are $u_{min} = -1.2$ and $u_{max} = 1.2$.

Solution

Using (12.18) and (12.19), we obtain $K_2 = 1.215$ and $K_1 = -1.2$ and the digital controller transfer function

$$C(z) = \frac{1.215z - 1.2}{z - 1}$$

The output obtained with the PI controller in incremental form (avoiding controller updating when the actuator saturates) is shown in Figure 12.14. Comparing the results to those shown in Figure 12.7, we note that the response is much faster both in rising to the reference value and in settling. In Figure 12.14, the effects of saturation are no longer noticeable.

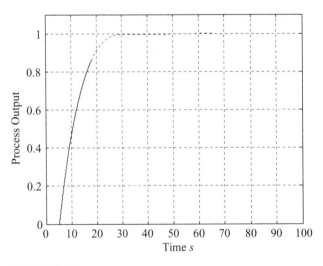

FIGURE 12.14

Process output with the PID in incremental form.

12.5 SAMPLING PERIOD SWITCHING

In many control applications, it is necessary to change the sampling period during operation to achieve the optimal usage of the computational resources. In fact, a single CPU usually performs many activities such as data storage, user interface, and communication, and possibly implements more than one controller. It is therefore necessary to optimize CPU utilization by changing the sampling period. For a given digital control law, on the one hand it is desirable to decrease the sampling period to avoid performance degradation; on the other hand decreasing it can overload the CPU and violate real-time constraints.

The problem of changing the control law when the sampling frequency changes can be solved by switching between controllers working in parallel, each with a different sampling period. This is a simple task provided that bumpless switching is implemented (see Section 12.4.3). However, using multiple controllers in parallel is computationally inefficient and is unacceptable if the purpose is to optimize CPU utilization. Thus, it is necessary to shift from one controller to another when the sampling period changes rather than operate controllers in parallel.

This requires computing the new controller parameters as well as past values of error and control variables that it requires to compute the control before switching. If the original sampling interval is T' and the new sampling interval is T, then we must switch from the controller

$$C'(z) = \frac{U(z)}{E(z)} = \frac{\mathbf{b}^T \mathbf{z}_m}{\mathbf{a}^T \mathbf{z}_n}$$

$$\mathbf{a} = [a_0(T') \quad a_1(T') \quad \cdots \quad 1]^T \qquad (12.22)$$

$$\mathbf{b} = [b_0(T') \quad b_1(T') \quad \cdots \quad b_m(T')]^T$$

$$\mathbf{z}_l = [1 \quad z \quad \cdots \quad z^l]$$

to the controller

$$C(z) = \frac{U(z)}{E(z)} = \frac{\mathbf{b}^T \mathbf{z}_m}{\mathbf{a}^T \mathbf{z}_n}$$

$$\mathbf{a} = [a_0(T) \quad a_1(T) \quad \cdots \quad 1]^T$$

$$\mathbf{b} = [b_0(T) \quad b_1(T) \quad \cdots \quad b_m(T)]^T$$

$$\mathbf{z}_l = [1 \quad z \quad \cdots \quad z^l]$$

Equivalently, we switch from the difference equation

$$u(kT') = -a_{n-1}(T')u((k-1)T') - \ldots - a_0(T')u((k-n)T') +$$
$$b_0(T')e((k-n+m)T') + \ldots + b_m(T')e((k-n)T')$$

to the difference equation

$$u(kT) = -a_{n-1}(T)u((k-1)T) - \ldots - a_0(T)u((k-n)T) +$$
$$b_0(T)e((k-n+m)T) + \ldots + b_m(T)e((k-n)T)$$

Thus, at the switching time instant, we must recompute the values of the parameter vectors **a** and **b** as well as the corresponding past $m + 1$ values of the tracking error e and the past n values of the control variable u.

We compute the new controller parameters using the controller transfer function, which explicitly depends on the sampling period. For example, if the PID controller of (12.11) is used, the new controller parameters can be easily computed using (12.12) with the new value of the sampling period T, or equivalently, using (12.10) where the sampling period T appears explicitly. To compute the past values of the tracking error e and of the control variable u, different techniques can be applied depending on whether the sampling period increases or decreases. For simplicity, we only consider the case where one sampling period is a divisor or multiple of the other. The case where the ratio between the previous and the new sampling periods (or vice versa) is not an integer is a simple extension, which is not considered here.

If the new sampling period T is a fraction of the previous sampling period T' (i.e., $T' = \lambda T$), the previous n values of the control variable $[u((k-1)T)$, $u((k-2)T), \ldots, u((k-n)T)]$ are determined with the control variable kept constant during the past λ periods. Otherwise, the $m + 1$ previous error values are computed using an **interpolator** such as a cubic polynomial. In particular, the coefficients c_3, c_2, c_1, and c_0 of a third-order polynomial $\tilde{e}(t) = c_3 t^3 + c_2 t^2 + c_1 t + c_0$ can be determined by considering the past three samples and the current value of the control error. The data yields the following linear system:

$$\begin{bmatrix} ((k-3)T')^3 & ((k-3)T')^2 & (k-3)T' & 1 \\ ((k-2)T')^3 & ((k-2)T')^2 & (k-2)T' & 1 \\ ((k-1)T')^3 & ((k-1)T')^2 & (k-1)T' & 1 \\ (kT')^3 & (kT')^2 & kT' & 1 \end{bmatrix} \begin{bmatrix} c_3 \\ c_2 \\ c_1 \\ c_0 \end{bmatrix} = \begin{bmatrix} e((k-3)T') \\ e((k-2)T') \\ e((k-1)T') \\ e(kT') \end{bmatrix} \quad \textbf{(12.23)}$$

Once the coefficients of the polynomial function have been determined, the previous values of the control error with the new sampling period are determined from the values of the polynomial functions evaluated at the required sampling instants. The procedure is illustrated in Figure 12.15, where the dashed line connecting the control error values between $(k-3)T$ and kT is the polynomial function $\tilde{e}(t)$.

If the new sampling period T is a multiple of the previous sampling period T' (i.e., $T = \lambda T'$), the previous m error samples are known. However, the data structure must be large enough to store them even if they are not necessarily with the sampling period T'. If the pole-zero difference of the process is equal to one, the equivalent n past control actions are approximately computed as the outputs estimated using the model of the control system with sampling period T. Specifi-

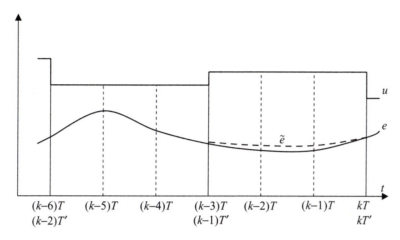

FIGURE 12.15

Switching controller sampling from a sampling period T' to a faster sampling rate with sampling period $T = T'/3$.

cally, let the process model obtained by discretizing the analog process with sampling period T be

$$G_{ZAS}(z) = \frac{Y(z)}{U(z)} = \frac{\beta_{h-1}z^{h-1} + \beta_{h-2}z^{h-2} + \ldots + \beta_0}{z^h + \alpha_{h-1}z^{h-1} + \ldots + \alpha_0}$$

where $h \geq n$. Then the equivalent past n control actions $u((k-1)T)$, $u((k-2)T)$, $\ldots$, $u((k-n)T)$ are determined by minimizing the difference between the measured output and that estimated by the model at the switching time:

$$\min |y(kT) - (-\alpha_{h-1}y((k-1)T) - \ldots - a_0 y((k-h)T) +$$
$$\beta_{h-1}u((k-1)T) + \ldots \beta_0 u((k-h)T))| \qquad (12.24)$$

We solve the optimization problem numerically using an appropriate approach such as the **simplex algorithm** (see Section 12.5.1). To initiate the search, initial conditions must be provided. The initial conditions can be selected as the values of the control signal at the same sampling instants determined earlier with the faster controller. The situation is depicted in Figure 12.16.

12.5.1 MATLAB Commands

When the sampling frequency is increased, the array **en** of the $m + 1$ previous error values can be computed using the following MATLAB command:

>> **en = interp1(ts, es, tn, 'cubic')**

where **ts** is a vector containing the last four sampling instants $[(k - 3)T'$, $(k - 2)T'$, $(k - 1)T'$, $kT']$ of the slower controller and **es** is a vector containing

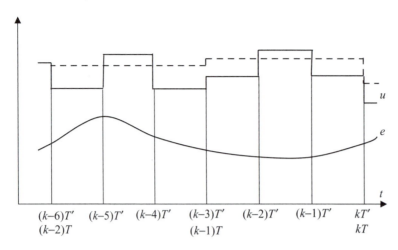

$$(k{-}6)T' \quad (k{-}5)T' \quad (k{-}4)T' \quad (k{-}3)T' \quad (k{-}2)T' \quad (k{-}1)T' \quad kT'$$
$$(k{-}2)T \qquad\qquad\qquad (k{-}1)T \qquad\qquad\qquad kT$$

FIGURE 12.16

Switching controller sampling from a sampling period T' to a slower sampling rate with sampling period $T = 3T'$.

the corresponding control errors $[e((k-3)T'),\, e((k-2)T'),\, e((k-1)T'),\, e(kT')]$. The array **tn** contains the $m+1$ sampling instants for which the past control errors for the new controller must be determined.

Alternatively, the vector of the coefficients $\mathbf{c} = [c_3,\, c_2,\, c_1,\, c_0]^T$ of the cubic polynomial can be obtained by solving the linear system (12.23) using the command

$$\gg \mathbf{c} = \mathbf{linsolve(M, e)}$$

where $\mathbf{M}$ is the matrix containing the time instants and $\mathbf{e}$ is the column vector containing the error samples such that (12.23) is written as $\mathbf{M}^*\mathbf{c} = \mathbf{e}$. Once the coefficients of the polynomial are computed, the values of the control error at previous sampling instants can be easily determined by interpolation.

To find the past control values using the simplex algorithm, use the command

$$\gg \mathbf{un = fminsearch(@(u)(abs(yk - ah1^*yk1 - \ldots -a0 \,^* \,ykh + bh1 \,^*}$$
$$\mathbf{u(1) + \ldots + b0^*u(n))), initcond)}$$

where **initcond** is the vector of n elements containing the initial conditions, **un** is the vector of control variables $[u((k-1)T),\, u((k-2)T),\, \ldots,\, u((k-n)T)]$, and the terms in the **abs** function are the values corresponding to expression (12.24), with the exception of $\mathbf{u(1)}, \ldots, \mathbf{u(n)}$ that must be written explicitly. Details are provided in the following examples.

EXAMPLE 12.10

Discuss the effect of changing the sampling rate on the step response of the process described in Example 12.7 and an analog PI controller with $K_p = 1.2$ and $T_i = 8$. Initially use a digital PI controller in incremental form with $T' = 0.4$ s; then switch at time $t = 22$ s to a faster controller with $T = 0.1$ s.

Solution

For PI control, we set T_d to zero in (12.18) to (12.20) to obtain the parameters

$$K_2 = K_p\left(1 + \frac{T'}{T_i}\right) = 1.2(1 + 0.4/8) = 1.26$$

$$K_1 = -K_p = -1.2$$

$$K_0 = 0$$

Using (12.21), the initial digital controller transfer function is

$$C(z) = \frac{U(z)}{E(z)} = \frac{1.26z - 1.2}{z - 1}$$

We simulate the system using MATLAB to compute the control and error signals. The control signal at $t = 20.8, 21.2$, and 21.6 s is plotted in Figure 12.17, whereas the error at time $t = 20.8, 21.2, 21.6$, and 22 s is plotted as circles in Figure 12.18. At time $t = 22$ s, the sampling period switches to $T = 0.1$ s and the control signal must be recomputed. The new PI controller parameters are

$$K_2 = K_p\left(1 + \frac{T'}{T_i}\right) = 1.2(1 + 0.1/8) = 1.215$$

$$K_1 = -K_p = -1.2$$

$$K_0 = 0$$

yielding the following transfer function:

$$C(z) = \frac{1.215z - 1.2}{z - 1}$$

The associated difference equation is therefore

$$u(k) = u(k-1) + 1.215e(k) - 1.2e(k-1) \tag{12.25}$$

Whereas at $t = 22$ s the value of $e(k) = -0.3139$ is unaffected by switching, the values of $u(k-1)$ and $e(k-1)$ must in general be recomputed for the new sampling period. For a first-order controller, the value of $u(k-1)$ is the value of the control signal at time $t = 22 - T = 22 - 0.1 = 21.9$ s. This value is the same as that of the control signal for sampling period T' over the interval 21.6 to 22 s. From the simulation results shown in Figure 12.17, we obtain $u(k-1) = 0.8219$ where the values of the control signal at $t = 21.7, 21.8$, and 21.9 s with the new controller are denoted by stars.

To calculate the previous value of the control error $e(k-1)$, we use a cubic polynomial to interpolate the last four error values with the slower controller (the interpolating function

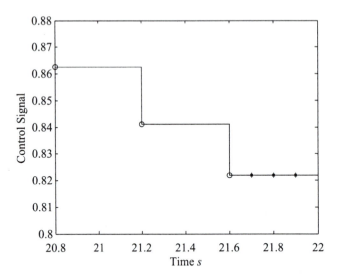

FIGURE 12.17

Control signal (*solid line*) obtained by simulating the system described in Example 12.10 with sampling period $T' = 0.4$ s. Circles denote the control values at the sampling instants, and stars denote the control values for $T = 0.1$ s.

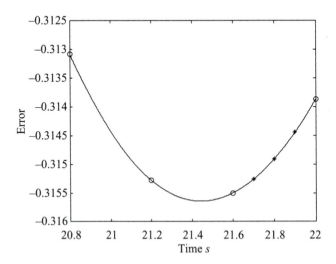

FIGURE 12.18

Control error interpolation $\tilde{e}(t)$ (*solid line*) obtained by simulating the system described in Example 12.10 with sampling period $T' = 0.4$ s. Circles denote the error values at the sampling instants, and stars denote the error values for $T = 0.1$ s.

is the solid line in Figure 12.18). Using the control errors at $t = 21.7$, 21.8, 21.9 s, $e(21.6) = -0.3155$, $e(21.2) = -0.3153$, and $e(20.8) = -0.3131$, the linear system (12.23) is solved for the coefficients $c_3 = -0.0003$, $c_2 = 0.0257$, $c_1 = -0.6784$, and $c_0 = 5.4458$. To solve the linear system, we use the MATLAB commands

```
>> M = [20.8^3 20.8^2 20.8 1; 21.2^3 21.2^2 21.2 1; 21.6^3 21.6^2
   21.6 1; 22.0^3 22.0^2 22.0 1];
>> e = [-0.3131 -0.3153 -0.3155 -0.3139]';
>> c = linsolve(M,e);  % Solve the linear system M c = e
>> en = c(1)*21.9^3 + c(2)*21.9^2 + c(3)*21.9 + c(4);
```

The value of the control error at time $t = 21.9$ s is therefore

$$e(k-1) = -0.0003 \cdot 21.9^3 + 0.0257 \cdot 21.9^2 - 0.6784 \cdot 21.9 + 5.4458 = -0.3144$$

Alternatively, we can use the MATLAB command for cubic interpolation

> **>> en = interp1([20.8 21.2 21.6 22], [-0.3131 -0.3153 -0.3155**
> **-0.3139],21.9,'cubic');**

The control error for the faster controller at $t = 21.7$, 21.8, and 21.9 s is denoted by stars in Figure 12.18. From (12.25), we observe that only the error at $t = 21.9$ is needed to calculate the control at $t = 22$ s after switching to the faster controller. We compute the control value:

$$u(k) = u(k-1) + 1.215e(k) - 1.2e(k-1)$$
$$= 0.8219 + 1.215 \times (-0.3139) - 1.2 \times (-0.3144) = 0.8178$$

We compute the control at time $t = 22.1$ s and the subsequent sampling instants using the same expression without the need for further interpolation. Note that the overall performance is not significantly affected by changing the sampling period, and the resulting process output is virtually the same as the one shown in Figure 12.8.

EXAMPLE 12.11

Design a digital controller for the DC motor speed control system described in Example 6.16 with transfer function

$$G(s) = \frac{1}{(s+1)(s+10)}$$

to implement the analog PI controller

$$C(s) = 47.2 \frac{s+1}{s}$$

with the sampling period switched from $T' = 0.01$ s to $T = 0.04$ s at time $t = 0.5$ s. Obtain the step response of the closed-loop system, and discuss your results.

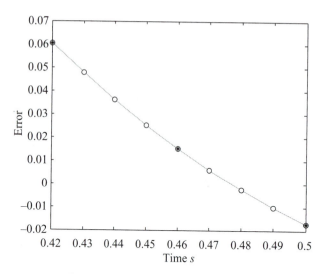

FIGURE 12.19

Control error for the two controllers described in Example 12.11. Circles denote the error values for the faster controller; the dark circles denote the error values for the slower controller.

Solution

Applying the bilinear transformation with $T' = 0.01$ to the controller transfer function $C(s)$, we obtain the initial controller transfer function

$$C(z) = \frac{47.44z - 46.96}{z - 1}$$

We simulate the system using MATLAB to compute the error and process output values. The error values at $t = 0.42, \ldots, 0.5$ s are shown in Figure 12.19. The process output for a unit step reference input step at the same sampling instants is shown in Figure 12.20.

Starting at $t = 0.5$ s, the controller transfer function obtained by bilinearly transforming the analog controller with sampling period $T = 0.04$ s becomes

$$C(z) = \frac{U(z)}{E(z)} = \frac{48.14z - 46.26}{z - 1}$$

The corresponding difference equation

$$u(k) = u(k-1) + 48.14e(k) - 46.26e(k-1)$$

is used to calculate the control variable starting at $t = 0.5$ s. From the MATLAB simulation results, the error values needed to compute the control at $t = 0.5$ s are $e(k) = -0.0172$ at $t = 0.5$ s and $e(k - 1) = 0.0151$ at $t = 0.5 - 0.04 = 0.46$ s. The control $u(k - 1)$ at $t = 0.46$ s must be determined by solving the optimization problem (12.24). The z-transfer function of the plant, ADC and DAC, with $T = 0.04$ is

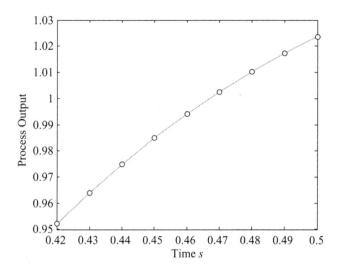

FIGURE 12.20

Process output for the faster controller described in Example 12.11.

$$G_{ZAS}(z) = \frac{Y(z)}{U(z)} = 10^{-4}\frac{6.936z + 5.991}{z^2 - 1.631z + 0.644}$$

$$= \frac{6.936 \times 10^{-4} z^{-1} + 5.991 \times 10^{-4} z^{-2}}{1 - 1.631z^{-1} + 0.644z^{-2}}$$

and the corresponding difference equation is

$$y(k) = 1.631y(k-1) - 0.644y(k-2) + 6.936 \cdot 10^{-4}u(k-1) + 5.991 \cdot 10^{-4}u(k-2)$$

Therefore, the optimization problem is

$$\min|y(0.5) - (1.631y(0.46) - 0.644y(0.42) + 6.936 \times 10^{-4}u(0.46) + 5.991 \times 10^{-4}u(0.42))|$$

Using the output values $y(0.5) = 1.0235$, $y(0.46) = 0.9941$, and $y(0.42) = 0.9522$, the values of $u(0.46)$ and $u(0.42)$ are computed by solving the optimization problem using the following MATLAB command:

```
>> u = fminsearch(@(un)(abs(1.0235-(1.631*0.9941-
0.644*0.9522+6.936e-4*un(1)+5.991e-4*un(2)))),[13.6 11.5]);
```

The initial conditions $u(k-1) = 13.6$ at $t = 0.46$ and $u(k-2) = 11.5$ at $t = 0.42$ are obtained from the values of the control variable at time $t = 0.42$ and $t = 0.46$ with the initial sampling period $T' = 0.01$ s (Figure 12.21). The optimization yields the control values $u(0.46) = 11.3995$ and $u(0.42) = 12.4069$. The resulting value of the objective function is

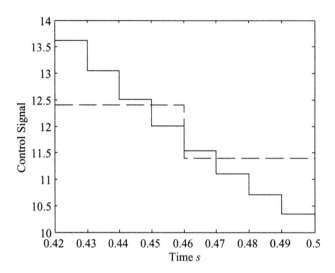

FIGURE 12.21

Control values for the two controllers described in Example 12.11. The solid line represents the control variable with the faster controller. The dashed line represents the equivalent control variable with the slower controller.

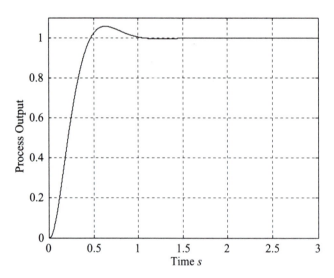

FIGURE 12.22

Step response described in Example 12.11 with sampling period switching.

zero (i.e., the measured output is equal to the model estimate at the switching time). Thus, the value of the control variable at the switching time $t = 0.5$ s is

$$u(0.5) = 11.3995 + 48.14 \times -0.0172 - 46.26 \times 0.0151 = 9.8730$$

The step response of the system, shown in Figure 12.22, has a small overshoot and time to first peak, and a short settling time. The response is smooth and does not have a discontinuity at the switching point.

12.5.2 Dual-Rate Control

In some industrial applications, samples of the process output are available at a rate that is slower than the sampling rate of the controller. If performance degrades significantly when the controller sampling rate is reduced to equal that of the process output, a dual-rate control scheme can be implemented. The situation is depicted in Figure 12.23, where it is assumed that the slow sampling period λT is a multiple of the fast sampling period T (i.e., λ is an integer). Thus, the ADC operates at the slower sampling rate λT, whereas the controller and the sample-and-hold operate at the faster sampling rate T.

To achieve the performance obtained when the output is sampled at the fast rate, a possible solution is to implement the so-called **dual-rate inferential control** scheme. It uses a fast-rate model of the process $\hat{G}_{ZAS}(z)$ to compute the missing output samples. The control scheme is shown in Figure 12.24, where a is an availability parameter for the output measurement defined by

FIGURE 12.23

Block diagram of dual-rate control. The controller and the ZOH operate with sampling period T; the process output sampling period is λT.

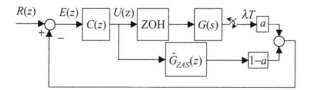

FIGURE 12.24

Block diagram of dual-rate inferential control.

$$a = \begin{cases} 0, t \neq k\lambda T \\ 1, t = k\lambda T \end{cases}$$

The controller determines the values of the control variable using the measured output at $t = k\lambda T$ and using the estimated output when $t = kT$ and $t \neq k\lambda T$. In the absence of disturbances and modeling errors, the dual-rate control scheme is equivalent to the fast single-rate control scheme. Otherwise, the performance of dual-rate control can deteriorate significantly. Because disturbances and modeling errors are inevitable in practice, the results of this approach must be carefully checked.

EXAMPLE 12.12

Design a dual-rate inferential control scheme with $T = 0.02$ and $\lambda = 5$ for the process (see Example 6.16)

$$G(s) = \frac{1}{(s+1)(s+10)}$$

and the controller

$$C(s) = 47.2 \frac{s+1}{s}$$

Solution

The fast rate model ($T = 0.02$) for the plant with DAC and ADC is

$$G_{ZAS}(z) = 10^{-4} \frac{1.86z + 1.729}{z^2 - 1.799z + 0.8025} = \frac{Y(z)}{U(z)} = \hat{G}_{ZAS}(z)$$

The difference equation governing the estimates of the output is

$$y(k) = 1.799\, y(k-1) - 0.8025\, y(k-2) + 1.86 \times 10^{-4} u(k-1) + 1.729 \times 10^{-4} u(k-2)$$

With $T = 0.02$, the controller transfer function obtained by bilinear transformation is

$$C(z) = \frac{47.67z - 46.73}{z - 1}$$

The controller determines the values of the control variable from the measured output at $t = 5kT$. When $t = kT$ and $t \neq 5kT$, we calculate the output estimates using the estimator difference equation.

The control scheme shown in Figure 12.24 yields the step response shown in Figure 12.25. The results obtained using a single-rate control scheme with $T = 0.1$ are shown in Figure 12.26. The step response for single-rate control has a much larger first peak and settling time.

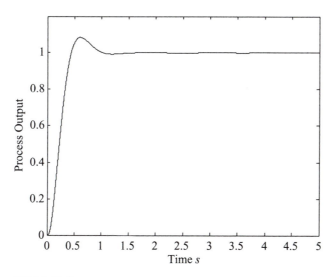

FIGURE 12.25

Step response described in Example 12.12 for a dual-rate controller with $T = 0.02$ and $\lambda = 5$.

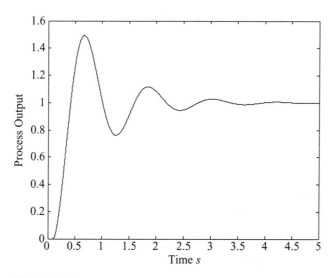

FIGURE 12.26

Step response described in Example 12.12 for a single-rate controller with $T = 0.1$.

RESOURCES

Albertos, P., M. Vallés, A. Valera, Controller transfer under sampling rate dynamic changes, *Proceedings European Control Conference*, Cambridge (UK), 2003.

Åström, K. J., and T. Hägglund, *Advanced PID Controllers*, ISA Press, 2006.

Cervin, A., D. Henriksson, B. Lincoln, J. Eker, and K.-E. Årzén, How does control timing affect performance? *IEEE Control Systems Magazine*, 23:16–30, 2003.

Gambier, A., Real-time control systems: A tutorial, *Proceedings 5th Asian Control Conference*, pp. 1024–1031, 2004.

Li, D. S., L. Shah, and T. Chen, Analysis of dual-rate inferential control systems, *Automatica*, 38:1053–1059, 2003.

Visioli, A., *Practical PID Control*, Springer, 2006.

PROBLEMS

12.1. Write pseudocode that implements the following controller:

$$C(z) = \frac{U(z)}{E(z)} = \frac{2.01z - 1.99}{z - 1}$$

12.2. Rewrite the pseudocode for the controller described in Problem 12.1 to decrease the execution time by assigning priorities to computational tasks.

12.3. Design an antialiasing filter for the position control system

$$G(s) = \frac{1}{s(s+10)}$$

with the analog controller (see Example 5.6)

$$C(s) = 50 \frac{s + 0.5}{s}$$

Select an appropriate sampling frequency and discretize the controller.

12.4. Determine the mean and variance values of the quantization noise when a 12-bit ADC is used to sample a variable in a range 0 to 10 V for (a) rounding and (b) truncation.

12.5. For the system and the controller described in Problem 12.3 with a sampling interval $T = 0.02$ s, determine the decrease in the phase margin due to the presence of the ZOH.

12.6. Consider an oven control system (Visioli, 2006) with transfer function

$$G(s) = \frac{1.1}{1300s + 1} e^{-25s}$$

and the PI controller

$$C(s) = 13\frac{200s + 1}{200s}$$

Let both the actuator and the sensor signals be in the range 0 to 5 V, and let 1° Celsius of the temperature variable correspond to 0.02 V. Design the hardware and software architecture of the digital control system.

12.7. Write the difference equations for the controller in (a) direct form, (b) parallel form, and (c) cascade form.

$$C(z) = 50\frac{(z - 0.9879)(z - 0.9856)}{(z - 1)(z - 0.45)}$$

12.8. For the PID controller that results by applying the Ziegler-Nichols tuning rules to the process

$$G(s) = \frac{1}{8s + 1}e^{-2s}$$

determine the discretized PID controller transfer functions (12.11) and (12.12) with $N = 10$ and $T = 0.1$.

12.9. Design a bumpless manual/automatic mode scheme for the PID controller ($T = 0.1$)

$$C(z) = \frac{252z^2 - 493.4z + 241.6}{(z - 1)(z - 0.13)}.$$

12.10. Design a bumpless manual/automatic mode scheme for the controller obtained in Example 6.18

$$C(z) = \frac{1.422(z - 0.8187)(z - 0.9802)(z + 1)}{(z - 1)(z + 0.9293)(z - 0.96)}$$

12.11. Determine the digital PID controller (with $T = 0.1$) in incremental form for the analog PID controller

$$C(s) = 3\left(1 + \frac{1}{8s} + 2s\right)$$

COMPUTER EXERCISES

12.12. Write a MATLAB script and design a Simulink diagram that implements the solution to Problem 12.8 with different filter parameter values N, and discuss the set-point step responses obtained by considering the effect of measurement noise on the process output.

12.13. Consider the analog process

$$G(s) = \frac{1}{8s+1} e^{-2s}$$

and the analog PI controller with $K_p = 3$ and $T_i = 8$. Obtain the set-point step response with a saturation limit of $u_{min} = -1.1$ and $u_{max} = 1.1$ and with a digital PI controller ($T = 0.1$) with
(a) No anti-windup
(b) A conditional integration anti-windup strategy
(c) A back-calculation anti-windup strategy.
(d) A digital PI controller in incremental form

12.14. Consider the analog process and the PI controller described in Problem 12.13. Design a scheme that provides a bumpless transfer between manual and automatic mode, and simulate it by applying a step set-point signal and by switching from manual mode, where the control variable is equal to one, to automatic mode at time $t = 60$ s. Compare the results with those obtained by a scheme without bumpless transfer.

12.15. Design and simulate a dual-rate inferential control scheme with $T = 0.01$ and $\lambda = 4$ for the plant

$$G(s) = \frac{1}{(s+1)(s+5)}$$

and the analog PI controller (see Problem 5.7). Then apply the controller to the process

$$\tilde{G}(s) = \frac{1}{(s+1)(s+5)(0.1s+1)}$$

to verify the robustness of the control system.

12.16. Consider the analog process and the analog PI controller described in Problem 12.13. Write a MATLAB script that simulates the step response with a digital controller when the sampling period switches at time $t = 0.52$ from $T = 0.04$ to $T = 0.01$.

12.17. Consider the analog process and the analog PI controller described in Problem 12.13. Write a MATLAB script that simulates the step response with a digital controller when the sampling period switches at time $t = 0.52$ from $T = 0.01$ to $T = 0.04$.

Table of Laplace and z-Transforms

I

No.	Continuous Time	Laplace Transform	Discrete Time*	Z-Transform
1	$\delta(t)$	1	$\delta(k)$	1
2	$1(t)$	$\dfrac{1}{s}$	$1(k)$	$\dfrac{z}{z-1}$
3	t	$\dfrac{1}{s^2}$	k^{**}	$\dfrac{z}{(z-1)^2}$
4	$\dfrac{t^2}{2!}$	$\dfrac{1}{s^3}$	k^2	$\dfrac{z(z+1)}{(z-1)^3}$
5	$\dfrac{t^3}{3!}$	$\dfrac{1}{s^4}$	k^3	$\dfrac{z(z^2+4z+1)}{(z-1)^4}$
6	$e^{-\alpha kT}$	$\dfrac{1}{s+\alpha}$	a^{k***}	$\dfrac{z}{z-a}$
7	$1-e^{-\alpha t}$	$\dfrac{\alpha}{s(s+\alpha)}$	$1-a^k$	$\dfrac{(1-a)z}{(z-1)(z-a)}$
8	$e^{-\alpha t}-e^{-\beta t}$	$\dfrac{\beta-\alpha}{(s+\alpha)(s+\beta)}$	a^k-b^k	$\dfrac{(a-b)z}{(z-a)(z-b)}$
9	$te^{-\alpha t}$	$\dfrac{1}{(s+\alpha)^2}$	ka^k	$\dfrac{az}{(z-a)^2}$
10	$\sin(\omega_n t)$	$\dfrac{\omega_n}{s^2+\omega_n^2}$	$\sin(\omega_n k)$	$\dfrac{\sin(\omega_n)z}{z^2-2\cos(\omega_n)z+1}$
11	$\cos(\omega_n t)$	$\dfrac{s}{s^2+\omega_n^2}$	$\cos(\omega_n k)$	$\dfrac{z[z-\cos(\omega_n)]}{z^2-2\cos(\omega_n)z+1}$

*The discrete time functions are similar to but not always a sampled form of the continuous time functions.

**Sampling t gives kT with a transform that is obtained by multiplying the transform of k by T.

***The function $e^{-\alpha kT}$ is obtained by setting $a = e^{-\alpha T}$.

No.	Continuous Time	Laplace Transform	Discrete Time*	Z-Transform
12	$e^{-\zeta\omega_n t}\sin(\omega_d t)$	$\dfrac{\omega_d}{(s+\zeta\omega_n)^2+\omega_d^2}$	$e^{-\zeta\omega_n k}\sin(\omega_d k)$	$\dfrac{e^{-\zeta\omega_n}\sin(\omega_d)z}{z^2-2e^{-\zeta\omega_n}\cos(\omega_d)z+e^{-2\zeta\omega_n}}$
13	$e^{-\zeta\omega_n t}\cos(\omega_d t)$	$\dfrac{s+\zeta\omega_n}{(s+\zeta\omega_n)^2+\omega_d^2}$	$e^{-\zeta\omega_n k}\cos(\omega_d k)$	$\dfrac{z[z-e^{-\zeta\omega_n}\cos(\omega_d)]}{z^2-2e^{-\zeta\omega_n}\cos(\omega_d)z+e^{-2\zeta\omega_n}}$
14	$\sinh(\beta t)$	$\dfrac{\beta}{s^2-\beta^2}$	$\sinh(\beta k)$	$\dfrac{\sinh(\beta)z}{z^2-2\cosh(\beta)z+1}$
15	$\cosh(\beta t)$	$\dfrac{s}{s^2-\beta^2}$	$\cosh(\beta k)$	$\dfrac{z[z-\cosh(\beta)]}{z^2-2\cosh(\beta)z+1}$

Properties of the z-Transform

No.	Name	Formula
1	Linearity	$\mathcal{Z}\{\alpha f_1(k) + \beta f_2(k)\} = \alpha F_1(z) + \beta F_2(z)$
2	Time Delay	$\mathcal{Z}\{f(k-n)\} = z^{-n}F(z)$
3	Time Advance	$\mathcal{Z}\{f(k+1)\} = zF(z) - zf(0)$
		$\mathcal{Z}\{f(k+n)\} = z^n F(z) - z^n f(0) - z^{n-1}f(1) - zf(n-1)$
4	Discrete-Time Convolution	$\mathcal{Z}\{f_1(k) * f_2(k)\} = \mathcal{Z}\left\{\sum_{i=0}^{k} f_1(i)\, f_2(k-i)\right\}$
		$\quad = F_1(z)\, F_2(z)$
5	Multiplication by Exponential	$\mathcal{Z}\{a^{-k}f(k)\} = F(az)$
6	Complex Differentiation	$\mathcal{Z}\{k^m f(k)\} = \left(-z\dfrac{d}{dz}\right)^m F(z)$
7	Final Value Theorem	$f(\infty) = \lim_{k\to\infty} f(k) = \lim_{z\to 1}(1 - z^{-1})F(z)$
		$\quad = \lim_{z\to 1}\left(\dfrac{z-1}{z}\right)F(z)$
8	Initial Value Theorem	$f(0) = \lim_{k\to 0} f(k) = \lim_{z\to\infty} F(z)$

Review of Linear Algebra

A.1 MATRICES

An $m \times n$ matrix is an array of entries[1] denoted by

$$A = [a_{ij}] = \begin{bmatrix} a_{11} & a_{12} & \cdots & a_{1n} \\ a_{21} & a_{22} & \cdots & a_{2n} \\ \vdots & \vdots & \ddots & \vdots \\ a_{m1} & a_{m2} & \cdots & a_{mn} \end{bmatrix}$$

with m rows and n columns. The matrix is said to be of order $m \times n$.

Rectangular matrix	$m \neq n$
Square matrix	$m = n$
Row vector	$m = 1$
Column vector	$n = 1$

EXAMPLE A.1: MATRIX REPRESENTATION

$$A = \begin{bmatrix} 1 & 2 & 3 \\ 4 & 5 & 6 \end{bmatrix} \quad \text{real } 2 \times 3 \text{ rectangular matrix}$$

>> A = [1 2 3; 4 5 6]

$$A = \begin{bmatrix} 1 & 2-j \\ 4+j & 5 \end{bmatrix} \quad \text{complex } 2 \times 2 \text{ square matrix}$$

>> A = [1, 2 − j; 4 + j, 5]

---

[1]Relevant MATLAB® commands are given as necessary and are preceded by ">>."

A.2 EQUALITY OF MATRICES

Equal matrices are matrices of the same order with equal corresponding entries:

$$A = B \Leftrightarrow [a_{ij}] = [b_{ij}], \quad i = 1, 2, \ldots, m \\ j = 1, 2, \ldots, n$$

EXAMPLE A.2: EQUAL MATRICES

$$A = \begin{bmatrix} 1 & 2 & 3 \\ 4 & 5 & 6 \end{bmatrix} \quad B = \begin{bmatrix} 1.1 & 2 & 3 \\ 4 & 5 & 6 \end{bmatrix} \quad C = \begin{bmatrix} 1 & 2 & 3 \\ 4 & 5 & 6 \end{bmatrix}$$

$$A \neq B, \quad C \neq B, \quad A = C$$

A.3 MATRIX ARITHMETIC

A.3.1 Addition and Subtraction

The sum (difference) of two matrices of the same order is a matrix with entries that are the sum (difference) of the corresponding entries of the two matrices:

$$C = A \pm B \Leftrightarrow [c_{ij}] = [a_{ij} \pm b_{ij}], \quad i = 1, 2, \ldots, m \\ j = 1, 2, \ldots, n$$

EXAMPLE A.3: MATRIX ADDITION/SUBTRACTION

$$>> C = A + B$$

$$>> C = A - B$$

Note: MATLAB accepts the command

$$>> C = A + b$$

if **b** is a scalar. The result is the matrix $C = [a_{ij} + b]$.

A.3.2 Transposition

Interchanging the rows and columns of a matrix:

$$C = A^T \Leftrightarrow [c_{ij}] = [a_{ji}], \quad i = 1, 2, \cdots, m \\ j = 1, 2, \cdots, n$$

EXAMPLE A.4: MATRIX TRANSPOSE

$$A = \begin{bmatrix} 1 & 2-j \\ 4+j & 5 \end{bmatrix}$$

$$>> B = A'$$

$$B = \begin{bmatrix} 1 & 4-j \\ 2+j & 5 \end{bmatrix}$$

Note: The (') command gives the complex conjugate transpose for a complex matrix.

Symmetric matrix $A = -A^T$
Hermitian matrix $A = A^*$ (*Denotes the complex conjugate transpose.)

EXAMPLE A.5: SYMMETRIC AND HERMITIAN MATRIX

$$A = \begin{bmatrix} 1 & 2 \\ 2 & 5 \end{bmatrix} \quad \text{symmetric}$$

$$A = \begin{bmatrix} 1 & 2-j \\ 2+j & 5 \end{bmatrix} \quad \text{hermitian}$$

Notation

Column vector $\quad \mathbf{x} = \begin{bmatrix} x_1 & x_2 & \cdots & x_n \end{bmatrix}^T = \begin{bmatrix} x_1 \\ x_2 \\ \vdots \\ x_n \end{bmatrix}$

Row vector $\quad \mathbf{x}^T = \begin{bmatrix} x_1 & x_2 & \cdots & x_n \end{bmatrix}$

EXAMPLE A.6: COLUMN AND ROW VECTORS

$$\mathbf{x} = \begin{bmatrix} 1 & 2 & 3 \end{bmatrix}^T \qquad \mathbf{y}^T = \begin{bmatrix} 1 & 2 & 3 \end{bmatrix}$$

$$>> x = [1; 2; 3]$$

$$x = 1$$
$$2$$
$$3$$

$$>> y = [1, 2, 3]$$

$$y = 1 \quad 2 \quad 3$$

A.3.3 Matrix Multiplication

A.3.3.1 *Multiplication by a Scalar*
Multiplication of every entry of the matrix by a scalar:

$$C = \alpha A \Leftrightarrow [c_{ij}] = [\alpha a_{ji}], \quad i = 1, 2, \ldots, m$$
$$j = 1, 2, \ldots, n$$

$$\gg C = a*A$$

Note: MATLAB is a case-sensitive mode that distinguishes between uppercase and lowercase variables.

A.3.3.2 *Multiplication by a Matrix*
The product of an $m \times n$ matrix and an $n \times l$ matrix is an $m \times l$ matrix—that is, $(m \times n) \cdot (n \times l) = (m \times l)$

$$C = AB \Leftrightarrow [c_{ij}] = \left[\sum_{k=1}^{n} a_{ik} b_{kj} \right], \quad i = 1, 2, \ldots, m$$
$$j = 1, 2, \ldots, l$$

Noncummutative $AB \neq BA$ (in general).
Normal matrix $A^*A = AA^*$ (commutative multiplication with its conjugate transpose).

Clearly, any symmetric (hermitian) matrix is also normal. But some normal matrices are not symmetric (hermitian).

Premultiplication by a row vector $m = 1$

$$(1 \times n) \cdot (n \times l) = (1 \times l) \quad C = \mathbf{c}^T = [c_1 \quad c_2 \quad \ldots \quad c_l]$$

Postmultiplication by a column vector $l = 1$

$$(m \times n) \cdot (n \times 1) = (m \times 1) \quad C = \mathbf{c} = [c_1 \quad c_2 \quad \ldots \quad c_l]^T$$

Multiplication of a row by a column $m = l = 1$

$$(1 \times n) \cdot (n \times 1) = (1 \times 1) \quad c = \text{scalar}$$

Note that this product is the same for any two vectors regardless of which vector is transposed to give a row vector because

$$c = \mathbf{a}^T \mathbf{b} = \mathbf{b}^T \mathbf{a} = \sum_{k=1}^{n} a_i b_i$$

This defines a dot product for any two real vectors and is often written in the form

$$<a,b>$$

Multiplication of a column by a row $n = 1$

$$(m \times 1) \cdot (1 \times l) = (m \times l) \quad C = m \times l \text{ matrix}$$

Positive integral power of a square matrix

$$A^s = AA \ldots A \quad (A \text{ repeated } s \text{ times})$$

$$A^s A^r = A^{s+r} = A^r A^s \quad \text{(commutative product)}$$

EXAMPLE A.7: MULTIPLICATION

$$A = \begin{bmatrix} 1 & 2 & 3 \\ 4 & 5 & 6 \end{bmatrix} \quad B = \begin{bmatrix} 1.1 & 2 \\ 4 & 5 \\ 1 & 2 \end{bmatrix}$$

1. Matrix by scalar

$$C = 4A = \begin{bmatrix} 4 \times 1 & 4 \times 2 & 4 \times 3 \\ 4 \times 4 & 4 \times 5 & 4 \times 6 \end{bmatrix}$$
$$= \begin{bmatrix} 4 & 8 & 12 \\ 16 & 20 & 24 \end{bmatrix}$$

>> C = 4*[1, 2, 3;4, 5, 6]

2. Matrix by matrix

$$C = AB = \begin{bmatrix} 1 & 2 & 3 \\ 4 & 5 & 6 \end{bmatrix} \begin{bmatrix} 1.1 & 2 \\ 4 & 5 \\ 1 & 2 \end{bmatrix}$$
$$= \begin{bmatrix} 1 \times 1.1 + 2 \times 4 + 3 \times 1 & 1 \times 2 + 2 \times 5 + 3 \times 2 \\ 4 \times 1.1 + 5 \times 4 + 6 \times 1 & 4 \times 2 + 5 \times 5 + 6 \times 2 \end{bmatrix}$$
$$= \begin{bmatrix} 12.1 & 18 \\ 30.4 & 45 \end{bmatrix}$$

>> C = A*B

3. Vector-matrix multiplication

$$C = \mathbf{x}^T B = \begin{bmatrix} 1 & 2 & 3 \end{bmatrix} \begin{bmatrix} 1.1 & 2 \\ 4 & 5 \\ 1 & 2 \end{bmatrix}$$
$$= \begin{bmatrix} 1 \times 1.1 + 2 \times 4 + 3 \times 1 & 1 \times 2 + 2 \times 5 + 3 \times 2 \end{bmatrix}$$
$$= \begin{bmatrix} 12.1 & 18 \end{bmatrix}$$

$$D = A\mathbf{y} = \begin{bmatrix} 1 & 2 & 3 \\ 4 & 5 & 6 \end{bmatrix} \begin{bmatrix} 1.1 \\ 4 \\ 1 \end{bmatrix}$$

$$= \begin{bmatrix} 1 \times 1.1 + 2 \times 4 + 3 \times 1 \\ 4 \times 1.1 + 5 \times 4 + 6 \times 1 \end{bmatrix} = \begin{bmatrix} 12.1 \\ 30.4 \end{bmatrix}$$

>> C = [1, 2, 3]*B;

>> D = A*[1.1; 4; 1];

4. Vector-vector

$$\mathbf{z} = \mathbf{x}^T \mathbf{y} = \begin{bmatrix} 1 & 2 & 3 \end{bmatrix} \begin{bmatrix} 1.1 \\ 4 \\ 1 \end{bmatrix}$$

$$= 1 \times 1.1 + 2 \times 4 + 3 \times 1 = 12.1$$

$$D = \mathbf{y}\mathbf{x}^T = \begin{bmatrix} 1.1 \\ 4 \\ 1 \end{bmatrix} \begin{bmatrix} 1 & 2 & 3 \end{bmatrix}$$

$$= \begin{bmatrix} 1.1 \times 1 & 1.1 \times 2 & 1.1 \times 3 \\ 4 \times 1 & 4 \times 2 & 4 \times 3 \\ 1 \times 1 & 1 \times 2 & 1 \times 3 \end{bmatrix}$$

$$= \begin{bmatrix} 1.1 & 2.2 & 3.3 \\ 4 & 8 & 12 \\ 1 & 2 & 3 \end{bmatrix}$$

>> z = [1, 2, 3]*[1.1; 4; 1]

>> D = [1.1; 4, 1]*[1, 2, 3]

5. Positive integral power of a square matrix

$$S = \begin{bmatrix} 1 & 2 \\ 0 & 4 \end{bmatrix} \quad S^3 = \begin{bmatrix} 1 & 2 \\ 0 & 4 \end{bmatrix} \begin{bmatrix} 1 & 2 \\ 0 & 4 \end{bmatrix} \begin{bmatrix} 1 & 2 \\ 0 & 4 \end{bmatrix} = \begin{bmatrix} 1 & 42 \\ 0 & 64 \end{bmatrix}$$

>> S^3

Diagonal of a matrix: The diagonal of a square matrix are the terms a_{ii}, $i = 1, 2, \ldots, n$.
Diagonal matrix: A matrix whose off-diagonal entries are all equal to zero.

$$A = \text{diag}\{a_{11}, a_{22}, \ldots, a_{nn}\}$$

$$= \begin{bmatrix} a_{11} & 0 & \cdots & 0 \\ 0 & a_{22} & \cdots & 0 \\ \vdots & \vdots & \ddots & \vdots \\ 0 & 0 & \cdots & a_{nn} \end{bmatrix}$$

EXAMPLE A.8: DIAGONAL MATRIX

$$A = \text{diag}\{1,5,7\} = \begin{bmatrix} 1 & 0 & 0 \\ 0 & 5 & 0 \\ 0 & 0 & 7 \end{bmatrix}$$

>> A = diag ([1, 5, 7])

Identity or unity matrix: A diagonal matrix with all diagonal entries equal to unity:

$$I = \text{diag}\{1,1,\ldots,1\} = \begin{bmatrix} 1 & 0 & \cdots & 0 \\ 0 & 1 & \cdots & 0 \\ \vdots & \vdots & \ddots & \vdots \\ 0 & 0 & \cdots & 1 \end{bmatrix}$$

We denote an $n \times n$ identity matrix by I_n. The identity matrix is a multiplicative identity because any $m \times n$ matrix A satisfies $AI_m = I_nA = A$. By definition, we have $A^0 = I_n$.

EXAMPLE A.9: IDENTITY MATRIX

$$I_3 = \text{diag}\{1,1,1\} = \begin{bmatrix} 1 & 0 & 0 \\ 0 & 1 & 0 \\ 0 & 0 & 1 \end{bmatrix}$$

>> eye(3)

Zero matrix: A matrix with all entries equal to zero

$$C = \mathbf{0}_{m \times n} \Leftrightarrow [c_{ij}] = [0], \quad i = 1, 2, \ldots, m$$
$$j = 1, 2, \ldots, n$$

For any $m \times n$ matrix A, the zero matrix has the properties

$$A\mathbf{0}_{n \times l} = \mathbf{0}_{n \times l}$$

$$\mathbf{0}_{l \times m} A = \mathbf{0}_{l \times n}$$

$$A \pm \mathbf{0}_{m \times n} = A$$

EXAMPLE A.10: ZERO MATRIX

$$\mathbf{0}_{2 \times 3} = \begin{bmatrix} 0 & 0 & 0 \\ 0 & 0 & 0 \end{bmatrix}$$

>> zeros (2,3)

A.4 DETERMINANT OF A MATRIX

The determinant of a square matrix is a scalar computed using its entries. For a 1×1 matrix, the determinant is simply the matrix itself. For a 2×2 matrix, the determinant is

$$\det(A) = |A| = a_{11}a_{22} - a_{12}a_{21}$$

For higher-order matrices, the following definitions are needed to define the determinant.

Minor: The ijth minor of an $n \times n$ matrix is the determinant of the $n - 1 \times n - 1$ matrix obtained by removing the ith row and the jth column and is denoted M_{ij}.

Cofactor of a matrix: The ijth cofactor of an $n \times n$ matrix is a signed minor given by

$$C_{ij} = (-1)^{i+j} M_{ij}$$

The sign of the ijth cofactor can be obtained from the ijth entry of the matrix

$$\begin{bmatrix} + & - & + & \cdots \\ - & + & - & \cdots \\ + & - & + & \cdots \\ \vdots & \vdots & \vdots & \ddots \end{bmatrix}$$

Determinant

$$\det(A) = |A| = \sum_{i=1}^{n} a_{is} C_{is} = \sum_{j=1}^{n} a_{sj} C_{sj}$$

that is, the determinant can be obtained by expansion along any row or column.

Singular matrix $\det(A) = 0$
Nonsingular matrix $\det(A) \neq 0$

Properties of Determinants

For an $n \times n$ matrix A,

$$\det(A) = \det(A^T)$$

$$\det(\alpha A) = \alpha^n \det(A)$$

$$\det(AB) = \det(A)\det(B)$$

EXAMPLE A.11: DETERMINANT OF A MATRIX

$$A = \begin{bmatrix} 1 & 2 & 3 \\ 4 & 5 & -1 \\ 1 & -5 & 0 \end{bmatrix}$$

$$|A| = 3 \times [4 \times (-5) - 5 \times 1] - (-1) \times [1 \times (-5) - 2 \times 1] + 0$$
$$= 3 \times (-25) + (-7) = -82$$

>> **det(A)**

Adjoint matrix: The transpose of the matrix of cofactors

$$\text{adj}(A) = \left[C_{ij} \right]^T$$

A.5 INVERSE OF A MATRIX

The inverse of a square matrix is a matrix satisfying

$$AA^{-1} = A^{-1}A = I_n$$

The inverse of the matrix is given by

$$A^{-1} = \frac{\text{adj}(A)}{\det(A)}$$

EXAMPLE A.12: INVERSE MATRIX

$$A = \begin{bmatrix} 1 & 2 & 3 \\ 2 & 4 & 5 \\ 0 & 6 & 7 \end{bmatrix} \quad A^{-1} = \frac{adj(A)}{\det(A)} = \begin{bmatrix} (28-30) & -(14-18) & (10-12) \\ -(14-0) & (7-0) & -(5-6) \\ (12-0) & -(6-0) & (4-4) \end{bmatrix} \Big/ 6$$

$$= \begin{bmatrix} -0.333 & 0.667 & -0.333 \\ -2.333 & 1.167 & 0.167 \\ 2 & -1 & 0 \end{bmatrix}$$

Use the command

>> **inv(A)**

>> **A\B**

to calculate $A^{-1}B$, and the command

>> **A/B**

to calculate AB^{-1}

Combinations of Operations

$$(ABC)^T = C^T B^T A^T$$

$$(ABC)^{-1} = C^{-1} B^{-1} A^{-1}$$

$$\left(A^T\right)^{-1} = \left(A^{-1}\right)^T = A^{-T}$$

Orthogonal matrix: A matrix whose inverse is equal to its transpose

$$A^{-1} = A^T$$

That is,

$$A^T A = AA^T = I_n$$

Using the properties of a determinant of a square matrix,

$$\det(I_n) = \det(A)\det(A^T) = \det(A)^2 = 1$$
$$\det(I_n) = \det(A)\det(A^T) = \det(A)^2 = 1$$

that is, $\det(A) = \pm 1$ for an orthogonal matrix.

EXAMPLE A.13: ORTHOGONAL MATRIX

The coordinate rotation matrix for a yaw angle (rotation about the z-axis) α is the orthogonal matrix

$$R(\alpha) = \begin{bmatrix} \cos(\alpha) & -\sin(\alpha) & 0 \\ \sin(\alpha) & \cos(\alpha) & 0 \\ 0 & 0 & 1 \end{bmatrix}$$

with $\det(R) = \cos^2(\alpha) + \sin^2(\alpha) = 1$.

Unitary matrix: A matrix whose inverse is equal to its complex conjugate transpose

$$A^{-1} = A^*$$

Trace of a matrix: The sum of the diagonal elements of a square matrix

$$\mathrm{tr}(A) = \sum_{i=1}^{n} a_{ii}$$

The trace satisfies the following properties:

$$\mathrm{tr}(A^T) = \mathrm{tr}(A)$$

$$\mathrm{tr}(AB) = \mathrm{tr}(BA)$$

$$\mathrm{tr}(A+B) = \mathrm{tr}(A) + \mathrm{tr}(B)$$

EXAMPLE A.14: TRACE OF A MATRIX

Find the trace of the matrix $R(\alpha)$ shown in Example A.13.

$$tr(R) = \cos(\alpha) + \cos(\alpha) + 1 = 1 + 2\cos(\alpha)$$

For $\alpha = \pi$, $\cos(\alpha) = -1$ and $tr(R) = -1$.

>> **trace(R)**

-1

Linearly independent vectors: A set of vectors $\{\mathbf{x}_i, i = 1, 2, \ldots n\}$ is linearly independent if

$$\alpha_1\mathbf{x}_1 + \alpha_2\mathbf{x}_2 + \cdots + \alpha_n\mathbf{x}_n = \mathbf{0} \Leftrightarrow \alpha_i = 0, i = 1, 2, \ldots, n$$

Otherwise, the set is said to be **linearly dependent.**

EXAMPLE A.15: LINEAR INDEPENDENCE

Consider the following row vectors: $\mathbf{a}^T = [3\ 4\ 0]$, $\mathbf{b}^T = [1\ 0\ 0]$, and $\mathbf{c}^T = [0\ 1\ 0]$.

The set $\{\mathbf{a}, \mathbf{b}, \mathbf{c}\}$ is linearly dependent because $\mathbf{a} = 3\mathbf{b} + 4\mathbf{c}$. But the sets $\{\mathbf{a},\mathbf{b}\}$, $\{\mathbf{b},\mathbf{c}\}$, and $\{\mathbf{a},\mathbf{c}\}$ are linearly independent.

Rank of a Matrix

Column rank: Number of linearly independent columns.

Row rank: Number of linearly independent rows.

The rank of a matrix is equal to its row rank, which is equal to its column rank.

For an $m \times n$ (rectangular) matrix A, the rank of the matrix is

$$r(A) \leq \min\{n, m\}$$

If equality holds, the matrix is said to be **full rank**. A full rank square matrix is nonsingular.

EXAMPLE A.16: RANK OF A MATRIX

The matrix

$$A = \begin{bmatrix} 3 & 4 & 0 \\ 1 & 0 & 0 \\ 0 & 1 & 0 \end{bmatrix}$$

has the row vectors considered in Example A.15. Hence, the matrix has 2 linearly independent row vectors (i.e., row rank 2). The first two columns of the matrix are also linearly independent (i.e., it has column rank 2). The largest square matrix with nonzero determinant is the 2×2 matrix:

$$\begin{bmatrix} 3 & 4 \\ 1 & 0 \end{bmatrix}$$

Clearly, the matrix has rank 2.

A.6 EIGENVALUES

The eigenvalues of an $n \times n$ matrix are the n roots of the characteristic equation

$$\det[\lambda I_n - A] = 0$$

Distinct eigenvalues: $\lambda_j \neq \lambda_i,\ i \neq j,\ i, j,\ = 1, 2, \ldots, n$

Repeated eigenvalues: $\lambda_i \neq \lambda_j$, for some $i \neq j$

Multiplicity of the eigenvalue: The number of repetitions of the repeated eigenvalue (also known as the **algebraic multiplicity**).

Spectrum of matrix A: The set of eigenvalues $\{\lambda_i,\ i = 1, 2, \ldots, n\}$.

Spectral radius of a matrix: Maximum absolute value over all the eigenvalues of the matrix.

Trace in terms of eigenvalues: $\mathrm{tr}(A) = \sum_{i=1}^{n} \lambda_i$

Upper triangular matrix

$$A = \begin{bmatrix} a_{11} & a_{12} & \cdots & a_{1n} \\ 0 & a_{22} & \cdots & a_{2n} \\ \vdots & \vdots & \ddots & \vdots \\ 0 & 0 & \cdots & a_{nn} \end{bmatrix}$$

Lower triangular matrix

$$A = \begin{bmatrix} a_{11} & 0 & \cdots & 0 \\ a_{21} & a_{22} & \cdots & 0 \\ \vdots & \vdots & \ddots & \vdots \\ a_{n1} & a_{n2} & \cdots & a_{nn} \end{bmatrix}$$

For lower triangular, upper triangular, and diagonal matrices,

$$\{\lambda_i, i = 1, 2, \ldots, n\} = \{a_{ii}, i = 1, 2, \ldots, n\}$$

A.7 EIGENVECTORS

The eigenvector of a matrix A are vectors that are mapped to themselves when multiplied by the matrix A:

$$A\mathbf{v} = \lambda\,\mathbf{v} \qquad [\lambda I_n - A]\mathbf{v} = 0$$

For a nonzero solution $\mathbf{v}$ to the preceding equation to exist, the premuliplying matrix must be rank deficient—that is, λ must be an eigenvalue of the matrix A.

The eigenvector is defined by a direction or by a specific relationship between its entries. Multiplication by a scalar changes the length but not the direction of the vector.

EXAMPLE A.17: EIGENVALUES AND EIGENVECTORS

Find the eigenvalues and eigenvectors of the matrix

$$A = \begin{bmatrix} 3 & 4 & 0 \\ 1 & 0 & 0 \\ 0 & 1 & 0 \end{bmatrix}$$

$$\lambda I_3 - A = \begin{bmatrix} \lambda - 3 & -4 & 0 \\ -1 & \lambda & 0 \\ 0 & -1 & \lambda \end{bmatrix}$$

$$\det[\lambda I_3 - A] = [(\lambda - 3)\lambda - 4]\lambda = [\lambda^2 - 3\lambda - 4]\lambda = (\lambda - 4)(\lambda + 1)\lambda$$

$$\lambda_1 = 4 \quad \lambda_2 = -1 \quad \lambda_3 = 0$$

$$AV = V\Lambda$$

$$\begin{bmatrix} 3 & 4 & 0 \\ 1 & 0 & 0 \\ 0 & 1 & 0 \end{bmatrix}\begin{bmatrix} v_{11} & v_{12} & v_{13} \\ v_{21} & v_{22} & v_{23} \\ v_{31} & v_{32} & v_{33} \end{bmatrix} = \begin{bmatrix} v_{11} & v_{12} & v_{13} \\ v_{21} & v_{22} & v_{23} \\ v_{31} & v_{32} & v_{33} \end{bmatrix}\begin{bmatrix} 4 & 0 & 0 \\ 0 & -1 & 0 \\ 0 & 0 & 0 \end{bmatrix}$$

1. $\lambda_3 = 0$
   $v_{13} = v_{23} = 0$ and v_{33} free. Let $v_{23} = 1$.

2. $\lambda_2 = -1$
   $v_{12} = -v_{22}$
   $v_{22} = -v_{32}$ Let $v_{12} = 1$.

3. $\lambda_1 = 4$
   $v_{11} = 4\,v_{21}$
   $v_{21} = 4\,v_{31}$ Let $v_{31} = 1$.

   Hence, the modal matrix of eigenvectors is

$$V = \begin{bmatrix} 16 & 1 & 0 \\ 4 & -1 & 0 \\ 1 & 1 & 1 \end{bmatrix}$$

The lengths or 2-norms of the eigenvectors are

$$\|\mathbf{v}_1\| = \left[16^2 + 4^2 + 1^2\right]^{1/2}$$

$$\|\mathbf{v}_2\| = \left[1^2 + 1^2 + 1^2\right]^{1/2}$$

$$\|\mathbf{v}_3\| = \left[0 + 0 + 1^2\right]^{1/2}$$

The three eigenvectors can be normalized using the vector norms to obtain the matrix

$$V = \begin{bmatrix} 0.9684 & 0.5774 & 0 \\ 0.2421 & -0.5774 & 0 \\ 0.0605 & 0.5774 & 1 \end{bmatrix}$$

>> A = [3, 4, 0; 1, 0, 0; 0, 1, 0]

>> [V, L]= eig(A)

V =

0	0.5774	-0.9684
0	-0.5774	-0.2421
1.0000	0.5774	-0.0605

L =

0	0	0
0	-1	0
0	0	4

The trace of the preceding matrix is

$$\text{tr}(A) = 3 + 0 + 0 = 3 = 0 + (-1) + 4 = \lambda_1 + \lambda_2 + \lambda_3$$

Normal matrix: Multiplication by its (conjugate) transpose is commutative

$$A^T A = A A^T \quad \left(A^* A = A A^*\right)$$

This includes symmetric (hermitian) matrices as a special case.

The matrix of eigenvectors of a normal matrix can be selected as an orthogonal (unitary) matrix:

$$A = V \Lambda V^T \quad \left(A = V \Lambda V^*\right)$$

Partitioned matrix: A matrix partitioned into smaller submatrices

$$\begin{bmatrix} A_{11} & A_{12} & \cdots \\ A_{21} & A_{22} & \cdots \\ \vdots & \vdots & \ddots \end{bmatrix}$$

Transpose of a partitioned matrix

$$\begin{bmatrix} A_{11}^T & A_{21}^T & \cdots \\ \hline A_{12}^T & A_{22}^T & \cdots \\ \hline \vdots & \vdots & \vdots & \ddots \end{bmatrix}$$

Sum/difference of partitioned matrices: $C = A \pm B \Leftrightarrow C_{ij} = A_{ij} \pm B_{ij}$

Product of partitioned matrices: Apply the rules of matrix multiplication with the products of matrix entries replaced by the noncommutative products of submatrices

$$C = AB \Leftrightarrow C_{ij} = \sum_{k=1}^{n} A_{ik} B_{kj}, \quad i = 1, 2, \ldots, r$$

$$i = 1, 2, \ldots, s$$

Determinant of a partitioned matrix

$$\begin{vmatrix} A_1 & A_2 \\ \hline A_3 & A_4 \end{vmatrix} = \begin{cases} |A_1||A_4 - A_3 A_1^{-1} A_2|, & A_1^{-1} \text{ exists} \\ |A_4||A_1 - A_2 A_4^{-1} A_3|, & A_4^{-1} \text{ exists} \end{cases}$$

Inverse of a partitioned matrix

$$\begin{bmatrix} A_1 & A_2 \\ \hline A_3 & A_4 \end{bmatrix}^{-1} = \begin{bmatrix} \left(A_1 - A_2 A_4^{-1} A_3\right)^{-1} & -A_1^{-1} A_2 \left(A_4 - A_3 A_1^{-1} A_2\right)^{-1} \\ \hline -A_4^{-1} A_3 \left(A_1 - A_2 A_4^{-1} A_3\right)^{-1} & \left(A_4 - A_3 A_1^{-1} A_2\right)^{-1} \end{bmatrix}$$

EXAMPLE A.18: PARTITIONED MATRICES

$$A = \begin{bmatrix} 1 & 2 & 5 \\ 3 & 4 & 6 \\ \hline 7 & 8 & 9 \end{bmatrix} \quad B = \begin{bmatrix} -3 & 2 & 5 \\ 3 & 1 & 7 \\ \hline -4 & 0 & 2 \end{bmatrix}$$

$$A + B = \begin{bmatrix} 1-3 & 2+2 & 5+5 \\ 3+3 & 4+1 & 6+7 \\ \hline 7-4 & 8+0 & 9+2 \end{bmatrix} = \begin{bmatrix} -2 & 4 & 10 \\ 6 & 5 & 13 \\ \hline 3 & 8 & 11 \end{bmatrix}$$

$$AB = \begin{bmatrix} \begin{bmatrix} 1 & 2 \\ 3 & 4 \end{bmatrix}\begin{bmatrix} -3 & 2 \\ 3 & 1 \end{bmatrix} + \begin{bmatrix} 5 \\ 6 \end{bmatrix}[-4 \quad 0] & \begin{bmatrix} 1 & 2 \\ 3 & 4 \end{bmatrix}\begin{bmatrix} 5 \\ 7 \end{bmatrix} + \begin{bmatrix} 5 \\ 6 \end{bmatrix}2 \\ [7 \quad 8]\begin{bmatrix} -3 & 2 \\ 3 & 1 \end{bmatrix} + 9[-4 \quad 0] & [7 \quad 8]\begin{bmatrix} 5 \\ 7 \end{bmatrix} + 9 \times 2 \end{bmatrix}$$

$$= \begin{bmatrix} -17 & 4 & 29 \\ -21 & 10 & 55 \\ \hline -33 & 22 & 109 \end{bmatrix}$$

>> A1 = [1, 2; 3, 4];

>> a2 = [5; 6];

>> a3 = [7, 8];

>> a4 = 9;

>> A = [A1, a2; a3, a4];

>> B = [[−3,2; 3,1], [5; 7]; [−4,0], 2];

>> A + B

−2	4	10
6	5	13
3	8	11

>> A * B

−17	4	29
−21	10	55
−33	22	109

Matrix Inversion Lemma

The following identity can be used in either direction to simplify matrix expressions:

$$\left[A_1 + A_2 A_4^{-1} A_3 \right]^{-1} = A_1^{-1} - A_1^{-1} A_2 \left[A_4 + A_3 A_1^{-1} A_2 \right]^{-1} A_3 A_1^{-1}$$

A.8 NORM OF A VECTOR

The norm is a measure of size or length of a vector. It satisfies the following axioms, which apply to the familiar concept of length in the plane.

Norm Axioms

1. $||\mathbf{x}|| = 0$ if and only if $\mathbf{x} = 0$
2. $||\mathbf{x}|| > 0$ for $\mathbf{x} \neq 0$
3. $||\alpha\,\mathbf{x}|| = |\alpha|\,||\mathbf{x}||$
4. $||\mathbf{x} + \mathbf{y}|| \leq ||\mathbf{x}|| + ||\mathbf{y}||$ (triangle inequality)

l_p Norms

$$l_\infty \textbf{ norm:} \quad ||\mathbf{x}||_\infty = \max_i |x_i|$$

$$l_2 \textbf{ norm:} \quad ||\mathbf{x}||_2^2 = \sum_{i=1}^{n} |x_i|^2$$

$$l_1 \textbf{ norm:} \quad ||\mathbf{x}||_1 = \sum_{i=1}^{n} |x_i|$$

Equivalent Norms

Norms that satisfy the inequality

$$k_1\|\mathbf{x}\|_i \le \|\mathbf{x}\|_j \le k_2\|\mathbf{x}\|_i$$

with finite constants k_1 and k_2. All norms for $n \times 1$ real vectors are equivalent. All equivalent norms are infinite if and only if any one of them is infinite.

EXAMPLE A.19: VECTOR NORMS

$$\mathbf{a}^T = [1, \quad 2, \quad -3]$$

$$\|\mathbf{a}\|_1 = |1| + |2| + |-3| = 6$$

$$\|\mathbf{a}\|_2 = \sqrt{1^2 + 2^2 + (-3)^2} = 3.7417$$

$$\|\mathbf{a}\|_\infty = \max\{|1|,|2|,|-3|\} = 3$$

```
>>a = [1; 2; -3]
>>norm(a, 2)        % 2 induced norm (square root of sum of squares)
3.7417
>>norm(a, 1)        % 1 induced norm (sum of absolute values)
6
>>norm(a, inf)      % infinity induced norm (max element)
3
```

A.9 MATRIX NORMS

Satisfy the norm axioms.

Frobenius Norm

$$\|A\|_F = \sqrt{\sum_{i=1}^{m}\sum_{j=1}^{n}|a_{ij}|^2} = \sqrt{tr\{A^T A\}}$$

Other matrix norms

$$\|A\| = \max_{i,j}|a_{ij}|$$

$$\|A\|_F = \sum_{i=1}^{m}\sum_{j=1}^{n}|a_{ij}|$$

Induced Matrix Norms

Norms that are induced from vector norms using the definition

$$\|A\|_i = \max_{\mathbf{x}}\frac{\|A\mathbf{x}\|}{\|\mathbf{x}\|} = \max_{\|\mathbf{x}\|=1}\|A\mathbf{x}\|$$

where $\|\bullet\|$ is any vector norm.

Submultiplicative Property

$$\|A\mathbf{x}\| \le \|A\|\|\mathbf{x}\|$$

$$\|AB\| \le \|A\|\|B\|$$

All induced norms are submultiplicative, but only some noninduced norms are.

l_1 **Norm** $\|A\|_1 = \max_j \sum_{i=1}^{m} |a_{ij}|$ (maximum absolute column sum)

L_∞ **Norm** $\|A\|_\infty = \max_j \sum_{ji=1}^{n} |a_{ij}|$ (maximum absolute row sum)

l_2 **Norm** $\|A\|_2 = \max_j \lambda_i^{1/2}\left(A^T A\right)$ (maximum singular value = maximum eigen-value of A^TA)

Note: The norms given earlier are **not** induced norms.

EXAMPLE A.20: NORM OF A MATRIX

$$A = \begin{bmatrix} 1 & 2 \\ 3 & -4 \end{bmatrix}$$

$$\|A\|_1 = \max\left\{|1|+|3|, |2|+|-4|\right\} = 6$$

$$\|A\|_2 = \lambda_{\max}^{1/2}\left\{\begin{bmatrix} 1 & 3 \\ 2 & -4 \end{bmatrix}\begin{bmatrix} 1 & 2 \\ 3 & -4 \end{bmatrix}\right\} = \lambda_{\max}^{1/2}\left\{\begin{bmatrix} 10 & -10 \\ -10 & 20 \end{bmatrix}\right\} = 5.1167$$

$$\|A\|_\infty = \max\left\{|1|+|2|, |3|+|-4|\right\} = 7$$

$$\|A\|_F = \sqrt{|1|^2 + |2|^2 + |3|^3 + |-4|^4} = 5.4772$$

```
>>norm(A, 1)        % 1 induced norm (max of column sums)
6
>>norm(A, 2)        % 2 induced norm (max singular value)
5.1167
>>norm(A, inf)      % infinity induced norm (max of row sums)
7
>>norm(A, 'fro')    % 2 (square root of sum of squares)
5.4772
```

A.10 QUADRATIC FORMS

A quadratic form is a function of the form

$$V(\mathbf{x}) = \mathbf{x}^T P \mathbf{x} = \sum_{i=1}^{n} \sum_{j=1}^{n} p_{ij} x_i x_j$$

where $\mathbf{x}$ is an $n \times 1$ real vector and P is an $n \times n$ matrix. The matrix P can be assumed to be symmetric without loss of generality. To show this, assume that P is not symmetric and rewrite the quadratic form in terms of the symmetric component and the skew-symmetric component of P as follows:

$$V(\mathbf{x}) = \mathbf{x}^T \left(\frac{P + P^T}{2} \right) \mathbf{x} + \mathbf{x}^T \left(\frac{P - P^T}{2} \right) \mathbf{x}$$

$$= \mathbf{x}^T \left(\frac{P + P^T}{2} \right) \mathbf{x} + \frac{1}{2} \left(\mathbf{x}^T P \mathbf{x} - (P\mathbf{x})^T \mathbf{x} \right)$$

Interchanging the row and column in the last term gives

$$V(\mathbf{x}) = \mathbf{x}^T \left(\frac{P + P^T}{2} \right) \mathbf{x} + \frac{1}{2} \left(\mathbf{x}^T P \mathbf{x} - \mathbf{x}^T P \mathbf{x} \right)$$

$$= \mathbf{x}^T \left(\frac{P + P^T}{2} \right) \mathbf{x}$$

Thus, if P is not symmetric, we can replace it with its symmetric component without changing the quadratic form.

The sign of a quadratic form for nonzero vectors $\mathbf{x}$ can be invariant depending on the matrix P. In particular, the eigenvalues of the matrix P determine the sign of the quadratic form. To see this, we examine the eigenvalues-eigenvector decomposition of the matrix P in the quadratic form

$$V(\mathbf{x}) = \mathbf{x}^T P \mathbf{x}$$

We assume, without loss of generality, that P is symmetric. Hence, its eigenvalues are real and positive, and its modal matrix of eigenvectors is orthogonal. The matrix can be written as

$$P = V_p \Lambda V_p^T$$

$$\Lambda = \text{diag}\{\lambda_1, \lambda_2, \dots, \lambda_n\}$$

Using the eigenvalues decomposition of the matrix, we have

$$V(\mathbf{x}) = \mathbf{x}^T V_p \Lambda V_p^T \mathbf{x}$$

$$= \mathbf{y}^T \Lambda \mathbf{y}$$

$$= \sum_{i=1}^{n} \lambda_i y_i^2 > 0$$

$$\mathbf{y} = [y_1 \quad y_2 \quad \cdots \quad y_n]$$

Because the modal matrix V_p is invertible, there is a unique $\mathbf{y}$ vector associated with each $\mathbf{x}$ vector. The expression for the quadratic form in terms of the eigenvalues allows us to characterize it and the associated matrix as follows.

Positive definite: A quadratic form is positive definite if

$$V(\mathbf{x}) = \mathbf{x}^T P \mathbf{x} > 0, \quad \mathbf{x} \neq 0$$

This is true if the eigenvalues of P are all positive, in which case we say that P is a positive definite matrix and we denote this by $P > 0 >$.

Negative definite: A quadratic form is negative definite if

$$V(\mathbf{x}) = \mathbf{x}^T P \mathbf{x} < 0, \quad \mathbf{x} \neq \mathbf{0}$$

This is true if the eigenvalues of P are all negative, in which case we say that P is a negative definite matrix and we denote this by $P < 0$.

Positive semidefinite: A quadratic form is positive semidefinite if

$$V(\mathbf{x}) = \mathbf{x}^T P \mathbf{x} \geq 0, \quad \mathbf{x} \neq \mathbf{0}$$

This is true if the eigenvalues of P are all positive or zero, in which case we say that P is a positive semidefinite matrix and we denote this by $P \geq 0$. Note that in this case, if an eigenvalues λ_i is zero then the nonzero vector $\mathbf{y}$ with its ith entry equal to 1 and all other entries zero gives a zero value for V. Thus, there is a nonzero vector $\mathbf{x} = V_p \mathbf{y}$ for which V is zero.

Negative semidefinite: A quadratic form is negative semidefinite if

$$V(\mathbf{x}) = \mathbf{x}^T P \mathbf{x} \geq 0, \quad \mathbf{x} \neq \mathbf{0}$$

This is true if the eigenvalues of P are all negative or zero, in which case we say that P is a negative semidefinite matrix and we denote this by $P \leq 0$. In this case, if an eigenvalues λ_i is zero, then V is zero for the nonzero $\mathbf{x} = V_p \mathbf{y}$ where $\mathbf{y}$ is a vector with its ith entry equal to 1 and all other entries zero.

Indefinite: If the matrix Q has some positive and some negative eigenvalues, then the sign of the corresponding quadratic form depends on the vector $\mathbf{x}$ and the matrix is called indefinite.

A.11 MATRIX DIFFERENTIATION/INTEGRATION

The derivative (integral) of a matrix is a matrix whose entries are the derivatives (integrals) of the entries of the matrix.

EXAMPLE A.21: MATRIX DIFFERENTIATION AND INTEGRATION

$$A(t) = \begin{bmatrix} 1 & t & \sin(2t) \\ t & 0 & 4+t \end{bmatrix}.$$

$$\int_0^t A(\tau)\,d\tau = \begin{bmatrix} \int_0^t 1\,d\tau & \int_0^t \tau\,d\tau & \int_0^t \sin(2\tau)\,d\tau \\ \int_0^t \tau\,d\tau & 0 & \int_0^t (4+\tau)\,d\tau \end{bmatrix}$$

$$= \begin{bmatrix} t & t^2/2 & \{1-\cos(t)\}/2 \\ t^2/2 & 0 & 4t+t^2/2 \end{bmatrix}$$

$$\frac{dA(t)}{dt} = \begin{bmatrix} \dfrac{d1}{dt} & \dfrac{dt}{dt} & \dfrac{d\sin(2t)}{dt} \\ \dfrac{dt}{dt} & 0 & \dfrac{d(4+t)}{dt} \end{bmatrix} = \begin{bmatrix} 0 & 1 & 2\cos(2t) \\ 1 & 0 & 1 \end{bmatrix}$$

Derivative of a product: $\dfrac{dAB}{dt} = A\dfrac{dB}{dt} + \dfrac{dA}{dt}B$

Derivative of the inverse matrix: $\dfrac{d(A^{-1})}{dt} = -A^{-1}\dfrac{dA}{dt}A^{-1}$

Gradient vector: The derivative of a scalar function $f(\mathbf{x})$ with respect to the vector $\mathbf{x}$ is known as the gradient vector and is given by the n by 1 vector

$$\frac{\partial f(\mathbf{x})}{\partial \mathbf{x}} = \left[\frac{\partial f(\mathbf{x})}{\partial x_i}\right]$$

Some authors define the gradient as a row vector.

Jacobian matrix: The derivative of an $n \times 1$ vector function $\mathbf{f}(\mathbf{x})$ with respect to the vector $\mathbf{x}$ is known as the Jacobian matrix and is given by the $n \times n$ matrix

$$\frac{\partial \mathbf{f}(\mathbf{x})}{\partial \mathbf{x}} = \left[\frac{\partial f_i(\mathbf{x})}{\partial x_j}\right]$$

Gradient of inner product

$$\frac{\partial \mathbf{a}^T\mathbf{x}}{\partial \mathbf{x}} = \frac{\partial \mathbf{x}^T\mathbf{a}}{\partial \mathbf{x}} = \left[\frac{\partial \sum\limits_{i=1}^{n} a_i x_i}{\partial x_i}\right] = [a_i] = \mathbf{a}$$

Gradient matrix of a quadratic form

$$\frac{\partial \mathbf{x}^T P \mathbf{x}}{\partial \mathbf{x}} = \mathbf{x}^T\frac{\partial P\mathbf{x}}{\partial \mathbf{x}} + \frac{\partial(P^T\mathbf{x})^T}{\partial \mathbf{x}}\mathbf{x}$$
$$= (P + P^T)\mathbf{x}$$

Because P can be assumed to be symmetric with no loss of generality, we write

$$\frac{\partial \mathbf{x}^T P \mathbf{x}}{\partial \mathbf{x}} = 2P\mathbf{x}$$

Hessian matrix of a quadratic form: The Hessian or second derivative matrix is given by

$$\frac{\partial^2 \mathbf{x}^T P \mathbf{x}}{\partial \mathbf{x}^2} = \frac{\partial 2P\mathbf{x}}{\partial \mathbf{x}} = 2\left[\frac{\partial \mathbf{p}_i^T\mathbf{x}}{\partial x_j}\right] = 2[p_{ij}]$$

where the ith entry of the vector $P\mathbf{x}$ is

$$\mathbf{p}_i^T\mathbf{x}$$

$$P = [p_{ij}] = \begin{bmatrix} \mathbf{p}_1^T \\ \mathbf{p}_2^T \\ \vdots \\ \mathbf{p}_n^T \end{bmatrix}$$

$$\frac{\partial^2 \mathbf{x}^T P \mathbf{x}}{\partial \mathbf{x}^2} = 2P$$

A.12 KRONECKER PRODUCT

The Kronecker product of two matrices A of order $m \times n$ and B of order $p \times q$ is denoted by $\otimes$ and is defined as

$$A \otimes B = \begin{bmatrix} a_{11}B & a_{12}B & \cdots & a_{1n}B \\ a_{21}B & a_{22}B & \cdots & a_{2n}B \\ \vdots & \vdots & \ddots & \vdots \\ a_{m1}B & a_{m2}B & \cdots & a_{mn}B \end{bmatrix}$$

The resulting matrix is of order $m.p \times n.q$.

EXAMPLE A.22: KRONECKER MATRIX PRODUCT

The Kronecker product of the two matrices:

$$A = \begin{bmatrix} 1 & 2 & 3 \\ 4 & 5 & 6 \end{bmatrix} \quad B = \begin{bmatrix} 1.1 & 2 & 3 \\ 4 & 5 & 6 \end{bmatrix}$$

$$A \otimes B = \begin{bmatrix} 1\begin{bmatrix} 1.1 & 2 & 3 \\ 4 & 5 & 6 \end{bmatrix} & 2\begin{bmatrix} 1.1 & 2 & 3 \\ 4 & 5 & 6 \end{bmatrix} & 3\begin{bmatrix} 1.1 & 2 & 3 \\ 4 & 5 & 6 \end{bmatrix} \\ 4\begin{bmatrix} 1.1 & 2 & 3 \\ 4 & 5 & 6 \end{bmatrix} & 5\begin{bmatrix} 1.1 & 2 & 3 \\ 4 & 5 & 6 \end{bmatrix} & 6\begin{bmatrix} 1.1 & 2 & 3 \\ 4 & 5 & 6 \end{bmatrix} \end{bmatrix}$$

$$= \left[\begin{array}{ccc|ccc|ccc} 1.1 & 2 & 3 & 2.2 & 4 & 6 & 3.3 & 6 & 9 \\ 4 & 5 & 6 & 8 & 10 & 12 & 12 & 15 & 18 \\ \hline 4.4 & 8 & 12 & 5.5 & 10 & 15 & 6.6 & 12 & 18 \\ 16 & 20 & 24 & 20 & 25 & 30 & 24 & 10 & 36 \end{array} \right]$$

```
>> kron(a,b)

ans =
    1.1000    2.0000    3.0000    2.2000    4.0000    6.0000    3.3000    6.0000    9.0000
    4.0000    5.0000    6.0000    8.0000   10.0000   12.0000   12.0000   15.0000   18.0000
    4.4000    8.0000   12.0000    5.5000   10.0000   15.0000    6.6000   12.0000   18.0000
   16.0000   20.0000   24.0000   20.0000   25.0000   30.0000   24.0000   30.0000   36.0000
```

RESOURCES

Barnett, S., *Matrices in Control Theory,* R. E. Krieger, 1984.

Brogan, W. L., *Modern Control Theory,* Prentice Hall, 1985.

Fadeeva, V. N., *Computational Methods of Linear Algebra*, Dover, 1959.

Gantmacher, F. R., *The Theory of Matrices,* Chelsea, 1959.

Noble, B., and J. W. Daniel, *Linear Algebra,* Prentice Hall, 1988.

Index